UNITEXT for Physics

T0177966

More information about this series at http://www.springer.com/series/13351

Attilio Rigamonti · Pietro Carretta

Structure of Matter

An Introductory Course with Problems and Solutions

Third Edition

 Springer

Attilio Rigamonti
University of Pavia
Pavia
Italy

Pietro Carretta
University of Pavia
Pavia
Italy

ISSN 2198-7882
UNITEXT for Physics
ISBN 978-3-319-37444-4
DOI 10.1007/978-3-319-17897-4

ISSN 2198-7890 (electronic)

ISBN 978-3-319-17897-4 (eBook)

*

Springer Cham Heidelberg New York Dordrecht London

Printed on acid-free paper

Springer International Publishing AG Switzerland is part of Springer Science+Business Media
(www.springer.com)

Die Wahrheit ist das Kind der Zeit, nicht der Autoritaet. Unsere Unwissenheit ist unendlich, tragen wir einen Kubikmillimeter ab!

—from B. Brecht, in *Leben des Galilei*

There is no end to this wonderful world of experimental discovery and mental constructions of reality as new facts become known. That is why physicists have more fun than most people.

—Miklos Gyulassy

To Luca and Margherita Strozzi Rigamonti,
with hope

—Attilio Rigamonti

To Gegia, Enri, Cate and Dario

—Pietro Carretta

Preface

Intended Audience, Approach and Presentation

This text is intended for a course of about six months for undergraduate students. It arises from the adaptation and the amendments to a text for a full-year course in Structure of Matter, written by one of the authors (A.R.) more than 30 years ago. At that time only a few (if any) textbooks having the suited form for introduction to the basic quantum properties of atoms, molecules, and crystals in a comprehensive and interrelated way, were available. Along the past 20 years many excellent books pursuing the aforementioned aim have been published (some of them are listed at the end of this preface). Still there are reasons, in our opinion, to attempt a further text devoted to the quantum roots of condensed matter properties. A practical aspect in this regard involves the organization of studies in Physics, after the huge scientific outburst of the various topics of fundamental and technological character in recent decades. In most universities there is now a first period of three or four years, common to all students and devoted to elementary aspects, followed by an advanced program in more specialized fields of Physics. The difficult task is to provide a common and formative introduction in the first period suitable as a basis for building up more advanced courses and to bridge the area between elementary physics and topics pertaining to research activities. The present attempt toward a readable book, hopefully presenting those desired characteristics, essentially is based on a mixture of simplified institutional theory with solved problems. The hope, in this way, is to provide physical insights, basic culture, and motivation, without deteriorating the possibility of advanced subsequent learning.

Organization

Structure of Matter is such a wide field involving so many subjects that a first task is to find a way to confine an introductory text. The present status of that discipline represents a key construction of the scientific knowledge, possibly equated only by the unitary description of the electromagnetic phenomena. Even by limiting attention to conventional topics of condensed matter only, namely atoms, molecules, and crystals, we are still left with an ample field. For instance, semiconductors or superconductors, the electric and magnetic properties of matter and its interaction with electromagnetic radiation, the microscopic mechanisms underlying solid-state devices as well as masers and lasers, are to be considered as belonging to the field of structure of matter (without mentioning the "artificial" matter involving systems such as nanostructures, photonic crystals, or special materials obtained by subtle manipulations of atoms by means of special techniques). In this text the choice has been to limit the attention to key concepts and to the typical aspects of atoms (Chaps. 1–5), molecules (Chaps. 7–10) and of crystalline solids (Chaps. 11–14), looking at the basic "structural" subjects without dealing with the properties that originate from them. This choice is exemplified by referring to crystals: electronic states and quantum motions of ions have been described without going into the details regarding the numerous properties related to these aspects. Only in a few illustrative cases favoring better understanding or comprehensive view, derivation of some related properties has been given (examples are some thermodynamical properties due to nuclear motions in molecules and crystals or some of the electric or magnetic properties). Chapter 6 has the particular aim to lead the reader to an illustrative overview of the quantum behaviors of angular momenta and magnetic moments, with an introduction to spin statistics, magnetic resonance, and spin motions and a mention of spin thermodynamics, through the description of adiabatic demagnetization. The four new Chaps. (15–18) introduced in the 3rd Edition deal with relevant phases of solid matter (magnetic, electric, and superconductive) and to the related phase transitions.

All along the text emphasis is given to the role of spectroscopic experiments giving access to the quantum properties by means of electromagnetic radiation. In the spirit to limit the attention to key subjects, frequent referring is given to the electric dipole moment and to selection rules, rather than to other aspects of the many experiments of spectroscopic character used to explore the matter at microscopic level. Other unifying concepts present along the text are the ones embedded in statistical physics and thermal excitations, as it is necessary in view of the many-body character of condensed matter in equilibrium with a thermal reservoir.

Prerequisite, Appendices, and Problems

Along the text the use of quantum mechanics, although continuous, only involves the basic background that the reader should have achieved in undergraduate courses. Knowledge of statistical physics is required based on Boltzmann, Fermi–Dirac and Bose–Einstein statistical distributions, with the relationships of thermodynamical quantities to the partition function (some of the problems work as proper recall, particularly for the physics of paramagnets or for black-body radiation). Finally the reader is assumed to have knowledge of classical electromagnetism and

Hamiltonian mechanics. Appendices are intended to provide *ad hoc* recalls, in some cases applied to appropriate systems or to phenomena useful for illustration. The Gaussian CGS units are used.

The problems should be considered entangled to the formal presentation of the subjects, being designed as an intrinsic part of the pathway the student should move by in order to grasp the key concepts. One of the reasons to entangle problems and institutional theory can be found in what Feynman wrote in the preface to his Lectures: "I think one way we could help the students more would be by putting more hard work into developing a set of problems which would elucidate some of the ideas in the lectures." Some of the problems are simple applications of the equations and in these cases the solutions are only sketched. Other problems are basic building blocks and possibly expansions of the formal description. Then the main steps of the solution are presented in some detail. The aim of the *mélange* intuition-theory-exercises pursued in the text is to favor the acquisition of the basic knowledge in the wide and wonderful field of condensed matter, emphasizing how phenomenological properties originate from the microscopic, quantum features of nature.

It should be obvious that a book of this size can present only a fraction of the present knowledge in the field. If the reader could achieve even an elementary understanding of the atoms, molecules, and crystals, how they react to electric and magnetic fields, how they interact with electromagnetic radiation, and respond to thermal excitation, the book will have fulfilled its purpose.

The fundamental blocks of the physical world are thought to be the subnuclear elementary particles. However, the beauty of the natural world rather originates from the architectural construction of the blocks occurring in the matter. Ortega Y Gasset wrote "If you wish to admire the beauty of a cathedral you have to respect for distance. If you go too close, you just see a brick." Furthermore, one could claim that the world of condensed matter more easily allows one to achieve a private discovery of phenomena. In this respect let us report what Edward Purcell wrote in his Nobel lecture: "To see the world for a moment as something rich and strange is the private reward of many a discovery."

Acknowledgments

The authors wish to acknowledge Giacomo Mauro D'Ariano, who has inspired and solved several problems and provided enlightening remarks with his collaboration to the former course "Structure of Matter" given by one of us (A.R.), along two decades. Acknowledgments for suggestions or indirect contributions through discussions or comments are due to G. Amoretti, A. Balzarotti, G. Benedek, G. Bonera, M. Bornatici, F. Borsa, L. Bossoni, G. Caglioti, R. Cantelli, L. Colombo, M. Corti, A. Debernardi, G. Grosso, A. Lascialfari, R. Mackeviciute, D. Magnani, N. Manini, F. Miglietta, G. Onida, G. Pastori Parravicini, M. Pieruccini, G. Prando, L. Romanó, S. Romano, A. Rosso, S. Sanna, G. Senatore, J. Spalek, F. Tedoldi, V. Tognetti, A.A. Varlamov, J. Villain.

M. Medici, N. Papinutto and G. Ventura are acknowledged for their help in printing preliminary versions of the manuscript and preparing some figures. The authors anticipate their gratitude to other students who, through vigilance and desire of learning will find errors and didactic mistakes.

This book has been written while receiving inspiration from a number of text-books dealing with particular items or from problems and exercises suggested or solved in them. The texts reported below are not recalled as a real "further-reading list," since it would be too ample and possibly useless. The list is rather the acknowledgment of the suggestions received when seeking inspiration, information, or advices.

A. Abragam, *L'effet Mossbauer et ses applications a l'etude des champs internes*, Gordon and Breach (1964).

M. Alonso and E.J. Finn, *Fundamental University Physics Vol.III- Quantum and Statistical Physics*, Addison Wesley (1973).

D.J. Amit and Y. Verbin, *Statistical Physics - An Introductory course*, World Scientific (1999).

J.F. Annett, *Superconductivity, Superfluids and Condensates*, Oxford University Press, Oxford (2004).

N.W. Ashcroft and N.D. Mermin, *Solid State Physics*, Holt, Rinehart and Winston (1976).

P.W. Atkins and R.S. Friedman, *Molecular Quantum Mechanics*, Oxford University Press, Oxford (1997).

A. Balzarotti, M. Cini, M. Fanfoni, *Atomi, Molecole e Solidi. Esercizi risolti*, Springer Verlag (2004).

A. Barone and G. Paternó, *Physics and Applications of the Josephson Effect*, John Wiley, New York (1982).

F. Bassani e U.M. Grassano, *Fisica dello Stato Solido*, Bollati Boringhieri (2000).

J.S. Blakemore, *Solid State Physics*, W.B. Saunders Co. (1974).

R. Blinc and B. Zeks, *Soft Modes in Ferroelectrics and Antiferroelectrics*, North-Holland Publishing Company, Amsterdam (1974).

S.J. Blundell, *Magnetism in Condensed Matter*, Oxford Master Series in Condensed Matter Physics, Oxford U.P. (2001).

S.J. Blundell and K.M. Blundell, *Concepts in Thermal Physics*, 2nd Edition, Oxford Master Series in Condensed Matter Physics, Oxford U.P. (2010).

S. Boffi, *Da Laplace a Heisenberg*, La Goliardica Pavese (1992).

B.H. Bransden and C.J. Joachain, *Physics of atoms and molecules*, Prentice Hall (2002).

W. Buckel, *Superconductivity- Fundamental and Applications*, VCH Weinheim (1991).

D. Budker, D.F. Kimball and D.P. De Mille, *Atomic Physics - An Exploration Through Problems and Solutions*, Oxford University Press (2004).

G. Burns, *Solid State Physics*, Academic Press, Inc. (1985).

G. Burns, *High Temperature Superconductivity - An Introduction*, Academic Press, Inc. (1992).

G. Caglioti, *Introduzione alla Fisica dei Materiali*, Zanichelli (1974).

B. Cagnac and J.C. Pebay - Peyroula, *Physique atomique, tome 2*, Dunod Université, Paris (1971).

P. Caldirola, *Istituzioni di Fisica Teorica*, Editrice Viscontea, Milano (1960).

M. Cini, *Topics and Methods in Condensed Matter Theory*, Springer (2010).

W. Cochran, *Lattice Vibrations*, Reports on Progress in Physics XXVI, 1 (1963).

J.M.D. Coey, *Magnetism and Magnetic Materials*, Cambridge University Press, Cambridge (2009).

L. Colombo, *Elementi di Struttura della Materia*, Hoepli (2002).

E.U. Condon and G.H. Shortley, *The Theory of Atomic Spectra*, Cambridge University Press, London (1959).

C.A. Coulson, *Valence*, Oxford Clarendon Press (1953).

J.A. Cronin, D.F. Greenberg, V.L. Telegdi, *University of Chicago Graduate Problems in Physics*, Addison-Wesley (1967).

M. Cyrot and D. Pavuna, *Introduction to Superconductivity and High-T_c Materials*, World Scientific, Singapore (1992).

G.M. D'Ariano, *Esercizi di Struttura della Materia*, La Goliardica Pavese (1989).

J.P. Dahl, *Introduction to the Quantum World of Atoms and Molecules*, World Scientific (2001).

P.G. de Gennes, *Superconductivity of Metals and Alloys*, Addison-Wesley (1989).

W. Demtröder, *Molecular Physics*, Wiley-VCH (2005).

W. Demtröder, *Atoms, Molecules and Photons*, Springer Verlag (2006).

R.N. Dixon, *Spectroscopy and Structure*, Methuen and Co LTD London (1965).

R. Eisberg and R. Resnick, *Quantum Physics of Atoms, Molecules, Solids, Nuclei and Particles*, J. Wiley and Sons (1985).

H. Eyring, J. Walter and G.E. Kimball, *Quantum Chemistry*, J. Wiley, New York (1950).

R.P. Feynman, R.B. Leighton and M. Sands,*The Feynman Lectures on Physics Vol. III*, Addison Wesley, Palo Alto (1965).

R. Fieschi e R. De Renzi, *Struttura della Materia*, La Nuova Italia Scientifica, Roma (1995).

A.P. French and E.F. Taylor, *An Introduction to Quantum Physics*, The M.I.T. Introductory Physics Series, Van Nostrand Reinhold (UK)(1986).

R. Gautreau and W. Savin, *Theory and Problems of Modern Physics*, (Schaum's series in Science) Mc Graw-Hill Book Company (1978).

M. Gitterman and V. Halperin, *Phase Transitions*, World Scientific (2013).

H.J. Goldsmid (Editor), *Problems in Solid State Physics*, Pion Limited London (1972).

H. Goldstein, *Classical Mechanics*, Addison-Wesley (1965).

D.L. Goodstein, *States of Matter*, Dover Publications Inc. (1985).

G. Grosso and G. Pastori Parravicini, *Solid State Physics*, 2nd Edition, Academic Press (2013).

A.P. Guimarães, *Magnetism and Magnetic Resonance in Solids*, J. Wiley and Sons (1998).

H. Haken and H.C. Wolf, *Atomic and Quantum Physics*, Springer Verlag Berlin (1987).

H. Haken and H.C. Wolf, *Molecular Physics and Elements of Quantum Chemistry*, Springer Verlag Berlin (2004).

W.A. Harrison, *Solid State Theory*, Dover Publications Inc., New York (1979).

G. Herzberg, *Molecular Spectra and Molecular Structure*, Vol. I, II and III, D. Van Nostrand, New York (1964-1966, reprint 1988-1991).

J.R. Hook and H.E. Hall, *Solid State Physics*, J. Wiley and Sons (1999).

H. Ibach and H. Lüth, *Solid State Physics: an Introduction to Theory and Experiments*, Springer Verlag (1990).

C.S. Johnson and L.G. Pedersen, *Quantum Chemistry and Physics*, Addison - Wesley (1977).

D.C. Johnston, *Handbook of Magnetic Materials Vol.10, Ed. K.H.J.Buschow, Chapter 10*, Elsevier (1997).

H. Kuzmany, *Solid-State Spectroscopy*, Springer-Verlag, Berlin (1998).

C. Kittel, *Elementary Statistical Physics*, J. Wiley and Sons (1958).

C. Kittel, *Introduction to Solid State Physics*, 8th Edition, J. Wiley and Sons (2005).

L.D. Landau and E.M. Lifshitz, *Statistical Physics*, Pergamon Press, Oxford (1959).

A. Larkin and A.A. Varlamov, *Theory of Fluctuations in Superconductors*, Oxford Science Publications, Clarendon Press, Oxford (2005).

Y.-K. Lim (Editor), *Problems and Solutions on Thermodynamic and Statistical Mechanics*, World Scientific, Singapore (2012).

M.E. Lines and A.M. Glass, *Principles and Applications of Ferroelectrics and Related Materials*, Clarendon Press, Oxford (1977).

N. Manini, *Introduction to the Physics of Matter*, Springer (2014).

J.D. Mc Gervey, *Quantum Mechanics - Concepts and Applications*, Academic Press, New York (1995).

L. Mihály and M.C. Martin, *Solid State Physics - Problems and Solutions*, J. Wiley (1996).

M.A. Morrison, T.L. Estle and N.F. Lane, *Quantum States of Atoms, Molecules and Solids*, Prentice - Hall Inc., New Jersey (1976).

K.A. Müller and A. Rigamonti (Editors), *Local Properties at Phase Transitions*, North-Holland Publishing Company, Amsterdam (1976).

C.P. Poole, H.A. Farach, R.J. Creswick, *Superconductivity*, Academic Press, San Diego (1995).

C.P. Poole, H.A. Farach, R.J. Creswick and R. Prozorov, *Superconductivity*, Academic Press, San Diego (2007).

E.M. Purcell, *Electricity and Magnetism, Berkeley Physics Course Vol.2*, Mc Graw-Hill (1965).

N.F. Ramsey, *Nuclear Moments*, J. Wiley Inc., New York (1953).

A. Rigamonti, *Introduzione alla Struttura della Materia*, La Goliardica Pavese (1977).

M.N. Rudden and J. Wilson, *Elements of Solid State Physics*, J. Wiley and Sons (1996).

M. Roncadelli, *Aspetti Astrofisici della Materia Oscura*, Bibliopolis, Napoli (2004).

H. Semat, *Introduction to Atomic and Nuclear Physics*, Chapman and Hall LTD (1962).

J.C. Slater, *Quantum Theory of Matter*, Mc Graw-Hill, New York (1968).

C.P. Slichter, *Principles of Magnetic Resonance*, Springer Verlag Berlin (1990).

H. Stanley, *Introduction to Phase Transitions and Critical Phenomena*, Oxford University Press, Oxford (1971).

S. Svanberg, *Atomic and Molecular Spectroscopy*, Springer Verlag, Berlin (2003).

D. Tabor, *Gases, liquids and solids*, Cambridge University Press (1993).

P.L. Taylor and O. Heinonen, *A Quantum Approach to Condensed Matter Physics*, Cambridge University Press (2002).

M. Tinkham, *Introduction to Superconductivity*, Dover Publications Inc., New York (1996).

M.A. Wahab, *Solid State Physics (Second Edition)*, Alpha Science International Ltd. (2005).

S. Weinberg, *The first three minutes: a modern view of the origin of the universe*, Amazon (2005).

M. White, *Quantum Theory of Magnetism*, McGraw-Hill (1970).

J.M. Ziman, *Principles of the Theory of Solids*, Cambridge University Press (1964).

Pavia   Attilio Rigamonti
December 2014   Pietro Carretta

Contents

1   **Atoms: General Aspects**   1
   1.1   Central Field Approximation   2
   1.2   Self-Consistent Construction of the Effective Potential.   5
   1.3   Degeneracy from Dynamical Equivalence.   6
   1.4   Hydrogenic Atoms: Illustration of Basic Properties   7
   1.5   Finite Nuclear Mass. Positron, Muonic
       and Rydberg Atoms   21
   1.6   Orbital and Spin Magnetic Moments and Spin-Orbit
       Interaction   25
   1.7   Spectroscopic Notation for Multiplet States   32
   Appendix 1.1 Electromagnetic Spectral Ranges
            and Fundamental Constants   36
   Appendix 1.2 Perturbation Effects in Two-Level Systems.   37
   Appendix 1.3 Transition Probabilities and Selection Rules   41
   Specific References and Further Reading   58

2   **Typical Atoms**   61
   2.1   Alkali Atoms   61
   2.2   Helium Atom   69
       2.2.1   Generalities and Ground State.   69
       2.2.2   Excited States and the Exchange Interaction   73
   2.3   Pauli Principle, Determinantal Eigenfunctions
       and Superselection Rule.   80
   Specific References and Further Reading   87

3   **The Shell Vectorial Model.**   89
   3.1   Introductory Aspects   89
   3.2   Coupling of Angular Momenta   91
       3.2.1   LS Coupling Model.   91
       3.2.2   The Effective Magnetic Moment.   95

      3.2.3   Illustrative Examples and the Hund Rules
               for the Ground State   97
3.3   jj Coupling Scheme.   105
3.4   Quantum Theory for Multiplets. Slater
        Radial Wavefunctions   110
3.5   Selection Rules.   114
Specific References and Further Reading   120

4 **Atoms in Electric and Magnetic Fields**.   121
4.1   Introductory Aspects   121
4.2   Stark Effect and Atomic Polarizability.   124
4.3   Hamiltonian in Magnetic Field   129
      4.3.1   Zeeman Regime   130
      4.3.2   Paschen-Back Regime   132
4.4   Paramagnetism of Non-interacting Atoms
        and Mean Field Interaction.   136
4.5   Atomic Diamagnetism   140
Appendix 4.1 Electromagnetic Units and Gauss System   143
Specific References and Further Reading   155

5 **Nuclear Moments and Hyperfine Interactions**   157
5.1   Introductory Generalities   157
5.2   Magnetic Hyperfine Interaction—F States   159
5.3   Electric Quadrupole Interaction.   166
Appendix 5.1 Fine and Hyperfine Structure in Hydrogen   172
Specific References and Further Reading   191

6 **Spin Statistics, Magnetic Resonance, Spin Motion and Echoes** . . .   193
6.1   Spin Statistics, Spin-Temperature and Fluctuations   193
6.2   The Principle of Magnetic Resonance
        and the Spin Motion   201
6.3   Spin and Photon Echoes   208
6.4   Ordering and Disordering in Spin Systems: Cooling
        by Adiabatic Demagnetization   210
Specific References and Further Reading   222

7 **Molecules: General Aspects**.   223
7.1   Born-Oppenheimer Separation and the Adiabatic
        Approximation   224
7.2   Classification of the Electronic States   228
      7.2.1   Generalities   228
      7.2.2   Schrödinger Equation in Cylindrical Symmetry.   229

        7.2.3   Separated-Atoms and United-Atoms Schemes
              and Correlation Diagram   231
   Further Reading   236

8 Electronic States in Diatomic Molecules   237
   8.1 H_2^+ as Prototype of MO Approach   237
        8.1.1   Eigenvalues and Energy Curves   237
        8.1.2   Bonding Mechanism and the Exchange
              of the Electron   245
   8.2   Homonuclear Molecules in the MO Scenario   248
   8.3 H_2 as Prototype of the VB Approach.   253
   8.4   Comparison of MO and VB Scenarios in H_2: Equivalence
       from Configuration Interaction   257
   8.5   Heteronuclear Molecules and the Electric Dipole Moment . . .   260
   Specific References and Further Reading   268

9 Electronic States in Selected Polyatomic Molecules   271
   9.1   Qualitative Aspects of NH_3 and H_2O Molecules   273
   9.2   Bonds Due to Hybrid Atomic Orbitals.   274
   9.3   Delocalization and the Benzene Molecule   278
   Appendix 9.1 Ammonia Molecule in Electric Field
              and the Ammonia Maser   281
   Further Reading   288

10 Nuclear Motions in Molecules and Related Properties   289
   10.1   Generalities and Introductory Aspects
        for Diatomic Molecules   289
   10.2   Rotational Motions   291
        10.2.1   Eigenfunctions and Eigenvalues   291
        10.2.2   Principles of Rotational Spectroscopy   292
        10.2.3   Thermodynamical Energy from Rotational
               Motions   295
        10.2.4   Orientational Electric Polarizability   296
        10.2.5   Extension to Polyatomic Molecules and Effect
               of the Electronic Motion in Diatomic Molecules   297
   10.3   Vibrational Motions.   301
        10.3.1   Eigenfunctions and Eigenvalues   301
        10.3.2   Principles of Vibrational Spectroscopy
               and Anharmonicity Effects   304
   10.4   Morse Potential.   308
   10.5   Roto-Vibrational Eigenvalues and Coupling Effects.   310
   10.6   Polyatomic Molecules: Normal Modes.   316
   10.7   Principles of Raman Spectroscopy.   320

10.8   Electronic Spectra and Franck—Condon Principle   324
10.9   Effects of Nuclear Spin Statistics in Homonuclear
        Diatomic Molecules.   327
Specific References and Further Reading   335

11 Crystal Structures.   337
11.1   Translational Invariance, Bravais Lattices
        and Wigner-Seitz Cell   338
11.2   Reciprocal Lattice and Brillouin Cell.   343
11.3   Typical Crystal Structures   345
Specific References and Further Readings.   351

12 Electron States in Crystals   353
12.1   Introductory Aspects and the Band Concept   353
12.2   Translational Invariance and the Bloch Orbital   356
12.3   Role and Properties of k   358
12.4   Periodic Boundary Conditions and Reduction
        to the First Brillouin zone   361
12.5   Density of States, Dispersion Relations and Critical Points. . .   363
12.6   The Effective Electron Mass.   365
12.7   Models of Crystals   368
        12.7.1   Electrons in Empty Lattice   368
        12.7.2   Weakly Bound Electrons   372
        12.7.3   Tightly Bound Electrons   375
Specific References and Further Reading   389

13 Miscellaneous Aspects Related to the Electronic Structure   391
13.1   Typology of Crystals.   391
13.2   Bonding Mechanisms and Cohesive Energies   394
        13.2.1   Ionic Crystals   394
        13.2.2   Lennard-Jones Interaction and Molecular Crystals . . .   396
13.3   Electron States of Magnetic Ions in a Crystal Field   401
13.4   Simple Picture of the Electric Transport.   405
Appendix 13.1 Magnetism from Itinerant Electrons   409
Specific References and Further Reading   415

14 Vibrational Motions of the Ions and Thermal Effects   417
14.1   Motions of the Ions in the Harmonic Approximation   417
14.2   Branches and Dispersion Relations   419
14.3   Models of Lattice Vibrations   419
        14.3.1   Monoatomic One-Dimensional Crystal.   420
        14.3.2   Diatomic One-Dimensional Crystal   422
        14.3.3   Einstein and Debye Crystals.   425
14.4   Phonons.   428

14.5   Thermal Properties Related to Lattice Vibrations.   430
14.6   The Mössbauer Effect   435
Specific References and Further Reading   444

15 Phase Diagrams, Response Functions and Fluctuations.   445
15.1   Phase Diagrams, Thermodynamic Responses
        and Critical Points: Introductory Remarks   446
15.2   Free Energy for Homogeneous Systems.   455
15.3   Non Homogeneous Systems and Fluctuations.   457
15.4   Time Dependence of the Fluctuations   460
15.5   Generalized Dynamical Susceptibility and Experimental
        Probes for Critical Dynamics   463
Appendix 15.1 From Single Particle to Collective Response   466
Specific References and Further Reading.   476

16 Dielectrics and Paraelectric-Ferroelectric Phase Transitions   477
16.1   Dielectric Properties of Crystals. Generalities   477
16.2   Clausius-Mossotti Relation and the Onsager
        Reaction Field   480
16.3   Dielectric Response for Model Systems   481
16.4   The Ferroelectric Transition in the Mean Field Scenario.   485
16.5   The Critical Dynamics Driving the Transition.   492
Appendix 16.1 Pseudo-Spin Dynamics for Order-Disorder
               Ferroelectrics   494
Appendix 16.2 Distribution of Correlation Times and Effects
               around the Transition.   498
Specific References and Further Reading   503

17 Magnetic Orders and Magnetic Phase Transitions   505
17.1   Introductory Aspects on Electronic Correlation   505
17.2   Mechanisms of Exchange Interaction.   510
17.3   Antiferromagnetism, Ferrimagnetism and Spin Glasses.   513
17.4   The Excitations in the Ordered States   517
17.5   Superparamagnetism and Frustrated Magnetism   520
Appendix 17.1 Phase Diagram and Related Effects in 2D Quantum
               Heisenberg Antiferromagnets (2DQHAF).   524
Appendix 17.2 Remarks on Scaling and Universality   530
Specific References and Further Reading   538

**18 Superconductors, the Superconductive Phase Transition
and Fluctuations**.   539
18.1   Historical Overview and Phenomenological Aspects   539
18.2   Microscopic Properties of the Superconducting State   545
        18.2.1   The Cooper Pair   545

	18.2.2	Some Properties of the Superconducting State	547
	18.2.3	The Particular Meaning of the Superconducting Wave Function	549
18.3	London Theory and the Flux Expulsion		551
18.4	Flux Quantization in Rings		553
18.5	The Josephson Junction		554
18.6	The SQUID Device		557
18.7	Type II Superconductors		559
18.8	High-Temperature Superconductors		560
18.9	Ginzburg-Landau (GL) Theory		564
	18.9.1	The GL Functional	564
	18.9.2	The GL Equations	566
	18.9.3	Uniform and Homogeneous SC and No Magnetic Field	567
	18.9.4	Surface Effects (in Bulk SC and in the Absence of Field)	568
	18.9.5	The London Penetration Length	570
18.10	The Parameter κ and the Vortex		571
18.11	Effects of Superconducting Fluctuations		574
	18.11.1	Introductory Remarks	574
	18.11.2	Paraconductivity and Fluctuating Diamagnetism	577
18.12	Superconducting Nanoparticles and the Zero-Dimensional Condition		580
Specific References and Further Reading			590

Index 591

Chapter 1
Atoms: General Aspects

Topics

Central Field Approximation
Effective Potential and One-Electron Eigenfunctions
Special Atoms (Hydrogenic, Muonic, Rydberg)
Magnetic Moments and Spin-Orbit Interaction
Electromagnetic Radiation, Matter and Transitions
Two-Level Systems and Related Aspects

The aim of this and of the following three chapters is the derivation of the main properties of the atoms and the description of their behavior in magnetic and electric fields. We shall begin with the assumption of point-charge nucleus with mass much larger than the electron mass and by taking into account only the Coulomb energy. Other interaction terms, of magnetic origin as well as the relativistic effects, will be initially disregarded.

In the light of the *central field approximation* it is appropriate to recall the results pertaining to one-electron atoms, namely the *hydrogenic atoms* (Sect. 1.4). When dealing with the properties of typical multi-electron atoms, such as *alkali atoms* or *helium atom* (Chap. 2) one shall realize that relevant modifications to that simplified framework are actually required. These are, for instance, the inclusion of the *spin-orbit interaction* (recalled at Sect. 1.6) and the effects due to the *exchange degeneracy* (Sect. 1.3, discussed in detail at Sect. 2.2).

The properties of a useful reference model, the two-level system, and some aspects of the electromagnetic radiation in interaction with matter are recalled in Appendices and/or in ad-hoc problems.

© Springer International Publishing Switzerland 2015
A. Rigamonti and P. Carretta, *Structure of Matter*,
UNITEXT for Physics, DOI 10.1007/978-3-319-17897-4_1

1.1 Central Field Approximation

The wave function $\psi(\mathbf{r}_1, \mathbf{r}_2, ..., \mathbf{r}_N)$ describing the stationary state of the N electrons in the atom follows from the Schrödinger equation

$$\left[\frac{-\hbar^2}{2m} \sum_i \nabla_i^2 - \sum_i \frac{Ze^2}{r_i} + \sum_{i \neq j}' \frac{e^2}{r_{ij}}\right] \psi(\mathbf{r}_1, \mathbf{r}_2, ..., \mathbf{r}_N) = E\psi(\mathbf{r}_1, \mathbf{r}_2, ..., \mathbf{r}_N)$$

$$\left[T_e + V_{ne} + V_{ee}\right] \psi(\mathbf{r}_1, \mathbf{r}_2, ..., \mathbf{r}_N) = E\psi(\mathbf{r}_1, \mathbf{r}_2, ..., \mathbf{r}_N) \quad (1.1)$$

where in the Hamiltonian one has the kinetic energy T_e, the potential energy V_{ne} describing the Coulomb interaction of the electrons with the nucleus of charge Ze and the electron-electron repulsive interaction V_{ee} (Fig. 1.1).

If the inter-electron interaction V_{ee} could be neglected, the total Hamiltonian would be $\mathcal{H} = \sum_i \mathcal{H}_i$, with $\mathcal{H}_i$ the one-electron Hamiltonian. Then $\psi(\mathbf{r}_1, \mathbf{r}_2, ...) = \prod_i \phi(\mathbf{r}_i)$, with $\phi(\mathbf{r}_i)$ one-electron eigenfunctions. V_{ee} does not allow one to separate the variables $\mathbf{r}_i$, in correspondence to the fact that the motion of a given electron does depend from the ones of the others. Furthermore V_{ee} is too large to be treated as a perturbation of $[T_e + V_{ne}]$. As we shall see (Sect. 2.2), even in the case of Helium atom, with only one pair of interacting electrons, the ground-state energy correction related to V_{ee} is about 30 % of the energy of the unperturbed state correspondent to $V_{ee} = 0$.

The search for an approximate solution of Eq. (1.1) can initiate by considering the form of the potential energy $V(\mathbf{r}_i)$, for a given electron, in the limiting cases of distances r_i from the nucleus much larger and much smaller than the average distance d of the other $(N - 1)$ electrons:

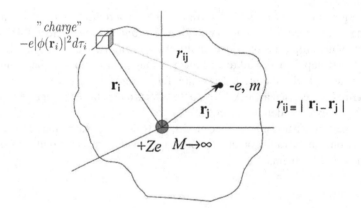

Fig. 1.1 Schematic view of multi-electron atom. The nucleus is assumed as a point charge Ze, with mass M much larger than the mass m of the electron, of charge $-e$

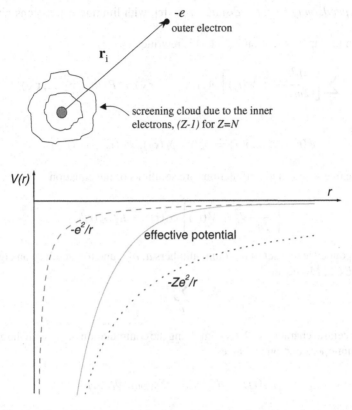

Fig. 1.2 Sketchy view of the electronic cloud screening the nuclear charge for an outer electron and correspondent forms of the potential energy in the limiting cases of large and small distances and of the effective central field potential energy (*solid line*). Details on the role of the screening cloud shall be given in describing the alkali atoms (Sect. 2.1)

$$r_i \gg d \qquad V(r_i) \simeq \frac{-e^2}{r_i}$$

$$r_i \ll d \qquad V(r_i) \simeq \frac{-Ze^2}{r_i} + \text{const.} \qquad (1.2)$$

having taken into account that for neutral atoms ($N = Z$) when $r_i \gg d$ the electrons screen ($Z - 1$) protons, while for $r_i \ll d$ ($N - 1$) electrons yield a constant effective potential, as expected for an average spherical charge distribution (Fig. 1.2). We shall discuss in detail the role of the screening cloud due to the inner electrons when dealing with alkali atoms (Sect. 2.1).

In the light of the form of the potential energy suggested by Eq. (1.2) and neglecting correlation effects in the electronic positions, one deals with the *central field approximation*, first considered by *Hartree* and *Slater*. In this context any electron is moving in an *effective average field*, due to the nucleus and to the other electrons, which *depends only from the distance* $r \equiv |\mathbf{r}|$, with limiting expressions given by Eq. (1.2).

Within this approximation Eq. (1.1) is rewritten

$$\sum_i \left[\frac{-\hbar^2}{2m} \nabla_i^2 + V(r_i) \right] \psi(\mathbf{r}_1, \mathbf{r}_2, ..., \mathbf{r}_N) = E \psi(\mathbf{r}_1, \mathbf{r}_2, ..., \mathbf{r}_N) \qquad (1.3)$$

implying

$$\psi(\mathbf{r}_1, \mathbf{r}_2, ..., \mathbf{r}_N) = \phi_a(\mathbf{r}_1)\phi_b(\mathbf{r}_2)...\phi_c(\mathbf{r}_i)...\phi_z(\mathbf{r}_N), \qquad (1.4)$$

where the one-electron eigenfunctions are solutions of the equation

$$\left\{ \frac{-\hbar^2}{2m} \nabla_i^2 + V(r_i) \right\} \phi_a(\mathbf{r}_i) = E_i^a \phi_a(\mathbf{r}_i) \qquad (1.5)$$

in correspondence to a set of quantum numbers $a, b...$, and to one-electron eigenvalues E_1^a, E_2^b.... Moreover

$$E = \sum_i E_i^{a...}. \qquad (1.6)$$

From the central character of $V(r_i)$, implying the commutation of $\mathcal{H}_i$ with the angular momentum operators, one deduces

$$\phi_a(\mathbf{r}_i) = R_{n(i)l(i)}(r_i) Y_{l(i)m(i)}(\theta_i, \varphi_i) \qquad (1.7)$$

where $Y_{l(i)m(i)}(\theta_i, \varphi_i)$ are the spherical harmonics and then the set of quantum numbers is $a \equiv n_i, l_i, m_i$.

Thus the one-electron states are labeled by the numbers (n_1, l_1), (n_2, l_2) etc. or by the equivalent symbols $(1s)$, $(2s)$, $(2p)$ etc.

The spherical symmetry associated with $\sum_i V(r_i)$ also implies that for the total angular momentum $\mathbf{L} = \sum_i \mathbf{l}_i$, $|\mathbf{L}|$ and L_z are constants of motion. Then one can label the atomic states with quantum numbers $L = 0, 1, 2...$ $L(L+1)\hbar^2$ is the square of the angular momentum of the whole atom, while the number M (the equivalent for the atom of the one-electron number m) characterizes the component $M\hbar$ of L along a given direction (usually indicated by z). It is noted that at this point we have no indication on how L and M result from the correspondent numbers l_i and m_i. The composition of the angular momenta will be discussed at Chap. 3. Anyway, since now we realize that the atomic states can be classified in the form S, P, D, F etc. in correspondence to the values $L = 0, 1, 2, 3$ etc.

1.2 Self-Consistent Construction of the Effective Potential

In the assumption that the one-electron wavefunctions $\phi_a(\mathbf{r}_i)$ have been found one can achieve a self-consistent construction of the effective potential energy $V(r_i)$. As it is known $-e|\phi_a(\mathbf{r})|^2 d\tau$ can be thought as the fraction of electronic charge in the volume element $d\tau$. Owing to the classical analogy, one can write the potential energy for a given jth electron as[1] (see Fig. 1.1)

$$V(r_j) = -\frac{Ze^2}{r_j} + \sum_{i \neq j} \int \frac{e^2|\phi_a(\mathbf{r}_i)|^2}{r_{ij}} d\tau_i \qquad (1.8)$$

This relationship between $V(r)$ and ϕ_a suggests that once a given $V(r)$ is assumed, Eq. (1.5) can be solved (by means of numerical methods) to obtain $\phi(\mathbf{r}_i)$ in the form (1.7). Then one can build up a new expression for $V(r_i)$ and iterate the procedure till the radial parts of the wavefunctions at the nth step differ from the ones at the (n − 1)th step in a negligible way. This is the physical content of the *self-consistent method* devised by *Hartree* to obtain the radial part of the one-electron eigenfunctions or, equivalently, the best *approximate* expression for $V(r_i)$. Here we only mention that a more appropriate procedure has to be carried out using eigenfunctions which include the spin variables and the dynamical equivalence (Sect. 1.3), with the *antisymmetry* requirement. Such a generalization of the Hartree method has been introduced by *Fock* and *Slater* and it is known as *Hartree-Fock* method. The appropriate many-electrons eigenfunctions have the *determinantal* form (see Sect. 2.3). A detailed derivation of the effective potential energy for the simplest case of two electrons on the basis of Eq. (1.8) is given in Problem 2.7.

The potential energy $V(r_i)$ can be conveniently described through an effective nuclear charge $Z_{eff}(r)$ by means of the relation

$$V(r) = -\frac{e^2}{r} Z_{eff}(r) \qquad (1.9)$$

(now the index i is dropped). The sketchy behavior of the effective nuclear charge is shown in Fig. 1.3. The dependence on r at intermediate distance has to be derived, for instance, by means of the self-consistent method or by other numerical methods.

---

[1] Equation (1.8) can also be derived by applying the variational principle to the energy function constructed on the basis of the ϕ_a's with the complete Hamiltonian, for a variation $\delta\phi_a$ leaving the one-electron eigenfunction normalized.

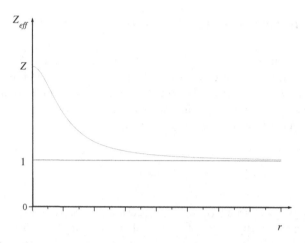

Fig. 1.3 Sketchy behavior of the effective nuclear charge acting on a given electron at the distance r from the nucleus of charge Ze, arising from the screening due to other electrons. The charge $(Z - N - 1)$ (1 for neutral atom with $Z = N$) is often called *residual charge* (for a quantitative estimate of $Z_{eff}(r)$ for the ground state of Helium see Problem 2.7)

1.3 Degeneracy from Dynamical Equivalence

From Eqs. (1.3), (1.5) and (1.7) the N-electron wavefunction implies the assignment of a set of quantum numbers a_i to each ith electron. This assignment cannot be done in a unique way, since the electrons are *indistinguishable*, the Hamiltonian $\mathcal{H} = \sum_i \mathcal{H}_i$ being invariant upon exchange of the indexes (*exchange symmetry*). Therefore, for a state of the atom correspondent to a given eigenvalue, one has to write an eigenfunction combining with equal weights all the possible configurations, with the quantum numbers a_i variously assigned to different electrons. Therefore

$$\psi(\mathbf{r}_1, \mathbf{r}_2, ..., \mathbf{r}_N) = \sum_P P\left[\phi_{a_1}(\mathbf{r}_1)\phi_{a_2}(\mathbf{r}_2)...\phi_{a_N}(\mathbf{r}_N)\right] \qquad (1.10)$$

where P is an operator permuting electrons and quantum numbers.

It should be stressed that this remark on the role of the dynamical equivalence is incomplete and somewhat misleading. In fact we shall reformulate it after the introduction of a further quantum number, the spin number. Moreover, we will have to take into account the *Pauli principle*, that limits the acceptable wavefunctions obtained upon permutation to the ones changing sign (*antisymmetric*). This topic will be discussed after the analysis of Helium atom, with two electrons (Sect. 2.2). The eigenfunction in form of the *Slater determinant* (Sect. 2.3) does take into account the exchange degeneracy and the antisymmetry requirement.

We conclude these preliminary aspects observing that a proper quantum treatment, within a perturbative approach, at least should take into account the modifications to the central field approximation due to the Hamiltonian

$$\mathcal{H}_P = -\sum_i \frac{Ze^2}{r_i} + \sum_{i \neq j}' \frac{e^2}{r_{ij}} - \sum_i V(r_i), \tag{1.11}$$

resulting from the difference between the Hamiltonian in (1.1) and the one in (1.3). This is the starting point of the *Slater theory* for *multiplets*.

1.4 Hydrogenic Atoms: Illustration of Basic Properties

The central field approximation allows one to reduce the Schrödinger equation to the form given by Eqs. (1.3) and (1.5). This latter suggests the opportunity to recall the basic properties for one-electron atoms, with Z protons at the nuclear site (Hydrogenic atoms). The Schrödinger equation is rewritten

$$\left[\frac{-\hbar^2}{2m} \nabla^2_{r,\theta,\varphi} - \frac{Ze^2}{r} \right] \phi_{n,l,m}(r, \theta, \varphi) = E_n \phi_{n,l,m}(r, \theta, \varphi) \tag{1.12}$$

with $\phi_{n,l,m}$ of the form in Eq. (1.7). To abide by the description for the Hydrogen atom, one can substitute everywhere the proton charge $(+e)$ by Ze in the eigenvalues and in the wavefunctions. Then

$$E_n = -\frac{m(Ze)^2 e^2}{2\hbar^2} \frac{1}{n^2} = -Z^2 R_H hc \frac{1}{n^2} \tag{1.13}$$

(with R_H Rydberg constant, given by $109{,}678\,\text{cm}^{-1}$, correspondent to $13.598\,\text{eV}$). The spherical harmonics entering the wavefunction $\phi_{n,l,m}$ (see Eq. (1.7)) are reported in Tables 1.1 and 1.2, up to $l = 3$.

The radial functions $R_{nl}(r)$ in Eq. (1.7) result from the solution of

$$\frac{d^2 R}{dr^2} + \frac{2}{r} \frac{dR}{dr} + \left[\frac{2m}{\hbar^2} (E + \frac{Ze^2}{r}) - \frac{l(l+1)}{r^2} \right] R = 0 \tag{1.14}$$

or

$$\frac{-\hbar^2}{2mr^2} \frac{d}{dr} r^2 \frac{dR}{dr} + \left[\frac{l(l+1)\hbar^2}{2mr^2} - \frac{Ze^2}{r} \right] R = ER, \tag{1.15}$$

namely a one-dimensional (1D) equation with an effective potential energy V_{eff} which includes the centrifugal term related to the non-inertial frame of reference of the radial axis. The shape of V_{eff} is shown in Fig. 1.4. In comparison to the Hydrogen atom, Eq. (1.12) shows that in Hydrogenic atoms one has to rescale the distances by the factor Z. Instead of $a_0 = \hbar^2/me^2 = 0.529\,\text{Å}$ (radius of the first orbit in the Bohr atom, corresponding to an energy $-R_H hc = -e^2/2a_0$), the characteristic length thus becomes (a_0/Z).

Table 1.1 Normalized spherical harmonics, up to $l = 3$

$s(l = 0)$	$Y_{00} = \frac{1}{\sqrt{4\pi}}$
$p(l = 1)$	$Y_{1-1} = \sqrt{\frac{3}{8\pi}} \frac{x-iy}{r} = \sqrt{\frac{3}{8\pi}} sin\theta e^{-i\phi}$
	$Y_{10} = \sqrt{\frac{3}{4\pi}} \frac{z}{r} = \sqrt{\frac{3}{4\pi}} cos\theta$
	$Y_{11} = -\sqrt{\frac{3}{8\pi}} \frac{x+iy}{r} = -\sqrt{\frac{3}{8\pi}} sin\theta e^{i\phi}$
$d(l = 2)$	$Y_{2-2} = \sqrt{\frac{15}{32\pi}} \frac{(x-iy)^2}{r^2} = \sqrt{\frac{15}{32\pi}} sin^2\theta e^{-2i\phi}$
	$Y_{2-1} = \sqrt{\frac{15}{8\pi}} \frac{z(x-iy)}{r^2} = \sqrt{\frac{15}{8\pi}} sin\theta cos\theta e^{-i\phi}$
	$Y_{20} = \sqrt{\frac{5}{16\pi}} \frac{3z^2-r^2}{r^2} = \sqrt{\frac{5}{16\pi}} (3cos^2\theta - 1)$
	$Y_{21} = -\sqrt{\frac{15}{8\pi}} \frac{z(x+iy)}{r^2} = -\sqrt{\frac{15}{8\pi}} sin\theta cos\theta e^{i\phi}$
	$Y_{22} = \sqrt{\frac{15}{32\pi}} \frac{(x+iy)^2}{r^2} = \sqrt{\frac{15}{32\pi}} sin^2\theta e^{2i\phi}$
$f(l = 3)$	$Y_{3-3} = \sqrt{\frac{35}{64\pi}} \frac{(x-iy)^3}{r^3} = \sqrt{\frac{35}{64\pi}} sin^3\theta e^{-3i\phi}$
	$Y_{3-2} = \sqrt{\frac{105}{32\pi}} \frac{z(x-iy)^2}{r^3} = \sqrt{\frac{105}{32\pi}} sin^2\theta cos\theta e^{-2i\phi}$
	$Y_{3-1} = \sqrt{\frac{21}{64\pi}} \frac{(5z^2-r^2)(x-iy)}{r^3} = \sqrt{\frac{21}{64\pi}} (5cos^2\theta - 1)sin\theta e^{-i\phi}$
	$Y_{30} = \sqrt{\frac{7}{16\pi}} \frac{(5z^2-3r^2)z}{r^3} = \sqrt{\frac{7}{16\pi}} (5cos^2\theta - 3)cos\theta$
	$Y_{31} = -\sqrt{\frac{21}{64\pi}} \frac{(5z^2-r^2)(x+iy)}{r^3} = -\sqrt{\frac{21}{64\pi}} (5cos^2\theta - 1)sin\theta e^{i\phi}$
	$Y_{32} = \sqrt{\frac{105}{32\pi}} \frac{z(x+iy)^2}{r^3} = \sqrt{\frac{105}{32\pi}} sin^2\theta cos\theta e^{2i\phi}$
	$Y_{33} = -\sqrt{\frac{35}{64\pi}} \frac{(x+iy)^3}{r^3} = -\sqrt{\frac{35}{64\pi}} sin^3\theta e^{3i\phi}$

Since $|\phi(r, \theta, \phi)|^2 d\tau$ corresponds to the probability to find the electron inside the volume element $d\tau = r^2 sin\theta dr d\theta d\phi$, from the form of the eigenfunctions the physical meaning of the spherical harmonics is grasped: $Y^*Y sin\theta d\theta d\phi$ yields the probability that the vector **r**, ideally following the electron in its motion, falls within the elemental solid angle $d\Omega$ around the direction defined by the polar angles θ and ϕ, as shown in below.

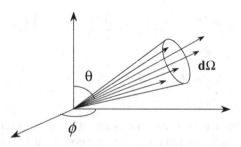

In the states labeled by the quantum numbers (n, l, m) the eigenvalue equations for the modulus square and for the z-component of the angular momentum are

Table 1.2 Normalized spherical harmonics in the real form (see text)

$s(l=0)$	$Y_{00} = \frac{1}{\sqrt{4\pi}}$
$p(l=1)$	$Y_x = \sqrt{\frac{3}{4\pi}}\frac{x}{r} = \sqrt{\frac{3}{4\pi}}sin\theta cos\phi$
	$Y_y = \sqrt{\frac{3}{4\pi}}\frac{y}{r} = \sqrt{\frac{3}{4\pi}}sin\theta sin\phi$
	$Y_z = \sqrt{\frac{3}{4\pi}}\frac{z}{r} = \sqrt{\frac{3}{4\pi}}cos\theta$
$d(l=2)$	$Y_{z^2} = \sqrt{\frac{5}{16\pi}}\frac{3z^2-r^2}{r^2} = \sqrt{\frac{5}{16\pi}}(3cos^2\theta - 1)$
	$Y_{zx} = \sqrt{\frac{15}{4\pi}}\frac{zx}{r^2} = \sqrt{\frac{15}{4\pi}}sin\theta cos\theta cos\phi$
	$Y_{zy} = \sqrt{\frac{15}{4\pi}}\frac{zy}{r^2} = \sqrt{\frac{15}{4\pi}}sin\theta cos\theta sin\phi$
	$Y_{x^2-y^2} = \sqrt{\frac{15}{16\pi}}\frac{x^2-y^2}{r^2} = \sqrt{\frac{15}{16\pi}}sin^2\theta cos2\phi$
	$Y_{xy} = \sqrt{\frac{15}{4\pi}}\frac{xy}{r^2} = \sqrt{\frac{15}{16\pi}}sin^2\theta sin2\phi$
$f(l=3)$	$Y_{z^3} = \sqrt{\frac{7}{16\pi}}\frac{(5z^2-3r^2)z}{r^3} = \sqrt{\frac{7}{16\pi}}(5cos^2\theta - 3)cos\theta$
	$Y_{z^2x} = \sqrt{\frac{21}{32\pi}}\frac{(5z^2-r^2)x}{r^3} = \sqrt{\frac{21}{32\pi}}(5cos^2\theta - 1)sin\theta cos\phi$
	$Y_{z^2y} = \sqrt{\frac{21}{32\pi}}\frac{(5z^2-r^2)y}{r^3} = \sqrt{\frac{21}{32\pi}}(5cos^2\theta - 1)sin\theta sin\phi$
	$Y_{z(x^2-y^2)} = \sqrt{\frac{105}{16\pi}}\frac{z(x^2-y^2)}{r^3} = \sqrt{\frac{105}{16\pi}}sin^2\theta cos\theta cos2\phi$
	$Y_{zxy} = \sqrt{\frac{105}{4\pi}}\frac{zxy}{r^3} = \sqrt{\frac{105}{16\pi}}sin^2\theta cos\theta sin2\phi$
	$Y_{x^2y} = \sqrt{\frac{35}{32\pi}}\frac{(3x^2y-y^3)}{r^3} = \sqrt{\frac{35}{32\pi}}sin^3\theta cos3\phi$
	$Y_{y^2x} = \sqrt{\frac{35}{32\pi}}\frac{(x^3-3y^2x)}{r^3} = \sqrt{\frac{35}{32\pi}}sin^3\theta sin3\phi$

$$\hat{l}^2\phi_{nlm} = R_{nl}(r)\hat{l}^2 Y_{lm}(\theta,\varphi) = R_{nl}(r)l(l+1)\hbar^2 Y_{lm}(\theta,\varphi);$$
$$\hat{l}_z\phi_{nlm} = R_{nl}(r)\hat{l}_z Y_{lm}(\theta,\varphi) = R_{nl}(r)\hat{l}_z \Theta_{lm}(\theta)e^{im\varphi} =$$
$$= R_{nl}(r)\Theta_{lm}(\theta)\hat{l}_z e^{im\varphi} = R_{nl}(r)Y_{lm}(\theta,\varphi)m\hbar \qquad (1.16)$$

Finally, from Eq. (1.13) it is noted that a given state of the Hydrogenic atom is Z^2 times more bound than the correspondent state in the Hydrogen atom. This happens because on the average, the electron is Z-times closer to a nuclear charge increased by a factor Z.

The normalized wave functions for Hydrogenic atoms are reported in Table 1.3. It is remarked that for $r \ll a_0/Z$ one has

$$(\phi_{nlm})_{r\to 0} \propto R_{nl} \propto r^l \qquad (1.17)$$

while for large distance

$$(\phi_{nlm})_{r\to\infty} \propto R_{nl} \propto e^{-\frac{r}{a_0}\frac{Z}{n}} \qquad (1.18)$$

Fig. 1.4 Effective potential
energy in the 1D
Schrödinger equation for
$R(r)$ (Eq. (1.15)), for the
low-energy states.
Horizontal lines indicate the
eigenvalues for $n = 1, 2$ and
3, given by $-Z^2 R_H hc/n^2$
$(a_0 = \hbar^2/me^2)$

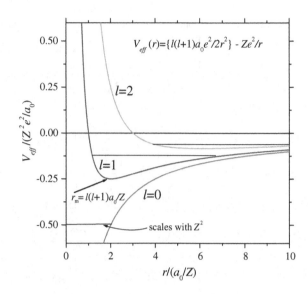

From the wavefunctions relevant properties of the states, such as the radial probability
density

$$P_{nl}(r) = \int d\varphi \int d\theta sin\theta r^2 |\phi_{nl}|^2, \qquad (1.19)$$

or the expectation values of any positional function $f(\mathbf{r})$

$$< f >_{nl} = \int |\phi_{nl}|^2 f(\mathbf{r}) d\tau \qquad (1.20)$$

can be derived. The radial probability densities for the $1s$, $2s$ and $2p$ states are
depicted in Fig. 1.5.

For spherical symmetry $P_{nl}(r)$ can be written as $4\pi r^2 |\phi_{nl}|^2$. It should be remarked
that for $Z = 1$ the maximum in P_{1s} occurs at $r = a_0$, corresponding to the radius
of the first orbit in the Bohr model (see Problem 1.4). For the states at $n = 2$ the
correspondence of the maximum in $P_{nl}(r)$ with the radius of the Bohr orbit pertains
to the $2p$ states.

The first excited state ($n = 2$), corresponding to $E_2 = -(Z^2e^2/2a_0)(1/4)$, is
the superposition of four degenerate states: $2s, 2p_1, 2p_0$ and $2p_{-1}$. To describe the
$2p$ states, instead of the functions $\phi_{2p,m=\pm1,0}$ (see Table 1.3) one may use the linear
combinations (see Table 1.2).

Table 1.3 Normalized eigenfunctions for Hydrogenic atoms, for $n = 1, 2$ and 3

n	l	m	Eigenfunctions
1	0	0	$\phi_{100} = \frac{1}{\sqrt{\pi}}(\frac{Z}{a_0})^{3/2} e^{-Zr/a_0}$
2	0	0	$\phi_{200} = \frac{1}{4\sqrt{2\pi}}(\frac{Z}{a_0})^{3/2}(2 - \frac{Zr}{a_0})e^{-Zr/2a_0}$
2	1	0	$\phi_{210} = \frac{1}{4\sqrt{2\pi}}(\frac{Z}{a_0})^{3/2}\frac{Zr}{a_0}e^{-Zr/2a_0}\cos\theta$
2	1	± 1	$\phi_{21\pm1} = \mp\frac{1}{8\sqrt{\pi}}(\frac{Z}{a_0})^{3/2}\frac{Zr}{a_0}e^{-Zr/2a_0}\sin\theta e^{\pm i\varphi}$
3	0	0	$\phi_{300} = \frac{1}{81\sqrt{3\pi}}(\frac{Z}{a_0})^{3/2}(27 - 18\frac{Zr}{a_0} + 2\frac{Z^2r^2}{a_0^2})e^{-Zr/3a_0}$
3	1	0	$\phi_{310} = \frac{\sqrt{2}}{81\sqrt{\pi}}(\frac{Z}{a_0})^{3/2}(6 - \frac{Zr}{a_0})\frac{Zr}{a_0}e^{-Zr/3a_0}\cos\theta$
3	1	± 1	$\phi_{31\pm1} = \mp\frac{1}{81\sqrt{\pi}}(\frac{Z}{a_0})^{3/2}(6 - \frac{Zr}{a_0})\frac{Zr}{a_0}e^{-Zr/3a_0}\sin\theta e^{\pm i\varphi}$
3	2	0	$\phi_{320} = \frac{1}{81\sqrt{6\pi}}(\frac{Z}{a_0})^{3/2}(\frac{Z^2r^2}{a_0^2})e^{-Zr/3a_0}(3\cos^2\theta - 1)$
3	2	± 1	$\phi_{32\pm1} = \mp\frac{1}{81\sqrt{\pi}}(\frac{Z}{a_0})^{3/2}(\frac{Z^2r^2}{a_0^2})e^{-Zr/3a_0}\sin\theta\cos\theta e^{\pm i\varphi}$
3	2	± 2	$\phi_{32\pm2} = \frac{1}{162\sqrt{\pi}}(\frac{Z}{a_0})^{3/2}(\frac{Z^2r^2}{a_0^2})e^{-Zr/3a_0}\sin^2\theta e^{\pm 2i\varphi}$

$$\phi_{2px} = \frac{1}{\sqrt{2}}[\phi_{2p,1} + \phi_{2p,-1}] \propto \sin\theta\cos\varphi \propto x$$

$$\phi_{2py} = \frac{i}{\sqrt{2}}[\phi_{2p,1} - \phi_{2p,-1}] \propto \sin\theta\sin\varphi \propto y$$

$$\phi_{2pz} = \phi_{2p,0} \propto \cos\theta \propto z \tag{1.21}$$

From these expressions, also in the light of the $P_{2p}(r)$ depicted in Fig. 1.5 and in view of the equivalence between the x, y and z directions, one can represent the *atomic orbitals* (the quantum equivalent of the classical orbits) in the form reported in Fig. 1.6.

The degeneracy in x, y, z is *necessary*, in view of the spherical symmetry of the potential. On the contrary the degeneracy in l, namely same energy for $s, p, d...$ states for a given n, is *accidental*, being the consequence of the Coulombic form of the potential. We shall see that when the potential takes a different radial dependence because of $Z_{eff}(r)$, then the degeneracy in l is removed (Sect. 2.1).

It is reminded that the difference between the $2p_{1,0,-1}$ and the $2p_{x,y,z}$ representation involves the eigenvalue for $\hat{l}_z$. The former are eigenfunctions of $\hat{l}_z$ while the latter are not, as shown for instance for ϕ_{2px}:

$$\hat{l}_z\phi_{2px} = -i\hbar\frac{\partial}{\partial\varphi}\phi_{2px} = -i\hbar\frac{\partial}{\partial\varphi}f(r)\sin\theta\cos\varphi = +i\hbar\phi_{2py} \tag{1.22}$$

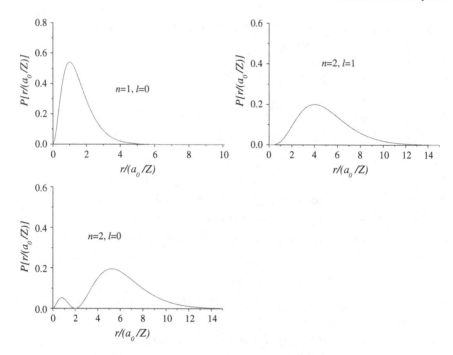

Fig. 1.5 Radial probability densities for $1s$, $2s$ and $2p$ states in Hydrogenic atoms

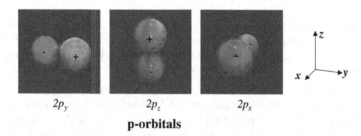

p-orbitals

Fig. 1.6 Illustrative plots, for the $2p$ states of Hydrogenic atoms, of the atomic orbitals, defined as the shape of the surfaces where $|\phi_{nl}|^2 = constant$, meantime with probability of presence of the electron in the internal volume given by 0.9. It should be remarked that the sign $+$ or $-$, related to the sign of Y_{2p}, can actually be interchanged. However the *relationship* of the sign along the different directions is relevant, since it fixes the parity of the state under the operation of reversing the direction of the axes or, equivalently, of bringing $\mathbf{r}$ in $-\mathbf{r}$

Obviously the difference is only in the description and no real modification occurs in regard of the measurements. This is inferred, for example, from the definition of ϕ_{2px} in terms of the basis of the eigenfunctions for $\hat{l}_z$ (see Eq. (1.21)).

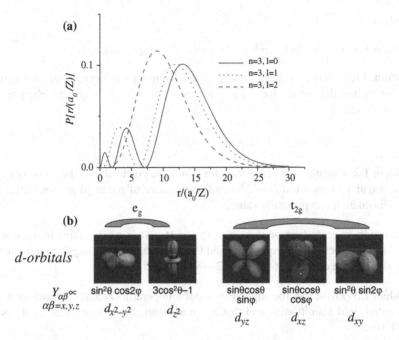

Fig. 1.7 Radial probability densities (**a**) for the $n = 3$ states of Hydrogenic atoms. In part (**b**) of the Figure the angular distribution of the $3d$ atomic orbitals is reported. The d_{z^2} and $d_{x^2-y^2}$, grouped together are commonly called e_g levels, while the d_{xy}, d_{xz} and d_{yz} are called t_{2g} levels (we shall return to these aspects at Sect. 13.3)

Finally in Fig. 1.7a the radial probability densities for the $n = 3$ states are plotted. The linear combinations of $3d$ states with different m's, leading to the most common representation, with the correspondent atomic orbitals are shown in Fig. 1.7b.

Some expectation values of current use are reported in Table 1.4.

Table 1.4 Expectation values of some quantities in Hydrogenic atoms

$$< r >_{nlm} \equiv \int \phi^*_{nlm}(\mathbf{r}) r \phi_{nlm}(\mathbf{r}) d\tau \equiv \int_0^\infty |R_{nl}|^2 r^3 dr =$$
$$= n^2 \frac{a_0}{Z} [1 + \frac{1}{2}(1 - \frac{l(l+1)}{n^2})] = \frac{a_0}{2Z}[3n^2 - l(l+1)]$$

$$< r^2 >_{nlm} = \frac{n^2}{2}(\frac{a_0}{Z})^2[5n^2 + 1 - 3l(l+1)]$$

$$< r^{-1} >_{nlm} = [n^2 \frac{a_0}{Z}]^{-1}$$

$$< r^{-2} >_{nlm} = \frac{Z^2}{a_0^2}[n^3(l + \frac{1}{2})]^{-1}$$

$$< V >_{nlm} = -\frac{Z^2 e^2}{a_0 n^2}$$

$$< T >_{nlm} = \frac{Z^2 e^2}{2a_0 n^2}$$

$$< r^{-3} >_{nlm} = \frac{Z^3}{a_0^3 n^3 [l(l+1)(l+\frac{1}{2})]} \qquad (l \neq 0)$$

For $l = 0$ one has the divergence in the lower limit of the integral, since in $< r^{-3} >_{nlm} = \int \phi^*_{nl} \frac{1}{r^3} \phi_{nl} r^2 \sin\theta dr d\theta d\varphi$
$\phi_{nl} \propto r^l$ for $r \to 0$ (see Eq. (1.17)).

Problems

Problem 1.1 For two independent electrons prove Eqs. (1.4) and (1.6).

Solution: From $\mathcal{H}_1\phi_1 = E_1\phi_1$ and $\mathcal{H}_2\phi_2 = E_2\phi_2$, by multiplying the first equation for ϕ_2 and the second for ϕ_1, one writes $\mathcal{H}_1\phi_1\phi_2 = E_1\phi_1\phi_2$ and $\mathcal{H}_2\phi_1\phi_2 = E_2\phi_1\phi_2$. Then

$$\mathcal{H}\phi = (\mathcal{H}_1 + \mathcal{H}_2)\phi_1\phi_2 = (E_1 + E_2)\phi_1\phi_2 = E\phi.$$

Problem 1.2 One electron is in a state for which the eigenvalue of the z-component of the angular momentum $\hat{l}_z$ is $3\hbar$, while the square of the angular momentum is $12\hbar^2$. Evaluate the expectation value of $\hat{l}_x^2$.

Solution: In $\hbar^2$ units, from $< \hat{l}_x^2 > + < \hat{l}_y^2 > = \hat{l}^2 - < \hat{l}_z^2 >$. By taking into account that x and y directions are equivalent and that the square of the angular momentum has to be 12, one deduces $< \hat{l}_x^2 > = (12 - 9)/2 = 1.5$.

Problem 1.3 Prove that the angular momentum operators $\hat{l}_z$ and $\hat{l}^2$ commute with the central field Hamiltonian and that a common set of eigenfunctions exists (see Eq. (1.16)).

Solution: In Cartesian coordinates, omitting $i\hbar$

$$\hat{l}_x\hat{l}_y - \hat{l}_y\hat{l}_x =$$

$$= \left(-y\frac{\partial}{\partial z} + z\frac{\partial}{\partial y}\right)\left(-z\frac{\partial}{\partial x} + x\frac{\partial}{\partial z}\right) - \left(-z\frac{\partial}{\partial x} + x\frac{\partial}{\partial z}\right)\left(-y\frac{\partial}{\partial z} + z\frac{\partial}{\partial y}\right) =$$

$$= y\frac{\partial}{\partial x} + yz\frac{\partial^2}{\partial z\partial x} - xy\frac{\partial^2}{\partial z^2} - z^2\frac{\partial^2}{\partial y\partial x} + xz\frac{\partial^2}{\partial y\partial z} -$$

$$- \left[zy\frac{\partial^2}{\partial z\partial x} - z^2\frac{\partial^2}{\partial x\partial y} - xy\frac{\partial^2}{\partial z^2} + x\frac{\partial}{\partial y} + xz\frac{\partial^2}{\partial z\partial y}\right] =$$

$$= \left(y\frac{\partial}{\partial x} - x\frac{\partial}{\partial y}\right) = \hat{l}_z$$

In analogous way the commutation rules for the components are found:

$$[\hat{l}_x, \hat{l}_y] = i\hbar\hat{l}_z, \qquad [\hat{l}_z, \hat{l}_x] = i\hbar\hat{l}_y, \qquad [\hat{l}_y, \hat{l}_z] = i\hbar\hat{l}_x.$$

In spherical polar coordinates, from

$$\hat{l}^2 = -\hbar^2\left[\frac{1}{sin\theta}\frac{\partial(sin\theta\frac{\partial}{\partial\theta})}{\partial\theta} + \frac{1}{sin^2\theta}\frac{\partial^2}{\partial\varphi^2}\right]$$

while $\hat{l}_z = -i\hbar\partial/\partial\varphi$, one finds $[\hat{l}^2, \hat{l}_z] = 0$.

For the central field Hamiltonian (omitting irrelevant constants) $\mathcal{H} = -\nabla^2 + V(r)$ in Cartesian coordinates for the kinetic energy

$$Tl_z = y\nabla^2\frac{\partial}{\partial x} - x\nabla^2\frac{\partial}{\partial y} = (y\frac{\partial}{\partial x} - x\frac{\partial}{\partial y})\nabla^2 = l_z T,$$

for the φ-independent $V(r)$ the commutation with $\hat{l}_z$ directly follows.

Now we prove that when an operator M commutes with the Hamiltonian a set of common eigenstates can be found, so that the two operators describe observables with well defined values (the statement holds for any pair of commuting operators).

From $M\mathcal{H} - \mathcal{H}M = 0$ any matrix element involving the Hamiltonian eigenfunctions reads

$$< i|M\mathcal{H} - \mathcal{H}M|j > = < i|M\mathcal{H}|j > - < i|\mathcal{H}M|j > = 0.$$

From the multiplication rule

$$\sum_l < i|M|l >< l|\mathcal{H}|j > - \sum_k < i|\mathcal{H}|k >< k|M|j > = 0.$$

$\mathcal{H}$ being diagonal one writes

$$< i|M|j >< j|\mathcal{H}|j > - < i|\mathcal{H}|i >< i|M|j > = 0,$$

namely $< i|M|j > (E_i - E_j) = 0$, that for $i \neq j$ proves the statement, when $E_j \neq E_i$ (for degenerate states the proof requires taking into account linear combinations of the eigenfunctions).

Problem 1.4 In the Bohr model for the Hydrogen atom the electron moves along circular orbits (*stationary states*) with no emission of electromagnetic radiation. The *Bohr-Sommerfeld condition* reads

$$\oint p_\theta d\theta = lh \qquad\qquad l = 1, 2, ...$$

p_θ being the moment conjugate to the polar angle θ in the plane of motion. Show that this quantum condition implies that the angular momentum is an integer multiple of $\hbar$ and derive the radius of the orbits and the correspondent energies of the atom.

Plot the energy levels in a scale of increasing energy and indicate the transitions allowed by the selection rule $\Delta l = \pm 1$, estimating the wavelengths of the first lines in the *Balmer spectroscopic series* (transitions $n'' \rightarrow n'$, with $n' = 2$).

Compare the energy levels for H with the ones for He^+ and for Li^{2+}.

Finally consider the motion of the electron in three-dimensions and by applying the quantum condition to the polar angles, by means of vectorial arguments obtain the quantization $\hat{l}_z = m\hbar$ for the z-component of the angular momentum.

Solution: From the Lagrangian $\mathcal{L} = I(\partial\theta/\partial t)^2/2 + e^2/r$ one has $p_\theta = I\partial\theta/\partial t$, with I moment of inertia and $\partial\theta/\partial t = \omega = constant$. The quantum condition becomes

$$I\frac{\partial\theta}{\partial t} \oint d\theta = lh$$

so that $mr^2\omega 2\pi = lh$ and $mvr = l\hbar$. From the latter equation and the equilibrium condition for the orbits, where $mv^2/r = e^2/r^2$, the radii turn out

$$r_n = \frac{m^2 v^2 r^2}{me^2} = \frac{n^2\hbar^2}{me^2} = n^2 a_0$$

with $a_0 = \hbar^2/me^2 = 0.529\,\text{Å}$. The energy is

$$E = T + V = \frac{1}{2}mv^2 - \frac{e^2}{r} = -\frac{e^2}{2r}$$

(in agreement with the virial theorem, $< T > = < V > n/2$, with n exponent in $V \propto r^n$) and thus

$$E_n = -\frac{e^2}{2r_n} = -\frac{e^4 m}{2\hbar^2}\frac{1}{n^2}$$

as from Eq. (1.13), for $Z = 1$. A pictorial view of the orbits is the following (not in scale):

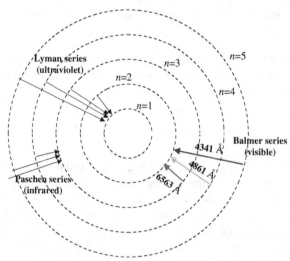

Below the energy levels are depicted:

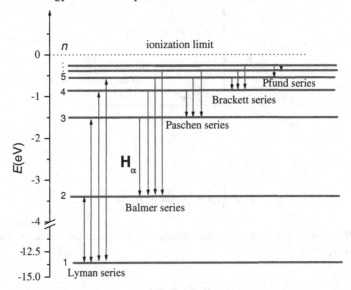

The Balmer series is shown below

Comparison of the energy levels with the ones in He$^+$ and in Li^{2+}:

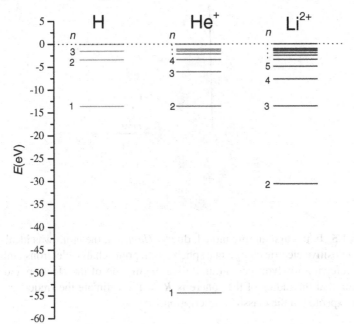

In three-dimensions

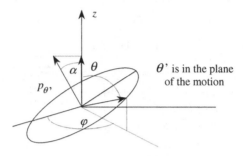

$$T = \frac{1}{2}[p_r\dot{r} + p_\theta\dot{\theta} + p_\varphi\dot{\varphi}]$$

and $p_{\theta'}d\theta' = p_\theta d\theta + p_\varphi d\varphi$ (since the energy is the same in the frame of reference (r, θ') and (r, θ, φ)). Thus

$$\oint p_\varphi d\varphi = mh$$

with m quantum number and p_φ constant, so that $p_\varphi = m\hbar$ and $p_{\theta'} = k\hbar$, while $cos\alpha = m/k$, with $k = 1, 2, 3...$ and m varies from $-k$ to $+k$.

A pictorial view of the quantization in terms of precession of the angular momentum for $l = 2$ (the "length" being $\sqrt{l(l+1)}\hbar$) is

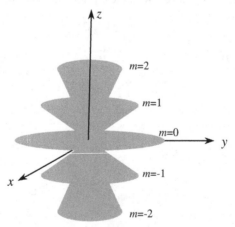

Problem 1.5 In the first atomic model, due to *Thomson*, the atom was idealized as a uniform positive electric charge in a sphere, with point-charge electrons embedded in it. By referring to Hydrogen atom, derive the motion of the electron and in the assumption that the radius of the sphere is $R = 1$ Å estimate the frequency of the radiation expected in the classical description.

Solution: The force at the distance r from the center of the sphere is

$$f(r) = -\frac{er^3}{R^3}\frac{e}{r^2}$$

and the electron motion is harmonic, with angular frequency $\omega = \sqrt{e^2/mR^3}$. For $m = 9.09 \times 10^{-28}$ g, $e = 4.8 \times 10^{-10}$ u.e.s. and $R = 1\,\text{Å}$ the frequency turns out $\nu = 2.53 \times 10^{15}\,\text{s}^{-1}$. In the classical picture the emission is at the same frequency (and at multiples) of the acceleration.

Problem 1.6 In the assumption that the proton can be thought as a sphere with homogeneous charge distribution and radius $R = 10^{-13}$ cm, evaluate the shift in the ground state energy of the Hydrogen atom due to the finite size of the nucleus in the perturbative approach (Note that $R \ll a_0$). Repeat the calculation for uniform distribution onto the surface of the sphere.

Solution: At the distance $r < R$ from the origin the potential energy is

$$V(r) = -\frac{e^2 r^3}{R^3 r} - \int_r^R \frac{e^2 4\pi r'^2}{r' 4\pi \frac{R^3}{3}} dr' = -\frac{3}{2}e^2 \left[\frac{1}{R} - \frac{r^2}{3R^3}\right]$$

The difference with respect to the energy for point charge nucleus implies an energy shift given by

$$< 1s|V_{diff}|1s > \; = \frac{1}{\pi a_0^3} \int_0^R e^{-\frac{2r}{a_0}} \left[\frac{e^2}{r} + \frac{e^2 r^2}{2R^3} - \frac{3e^2}{2R}\right] 4\pi r^2 dr$$

and for $r < R \ll a_0$

$$< 1s|V_{diff}|1s > \; = -\frac{2e^2}{a_0^3}\left[R^2 - \frac{R^2}{5} - R^2\right] = \frac{4}{5}\frac{e^2 R^2}{2a_0^3}$$

corresponding to about 3.9×10^{-9} eV.

For a uniform distribution onto the surface the perturbation Hamiltonian is $\mathcal{H}_P = +e^2/r - e^2/R$, for $0 \le r \le R$. The first-order energy correction is

$$< 1s|\mathcal{H}_P|1s > \; = \frac{e^2}{\pi a_0^3} \int_0^R e^{-\frac{2r}{a_0}} [\frac{1}{r} - \frac{1}{R}] 4\pi r^2 dr =$$

$$= \frac{4e^2}{a_0^3} \int_0^R [r - \frac{r^2}{R}] dr = \frac{2e^2 R^2}{a_0^3}\frac{1}{3} \simeq 6.5 \times 10^{-9}\,\text{eV}$$

Problem 1.7 For a Hydrogenic atom in the ground state evaluate the radius R of the sphere inside which the probability to find the electron is 0.9.

Solution: From

$$\int_0^R 4\pi r^2 |\phi_{1s}|^2 dr = 0.9$$

with $\phi_{1s} = \sqrt{1/\pi}(Z/a_0)^{3/2} exp(-Zr/a_0)$, since

$$\int_0^R r^2 e^{-2Zr/a_0} dr = -e^{-2ZR/a_0}\left[\frac{R^2 a_0}{2Z} + \frac{R a_0^2}{2Z^2} + \frac{a_0^3}{4Z^3}\right] + \frac{a_0^3}{4Z^3}$$

a trial and error numerical estimate yields $R \simeq 2.66a_0/Z$.

Problem 1.8 In the assumption that the ground state of Hydrogenic atoms is described by an eigenfunction of the form $exp(-ar^2/2)$, derive the best approximate eigenvalue by means of variational procedure.

Solution: The energy function is $E(a) = < \phi|\mathcal{H}|\phi > / < \phi|\phi >$, with

$$\mathcal{H} = -(\hbar^2/2m)[(d^2/dr^2) + (2/r)d/dr] - (Ze^2/r)$$

(see Eqs. (1.14) and (1.15)).
One has $< \phi|\phi > = 4\pi(1/4a)\sqrt{\pi/a}$, while

$$< \phi|\mathcal{H}|\phi > = 4\pi[(3\hbar^2/16m)\sqrt{\pi/a} - (Ze^2/2a)].$$

Then

$$E(a) = \frac{3\hbar^2}{4m}a - 2Ze^2\sqrt{\frac{a}{\pi}}$$

From $dE/da = 0$ one obtains $a_{min}^{1/2} = 4mZe^2/3\hbar^2\sqrt{\pi}$ and $E_{min} = -4e^4 Z^2 m/3\pi\hbar^2$ $\simeq 0.849 E_{1s}^H$ (for $Z = 1$).

Problem 1.9 Prove that on the average the electronic charge distribution associated with $n = 2$ states in Hydrogenic atoms is spherically symmetric. Observe how this statement holds for multi-electron atoms in the central field approximation.

Solution: The charge distribution is controlled by

$$\frac{1}{4}|\phi_{2,0,0}|^2 + \frac{1}{4}[|\phi_{2,1,-1}|^2 + |\phi_{2,1,0}|^2 + |\phi_{2,1,1}|^2]$$

where the latter term (see Table 1.1) is proportional to $[(1/2)sin^2\theta + cos^2\theta + (1/2)sin^2\theta] = 1$.

   In the central field approximation the statement holds, the wavefunctions being described in their angular dependence by spherical harmonics (This is a particular case of the *Unsold theorem* $\sum_{m=-l}^{m=+l} Y_{l,m}^* Y_{l,m} = (2l + 1)/4\pi$).

Problem 1.10 On the basis of a perturbative approach evaluate the correction to the ground state energy of Hydrogenic atoms when the nuclear charge is increased from Z to $(Z + 1)$ ($\int_0^\infty x^n exp(-ax)dx = n!/a^{n+1}$).

Solution: The exact result is $E_{Z+1} = -(e^2/2a_0)(Z + 1)^2$.
  The perturbative correction reads

$$E_{per}^{(1)} = -\int (\phi_Z^{1s})^* \frac{e^2}{r}(\phi_Z^{1s})d\tau = -\frac{4\pi e^2 Z^3}{\pi a_0^3}\int e^{-\frac{2Zr}{a_0}}rdr = -\frac{e^2 Z}{a_0}$$

In $(-e^2/2a_0)$ units the energy difference $E_{Z+1} - E_Z = 2Z + 1$ and for large Z this would practically coincide with $2Z$. It is noted that the fractional correction goes as $1/Z$, since $E^0 \propto Z^2$.

1.5 Finite Nuclear Mass. Positron, Muonic and Rydberg Atoms

To take into account the finite nuclear mass M in Hydrogenic atoms one can substitute the electron mass m with the reduced mass $\mu = Mm/(M + m)$. In fact this results from the very beginning, namely from the classical two-body Hamiltonian, the kinetic energy being

$$T = \frac{1}{2}\omega^2(Ma^2 + mb^2) = \frac{1}{2}\frac{Mm}{(M + m)}\omega^2 r^2 = \frac{1}{2}\mu\omega^2 r^2,$$

namely the one for a single mass μ rotating with angular velocity ω at the distance r:

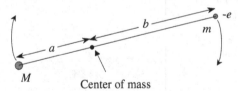

Center of mass

The potential energy does not change even though the nucleus is moving and therefore in order to account for the effects of finite nuclear mass, one simply substitutes m for μ in the eigenvalues and in the eigenfunctions. Then

$$E_n = -Z^2\frac{\mu e^4}{2\hbar^2}\frac{1}{n^2} = -\frac{e^2}{2a_0^*}\frac{Z^2}{n^2} \tag{1.23}$$

with $a_0^* = \hbar^2/\mu e^2$. In particular, the wavenumbers of the spectral lines (see Problem 1.4) are corrected according to

$$\bar{\nu} = Z^2 R_H \frac{1}{\left(1 + \frac{m}{M}\right)} \left(\frac{1}{n_f^2} - \frac{1}{n_i^2}\right) \tag{1.24}$$

where R_H is the Rydberg constant for the Hydrogen in the assumption of infinite nuclear mass (see Eq. (1.13)).

The *Deuterium* has been discovered (1932) from slightly shifted weak spectroscopic lines (*isotopic shift*), related to the correction to the eigenvalues in Eq. (1.23), due to the different nuclear masses for H and D.

A two particle system where the correction due to finite "nuclear" mass is strongly marked is obviously the *positronium* i.e. the Hydrogen-like "atom" where the proton is substituted by the positron. The reduced mass in this case is $\mu = m/2$, implying strong corrections to the eigenvalues and to the correspondent spectral lines (and to other effects that we shall discuss in subsequent chapters).

In Hydrogenic atoms it is possible to substitute the electron with a negative muon. From high energy collisions of protons on a target, two neutrons and a negative pion are produced. The pion decays into an antineutrino and a negative muon, of charge $-e$ and mass about 206.8 times the electron mass. The muon decays into an electron and two neutrinos, with life-time $\tau \simeq 2.2\,\mu$s. Before the muon decays it can be trapped by atoms in "electron-like orbits", thus generating the so called *muonic atoms*.

Most of the results derived for Hydrogenic atoms can be transferred to muonic atoms by the substitution of the electron mass with the muon mass $m_\mu = 206.8\,m$. Thus the distances have to be rescaled by the same amount and the muonic atoms are very "small", the dimension being of the order or less than the nuclear size (see Fig. 1.8). It is obvious that in this condition the approximation of nuclear point charge and Coulomb potential must be abandoned.

Fig. 1.8 Sketch of $|\phi_{1s}|^2$ for a muon in Pb ($Z = 82$) in the $1s$ state in the assumption of point charge nucleus (*solid line*), in comparison with the charge distribution of the nucleus itself, of radius around 6 Fermi (*dashed area*)

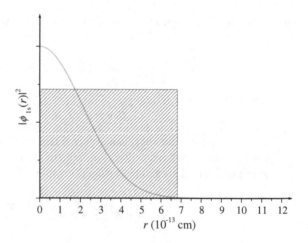

However, qualitatively, in muonic atoms the eigenvalues can still be obtained from the ones in Eq. (1.13) by multiplying for 206.8. Under this approximation the wavenumbers of the correspondent spectral lines become $\overline{\nu_\mu} = 206.8\,\overline{\nu_H}$ and the emission falls in the X-ray spectral range. The ionization potential is increased up to several MeV.[2]

Somewhat opposite to the muonic atoms are the "gigantic" *Rydberg atoms*, in which the electron, usually the one outside the inner shells (see the alkali atoms at Sect. 2.1) on the average is at very large distance from the nucleus. These atoms are found in interstellar spaces or can be produced in laboratory by irradiating atomic beams with lasers. The Rydberg atoms are therefore similar to Hydrogen atoms in excited states, the effective charge Z_{eff} (Fig. 1.3) being close to unit. Typically the quantum number n can reach several tens, hundreds in cosmic space. Since the expectation value of the distance (Table 1.4) increases with n^2, the "dimension" of the Rydberg atoms can reach $10^3 - 10^4$ Å. In these states the life time is very long (we shall see in Appendix 1.3 how the life time is related to the spontaneous emission of radiation) of the order of one second instead of the typical 10^{-8} s for inner levels in the Hydrogenic atoms. The eigenvalues scale with n^2 (Eq. (1.13)) and become of the order of 10^{-2} eV. Thus the Rydberg atoms are easily ionized and highly polarizable, the electric polarizability increasing approximately with the seventh power of the quantum number n (see Sect. 4.2 and Problem 4.21).

Problems

Problem 1.11 In the Hydrogen atom the H_α line (see Problem 1.4) has a wavelength 6562.80 Å. In Deuterium the H_α line shifts to 6561.01 Å. Estimate the ratio of the proton to deuteron mass.

Solution: From Eq. (1.24), $\lambda_D/\lambda_H = (1 + m/M_D)/(1 + m/M_H)$ and then

$$\frac{\Delta\lambda}{\lambda_H} = \frac{m(M_H - M_D)}{M_H M_D(1 + \frac{m}{M_H})} \simeq \frac{m(M_H - M_D)}{M_H M_D} = \frac{-1.79}{6562.8} \simeq \frac{\frac{M_H}{M_D} - 1}{1836},$$

yielding $M_H/M_D \simeq 0.4992$, i.e. $M_D = 2.0032 M_H$.

Problem 1.12 Show that in Rydberg atoms the frequency of the photon emitted from the transition between adjacent states at large quantum numbers n is close to the rotational frequency of the electron in the circular orbit of the Bohr atom (particular case of the *correspondence principle*).

Solution From Eq. (1.24), by neglecting the reduced mass correction, the transition frequency turns out

---

[2]It should be remarked that dramatic effects in muonic atoms involve also other quantities or interactions, for instance the spin-orbit interaction and the hyperfine field (see Sect. 5.1).

$$\nu = R_H c \frac{(n_i - n_f)(n_i + n_f)}{n_i^2 n_f^2}$$

which for $n_i, n_f \gg 1$ and $n_i - n_f = 1$ becomes $\nu \simeq 2R_H c/n^3$.

The Bohr rotational frequency, (see Problem 1.4) by taking into account the equilibrium condition $mv^2/r = e^2/r^2$, results

$$\nu_{rot} = \frac{mvr}{2\pi m r^2} = \frac{n\hbar m^2 e^4}{2\pi m n^4 \hbar^4} = \frac{2R_H c}{n^3}.$$

Problem 1.13 By using scaling arguments estimate the order of magnitude of the correction to the wavefunctions and eigenvalues in Hydrogen when the electron is replaced by a negative muon.

Solution: Since $\mu^{-1} = (m_p^{-1} + m_\mu^{-1})$, a_0 in the wavefunction is corrected by a factor $\simeq 186$. The eigenvalue depends linearly on the mass, then the energy is increased by a factor $\simeq 186$. These estimates neglect any modification in the potential energy. This is somewhat possible since $Z = 1$, while for heavy atoms (see Fig. 1.8) one should take into account the modification in the potential energy (see Problem 1.6). Similar considerations hold for *Protonium* (i.e. the "atom" with one positive and one negative proton), where only the states at small n are sizeably affected by the modified nuclear potential.

Problem 1.14 By scaling arguments evaluate how the ground state energy, the wavelength of $(2p \rightarrow 1s)$ transition and the life time of the $2p$ state are modified from Hydrogen atom to Positronium (for the life time see Appendix 1.3 and neglect the *annihilation process* related to the overlap of the wavefunctions in the $1s$ state).

Solution: The reduced mass is about half of the one in Hydrogen. Therefore the eigenvalue for the ground state is 6.8 eV. The transition frequency is at wavelength 2430 Å. For the life-time, from Appendix 1.3 one notices that the decay rate is proportional to the third power of the energy separation and to the second power of the dipole matrix element. The energy separation is one half while the length scale is twice, the decay rate is $1/2$. Then the life time is increased by a factor 2, namely from 1.6 to 3.2 ns. One could remark that nuclear-size effects, relevant in Hydrogen high-resolution spectroscopy (Appendix 5.1), are absent for positronium.

Problem 1.15 In experiments with radiation in cavity interacting with atoms, collimated beams of ^{85}Rb atoms in the $63p$ state are driven to the $61d$ state. On the basis of the classical analogy (see Problem 1.12) estimate the frequency required for the transition, the "radius" of the atom (for $n = 63$) and the order of magnitude of the electric dipole matrix element.

Solution: $\nu \simeq 2R_H c \Delta n/n^3 = 55.2$ GHz; $< r > \simeq n^2 a_0 = 2100.4$ Å; dipole matrix element (see Appendix 1.3) $\delta \simeq e < r > = 1.009 \times 10^{-14}$ u.e.s. cm.

Problem 1.16 In a Rydberg atom the outer electron is in the $n = 50$ state. Evaluate the electric field $\mathcal{E}$ required to ionize the atom (Hint: assume a potential energy of the form $V(r) = -e^2/r - er\mathcal{E}\cos\theta$ and disregard the possibility of quantum tunneling).

Solution: From $dV/dr = 0$ the maximum in the potential energy is found at $r_m = \sqrt{e/\mathcal{E}}$, where $V(r_m) = -2e^{3/2}\sqrt{\mathcal{E}}$.

The energy of the Rydberg atom is approximately $E_n \simeq (-e^2/2a_0)(1/n^2)$ and equating it to $V(r_m)$ one obtains $(e^2/2a_0)(1/n^2) = 2e^{3/2}\sqrt{\mathcal{E}}$, i.e. $\mathcal{E} = e/16a_0^2 n^4$, corresponding to

$$\mathcal{E} \simeq 51 \text{ V/cm}$$

1.6 Orbital and Spin Magnetic Moments and Spin-Orbit Interaction

As we shall see in detail in Chap. 2, the spectral lines observed in moderate resolution (e.g. the yellow doublet resulting from the $3p \leftrightarrow 3s$ transition in the Na atom) indicate that also interactions of magnetic character have to be taken into account in dealing with the electronic structure of the atoms.

The magnetic moment associated with the orbital motion, somewhat corresponding to a current, can be derived from the Hamiltonian for an electron in a static magnetic field H along the z direction, with vector potential

$$\mathbf{A} = \frac{1}{2}\mathbf{H} \times \mathbf{r} \qquad (1.25)$$

and scalar potential $\phi = 0$.

The one-electron Hamiltonian[3] is

$$\mathcal{H} = \frac{1}{2m}\left(\mathbf{p} + \frac{e}{c}\mathbf{A}\right)^2 + V - e\phi \qquad (1.26)$$

yielding, to the first order in $\mathbf{A}$, the operator

$$\mathcal{H} = \mathcal{H}_0 - i\frac{e\hbar}{mc}\mathbf{A}.\nabla \qquad (1.27)$$

where $\mathcal{H}_0$ is the Hamiltonian in the absence of magnetic or electric fields and where it has been taken into account that $\mathbf{A}$ and ∇ are commuting operators (*Lorentz gauge*). Therefore, in the light of Eq. (1.25) the Hamiltonian describing the effect of the magnetic field is

---

[3] This form of classical Hamiltonian associated with the force $\mathbf{F} = -e\mathcal{E} - e(\mathbf{v}/c) \times \mathbf{H}$ is required in order to have the kinetic energy expressed in terms of the generalized moment $\mathbf{p} = m\mathbf{v} - e\mathbf{A}/c$ (see the text by *Goldstein*) so that, in the quantum mechanical description, $\mathbf{p} = -i\hbar\nabla$.

$$\mathcal{H}_{mag} = -i\frac{e\hbar}{2mc}\mathbf{H} \times \mathbf{r}.\mathbf{\nabla} = \frac{e}{2mc}\mathbf{l}.\mathbf{H}. \tag{1.28}$$

Compared to the classical Hamiltonian $-\boldsymbol{\mu} \cdot \mathbf{H}$ for a magnetic moment in a field, $\mathcal{H}_{mag}$ allows one to assign to the angular momentum $\mathbf{l}$ a magnetic moment operator given by

$$\boldsymbol{\mu}_l = -\frac{e}{2mc}\hbar\mathbf{l} = -\mu_B\mathbf{l} \tag{1.29}$$

where $\mu_B = e\hbar/2mc$ is called *Bohr magneton*, numerically 0.927×10^{-20} erg/G. Equation 1.29 can be obtained even classically in the framework of the Bohr model for the Hydrogen atom (See Problem 1.18).

Experimental evidences, such as spectral lines from atoms in magnetic field (see Chap. 4) as well the quantum electrodynamics developed by *Dirac*, indicate that an *intrinsic* angular momentum, the *spin* s, has to be assigned to the electron.

By extending the eigenvalue equations for the orbital angular momentum to spin, one writes

$$\mathbf{s}^2|\alpha> = s(s+1)\hbar^2|\alpha>, \qquad s_z|\alpha> = \frac{\hbar}{2}|\alpha>$$

$$\mathbf{s}^2|\beta> = s(s+1)\hbar^2|\beta>, \qquad s_z|\beta> = -\frac{\hbar}{2}|\beta> \tag{1.30}$$

$|\alpha>$ and $|\beta>$ being the spin eigenfunctions corresponding to quantum spin numbers $m_s = 1/2$ and $m_s = -1/2$, respectively, while $s = 1/2$.

As a first consequence of the spin, in the one-electron eigenfunction (*spin-orbital*) one has to include the spin variable, labeling the value of s_z. When the coupling between orbital and spin variables (the *spin-orbit interaction* that we shall estimate in the following) is weak, one can factorize the function in the form

$$\psi(r, \theta, \varphi, \mathbf{s}) = \phi(r, \theta, \varphi)\chi_{spin} \tag{1.31}$$

where χ_{spin} is $|\alpha>$ or $|\beta>$ depending on the value of the quantum number m_s.

To express the magnetic moment associated with s without resorting to quantum electrodynamics, one has to make an *ansatz* based on the experimental evidence. In partial analogy to Eq. (1.29) we write

$$\boldsymbol{\mu}_s = -2\mu_B\mathbf{s} \tag{1.32}$$

Due to the existence of elementary magnetic moments, an external magnetic field can be expected to remove the degeneracy in the z-component of the angular momenta. For instance for s_z, two sublevels are generated by the magnetic field, with energy separation $\Delta E = (e\hbar/mc)H$, a phenomenon that can be called *magnetic splitting* (Problem 1.17).

Now we are going to derive the Hamiltonian describing the interaction between the orbital and the spin magnetic moments. This will be done in the semiclassical model

Fig. 1.9 Definition of the magnetic field **H** acting on the electron due to the relative motion of the nucleus of charge Ze creating an electric field at the position **r**, in view of the relativistic transformation

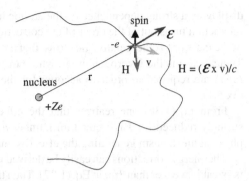

first used by *Thomas* and *Frenkel*, assuming classical expressions for the electric and magnetic fields acting on the electron. By referring to Fig. 1.9, the electric field at the electron is $E = (1/er)(dV/dr)\mathbf{r}$ (where V is the central field energy). From the relativistic transformation and by adding a factor $1/2$ introduced by *Thomas* to account for the non-inertial motion, one has

$$\mathbf{H} = \frac{1}{2cer}\frac{dV}{dr}\mathbf{r} \times \mathbf{v}. \tag{1.33}$$

Thus the magnetic Hamiltonian becomes

$$\mathcal{H}_{spin\text{-}orbit} = -\boldsymbol{\mu}_s \cdot \mathbf{H} = \frac{1}{2m^2c^2r}\frac{dV}{dr}(\mathbf{l.s}) \equiv \xi(r)\mathbf{l.s} \tag{1.34}$$

which can be viewed as an effective r-dependent magnetic field along **l** direction, acting on the spin magnetic moment when the electron is at the position **r**. It is noted that the function $\xi(r)$, of central character, is *essentially positive* and includes $\hbar^2$ from **l** and **s**.

An immediate physical interpretation of the Hamiltonian in Eq. (1.34) can be achieved by referring to Hydrogenic atoms, where

$$\xi(r) = \frac{Ze^2\hbar^2}{2m^2c^2r^3} \tag{1.35}$$

Then the energy associated with $\mathcal{H}_{spin\text{-}orbit}$ is of the order of

$$E_{SO} \simeq (Ze^2/2m^2c^2) < r^{-3} > n\hbar.(1/2)\hbar$$

and from Table 1.4, where $< r^{-3} > \simeq Z^3/a_0^3 n^3 l^3$, one has

$$E_{SO} \simeq \frac{e^2\hbar^2 Z^4}{4m^2c^2a_0^3 n^5} \tag{1.36}$$

displaying a strong dependence on the atomic number Z. For small Z the spin-orbit interaction turns out of the order of the correction related to the velocity dependence of the mass or to other relativistic terms, that have been neglected (see Problem 1.38). Typical case is the Hydrogen atom, where the relativistic corrections of *Dirac* and *Lamb* are required in order to account for the detailed fine structure (see Appendix 5.1).

From Eq. (1.36) one realizes that the effects of the spin-orbit interaction are strongly reduced for large quantum number n, as it is conceivable in view of the physical mechanism generating the effective magnetic field on the electron.

The energy corrections can easily be derived within the assumption that $\mathcal{H}_{spin-orbit}$ is sizeably weaker than $\mathcal{H}_0$, in Eq. (1.27). Then the perturbation theory can be applied to spin-orbital eigenfunctions, $\psi(r, \theta, \varphi, \mathbf{s}) = \phi(r, \theta, \varphi)\chi_{spin} \equiv \phi_{n,l,m,m_s}$, the operators $\hat{l}^2, \hat{s}^2, l_z, s_z$ being diagonal for the unperturbed system. Since the energy terms are often small in comparison to the energy separation between unperturbed states at different quantum numbers n and l, one can evaluate the energy corrections due to $\mathcal{H}_{spin-orbit}$ within the (nl) representation:

$$\Delta E_{SO} = \int R_{nl}^*(r)\xi(r)R_{nl}(r)r^2 dr \sum_{spin} \int \chi_{m_s'}^* Y_{lm'}^* (\mathbf{l.s})\chi_{m_s} Y_{lm} sin\theta d\theta d\varphi \quad (1.37)$$

that can be written in the form[4]

$$(\Delta E_{SO})_{m',m_s',m,m_s} = \xi_{nl} < m'm_s'|\mathbf{l} \cdot \mathbf{s}|mm_s > . \quad (1.38)$$

The spin orbit constant

$$\xi_{nl} = \int R_{nl}^*(r)\xi(r)R_{nl}(r)r^2 dr \quad (1.39)$$

can be thought as a measure of the "average" magnetic field on the electron in the nl state. This average field is again along the direction of $\mathbf{l}$ and acting on $\boldsymbol{\mu}_s$ implies an interaction of the form $\mathcal{H}_{spin-orbit} \propto -\mathbf{h}_{eff}.\boldsymbol{\mu}_s$.

To evaluate the energy corrections due to the Hamiltonian $\xi_{nl}\mathbf{l.s}$ instead of the formal diagonalization one can proceed with a first step of a more general approach (the so-called *vectorial model*) that we will describe in detail at Chap. 3. Let us define

$$\mathbf{j} = \mathbf{l} + \mathbf{s} \quad (1.40)$$

---

[4]It could be remarked that the ϕ_{n,l,m,m_s} are not the proper eigenfunctions since $(\mathbf{l.s})$ does not commute with l_z and s_z. However, when $(\mathbf{l} \cdot \mathbf{s})$ is replaced by the linear combination of $\hat{j}^2, \hat{l}^2$ and $\hat{s}^2$ (see Eq. (1.41)) and the eigenvalues are derived on the basis of the eigenfunctions of $\hat{j}^2$ and j_z, the appropriate ΔE_{SO} are obtained.

as the total, single-electron, angular momentum. For analogy with $\mathbf{l}$ and $\mathbf{s}$, $\mathbf{j}$ is specified by a quantum number j (integer or half-integer) and by the magnetic quantum number m_j taking the $(2j+1)$ values from $-j$ to $+j$, with the usual meaning in terms of quantization of the modulus and of the z-component of $\mathbf{j}$, respectively.

The operators $\mathbf{l}$ and $\mathbf{s}$ commute since they act on different variables, so that $\mathbf{l.s}$ can be substituted by

$$\mathbf{l} \cdot \mathbf{s} = \frac{1}{2}(\hat{j}^2 - \hat{l}^2 - \hat{s}^2) \tag{1.41}$$

involving only the modula, with eigenvalues $j(j+1)$, $l(l+1)$ and $s(s+1)$ [4].

Therefore, for $l \neq 0$ one has the two cases, $j = l + 1/2$ and $j = l - 1/2$, that in a vectorial picture correspond to spin parallel and antiparallel to $\mathbf{l}$.

Then the energy corrections due to $\mathcal{H}_{spin\text{-}orbit}$ are

$$\Delta E_{SO} = \xi_{nl} \hbar^2 l/2,$$

for $j = l + 1/2$, and

$$\Delta E_{SO} = -\xi_{nl} \hbar^2 (l+1)/2$$

for $j = (l - 1/2)$. Being ξ_{nl} *positive* the doublet sketched below is generated.

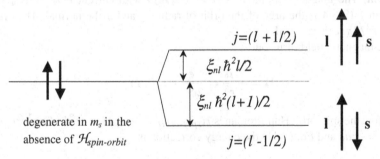

For s state only a *shift*, of relativistic origin, has to be associated with $\mathcal{H}_{spin\text{-}orbit}$ (see Problems 1.19 and 1.38).

Problems

Problem 1.17 Show that because of the spin magnetic moment, a magnetic field removes the degeneracy in m_s and two sublevels with energy separation $(e\hbar/mc)H$ are induced (*magnetic splitting*).

Solution: From the Hamiltonian

$$\mathcal{H}_{mag} = -\boldsymbol{\mu}_s.\mathbf{H} = -(-2\mu_B s_z)H = \frac{e\hbar}{mc}H s_z$$

and the s_z eigenvalues $\pm 1/2$, one has

$$\Delta E = E_{1/2} - E_{-1/2} = \frac{e\hbar}{mc} H \equiv 2\mu_B H$$

with the splitting sketched below

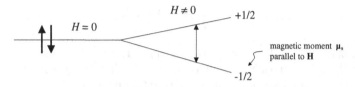

Numerically, for $H = 1\,\mathrm{T}$, $\Delta E \simeq 1.16 \times 10^{-4}\,\mathrm{eV} \simeq k_B T$ for $T \simeq 1.34$ K.

Problem 1.18 For the electron in the circular orbit of the Bohr model, derive the relationship between angular momentum and magnetic moment. By assigning to the electron the spin magnetic moment derive the correction to the energy levels due to spin-orbit interaction, comparing the results for $n = 2$ and $n = 3$ to the estimates in the Thomas-Frenkel approach (Sect. 1.6).

Solution: The magnetic moment is $\boldsymbol{\mu} = (iA/c)\hat{n}$ with current $i = -e\nu_{rot}$ (see Problem 1.12). A is the area of the orbit of radius r and $\hat{n}$ the normal. Thus $\boldsymbol{\mu} = -(e\nu\pi r^2/2\pi rc)\hat{n} \equiv -\mu_B \mathbf{l}$.

The magnetic field turns out

$$\mathbf{H} = -\frac{\boldsymbol{\mu}}{r^3} = \frac{e}{2cr^3}\mathbf{v} \times (-\mathbf{r})$$

Therefore the spin-orbit Hamiltonian is $\mathcal{H}_{spin\text{-}orbit} = -\boldsymbol{\mu}_s.\mathbf{H} = (e^2\hbar^2/2m^2c^2r^3)\mathbf{l}.\mathbf{s}$. For $r_n = n^2 a_0$ and Eq. (1.41) the energy correction is

$$E_{SO} = \frac{e^2\hbar^2}{2m^2c^2n^6a_0^3}\frac{1}{2}[j(j+1) - l(l+1) - s(s+1)]$$

By using for r_n^{-3} the expectation value

$$<r^{-3}> = \frac{1}{a_0^3 n^3 l(l+1)(l+\frac{1}{2})}$$

and indicating $e^2\hbar^2/4m^2c^2a_0^3 = 3.62 \times 10^{-4}\,\mathrm{eV}$ with E_0, one writes

$$E_{SO} = E_0\frac{1}{n^3 l(l+1)(l+\frac{1}{2})}[j(j+1) - l(l+1) - s(s+1)]$$

and

$$n = 2, l = 1, j = 1/2 \qquad E_{SO} = -E_0/12$$
$$n = 2, l = 1, j = 3/2 \qquad E_{SO} = E_0/24$$
$$n = 3, l = 1, j = 1/2 \qquad E_{SO} = -2E_0/81$$
$$n = 3, l = 1, j = 3/2 \qquad E_{SO} = E_0/81$$

Problem 1.19 By referring to one-electron s states try to derive the correction to the unperturbed energy value due to $\mathcal{H}_{spin\text{-}orbit}$, making a remark on what has to be expected.

Solution: $\mathcal{H}_{spin\text{-}orbit} = \xi_{nl}(\mathbf{l.s})$ with $\xi_{nl} \propto \int R_{nl}^*(r)\xi(r)R_{nl}(r)r^2dr$.

Since $R_{nl}(r) \propto r^l$, for $l = 0$, ξ_{nl} diverges for $r \to 0$, while $\mathbf{l.s} = 0$.

The final result is an energy shift that cannot be derived along the procedure neglecting relativistic effects (see Problem 1.38). A discussion of the fine and hyperfine structure in the Hydrogen atom, including the relativistic effects, is given in Appendix 5.1.

Problem 1.20 Evaluate the effective magnetic field that can be associated with the orbital motion of the optical electron in the Na atom, knowing that the transition $3p \to 3s$ yields a doublet with two lines at wavelenghts 5889.95 and 5895.92 Å.

Solution: From the difference in the wavelengths the energy separation of the $3p$ levels turns out

$$|\Delta E| = \frac{hc|\Delta\lambda|}{\lambda^2} = 2.13 \times 10^{-3}\text{eV}$$

ΔE can be thought to result from an effective field $H = \Delta E/2\mu_B$ (see Problem 1.17). Thus, $H = 2.13 \times 10^{-3}/2 \times 5.79 \times 10^{-5}\,\text{T} = 18.4\,\text{T}$.

Problem 1.21 The ratio (magnetic moment μ/angular momentum $\mathbf{L}$), often expressed as $(\mu/\mu_B)/(\mathbf{L}/\hbar)$, is called *gyromagnetic ratio*. Assuming that the electron is a sphere of mass m and charge $-e$ homogeneously distributed onto the surface, rotating at constant angular velocity, show that the gyromagnetic ratio turns out $\gamma = \mu/L = -5e/6mc$.

Solution: $m = (4\pi/3)\rho R^3$ while the angular momentum is
$$L = \int_0^{2\pi} \int_0^{\pi} \int_0^R \rho w r^4 \sin^3\theta d\theta d\varphi dr = (2/5)mR^2\omega, \rho \text{ being the specific mass.}$$

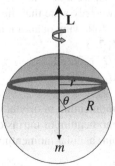

The surface charge density is $\sigma = -e/4\pi R^2$ and from
$\mu = Ai/c = (\pi R^4 \sigma w/c) \int sin^3\theta d\theta = -5eL/6mc$ one has $\gamma = -5e/6mc$.

Problem 1.22 Express numerically the spin-orbit constant ξ_{nl} for the $3p$, $3d$ and $4f$ states of the Hydrogen atom.

Solution: From Eq. (1.35) and the expectation values of $< r^{-3} >$ (Table 1.4)

$$\xi_{3p} = 1.29 \times 10^{37} \ erg^{-1}s^{-2} \ \hbar^2 = 8.94 \times 10^{-6} \ eV,$$
$$\xi_{3d} = 2.58 \times 10^{36} \ erg^{-1}s^{-2} \ \hbar^2 = 1.79 \times 10^{-6} \ eV,$$
$$\xi_{4f} = 3.88 \times 10^{35} \ erg^{-1}s^{-2} \ \hbar^2 = 0.27 \times 10^{-6} \ eV.$$

Problem 1.23 Show that when the spin-orbit interaction is taken into account the effective magnetic moment of an electron can be written $\mu_{\pm} = (-e/2mc)g_{\pm}(l+s)$

with $g_{\pm} = 1 \pm [1/(2l+1)]$.

Solution: Here g is a particular case of the *Lande' g factor*, to be discussed at Sect. 3.2. $\pm$ means spin parallel or antiparallel to **l**.
    For **s** $\parallel$ **l**

$$g_+ = 1 + \frac{(2l+1)(2l+3)+3-2l(2l+2)}{2(2l+1)(2l+3)} = 1 + \frac{1}{(2l+1)}$$

while for g_-, **s** antiparallel to **l**

$$g_- = 1 + \frac{(2l-1)(2l+1)+3-2l(2l+2)}{2(2l+1)(2l-1)} = 1 - \frac{1}{(2l+1)}.$$

1.7 Spectroscopic Notation for Multiplet States

In the light of spin-orbit interaction the one-electron states have to be labeled by quantum numbers n, l, j and m_j, with $s = 1/2$. Accordingly, a *fine structure* of the levels is induced, in form of *doublets*.
    As we shall see in detail in Chaps. 2 and 3, in the atom other couplings between l_i and s_i occur. At the moment we only state that the whole electronic structure of the atom can be described by the following quantum numbers:
    L, taking possible values 0, 1, 2, 3...
    S, taking possible values 0, 1/2, 1, 3/2, 2...
    J, taking possible values 0, 1/2, 1, 3/2 , 2...
to be associated with
    $\mathbf{L} = \sum_i \mathbf{l}_i$, the total angular momentum of orbital character,
    $\mathbf{S} = \sum_i \mathbf{s}_i$, the total angular momentum of intrinsic character
and with the total (orbital and spin) angular momentum $\mathbf{J} = \mathbf{L} + \mathbf{S}$, or to $\mathbf{J} = \sum_i \mathbf{j}_i$.

It is customary to use the following notation for the multiplet state of the atom
$$^{2S+1}Letter_J$$
where *Letter* means S, P, D, F, etc. for $L = 0, 1, 2, 3$ etc., $(2S+1)$ is the total number of the fine structure levels when $S < L$ $((2L + 1)$ the analogous when $L < S$).

The electronic configurations and the spectroscopic notations for the ground-state of the atoms are reported in the following.

Z	Element	Symbol	Configuration	Term
1	Hydrogen	H	$1s^1$	$^2S_{1/2}$
2	Helium	He	$1s^2$	$^1S_0$
3	Lithium	Li	$1s^2 2s^1$	$^2S_{1/2}$
4	Beryllium	Be	$1s^2 2s^2$	$^1S_0$
5	Boron	B	$1s^2 2s^2 2p^1$	$^2P_{1/2}$
6	Carbon	C	$1s^2 2s^2 2p^2$	$^3P_0$
7	Nitrogen	N	$1s^2 2s^2 2p^3$	$^4S_{3/2}$
8	Oxygen	O	$1s^2 2s^2 2p^4$	$^3P_2$
9	Fluorine	F	$1s^2 2s^2 2p^5$	$^2P_{3/2}$
10	Neon	Ne	$1s^2 2s^2 2p^6$	$^1S_0$
11	Sodium	Na	$[Ne]3s^1$	$^2S_{1/2}$
12	Magnesium	Mg	$[Ne]3s^2$	$^1S_0$
13	Aluminum	Al	$[Ne]3s^2 3p^1$	$^2P_{1/2}$
14	Silicon	Si	$[Ne]3s^2 3p^2$	$^3P_0$
15	Phosphorus	P	$[Ne]3s^2 3p^3$	$^4S_{3/2}$
16	Sulfur	S	$[Ne]3s^2 3p^4$	$^3P_2$
17	Chlorine	Cl	$[Ne]3s^2 3p^5$	$^2P_{3/2}$
18	Argon	Ar	$[Ne]3s^2 3p^6$	$^1S_0$
19	Potassium	K	$[Ar]4s^1$	$^2S_{1/2}$
20	Calcium	Ca	$[Ar]4s^2$	$^1S_0$
21	Scandium	Sc	$[Ar]3d^1 4s^2$	$^2D_{3/2}$
22	Titanium	Ti	$[Ar]3d^2 4s^2$	$^3F_2$
23	Vanadium	V	$[Ar]3d^3 4s^2$	$^4F_{3/2}$
24	Chromium	Cr	$[Ar]3d^5 4s^1$	$^7S_3$
25	Manganese	Mn	$[Ar]3d^5 4s^2$	$^6S_{5/2}$
26	Iron	Fe	$[Ar]3d^6 4s^2$	$^5D_4$
27	Cobalt	Co	$[Ar]3d^7 4s^2$	$^4F_{9/2}$
28	Nickel	Ni	$[Ar]3d^8 4s^2$	$^3F_4$
29	Copper	Cu	$[Ar]3d^{10} 4s^1$	$^2S_{1/2}$
30	Zinc	Zn	$[Ar]3d^{10} 4s^2$	$^1S_0$
31	Gallium	Ga	$[Ar]3d^{10} 4s^2 4p^1$	$^2P_{1/2}$
32	Germanium	Ge	$[Ar]3d^{10} 4s^2 4p^2$	$^3P_0$
33	Arsenic	As	$[Ar]3d^{10} 4s^2 4p^3$	$^4S_{3/2}$
34	Selenium	Se	$[Ar]3d^{10} 4s^2 4p^4$	$^3P_2$
35	Bromine	Br	$[Ar]3d^{10} 4s^2 4p^5$	$^2P_{3/2}$
36	Krypton	Kr	$[Ar]3d^{10} 4s^2 4p^6$	$^1S_0$

Z	Element	Symbol	Configuration	Term
37	Rubidium	Rb	$[Kr]5s^1$	$^2S_{1/2}$
38	Strontium	Sr	$[Kr]5s^2$	$^1S_0$
39	Yttrium	Y	$[Kr]4d^15s^2$	$^2D_{3/2}$
40	Zirconium	Zr	$[Kr]4d^25s^2$	$^3F_2$
41	Niobium	Nb	$[Kr]4d^45s^1$	$^6D_{1/2}$
42	Molybdenum	Mo	$[Kr]4d^55s^1$	$^7S_3$
43	Technetium	Tc	$[Kr]4d^55s^2$	$^6S_{5/2}$
44	Ruthenium	Ru	$[Kr]4d^75s^1$	$^5F_5$
45	Rhodium	Rh	$[Kr]4d^85s^1$	$^4F_{9/2}$
46	Palladium	Pd	$[Kr]4d^{10}$	$^1S_0$
47	Silver	Ag	$[Kr]4d^{10}5s^1$	$^2S_{1/2}$
48	Cadmium	Cd	$[Kr]4d^{10}5s^2$	$^1S_0$
49	Indium	In	$[Kr]4d^{10}5s^25p^1$	$^2P_{1/2}$
50	Tin	Sn	$[Kr]4d^{10}5s^25p^2$	$^3P_0$
51	Antimony	Sb	$[Kr]4d^{10}5s^25p^3$	$^4S_{3/2}$
52	Tellurium	Te	$[Kr]4d^{10}5s^25p^4$	$^3P_2$
53	Iodine	I	$[Kr]4d^{10}5s^25p^5$	$^2P_{3/2}$
54	Xenon	Xe	$[Kr]4d^{10}5s^25p^6$	$^1S_0$
55	Cesium	Cs	$[Xe]6s^1$	$^2S_{1/2}$
56	Barium	Ba	$[Xe]6s^2$	$^1S_0$
57	Lanthanum	La	$[Xe]5d^16s^2$	$^2D_{3/2}$
58	Cerium	Ce	$[Xe]4f^15d^16s^2$	$^1G_4$
59	Praseodymium	Pr	$[Xe]4f^36s^2$	$^4I_{9/2}$
60	Neodymium	Nd	$[Xe]4f^46s^2$	$^5I_4$
61	Promethium	Pm	$[Xe]4f^56s^2$	$^6H_{5/2}$
62	Samarium	Sm	$[Xe]4f^66s^2$	$^7F_0$
63	Europium	Eu	$[Xe]4f^76s^2$	$^8S_{7/2}$
64	Gadolinium	Gd	$[Xe]4f^75d^16s^2$	$^9D_2$
65	Terbium	Tb	$[Xe]4f^96s^2$	$^6H_{15/2}$
66	Dysprosium	Dy	$[Xe]4f^{10}6s^2$	$^5I_8$
67	Holmium	Ho	$[Xe]4f^{11}6s^2$	$^4I_{15/2}$
68	Erbium	Er	$[Xe]4f^{12}6s^2$	$^3H_6$
69	Thulium	Tm	$[Xe]4f^{13}6s^2$	$^2F_{7/2}$
70	Ytterbium	Yb	$[Xe]4f^{14}6s^2$	$^1S_0$
71	Lutetium	Lu	$[Xe]4f^{14}5d^16s^2$	$^2D_{3/2}$
72	Hafnium	Hf	$[Xe]4f^{14}5d^26s^2$	$^3F_2$

Z	Element	Symbol	Configuration	Term
73	Tantalum	Ta	$[Xe]4f^{14}5d^36s^2$	$^4F_{3/2}$
74	Tungsten	W	$[Xe]4f^{14}5d^46s^2$	$^5D_0$
75	Rhenium	Re	$[Xe]4f^{14}5d^56s^2$	$^6S_{5/2}$
76	Osmium	Os	$[Xe]4f^{14}5d^66s^2$	$^5D_4$
77	Iridium	Ir	$[Xe]4f^{14}5d^76s^2$	$^4F_{9/2}$
78	Platinum	Pt	$[Xe]4f^{14}5d^96s^1$	$^3D_3$
79	Gold	Au	$[Xe]4f^{14}5d^{10}6s^1$	$^2S_{1/2}$
80	Mercury	Hg	$[Xe]4f^{14}5d^{10}6s^2$	$^1S_0$
81	Thallium	Tl	$[Xe]4f^{14}5d^{10}6s^26p^1$	$^2P_{1/2}$
82	Lead	Pb	$[Xe]4f^{14}5d^{10}6s^26p^2$	$^3P_0$
83	Bismuth	Bi	$[Xe]4f^{14}5d^{10}6s^26p^3$	$^4S_{3/2}$
84	Polonium	Po	$[Xe]4f^{14}5d^{10}6s^26p^4$	$^3P_2$
85	Astatine	At	$[Xe]4f^{14}5d^{10}6s^26p^5$	$^2P_{3/2}$
86	Radon	Rn	$[Xe]4f^{14}5d^{10}6s^26p^6$	$^1S_0$
87	Francium	Fr	$[Rn]7s^1$	$^2S_{1/2}$
88	Radium	Ra	$[Rn]7s^2$	$^1S_0$
89	Actinium	Ac	$[Rn]\,6d^17s^2$	$^2D_{3/2}$
90	Thorium	Th	$[Rn]\,6d^27s^2$	$^3F_2$
91	Protactinium	Pa	$[Rn]5f^26d^17s^2$	$^4K_{11/2}$
92	Uranium	U	$[Rn]5f^36d^17s^2$	$^5L_6$
93	Neptunium	Np	$[Rn]5f^46d^17s^2$	$^6L_{11/2}$
94	Plutonium	Pu	$[Rn]5f^67s^2$	$^7F_0$
95	Americium	Am	$[Rn]5f^77s^2$	$^8S_{7/2}$
96	Curium	Cm	$[Rn]5f^76d^17s^2$	$^9D_2$
97	Berkelium	Bk	$[Rn]5f^97s^2$	$^6H_{15/2}$
98	Californium	Cf	$[Rn]5f^{10}7s^2$	$^5I_8$
99	Einsteinium	Es	$[Rn]5f^{11}7s^2$	$^4I_{15/2}$
100	Fermium	Fm	$[Rn]5f^{12}7s^2$	$^3H_6$
101	Mendelevium	Md	$[Rn]5f^{13}7s^2$	$^2F_{7/2}$
102	Nobelium	No	$[Rn]5f^{14}7s^2$	$^1S_0$
103	Lawrencium	Lr	$[Rn]5f^{14}7s^27p^1$	$^2P_{1/2}$
104	Rutherfordium	Rf	$[Rn]\,5f^{14}6d^27s^2$	$^3F_2$
105	Dubnium	Db	$[Rn]\,5f^{14}6d^37s^2$	$^4F_{3/2}$
106	Seaborgium	Sg	$[Rn]\,5f^{14}6d^47s^2$	$^5D_0$
107	Bohrium	Bh	$[Rn]\,5f^{14}6d^57s^2$	$^6S_{5/2}$
108	Hassium	Hs	$[Rn]\,5f^{14}6d^67s^2$	$^5D_4$

Appendix 1.1 Electromagnetic Spectral Ranges and Fundamental Constants

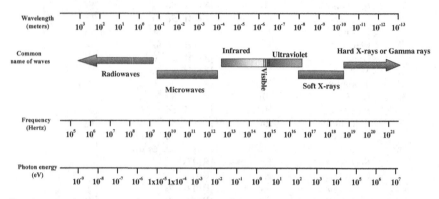

Fundamental constants (for magnetic quantities see Appendix 4.1)

Speed of light in vacuum	$c=2.99792 \times 10^{10}$ cm/s
Electron charge	$e=-1.60218 \times 10^{-19}$ Coulomb$=-4.8 \times 10^{-10}$ u.e.s.
Electron mass (at rest)	$m=9.10938 \times 10^{-28}$ g
Proton mass	$M=1.67262 \times 10^{-24}$ g
Neutron mass	$M_n=1.675 \times 10^{-24}$ g
Atomic mass unit (m(^{12}C)/12)	$u=1.661 \times 10^{-24}$ g
Planck constant	$h=6.62607 \times 10^{-27}$ erg.s$=4.1357 \times 10^{-15}$ eV.s
	$\hbar=(h/2\pi)=1.05457 \times 10^{-27}$ erg.s
Boltzmann constant	$k_B=1.38065 \times 10^{-16}$ erg/K
Stefan-Boltzmann constant (total emittance)	$\sigma=5.67 \times 10^{-5}$ erg/(s.cm^2.K^4)$= 5.67 \times 10^{-8}$ W/(m^2.K^4)
Bohr radius for atomic hydrogen (infinite nuclear mass)	$a_0=0.52918 \times 10^{-8}$ cm$= 0.52918$ Å
Rydberg constant (or Bohr energy e/2a$_0$) (infinite nuclear mass)	$R_H=109737$ cm$^{-1}= 13.606$ eV$=h.(3.29 \times 10^{15}$ Hz)
Bohr magneton	$\mu_B=e\hbar/2mc=0.9274 \times 10^{-20}$ erg/Gauss$= 0.9274 \times 10^{-23}$ A.m$^2 = 0.9274 \times 10^{-23}$ J/Tesla (see App.IV.1)
Nuclear magneton	$M_N=\mu_B m/M= \mu_B/1836.15= =e\hbar/2Mc=5.0508 \times 10^{-24}$ erg/Gauss
Proton magnetic moment (maximum component)	$\mu_P=M_N g_N I=M_N(5.586)(1/2)=1.4106 \times 10^{-23}$ erg/Gauss
Neutron magnetic moment	$\mu_n=-1.91315$ M_N
Avogadro number	$N_A=6.022 \times 10^{23}$ mol^{-1}
Electron volt	1 eV$= 1.602 \times 10^{-12}$ erg$=$ h.(2.418$\times 10^{14}$ Hz)
	1 erg$= 6.242 \times 10^{11}$ eV
Gas constant	$R=N_A k_B= 8.31447 \times 10^7$ erg/(mol.K)
k$_B$T at room temperature	0.0259 eV$\approx 1/40$ eV
Fine structure constant	$\alpha=e^2/\hbar c=1/137.036$

Appendix 1.2 Perturbation Effects in Two-Level Systems

We shall refer to a model system with two eigenstates, labelled $|1>$ and $|2>$, and correspondent eigenfunctions ϕ_1^0 and ϕ_2^0 forming a complete orthonormal basis. The model Hamiltonian is $\mathcal{H}_0$ and $\mathcal{H}_0\phi_m^0 = E_m\phi_m^0$, with $m = 1, 2$. In real systems the Hamiltonian $\mathcal{H}$ can differ from $\mathcal{H}_0$ owing to a small perturbation $\mathcal{H}_P$. Following a rapid transient (after turning on the perturbation) the stationary states are described by eigenfunctions that differ from the ones of the model system by a small amount, that can be written in terms of the unperturbed basis. This is equivalent to state that the eigenfunctions of the equation

$$\mathcal{H}\phi = E\phi \qquad (A.1.2.1)$$

are

$$\phi = c_1\phi_1^0 + c_2\phi_2^0 \qquad (A.1.2.2)$$

with c_1 and c_2 constants. By inserting ϕ in A.1.2.1 and multiplying by $< \phi_1^0|$ and by $< \phi_2^0|$ in turn, in the light of the orthonormality of the states, one derives for $c_{1,2}$

$$c_1(\mathcal{H}_{11} - E) + c_2\mathcal{H}_{12} = 0$$
$$c_1\mathcal{H}_{21} + c_2(\mathcal{H}_{22} - E) = 0$$

with $\mathcal{H}_{mn} = < m|\mathcal{H}|n >$. Non-trivial solutions imply

$$det \begin{pmatrix} \mathcal{H}_{11} - E & \mathcal{H}_{12} \\ \mathcal{H}_{21} & \mathcal{H}_{22} - E \end{pmatrix} = 0$$

and the eigenvalues turn out

$$E_{\mp} = \frac{1}{2}(\mathcal{H}_{11} + \mathcal{H}_{22}) \pm \frac{1}{2}\sqrt{(\mathcal{H}_{11} - \mathcal{H}_{22})^2 + 4\mathcal{H}_{12}\mathcal{H}_{21}} \qquad (A.1.2.3)$$

When the diagonal elements of $\mathcal{H}_P$ are zero, A.1.2.3 reduces to

$$E_{\mp} = \frac{1}{2}(E_1 + E_2) \pm \frac{1}{2}\sqrt{(E_1 - E_2)^2 + 4\varepsilon^2} \qquad (A.1.2.4)$$

where $\varepsilon^2 = | < 2|\mathcal{H}_P|1 > |^2$, $\mathcal{H}_P$ being Hermitian.

The perturbation effects strongly depend on the energy separation $\Delta E = E_2 - E_1$. For degenerate energy levels ($\Delta E = 0$) the largest energy correction occurs, the separation being given by 2ε. For perturbation much weaker than ΔE, Eq. (A.1.2.4) can be expanded, to yield the second-order corrections

$$E_{\mp} = E_{2,1} \pm \frac{\varepsilon^2}{\Delta E} \qquad (A.1.2.5)$$

The corrections to the unperturbed eigenvalues as a function of ΔE are illustrated below

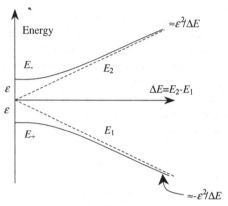

The eigenfunctions in the presence of $\mathcal{H}_P$ can be obtained by deriving the coefficients $c_{1,2}$ in A.1.2.2 in correspondence to $E = E_+$ and $E = E_-$. For widely separated unperturbed states one obtains

$$\phi_+ \simeq \phi_1^0 - \frac{\mathcal{H}_{12}}{\Delta E} \phi_2^0, \qquad \phi_- \simeq \phi_2^0 + \frac{\mathcal{H}_{12}}{\Delta E} \phi_1^0$$

while for degenerate eigenstates

$$\phi_+ = \frac{1}{\sqrt{2}} \left(\phi_1^0 + \frac{\mathcal{H}_{12}}{|\mathcal{H}_{12}|} \phi_2^0 \right), \qquad \phi_- = \frac{1}{\sqrt{2}} \left(\phi_1^0 - \frac{\mathcal{H}_{12}}{|\mathcal{H}_{12}|} \phi_2^0 \right) \qquad (A.1.2.6)$$

Now we turn to the *time evolution* of the system, by considering two cases. The first is the evolution of the system after a static, time-independent perturbation has been turned on, the second (to be discussed as Appendix 1.3 when a periodic time-dependent perturbation is applied.

To deal with the time dependence one has to refer to the complete unperturbed eigenfunctions and to the time-dependent Schrodinger equation:

$$[\mathcal{H}_0 + \mathcal{H}_P(t)]\psi = i\hbar \frac{\partial \psi}{\partial t} \qquad (A.1.2.7)$$

The eigenfunction A.1.2.2 is now written with time dependent coefficients

$$\psi = c_1(t)\psi_1^0 + c_2(t)\psi_2^0 \qquad (A.1.2.8)$$

with $|c_1|^2 + |c_2|^2 = 1$. Let us assume the initial condition $c_1(t = 0) = 1$ and $c_2(t = 0) = 0$. The probability that at the time t after turning on the perturbation the system is found in the state $|2>$ is given by

$$P_2(t) = |c_2(t)|^2 \tag{A.1.2.9}$$

The equation for $c_2(t)$ is obtained by inserting A.1.2.8 into A.1.2.7. Recalling that

$$\mathcal{H}_0 \psi^0_{1,2} = i\hbar \frac{\partial \psi^0_{1,2}}{\partial t}$$

one has

$$\mathcal{H}_P(c_1 \psi^0_1 + c_2 \psi^0_2) = i\hbar \left(\psi^0_1 \frac{dc_1}{dt} + \psi^0_2 \frac{dc_2}{dt} \right) \tag{A.1.2.10}$$

By multiplying this equation by $(\psi^0_1)^*$, integrating over the spatial coordinates and by taking into account that $\psi^0_{1,2}(t) = \phi^0_{1,2} exp(-i E^o_{1,2} t/\hbar)$ one finds

$$c_1 < 1|\mathcal{H}_P|1 > + c_2 < 1|\mathcal{H}_P|2 > e^{-i\omega_{21}t} = i\hbar \frac{dc_1}{dt} \tag{A.1.2.11}$$

where $\omega_{21} = (E^0_2 - E^0_1)/\hbar$; $< 1|\mathcal{H}_P|1 > \equiv \mathcal{H}_{11} \equiv \int (\phi^0_1)^* \mathcal{H}_P \phi^0_1 d\tau$ and $< 1|\mathcal{H}_P|2 > \equiv \mathcal{H}_{12} \equiv \int (\phi^0_1)^* \mathcal{H}_P \phi^0_2 d\tau$ are the matrix elements of the perturbation between the stationary states of the unperturbed system.[5]

In analogous way, from A.1.2.10, multiplying by $(\psi^0_2)^*$ one derives

$$c_1 \mathcal{H}_{21} e^{+i\omega_{21}t} + c_2 \mathcal{H}_{22} = i\hbar \frac{dc_2}{dt} \tag{A.1.2.12}$$

In order to illustrate these equations for $c_{1,2}$, let us refer to a perturbation which is constant in time, with no diagonal elements. Then $(\mathcal{H}_P)_{11} = (\mathcal{H}_P)_{22} = 0$ and $(\mathcal{H}_P)_{12} = \hbar\Gamma$, $(\mathcal{H}_P)_{21} = \hbar\Gamma^*$. Equations (A.1.2.11) and (A.1.2.12) become

$$\frac{dc_1}{dt} = -i\Gamma e^{-i\omega_{21}t} c_2 \qquad\qquad \frac{dc_2}{dt} = -i\Gamma^* e^{+i\omega_{21}t} c_1$$

By taking the derivative of the second and by using the first one, one has

$$\frac{d^2 c_2}{dt^2} = i\omega_{21} \frac{dc_2}{dt} - c_2 \Gamma^2$$

of general solution

$$c_2(t) = (A e^{i\Omega t} + B e^{-i\Omega t}) e^{\frac{i\omega_{21}t}{2}}$$

---

[5] In the *Feynman* formulation the coefficients $c_i = < i|\psi(t) >$ are the *amplitudes* that the system is in state $|i >$ at the time t and one has $i\hbar(dc_i/dt) = \sum_j \mathcal{H}_{ij}(t) c_j(t)$, $\mathcal{H}_{ij}$ being the elements of the *matrix Hamiltonian*.

with $\Omega = (1/2)\sqrt{w_{21}^2 + 4\Gamma^2}$. The constants A and B are obtained from the initial conditions already considered, yielding

$$c_2(t) = -\frac{i\Gamma}{\Omega}\sin\Omega t e^{\frac{iw_{21}t}{2}}$$

and therefore

$$P_2(t) = |c_2(t)|^2 = \frac{4\Gamma^2}{w_{21}^2 + 4\Gamma^2}\sin^2\frac{(\sqrt{w_{21}^2 + 4\Gamma^2})t}{2}, \qquad (A.1.2.13)$$

known as *Rabi* equation. $P_1(t) = 1 - P_2(t)$.

It is worthy to illustrate the Rabi equation in the case of equivalent states, so that $E_1^0 = E_2^0$. We shall refer to such a situation in discussing the molecular Hydrogen ion H_2^+ where an electron is shared between two protons (Sect. 8.1). Then $w_{21} = 0$ and Eq. (A.1.2.13) becomes

$$P_2(t) = \sin^2\Gamma t \qquad (A.1.2.14)$$

namely the system oscillates between the two states. After the time $t = \pi/2\Gamma$ the system is found in state $|2>$, even though the perturbation is weak. For H_2^+ one can say that the electron is being exchanged between the two protons.

For widely separated states so that $w_{21}^2 \gg 4\Gamma^2$ Eq. (A.1.2.13) yields

$$P_2(t) = \left(\frac{2\Gamma}{w_{21}}\right)^2 \sin^2\frac{w_{21}t}{2} \qquad (A.1.2.15)$$

predicting fast oscillations but very small probability to find the system in state $|2>$.

Pulse resonance techniques (see Chap. 6) can be thought as an application of the Rabi formula once that the two spin states (spin up and spin down in a magnetic field) are "forced to become degenerate" by the on-resonance irradiation at the separation frequency $(E_2^0 - E_1^0)/h$.

In the presence of a relaxation mechanism driving the system to the low-energy state, a term $-i\hbar\gamma$ (with γ the relaxation rate) should be included in the matrix element $\mathcal{H}_{22}$. In this case, from the solution of the equations for the coefficients $c_{1,2}(t)$ the probability $P_2(t)$ corrects Eq. (A.1.2.14) for the Rabi oscillations with a damping effect. For strong damping the oscillator crosses to the overdamped regime: after an initial raise $P_2(t)$ decays to zero without any oscillation (see the book by *Budker, Kimball* and *De Mille*). Some more detail on the relaxation mechanism for spins in a magnetic field will be given at Chap. 6.

Appendix 1.3 Transition Probabilities and Selection Rules

The phenomenological transition probabilities induced by electromagnetic radiation are defined in Problem 1.24, where the *Einstein relations* are also derived. To illustrate the mechanism underlying the effect of the radiation one has to express the absorption probability W_{12} between two levels $|1 >$ and $|2 >$ in terms of the Hamiltonian describing the interaction of the radiation with the system. Here this description is carried out by resorting to the time-dependent perturbation theory.

The perturbation Hamiltonian $\mathcal{H}_P(t)$, introduced in Appendix 1.2, is then specified in the form

$$\mathcal{H}_P(t) = H_1 e^{i\omega t}. \qquad (A.1.3.1)$$

In fact, from the one-electron Hamiltonian in e.m. field (see Eq. (1.26))

$$\mathcal{H} = \frac{(\mathbf{p} + e\mathbf{A}/c)^2}{2m} - e\varphi \qquad (A.1.3.2)$$

($\mathbf{A}$ and φ vector and scalar potentials), recalling that $[\mathbf{p}, \mathbf{A}] = -i\hbar(\nabla.\mathbf{A}) \propto div\,A = 0$ in the Lorentz gauge and that $\mathbf{A}(\mathbf{r}, t) = \mathbf{A}_0\,exp[i(\mathbf{k} \cdot \mathbf{r} - \omega t)]$, the first order perturbation Hamiltonian is

$$\mathcal{H}_{rad} = -\frac{i\hbar e}{mc}\mathbf{A}.\nabla \qquad (A.1.3.3)$$

By expanding $\mathbf{A}(\mathbf{r}, t)$

$$\mathbf{A}(\mathbf{r}, t) = \mathbf{A}_0 e^{-i\omega t}[1 + i(\mathbf{k} \cdot \mathbf{r}) + \cdots] \qquad (A.1.3.4)$$

and limiting the attention to the site-independent term (*electric dipole approximation* or *long-wave length approximation*) one can show that[6]

$$\mathcal{H}_{rad} \propto \mathbf{A}_0.\nabla \propto \mathbf{A}_0.\mathbf{r} \propto \mathbf{E}_0.\mathbf{r}\frac{c}{\omega_{21}}$$

Therefore H_1 in A.1.3.1 takes the form $H_1 = -e\mathbf{r}.\mathbf{E}_0$, with $\mathbf{E}_0$ amplitude of the e.m. field (*electric dipole mechanism of transition*).

Now we use the results obtained in Appendix 1.2, again considering that $(\mathcal{H}_{rad})_{11} = (\mathcal{H}_{rad})_{22} = 0$ and $(\mathcal{H}_{rad})_{12} = (\mathcal{H}_{rad})_{21}^*$. The equations for the coefficients $c_{1,2}$ become

---

[6]It is recalled that $\mathbf{E} = -(1/c)\partial\mathbf{A}/\partial t$ and that the matrix element of the ∇ operator can be expressed in terms of the one for $\mathbf{r}$:

$$< 2|\nabla|1 > = -m\omega_{21} < 2|\mathbf{r}|1 > /\hbar.$$

$$i\hbar\frac{dc_1}{dt} = c_2 e^{-i\omega_{21}t} \cos\omega t < 1|x|2 > eE_0$$

$$i\hbar\frac{dc_2}{dt} = c_1 e^{+i\omega_{21}t} \cos\omega t < 2|x|1 > eE_0 \qquad\text{(A.1.3.5)}$$

for a given x-component of the operator $\mathbf{r}$. For Eq. (A.1.3.5) only approximate solutions are possible, essentially based on the perturbation condition

$\mathcal{H}_P \ll \mathcal{H}_0$ (while they are solved exactly for $\omega = 0$, as seen in Appendix 1.2). For ω around ω_{21} one finds

$$|c_2(t)|^2 = \frac{t^2}{\hbar^2}\frac{sin^2\left(\frac{(\omega-\omega_{21})t}{2}\right)}{(\omega - \omega_{21})^2 t^2}| < 2|\mathcal{H}_1|1 > |^2. \qquad\text{(A.1.3.6)}$$

$|c_2(t)|^2$ has the time dependence depicted below, with a maximum at $\omega = \omega_{21}$ proportional to t^2.

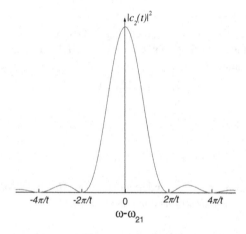

On increasing t the zeroes of the function tend to the origin while the maximum increases with t^2. Thus for $t \rightarrow \infty$ one has $|c_2(t)|^2 \propto \delta(\omega - \omega_{21})$, δ being the *Dirac delta function*. By taking into account the spread of the excited state due to the finite width (see Problem 1.24) or by resorting to the non-monocromatic character of the radiation, one writes

$$|c_2(t)|^2 \propto \frac{t^2}{\hbar^2}\int \rho(\omega)\frac{sin^2\frac{(\omega-\omega_{21})t}{2}}{(\omega - \omega_{21})^2\frac{t^2}{4}}d\omega \qquad\text{(A.1.3.7)}$$

where the frequency distribution $\rho(\omega)$ of the radiation is a slowly varying function around ω_{21}. Then one can set $\rho(\omega) \simeq \rho(\omega_{21})$. The integration over ω yields $2\pi/t$, and thus the transition probability per unit time becomes

$$W_{12} = |c_2(t)|^2/t = \frac{2\pi}{\hbar^2}| < 2|\mathcal{H}_1|1 > |^2\delta(\omega - \omega_{21}).$$

For linear polarization of the radiation along $\hat{\varepsilon}$ this equation reads

$$W_{12} = \frac{2\pi}{\hbar^2}| < 2| - e\mathbf{r}.\hat{\varepsilon}|1 > |^2 E_0^2\delta(\omega - \omega_{21}) \qquad (A.1.3.8)$$

For random orientation of $\mathbf{r}$ with respect to the e.m. wave one has to average $cos^2\theta$ over θ, to obtain 1/3. By introducing the energy density $\rho(\omega_{21})$ or $\rho(\nu_{21})$ ($\rho = < E^2 > /4\pi$) one finally has

$$W_{12} = \frac{2\pi}{3\hbar^2}\rho(\nu_{21})|\mathbf{R}_{21}|^2 \qquad (A.1.3.9)$$

where $|\mathbf{R}_{21}|^2 = | < 2| - ex|1 > |^2 + | < 2| - ey|1 > |^2 + | < 2| - ez|1 > |^2$.

$\mathbf{R}_{21}$ represents an effective *quantum electric dipole associated with a pair of states*. It can be defined $\mathbf{R}_{21} = < 2| - e\mathbf{r}|1 > exp(-i\omega_{21}t)$ and thus it can be thought a kind of electric dipole oscillating at the frequency of the transition. Since the power irradiated by a classical dipole is
$P = 2 < \ddot{\mu}^2 > /3c^3 \propto \omega^4 < \mu^2 >$, if one writes for the spontaneous emission (see Problem 1.24) $P = A_{21}h\nu_{21} \propto \nu_{21}^3 h\nu_{21} < |\mathbf{R}_{21}|^2 > \propto \omega^4 < |\mathbf{R}_{21}|^2 >$, that heuristic definition of $\mathbf{R}_{21}$ is justified.

The selection rules arise from the condition

$$\mathbf{R}_{21} \equiv < 2| - e\mathbf{r}|1 > \neq 0.$$

In the central field approximation the selection rules are[7]
(i) each electron makes a transition independently from the others;
(ii) neglecting the spin, the electric dipole transitions are possible when $\Delta l = \pm 1$ and $\Delta m = 0, \pm 1$ (according to parity arguments involving the spherical harmonics).

When the spin-orbit interaction is taken into account the selection rules are
$\Delta j = 0, \pm 1$ and $j = 0 \leftrightarrow j = 0$ transition not allowed;
$\Delta m = 0, \pm 1$ and no transition from $m = 0 \leftrightarrow m = 0$, when $\Delta j = 0$.

The *magnetic dipole* transitions (mechanism associated with the term ($i\mathbf{k}.\mathbf{r}$) in A.1.3.4) are controlled by the selection rules
$\Delta l = 0$ and $\Delta m = 0, \pm 1$
while for the transition driven by the *electric quadrupole* mechanism
$\Delta l = 0, \pm 2$ and $\Delta m = 0, \pm 1, \pm 2 \quad (l = 0 \leftrightarrow l' = 0$ forbidden)

The transition probabilities associated with the magnetic dipole or with the electric quadrupole mechanisms in the visible spectral range are smaller than W_{12} in A.1.3.9 by a factor of the order of the square of the *fine structure constant* $\alpha = e^2/\hbar c \simeq 1/137$. Further details on the selection rules will be given at Sect. 3.5.

---

[7]To derive the selection rules remind that $< Y_{l_2,m_2}|x|Y_{l_1,m_1} > = \delta_{l_2,l_1\pm1}\delta_{m_2,m_1\pm1}$ (and similar for y) while $< Y_{l_2,m_2}|z|Y_{l_1,m_1} > = \delta_{l_2,l_1\pm1}\delta_{m_2,m_1}$.

Problems

Problem 1.24 Refer to an ensemble of non-interacting atoms, each with two levels of energy E_1 (ground state) and E_2 (excited state). By applying the conditions of statistical equilibrium in a black-body radiation bath, derive the relationships among the probabilities of *spontaneous emission A_{21}*, of *stimulated emission W_{21}* and of *absorption W_{12}* (*Einstein relations*). Then assume that at $t = 0$ all the atoms are in the ground state and derive the evolution of the statistical populations $N_1(t)$ and $N_2(t)$ as a function of the time t at which electromagnetic radiation at the transition frequency is turned on (non-degenerate ground and excited states).

Discuss some aspects of the Einstein relations in regards of the possible *maser* and *laser actions* and about the finite width of the spectral line (*natural broadening*), by comparing the result based on the *Heisenberg principle* with the classical description of emission from harmonic oscillator (*Lorentz* model).

Solution: From the definition of transition probabilities,

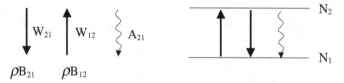

the time dependence of the *statistical populations* are written

$$\frac{dN_1}{dt} = -N_1 W_{12} + N_2 W_{21} + N_2 A_{21}$$

$$\frac{dN_2}{dt} = +N_1 W_{12} - N_2 W_{21} - N_2 A_{21}$$

In terms of the e.m. energy density at the transition frequency one has $W_{12} = B_{12}\rho(\nu_{12})$, $W_{21} = B_{21}\rho(\nu_{12})$, B_{12} and B_{21} being the *absorption* and *emission coefficients*, respectively.

One can assume that the system attains the equilibrium at a given temperature T inside a cavity where the black-body radiation implies the energy density (see Problem 1.25)

$$\rho(\nu_{12}) = \frac{8\pi h \nu_{12}^3}{c^3} \frac{1}{e^{\frac{h\nu_{12}}{k_B T}} - 1}. \qquad \text{a)}$$

At equilibrium $(dN_1/dt) = (dN_2/dt) = 0$. Then

$$\frac{N_1}{N_2} = \frac{W_{21} + A_{21}}{W_{12}} = \frac{\rho B_{21} + A_{21}}{\rho B_{12}} \quad \text{and} \quad \rho = \frac{A_{21}/B_{21}}{(N_1/N_2)(B_{21}/B_{12}) - 1}. \qquad \text{b)}$$

On the other hand, in accordance to Boltzmann statistics

$$\frac{N_1}{N_2} = e^{\frac{h\nu_{12}}{k_B T}} \qquad \text{c)}$$

The three equations a, b and c are satisfied for

$$B_{21} = B_{12} \qquad \text{and} \qquad A_{21} = \frac{8\pi h\nu_{12}^3}{c^3} B_{21}$$

These *Einstein relations*, derived in equilibrium condition are assumed to hold also out of equilibrium.

For levels 1 and 2 with statistical weights g_1 and g_2 respectively, $N_1/N_2 = \frac{g_1}{g_2} e^{\frac{h\nu_{12}}{k_B T}}$ and from the equilibrium condition

$$A_{21} = \frac{8\pi h\nu_{12}^3}{c^3} \frac{g_1}{g_2} B_{12} \qquad \text{and} \qquad A_{21} = \frac{8\pi h\nu_{12}^3}{c^3} B_{21}$$

so that $g_1 B_{12} = g_2 B_{21}$.

Now the system in the presence of radiation at the transition frequency (with initial condition $N_1(t = 0) = N$ and $N_2(t = 0) = 0$) is considered. Since

$$\frac{dN_1}{dt} = -N_1 W_{12} + N_2 W_{21} + N_2 A_{21} \equiv -N_1 W + N_2(W + A) =$$

$$= -N_1(2W + A) + N(W + A)$$

one derives

$$N_1(t) = \frac{N}{2W + A}(A + W + We^{-(2W+A)t})$$

plotted below:

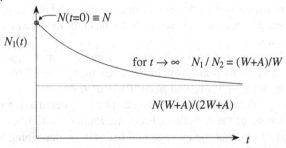

For $A \ll W \equiv W_{12} = W_{21}$ the saturation condition $N_1 = N_2 = N/2$ is achieved. It is noted that for $A \ll W$, by means of selective irradiation at the transition frequency the equilibrium condition implies a statistical temperature (describing N_1/N_2) different from the one of the thermostat. For $N_1 < N_2$ the *statistical temperature* would be negative (further discussion of these concepts is given at Chap. 6).

The condition of *negative temperature* (or *population inversion*) is a pre-condition for having radiation amplification in masers or in lasers. In the latter the *spontaneous emission* (i.e. A) acts as a disturbance, the correspondent "signal" being unrelated to an input.

Since $A_{21} \propto \nu_{12}^3$ the spontaneous emission can be negligible with respect to the *stimulated emission* $B_{12}\rho(\nu_{12})$ in the Microwave (MW) or in the Radiofrequency (RF) ranges, while it is usually rather strong in the visible range.

For the finite linewidth of a transition line the following is remarked. According to the uncertainty principle, because of the finite life-time τ of the excited state, of the order of A^{-1}, the uncertainty in the energy E_2 is $\Delta E \simeq A\hbar$ and then the linewidth is at least $\Delta\nu_{12} \simeq \tau^{-1}$. In the classical Lorentz description, the electromagnetic emission is related to a charge (the electron) in harmonic oscillation, with damping (*radiation damping*). The one-dimensional equation of motion of the charge can be written

$$m\frac{d^2x}{dt^2} + 2\Gamma m\frac{dx}{dt} + m\omega_0^2 x = 0$$

with solution $x(t) = x_0\, exp(-\Gamma t)exp(-i\omega_0 t)$ (for $\Gamma \ll \omega_0$). The Fourier transform is $FT[x(t)] = 2x_0/[\Gamma - i(\omega - \omega_0)]$ implying an intensity of the emitted radiation proportional to

$$I(\omega, \Gamma) \propto |FT[x(t)]|^2 \propto \frac{\Gamma}{\Gamma^2 + (\omega - \omega_0)^2}$$

namely a Lorenztian curve centered at ω_0 and of width Γ.

One can identify Γ with $\tau^{-1} \sim A$ and a certain equivalence of the classical description with the semi-classical theory of radiation is thus established.

Problem 1.25 (*The black-body radiation*).

Black-body radiation is the one present in a cavity of a body (e.g. a hot metal) brought to a given temperature T. It is related to the emission of e.m. energy over a wide frequency range.

The energy density $u(\nu, T)$ per unit frequency range around ν can be measured from the radiation $\rho_S(\nu, T)$ coming out from a small hole of area S (the *black-body*), per unit time and unit area. Prove that $\rho_S(\nu, T) = u(\nu, T)c/4$.

The electromagnetic field inside the cavity can be considered as a set of harmonic oscillators (the *modes* of the radiation). From the *Planck*'s estimate of the *thermal statistical energy*, prove that the average number $< n >$ describing the degree of excitation of one oscillator is

$$< n > = 1/[exp(h\nu/k_B T) - 1].$$

Then derive the *number of modes* $D(\omega)d\omega$ in the frequency range $d\omega$ around ω. (Note that $D(\omega)$ does not depend on the shape of the cavity).

By considering the photons as *bosonic* particles derive the *Planck distribution function*, the *Wien* law, the total energy in the cavity and the number of photons per unit volume.

Then consider the radiation as a thermodynamical system, imagine an expansion at constant energy and derive the exponent γ in the adiabatic transformation $TV^{\gamma-1} = const$. Evaluate how the entropy changes during the expansion.

Finally consider the e.m. radiation in the universe. During the expansion of the universe by a factor f each frequency is reduced by the factor $f^{1/3}$. Show that the Planck distribution function is retained along the expansion and derive the f-dependence of the temperature.

Solution:

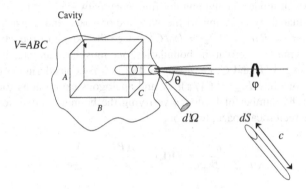

The energy emitted in $d\Omega$ from the element dS is

$$\rho_S d S d\nu = u(\nu, T) c \, cos\theta d\nu \frac{d\Omega}{4\pi} dS$$

Then

$$\rho_S(\nu, T) = \frac{u(\nu, T)c}{4\pi} \int_0^{2\pi} d\varphi \int_0^{\pi/2} cos\theta sin\theta d\theta = \frac{u(\nu, T)c}{4\pi} \frac{2\pi}{2} = \frac{u(\nu, T)c}{4}$$

In the *Planck* estimate the average energy per oscillator instead of being $< \varepsilon > = k_B T$ (as from the *equipartition principle* in the *Maxwell-Boltzmann* statistics) is evaluated according to

$$< \varepsilon > = \frac{\sum_{n=0}^{\infty} n\varepsilon_0 e^{-\frac{n\varepsilon_0}{k_B T}}}{\sum_{n=0}^{\infty} e^{-\frac{n\varepsilon_0}{k_B T}}}$$

where $\varepsilon_0 = h\nu$ is the quantum grain of energy for the oscillator at frequency ν. By defining $x = exp(-\varepsilon_0/k_B T)$ one writes

$$< \varepsilon >= \varepsilon_0 \frac{\sum_{n=0}^{\infty} n x^n}{\sum_{n=0}^{\infty} x^n}$$

and since $\sum_{n=0}^{\infty} x^n = 1/(1-x)$ for $x < 1$, while $\sum_{n=0}^{\infty} n x^n = x d(\sum_{n=0}^{\infty} x^n)/dx$, one obtains

$$< \varepsilon >= h\nu < n >, \qquad \text{with} \qquad < n >= \frac{1}{e^{\frac{h\nu}{k_B T}} - 1}$$

It is noted that for $k_B T \gg h\nu$, $< n > \rightarrow k_B T / h\nu$ and $< \varepsilon > \rightarrow k_B T$, the classical result for the average statistical energy of one-dimensional oscillator, i.e. for one of the modes of the e.m. field.

The number of modes having angular frequency between ω and $\omega + d\omega$ is conveniently evaluated by referring to the wavevector space and considering $k_x = (\pi/A)n_x$, $k_y = (\pi/B)n_y$ and $k_z = (\pi/C)n_z$ (see the sketch of the cavity). Since the e.m. waves must be zero at the boundaries, one must have an integer number of half-waves along A, B and C, i.e. $n_{x,y,z} = 1, 2, 3...$ By considering n as a continuous variable one has $dn_x = (A/\pi)dk_x$ and analogous expressions for the y and z directions. The number of $\mathbf{k}$ modes verifying the boundary conditions per unit volume of the reciprocal space, turns out

$$\frac{dn_x dn_y dn_z}{dk_x dk_y dk_z} \equiv D(\mathbf{k}) = \frac{ABC}{8\pi^3} = \frac{V}{8\pi^3}.$$

$D(\mathbf{k})$ is the *density of k-modes* or *density of k-states*.[8]

For photons the dispersion relation is $\omega = ck$ and the *number of modes $D(\omega)$* in $d\omega$ can be estimated from the volume $d\mathbf{k}$ in the reciprocal space in between the two surfaces at constant frequency ω and $\omega + d\omega$:

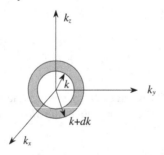

---

[8]This concept will be used for the electronic states and for the vibrational states in crystals, Chaps. 12 and 14. It is noted that the factor 8 is due to the fact that in this method of counting only positive components of the wave vectors have to be considered. For running waves, in the *Born-Von Karmann periodical conditions* (Sect. 12.4), the same number of excitations in the reciprocal space is obtained.

Then

$$D(\omega)d\omega = 2D(\mathbf{k})d\mathbf{k} = 2\frac{V}{8\pi^3}4\pi k^2 dk = \frac{V}{\pi^2 c^3}\omega^2 d\omega$$

and

$$D(\omega) = \frac{V}{\pi^2 c^3}\omega^2 \qquad \text{or} \qquad D(\nu) = \frac{8\pi V}{c^3}\nu^2,$$

the factor 2 being introduced to take into account the polarization states.

Photons are *bosonic* particles and therefore, by referring to the *Bose-Einstein statistical distribution function*

$$f_{BE} = 1/[exp(h\nu/k_B T) - 1]$$

one derives the *Planck distribution function* $\rho(\nu)$ (energy per unit volume in the unit frequency range) as follows. The energy related to the number of photons dn_ν within $d\nu$ around the frequency ν is

$$dE(\nu) = h\nu dn_\nu \qquad \text{and} \qquad dn_\nu = f_{BE}D(\nu)d\nu = \frac{8\pi V \nu^2}{c^3}\frac{d\nu}{e^{\frac{h\nu}{k_B T}} - 1}$$

By definition $\rho(\nu)d\nu = dE(\nu)/V$ and then

$$\rho(\nu) = \frac{8\pi h \nu^3}{c^3}\frac{1}{e^{\frac{h\nu}{k_B T}} - 1}$$

The *Wien law* can be obtained by looking for the maximum in $\rho(\nu)$: $d\rho/d\nu = 0$ for $h\nu_{max}/k_B T \simeq 2.8214$, corresponding to $\nu_{max} \simeq T \times 5.88 \times 10^{10}$ Hz (for T in Kelvin). It can be remarked that $\lambda_{max} = hc/5k_B T = (0.2898/T)$ cm is different from c/ν_{max}.

The total energy per unit volume $U(T)$ is obtained by integrating over the frequency and taking into account the number of modes in $d\nu$ and the average energy per mode $h\nu < n >$:

$$U(T) = \frac{1}{V}\int d\nu \frac{D(\nu)h\nu}{e^{\frac{h\nu}{k_B T}} - 1} = 3!\zeta(4)\frac{k_B^4 T^4}{\pi^2 c^3 \hbar^3}$$

where ζ is the Riemann zeta function, yielding $U(T) = \sigma T^4$, with $\sigma = 7.566 \times 10^{-15}$ erg $\cdot$ cm^{-3} K^{-4}, corresponding to the total emittance (*Stefan-Boltzmann law*) with the constant 5.6×10^{-12} W/cm^2 K^4.

The density of photons is obtaineijd by omitting in $U(T)$ the one-photon energy:

$$n_{tot}(T) = \frac{1}{V}\int d\nu \frac{D(\nu)}{e^{\frac{h\nu}{k_B T}} - 1} = 2\zeta(3)\frac{k_B^3 T^3}{\pi^2 c^3 \hbar^3} = 20.29 \times T^3 \, \text{cm}^{-3}$$

To derive the coefficient γ for expansion without exchange of energy, the radiation in the cavity is considered as a thermodynamical system of volume $V = ABC$ and temperature T (the temperature entering the energy distribution function). From $VU(T) = \sigma T^4 V = const$, one has $4V dT = -T dV$, i.e. $TV^{1/4} = const$ and therefore $\gamma = 5/4$.

During the expansion, since $N = n_{tot} V \propto T^3 V$, while $TV^{1/4} = const$, if the volume is increased by a factor f one has

$$T_{final} = T_{initial}\left(\frac{V_{initial}}{V_{initial} f}\right)^{\frac{1}{4}} = T_{initial} f^{-\frac{1}{4}}$$

and the number of photons becomes

$$N_{final} = N_{initial} f \left(\frac{T_{final}}{T_{initial}}\right)^3 = N_{initial} f^{\frac{1}{4}}$$

To evaluate the entropy the equation of state is required. The pressure of the radiation is obtained by considering the transfer of moment of the photons when they hit the surface and the well-known result $P = U/3$ is derived. For the entropy

$$dS = \frac{1}{T}d(UV) + \frac{PdV}{T} = \frac{V}{T}\frac{dU}{dT}dT + \frac{4U}{3T}dV$$

and since it has to be an exact differential $dU/dT = 4U/T$. Thus $PV = U_{tot}/3$, where $U_{tot} = UV$ and then

$$dS = \frac{4U}{3T}dV + \frac{V}{T}4\sigma T^3 dT$$

From the condition of exact differential $S = 4U_{tot}/3T$. The decrease of T yields an increase of the entropy because the number of photons increases. It is noted that for $T \to 0$, S, P and U tend to zero.

For transformation at constant entropy, assumed reversible, one would have $dS = 0$ and then

$$\frac{4\sigma T^3}{3}dV + 4\sigma V T^2 dT = 0,$$

so that $TV^{1/3} = const$.

In the expansion of the universe by a factor f the *cosmological principle* (each galaxy is moving with respect to any other by a velocity proportional to the distance) implies that each frequency ν_i is shifted to $\nu_f = \nu_i/f^{1/3}$. As a consequence of the expansion the energy density $du(\nu_i, T_i)$ in a given frequency range $d\nu_i$ is decreased by a factor f because of the increase in the volume and by a factor $f^{1/3}$ because of the energy shift for each photon. Then

$$du_f = \frac{du_i}{ff^{\frac{1}{3}}} = \frac{8\pi h\nu_i^3}{c^3} \frac{d\nu_i}{e^{\frac{h\nu_i}{k_B T}} - 1} \frac{1}{f^{\frac{4}{3}}}$$

which can be rewritten in terms of the new frequency $\nu_f = \nu_i / f^{1/3}$

$$du_f = \frac{8\pi h\nu_f^3}{c^3} \frac{d\nu_f}{e^{\frac{h\nu_f f^{1/3}}{k_B T}} - 1},$$

namely the same existing before the expansion, with the temperature scaled to $T/f^{1/3}$. Since $U_{tot} = UV \propto T^4 V$ the entropy of the universe is constant during the expansion, while the energy decreases by a factor $f^{1/3}$. The number of photons is constant.

In the figure the Planck distribution function (solid line) for the cosmic background radiation, resulting from a series of experimental detections, is evidenced.

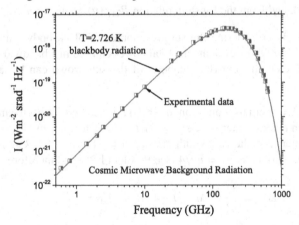

Problem 1.26 Derive the life-time of the Hydrogen atom in the $2p$ state and the *natural broadening* of the line resulting from the transition to the ground state. By neglecting relativistic effects (see Problem 1.38 and Appendix 5.1) evaluate the energy split due to the spin-orbit interaction (Sect. 1.6) and the effective field of orbital origin acting on the electron in the $2p$ state.

Solution: The life time (Problem 1.24) is $\tau = 1/A_{2p \to 1s}$, with $A_{2p \to 1s}$ the spontaneous emission transition probability. Then

$$A_{2p \to 1s} = \frac{32\pi^3 \nu^3}{3c^3 \hbar} \frac{1}{3} \sum_\alpha | < \phi_{2p\alpha}| - e\mathbf{r}|\phi_{1s} > |^2 \quad (\alpha \equiv 0, \pm 1)$$

From the evaluations of the matrix elements of the electric dipole components (See Table 1.3) one obtains

$$A_{2p\to 1s} = \left(\frac{2}{3}\right)^8 \frac{e^8}{c^3 a_0 \hbar^4} = 6.27 \times 10^8 \, s^{-1},$$

or $\tau = 1.6 \times 10^{-9}$ s.

Then the natural line-width can be written $\Delta\nu = (2.54 \times 10^{-7}/2\pi)\nu_{2p\to 1s}$.

The energy split due to the spin-orbit interaction is $\Delta E = (3/2)\xi_{2p}$, with $\xi_{2p} = e^2\hbar^2/(2m^2c^2a_0^3 \times 24)$, so that $\Delta E = 4.53 \times 10^{-5}$ eV and therefore $H = (\Delta E/2\mu_B) \simeq 4$ k Gauss (see also Problems 2.2 and 1.20).

Problem 1.27 Show that the stimulated emission probability W_{21} due to thermal radiation is equivalent to the spontaneous emission probability A_{21} times the average number of photons (Problem 1.24).

Solution: From $< n > = 1/[exp(h\nu/k_BT) - 1]$, while

$$\rho(\nu) = (8\pi h\nu^3/c^3)/[exp(h\nu/k_BT) - 1]$$

(Problem 1.25), and from the Einstein relation $B_{21}\rho(\nu) = A_{21} < n >$.

Problem 1.28 By considering the sun as a source of black-body radiation at the temperature $T \simeq 6000$ K, evaluate the total power emitted in a bandwidth of 1 MHz around the wavelength 3 cm (the diameter of the sun crown can be taken $2R = 10^6$ km).

Solution: For $\lambda = 3$ cm the condition $h\nu \ll k_BT$ is verified. To each e.m. mode one can attribute an average energy $< \varepsilon > = k_BT$. The density of modes is $(8\pi/c^3)\nu^2$ and thus the energy in the bandwidth $\Delta\nu$ is $\Delta u = (8\pi/c^3)\nu^2\Delta\nu k_BT$. The power emitted per unit surface is $\rho_S = uc/4$ (see Problem 1.25) and therefore

$$\Delta P = \frac{8\pi}{c^3}\nu^2\Delta\nu k_BT\frac{c}{4}4\pi R^2 \simeq 1.8 \times 10^9 \, W.$$

Problem 1.29 The energy flow from the sun arriving perpendicularly to the earth surface (neglecting atmospheric absorption) is $\Phi = 0.14$ W/cm^2. The distance from the earth to the sun is about 480 second-light. In the assumption that the sun can be considered as a black-body emitter, derive the temperature of the external crown.

Solution: The flow scales with the square of the distances. Thus the power emitted per unit surface from the sun can be written $\Phi_{tot} = (d/R)^2\Phi$ (d average distance, R radius of the sun). Then $\Phi_{tot} = 8 \times 10^3$ W/cm^2 and since (Problem 1.25) $\Phi = \sigma cT^4/4 = (5.67 \times 10^{-12} \times T^4)$ W/cm^2, one obtains $T_{Sun} \simeq 6129$ K.

Problem 1.30 Because of the thermal motions of the atoms the emission line from a lamp usually has a Gaussian shape. By referring to the yellow line at about 5800 Å by Sodium atom, neglecting the life-time broadening and assuming the Maxwellian distribution of the velocities, prove that statement. Estimate the order of magnitude of the broadening, for a temperature of the lamp of 500 K.

Show that the shift due to the recoil of the atom upon photon emission is negligible in comparison to the motional broadening. Comment on the possibility of *resonance absorption* by atoms in the ground state. At which wavelength one could expect that the resonance absorption would hardly be achieved?

Solution: Along the direction x of the motion the Doppler shift is

$$\lambda = \lambda_0 \left(1 \pm \frac{v_x}{c}\right)$$

The number of atoms $dn(v_x)$ moving with velocity between v_x and $v_x + dv_x$ is

$$dn(v_x) = N\sqrt{\frac{M}{2\pi k_B T}} e^{-\frac{Mv_x^2}{2k_B T}} dv_x,$$

N the number of atoms, with mass M (see Problem 1.34). The atoms emitting in the range $d\lambda$ around λ then are

$$dn(\lambda) = N\sqrt{\frac{Mc^2}{2\pi \lambda_0^2 k_B T}} e^{-\frac{Mc^2(\lambda-\lambda_0)^2}{2k_B T \lambda_0^2}} d\lambda$$

The intensity $I(\lambda)$ in the emission spectrum is proportional to $dn(\lambda)$

$$I(\lambda) \propto \sqrt{\frac{1}{\pi\delta^2}} e^{-\frac{(\lambda-\lambda_0)^2}{\delta^2}}$$

with $\delta = \sqrt{2k_B T/M}(\lambda_0/c)$.

Numerically, for the Na yellow line one has a broadening of about 1000 MHz, in wave-numbers, $1/\delta \simeq 0.034$ cm^{-1}.

The photon momentum being $h\nu/c$, the recoil energy is $E_R = (h\nu/c)^2/2M \simeq 9 \times 10^{-11}$ eV and the resonance absorption is not prevented. For wavelength in the range of the γ-rays the recoil energy would be larger than the Doppler broadening and without the *Mossbauer effect* (see Sect. 14.6) the resonance absorption would hardly be possible (Fig. 1.10).

Problem 1.31 *X-ray emission* can be obtained by removing an electron from inner states of atoms, with the subsequent transition of another electron from higher energy states to fill the vacancy. The X-Ray frequencies vary smoothly from element to element, increasing with the atomic number Z (see plot). Qualitatively justify the *Moseley law* $\lambda^{-1} \propto (Z - \sigma)^2$ (σ screening constant):

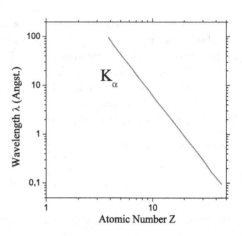

Fig. 1.10 Wavelength of the K_α line as a function of the atomic number

Solution: From the one-electron eigenvalues in central field with effective nuclear charge $(Z - \sigma)$ (σ reflecting the screening from other electrons, see Sects. 2.1 and 2.2), transitions between n_i and n_f imply the emission of a photon at energy

$$h\nu_{i \to f} = R_H hc(Z - \sigma)^2 \left(\frac{1}{n_f^2} - \frac{1}{n_i^2} \right)$$

The K-lines are attributed to the transitions to the final state $n_f = 1$. The K_α line corresponds to the longest wavelength ($n_i = 2$).

Problem 1.32 Estimate the order of magnitude of the voltage in an X-ray generator with Fe anode yielding the emission of the K_α line and the wavelength of the correspondent photon.

Solution: The energy of the K term is $E_K \simeq 13.6(Z - \sigma_K)^2(3/4)$ eV. For $\sigma_K \simeq 2$ one would obtain for the voltage $V \simeq 5800$ Volts. The wavelength of the K_α line turns out around 1.8 Å.

Problem 1.33 An electron is inside a sphere of radius $R_s = 1$ Å, with zero angular momentum. From the Schrödinger equation for the radial part of the wavefunction derive the lowest eigenvalue $E_{n=1}$ and the quantum pressure $P = -dE_{n=1}/dV$.

Solution: The equation for $r R(r)$ reads

$$-\frac{\hbar^2}{2m} \frac{d^2}{dr^2}(rR) = E(rR)$$

(see Eq. (1.14)). From the boundary condition $R(R_s) = 0$ one has $R \propto [sin(kr)]/kr$, with $k_n R_s = n\pi$ for $n = 1, 2, 3...$. Then

$$E_{n=1} = \frac{\hbar^2 k_{n=1}^2}{2m} = \frac{\pi^2 \hbar^2}{2m R_s^2}$$

and

$$P = \frac{\pi \hbar^2}{4m R_s^5}$$

For $R = 1\,\text{Å}$ one has $P = 9.6 \times 10^{12}$ dyne/cm^2. Compare this value with the one of the Fermi gas in a metal (Sect. 12.7).

Problem 1.34 From the Boltzmann distribution of the molecular velocities in ideal gas, show that the number of molecules n_c that hit the unit surface of the container per second is given by $n < v > /4$ (n number of molecules per cm^3) with $< v >$ the average velocity. Then numerically estimate n_c for molecular Hydrogen at ambient temperature and pressure (see Problem 1.30).

Solution: From the statistical distribution of the velocities the number of molecules moving along a given direction x with velocity between v_x and $v_x + dv_x$ is

$$dn(v_x) = n \left(\frac{M}{2\pi k_B T} \right)^{1/2} e^{-Mv_x^2/2k_B T} dv_x$$

The molecules colliding against the unit surface in a second are

$$n_c = \int_0^\infty v_x dn(v_x) = n \left(\frac{M}{2\pi k_B T} \right)^{1/2} \left(-\frac{k_B T}{M} \right) \left[e^{-Mv_x^2/2k_B T} \right]_0^\infty = n \left(\frac{k_B T}{2\pi M} \right)^{1/2}$$

The average velocity is

$$< v > = \frac{1}{n} \int_0^\infty v dn(v) = \frac{1}{n} \int_0^\infty v \left[4\pi n \left(\frac{M}{2\pi k_B T} \right)^{3/2} v^2 e^{-Mv^2/2k_B T} \right] dv =$$

$$= \left(\frac{8k_B T}{\pi M} \right)^{1/2} = \frac{4n_c}{n}$$

Numerically, for molecular Hydrogen H$_2$, $n_c \simeq 1.22 \times 10^{24}$ molecules/(s $\cdot$ cm^2).

Problem 1.35 Hydrogen atoms in the ground-state are irradiated at the resonance frequency $(E_{n=2} - E_{n=1})/h$, with e.m. radiation having the following polarization: (a) linear; (b) circular; (c) unpolarized.

By considering only electric dipole transitions, discuss the polarization of the fluorescent radiation emitted when the atoms return to the ground-state.

Solution: (a) The only possible transition is to the $2p_z$ state, with z the polarization axis ($\Delta m = 0$). No radiation is re-emitted along z while it is emitted in the xy plane, with polarization of the electric field along z.

(b) Only transitions to $2p_{\pm 1}$ state are possible ($\Delta m = \pm 1$). The fluorescent radiation when observed along the z direction is circularly polarized. By turning the observation axis from the z axis to the xy plane, the fluorescent radiation will progressively turn to the elliptical polarization, then to linearly polarized when the observation axis is in the xy plane.

(c) Any transition $1s \rightarrow 2p_{\pm 1,0}$ is possible, with uniform distribution over all the solid angle. The atom will be brought in the superposition state and the fluorescent radiation will have random wave-vector orientation and no defined polarization state.

Problem 1.36 An electron is moving along the x-axis under a potential energy $V(x) = (1/2)kx^2$, with $k = 5 \times 10^4$ dyne/cm. From the *Sommerfield quantization* (see Problem 1.4) obtain the amplitudes A of the motion in the lowest quantum states.

Solution: From $x(t) = A \, sin[(\sqrt{k/m})t + \varphi]$ the quantum condition in terms of the period $T = 2\pi\sqrt{m/k}$ reads

$$\oint m\dot{x}dx = \int_0^T m\dot{x}\ddot{x}dt = A^2k \int_0^T cos^2\left(\sqrt{\frac{k}{m}}t - \varphi\right)dt = A^2k\frac{T}{2} = nh$$

Thus $A_0 = 0$ (the zero-point energy is not considered here),
$A_1 = 1.47 \times 10^{-8}$ cm, $A_2 = 2.07 \times 10^{-8}$ cm.

Problem 1.37 The emission of radiation from *intergalactic Hydrogen* occurs at a wavelength $\lambda' = 21$ cm (see Sect. 5.2). The galaxy, that can be idealized as a rigid disc with homogeneous distribution of Hydrogen, is rotating. Estimate the Doppler broadening $\Delta\nu_{rot}$ of the radiation, assuming a period of rotation of 10^8 years and radius of the galaxy $R = 10$ kps $\simeq 3 \times 10^{18}$ cm. Prove that $\Delta\nu_{rot}$ is much larger than the broadening $\Delta\nu_T$ due to the thermal motion of the Hydrogen gas (assumed at a temperature $T = 100$ K) and larger than the shift due to the drift motion of the galaxy itself (at a speed of approximately $v_d = 10^7$ cm/s).

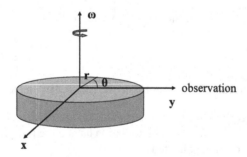

Solution: The Doppler shift at large distance along the y direction is

$$\nu(r, \theta) \simeq \nu_0\left(1 + \frac{\nu}{c}cos\theta\right)$$

with $\nu_o = \lambda'/c$. The mean-square average frequency is

$$\bar{\nu}_2 = \frac{1}{\pi R^2} \int_0^R r dr \int_0^{2\pi} \nu^2(r, \theta) d\theta = \nu_0^2 + \frac{\omega^2 \nu_0^2}{c \pi R^2} \int_0^R r^3 dr \int_0^{2\pi} \cos^2\theta d\theta$$

$$= \nu_0^2 \left(1 + \frac{\omega^2 R^2}{4c^2}\right)$$

Therefore

$$\Delta \nu_{rot} = \nu_0 \frac{\omega R}{2c} = \nu_0 \frac{\pi R}{cT} \simeq \nu_0 \cdot 10^{-3}$$

From Problem 1.30 one deduces the order of magnitude of the thermal broadening:

$$\Delta \nu_T = \frac{\nu_0}{\lambda'} \Delta \lambda = \frac{\nu_0}{\lambda'} \sqrt{\frac{2k_B T}{m_H}} \frac{\lambda'}{c} = \frac{\nu_0}{c} \sqrt{\frac{2k_B T}{m_H}} \simeq \nu_0 (4 \times 10^{-6})$$

For the drift associated with the linear motion of the galaxy one can approximately estimate the frequency shift of the order of $\Delta \nu_d = (\nu_d/c)\nu_0 \simeq 3.3 \times 10^{-4}\nu_0$.

Problem 1.38 In a description of the relativistic effects more detailed than the *Thomas-Frenkel* model (Sect. 1.6) to derive the one-electron spin-orbit Hamiltonian, the *Darwin term*

$$\mathcal{H}_D = \frac{\pi \hbar^2}{2m^2c^2} Z e^2 \delta(\mathbf{r}) \equiv \pi \alpha^2 \frac{Ze^2}{2a_0} a_0^3 \delta(\mathbf{r})$$

(with $\alpha = e^2/\hbar c = 1/137.036$ the *fine structure constant*) is found to be present.

Discuss the effects of $\mathcal{H}_D$ in Hydrogenic atoms, numerically comparing the corrections to the eigenvalues with the ones due to the spin-orbit Hamiltonian $\xi_{nl}\mathbf{l}.\mathbf{s}$.

Solution: From

$$< \phi_{nl} | \mathcal{H}_D | \phi_{nl} > \equiv D \int \phi_{nl}^*(\mathbf{r}) \delta(\mathbf{r}) \phi_{nl}(\mathbf{r}) d\mathbf{r} = D |\phi_{nl}(0)|^2,$$

with $D = \pi \alpha^2 Z e^2 a_0^2/2$, one sees that no effects due to $\mathcal{H}_D$ are present for non-s states (within the approximation of nuclear point-charge).

The shift for s states can be written (see Table 1.3)

$$\Delta E_D = \frac{Z^2\alpha^2}{n} \left(\frac{e^2 Z^2}{2a_0 n^2}\right) \equiv -\frac{Z^2\alpha^2}{n} E_n^0$$

with $E_n^0 = -Z^2 e^2/2a_0 n^2$ the unperturbed eigenvalues.

From

$$\xi_{nl} = (Ze^2/2m^2c^2) < r^{-3} >_{nlm}$$

and $< r^{-3} >_{nlm} = Z^3/[a_0^3 n^3 l(l+1/2)(l+1)]$ (see Table 1.4), $(l \neq 0)$

$$\Delta E_{SO} = \frac{Z^2\alpha^2(-E_n^0)}{2nl(l+\frac{1}{2})(l+1)}[j(j+1)-l(l+1)-3/4].$$

The relativistic kinetic energy is $c(p^2+m^2c^2)^{1/2}-mc^2 = (p^2/2m)-(p^4/8m^3c^2)$ + ... Then the energy correction reads

$$\Delta E_{kin} = \ <nlj|-p^4/8m^3c^2|nlj> \ = -\frac{1}{2mc^2} <nlj|(\frac{p^2}{2m})^2|nlj> =$$

$$= -\frac{1}{2mc^2} <nlj|(\mathcal{H}^0+e^2/r)^2|nlj>$$

and from the expectation value of $<r^{-2}>$ (see Table 1.4) one obtains

$$\Delta E_{kin} = -E_n^0\frac{Z^2\alpha^2}{n^2}\left[\frac{3}{4}-\frac{n}{l+\frac{1}{2}}\right].$$

From $\Delta E_D + \Delta E_{SO} + \Delta E_{kin}$ the eigenvalues of the *Dirac theory*,

$$E_{n,j}^{fs} = -E_n^0\frac{Z^2\alpha^2}{n^2}\left[\frac{3}{4}-\frac{n}{j+\frac{1}{2}}\right]$$

are obtained (see Appendix 5.1).

Specific References and Further Reading

1. H. Goldstein, *Classical Mechanics*, (Addison-Wesley, 1965).
2. D. Budker, D.F. Kimball and D.P. De Mille, *Atomic Physics - An Exploration Through Problems and Solutions*, (Oxford University Press, 2004).
3. B.H. Bransden and C.J. Joachain, *Physics of atoms and molecules*, (Prentice-Hall Inc., 2002).
4. B. Cagnac and J.C. Pebay - Peyroula, *Physique atomique, tome 2*, (Dunod Université, Paris, 1971).
5. E.U. Condon and G.H. Shortley, *The Theory of Atomic Spectra*, (Cambridge University Press, London, 1959).
6. J.P. Dahl, *Introduction to the Quantum World of Atoms and Molecules*, (World Scientific, 2001).
7. W. Demtröder, *Atoms, Molecules and Photons*, (Springer Verlag, 2006).
8. R. Eisberg and R. Resnick, *Quantum Physics of Atoms, Molecules, Solids, Nuclei and Particles*, (J. Wiley and Sons, 1985).
9. H. Eyring, J. Walter and G.E. Kimball, *Quantum Chemistry*, (J. Wiley, New York, 1950).
10. R.P. Feynman, R.B. Leighton and M. Sands, *The Feynman Lectures on Physics Vol. III*, (Addison Wesley, Palo Alto, 1965).
11. A.P. French and E.F. Taylor, *An Introduction to Quantum Physics*, (The M.I.T. Introductory Physics Series, Van Nostrand Reinhold (UK), 1986).

12. H. Haken and H.C. Wolf, *Atomic and Quantum Physics*, (Springer Verlag, Berlin, 1987).
13. M.A. Morrison, T.L. Estle and N.F. Lane, *Quantum States of Atoms, Molecules and Solids*, (Prentice-Hall Inc., New Jersey, 1976).
14. H. Semat, *Introduction to Atomic and Nuclear Physics*, (Chapman and Hall LTD, 1962).
15. J.C. Slater, *Quantum Theory of Matter*, (Mc Graw-Hill, New York, 1968).
16. S. Weinberg, *The first three minutes: a modern view of the origin of the universe*, (Amazon, 2005).

Chapter 2
Typical Atoms

Topics

Effects on the Outer Electron from the Inner Core
Helium Atom and the Electron-Electron Interaction
Exchange Interaction
Pauli Principle and Antisymmetry
Slater Determinantal Eigenfunctions

2.1 Alkali Atoms

Li, Na, K, Rb, Cs and Fr are a particular group of atoms characterized by one electron (often called *optical* being the one involved in optical spectra) with expectation value of the distance from the nucleus $< r >$ considerably larger than the one of the remaining $(N - 1)$ electrons, forming the internal *"core"*. The alkali atoms are suited for analyzing the role of the core charge in modifying the Coulomb potential $(-Ze^2/r)$ pertaining to Hydrogenic atoms (Sect. 1.4), as well as to illustrate the effect of the spin-orbit interaction (Sect. 1.6).

From spectroscopy one deduces the diagram of the energy levels for Li atom reported in Fig. 2.1, in comparison to the one for Hydrogen.

In Fig. 2.2 the analogous level scheme for Na atom is shown, with the main electric-dipole transitions yielding the emission spectrum.

The quantum numbers for the energy levels in Fig. 2.1 are the ones pertaining to the outer electron. At first we shall neglect the fine structure related to the spin-orbit interaction, which causes the splitting in doublets of the states at $l \neq 0$, as indicated for Na in Fig. 2.2.

A summarizing collection of the energy levels for alkali atoms is reported in Fig. 2.3. It should be remarked that because of the different extent of penetration in

© Springer International Publishing Switzerland 2015
A. Rigamonti and P. Carretta, *Structure of Matter*,
UNITEXT for Physics, DOI 10.1007/978-3-319-17897-4_2

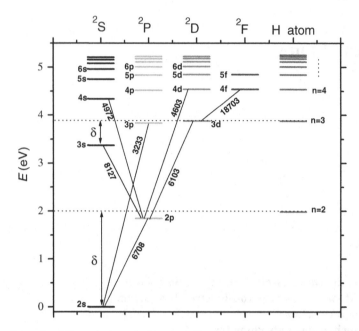

Fig. 2.1 Energy level diagram (*Grotrian diagram*) of Li atom, in term of the quantum numbers *nl* of the optical electron and comparison with the correspondent levels ($n > 1$) for H atom. The *quantum defect δ* (or *Rydberg* defect) indicated for 2s and 3s states, is a measure of the additional (negative) energy of the state in comparison to the correspondent state in Hydrogen. The wavelengths (in Å) for some transitions are reported

the core (as explained in the following) an inversion of the order of the energy levels in terms of the quantum number n (namely $|E_n| > |E_{n-1}|$) can occur.

From the Grotrian diagrams one deduces the following:

(i)   the sequence of the energy levels is similar to the one for H, with more bound and no more *l*-degenerate states;

(ii) the *quantum defect δ* for a given n-state (see Fig. 2.1) increases on decreasing the quantum number l;

(iii) the ground state for Li is 2s (3s for Na, etc.), with $L = l$ (and not the 1s state);

(iv) the transitions yielding the spectral lines obey the selection rule $\Delta l = \pm 1$.

These remarkable differences with respect to Hydrogen are related to an effective charge $Z_{eff}(r)$ for the optical electron (see Sect. 1.2) different from unit over a sizeable range of distance r from the nucleus.

In order to give a simple description of these effects we shall assume an *ad hoc* effective charge, of the form $Z_{eff} = (1+b/r)$, depicted in Fig. 2.4. The characteristic length b depends from the particular atom, it can be assumed constant over a large range of distance while for $r \to 0$ it must be such that $Z_{eff}(r) \to Z$.

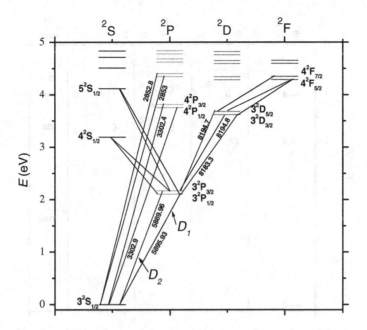

Fig. 2.2 Energy levels for Na atom with the electric dipole transitions ($\Delta l = \pm 1$) generating some spectral lines and correspondent wavelengths (in Å). The doublets related to spin-orbit interaction and resulting in states at different $j \equiv J$, are indicated (not in scale). The *yellow emission line* (a doublet) is due to the transition from the $^2P_{3/2}$ and $^2P_{1/2}$ states to the ground state $^2S_{1/2}$ with the optical electron in the 3s state

As a consequence of that choice for $Z_{eff}(r)$ the radial part of the Schrodinger equation for the optical electron takes a form strictly similar to the one in Hydrogen (see Sect. 1.4):

$$\frac{-\hbar^2}{2mr^2}\frac{d}{dr}r^2\frac{dR}{dr} + \left[\frac{l(l+1)\hbar^2 - 2me^2b}{2mr^2} - \frac{e^2}{r}\right]R = ER. \qquad (2.1)$$

It is remarked that for $b = 0$ the eigenvalues associated with Eq. (2.1) are $E_n = -R_Hhc/n^2$ (Eq. (1.13), for $Z = 1$).

If an effective quantum number l^* such that

$$l^*(l^* + 1) = l(l+1) - \frac{2me^2b}{\hbar^2} \equiv l(l+1) - \frac{2b}{a_0}$$

is introduced, then in the light of the formal treatment for Hydrogen, from Eq. (2.1) one derives the eigenvalues

$$E_n = -\frac{R_Hhc}{(n^*)^2}, \qquad (2.2)$$

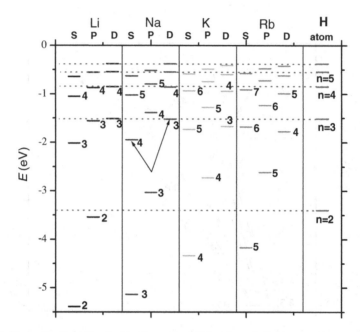

Fig. 2.3 Energy levels (neglecting the fine structure) for some alkali atoms, again compared with the states for Hydrogen at $n > 1$. The $4s$ state is more bound than the $3d$ state (see *arrows*), typical inversion of the order of the energies due to the extent of penetration of the s-electrons in the core, where the screening is not fully effective (see text and Fig. 2.6)

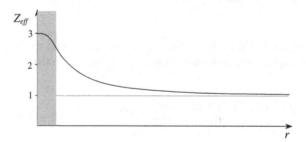

Fig. 2.4 Sketchy behavior of a plausible effective charge for the optical electron in Li atom. The dashed part of the Figure (not in scale) corresponds to the region of r not taken into account in the derivation of the energy levels. For Na, K, etc. atoms $Z_{eff}(r \rightarrow 0) \rightarrow Z$. A similar form of effective charge experimented by one electron because of the partial screening of the nuclear charge by the second electron is derived in Problem 2.7 for He atom

with n^* *not integer.* To evidence in these energy levels the numbers n and l pertaining to Hydrogen atom, we write $n^* = n - \delta l$, with $\delta l = l - l^*$, thus obtaining

$$E_{n,l} = -\frac{R_H hc}{(n - \delta l)^2} .$$

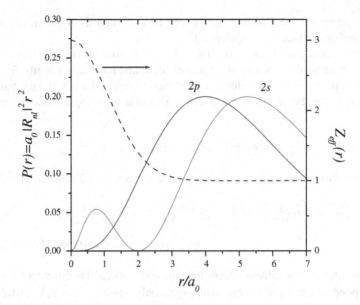

Fig. 2.5 Radial probability of presence for $2s$ and $2p$ electrons in Hydrogen and sketchy behavior of the effective charge for Li (see Fig. 2.4)

By neglecting the term in δl^2

$$E_{n,l} = -\frac{R_H hc}{(n - 2b/[a_0(2l+1)])^2} \equiv -\frac{R_H hc}{(n - \delta_{n,l})^2}. \qquad (2.3)$$

The eigenvalues are l-dependent, through a term that is atom-dependent (via b) and that decreases on increasing l, in agreement with the phenomenological findings.

The physical interpretation of the result described by Eq. (2.3) involves the amount of penetration of the optical electron within the core. In Fig. 2.5 it is shown that for $r \leq a_0$ the electron described by the $2s$ orbital has a radial probability of presence sizeably larger than the one for the $2p$ electron. This implies a reduced screening of the nuclear charge and then more bound state.

As a general rule one can state that the penetration within the core increases on decreasing l. In Fig. 2.6 it is shown how it is possible to have a more penetrating state for $n = 4$ rather than for $n = 3$, in spite of the fact that on the average the $3d$ electron is closer to the nucleus than the $4s$ electron. This effect is responsible of the inversion of the energy levels, with $|E_{4s}| > |E_{3d}|$, as already mentioned.

At the sake of illustration we give some quantum defects $\delta_{n,l}$ to be included in Eq. (2.3), for Na atom:

$\delta_{3s} = 1.373$	$\delta_{3p} = 0.883$	$\delta_{3d} = 0.01$
$\delta_{4s} = 1.357$	$\delta_{4p} = 0.867$	$\delta_{4d} = 0.011$
$\dots$	$\dots$	$\delta_{4f} \simeq 0$

These values for the quantum defects can be evaluated from the energy levels reported in Fig. 2.2 (see also Problem 2.1).

Finally a comment on the selection rule $\Delta l = \pm 1$ is in order. This rule is consistent with the statement that each electron makes the transition independently from the others, with the one-electron selection rule given in Appendix 1.3. In fact, the total wavefunction for the alkali atom, within the central field approximation, can be written

$$\phi(\mathbf{r}_1, \mathbf{r}_2, ..., \mathbf{r}_N) = \phi_{core}\phi_{optical}.$$

The electric dipole matrix element associated to a given $1 \leftrightarrow 2$ transition becomes

$$\mathbf{R}_{1\leftrightarrow 2} = -e \int (\phi_{core}^{(2)})^*(\mathbf{r}_1, \mathbf{r}_2, ...)(\phi^{(2)}(\mathbf{r}_n))^*[\mathbf{r}_1 + \mathbf{r}_2 + \cdots \mathbf{r}_n + \cdots + \mathbf{r}_N]$$
$$\times \phi_{core}^{(1)}(\mathbf{r}_1, \mathbf{r}_2, ...)\phi^{(1)}(\mathbf{r}_n)d\tau_1 d\tau_2...d\tau_N$$

Because of the orthogonality conditions the above integral is different from zero in correspondence to a given term involving $\mathbf{r}_n$ only when $\phi_{core}^{(2)} = \phi_{core}^{(1)}$, while

$$\int (\phi^{(2)}(\mathbf{r}_n))^*[\mathbf{r}_n]\phi^{(1)}(\mathbf{r}_n)d\tau_n$$

yields the selection rule $(\Delta l)_n = \pm 1$ and $(\Delta m)_n = 0, \pm 1$.

Fig. 2.6 Radial probability of presence for $3d$ and $4s$ electrons in Hydrogen. From the *dashed area* it is noted how the bumps in $P(r)$ for $r \leq 2a_0$ grant the presence of the $4s$ electron in the vicinity of the nucleus larger than the one pertaining to the $3d$ state

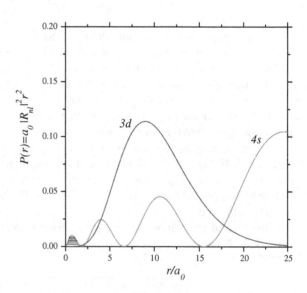

Now we take into account the doublet structure of each of the states at $l \neq 0$ (see the illustrative diagram in Fig. 2.2). The doublets result from spin-orbit interaction, as discussed at Sect. 1.6. The splitting of the np states of the optical electron turns out

Li	Na	K	Rb	Cs	
$2p$	$3p$	$4p$	$5p$	$6p$	
0.337	17.2	57.7	238	554	cm^{-1}
0.042	2.1	7.2	29.5	68.7	meV

supporting the energy corrections derived in terms of the spin-orbit constant ξ_{nl} (see for instance Problem 2.2). It can be observed that because of the selection rule $\Delta j = 0, \pm 1$ ($0 \leftrightarrow 0$ forbidden) (see Appendix 1.3) the spectral lines involving transitions between two non-S states in alkali atoms can display a fine structure in the form of three components (*compound doublets*).

Problems

Problem 2.1 The empirical values of the quantum defects $\delta_{n,l}$ (see Eq. (2.3)) for the optical electron in the Na atom are

Term	$n = 3$	$n = 4$	$n = 5$	$n = 6$
$l = 0$ s	1.373	1.357	1.352	1.349
$l = 1$ p	0.883	0.867	0.862	0.859
$l = 2$ d	0.010	0.011	0.013	0.011
$l = 3$ f	–	0.000	−0.001	−0.008

By neglecting the spin-orbit fine structure, write the wavenumbers of the main spectral series (see Fig. 2.2).

Solution: The spectral series are
principal (transitions from p to s terms), at wave numbers

$$\bar{\nu}_p = R_H \left[\frac{1}{[n_0 - \delta(n_0, 0)]^2} - \frac{1}{[n - \delta(n, 1)]^2} \right], \quad n \geq n_0, \ n_0 = 3;$$

sharp (transitions from s to p electron terms)

$$\bar{\nu}_s = R_H \left[\frac{1}{[n_0 - \delta(n_0, 1)]^2} - \frac{1}{[n - \delta(n, 0)]^2} \right], \quad n \geq n_0 + 1;$$

diffuse (transitions from d to p electron terms)

$$\bar{\nu}_d = R_H \left[\frac{1}{[n_0 - \delta(n_0, 1)]^2} - \frac{1}{[n - \delta(n, 2)]^2} \right], \quad n \geq n_0;$$

fundamental (transitions from f to d terms):

$$\bar{\nu}_f = R_H \left[\frac{1}{[n_0 - \delta(n_0, 2)]^2} - \frac{1}{[n - \delta(n, 3)]^2} \right], \quad n \geq n_0 + 1.$$

Problem 2.2 The spin-orbit splitting of the $6^2 P_{1/2}$ and $6^2 P_{3/2}$ states in Cesium atom causes a separation of the correspondent spectral line (transition to the $^2S_{1/2}$ ground-state) of 422 $\overset{\circ}{A}$, at wavelength around 8520 $\overset{\circ}{A}$. Evaluate the spin-orbit constant ξ_{6p} and the effective magnetic field acting on the electron in the $6p$ state.

Solution: From $\lambda'' - \lambda' = \Delta\lambda = 422 \overset{\circ}{A}$ and $\nu d\lambda = -\lambda d\nu$ one writes

$$\Delta E = h\Delta\nu \simeq h \cdot \frac{c}{\lambda'^2} \cdot \Delta\lambda \simeq 0.07 \, \text{eV}.$$

From

$$\Delta E_{SO} = \frac{\xi_{6p}}{2}\{j(j+1) - l(l+1) - s(s+1)\}$$

one has

$$\Delta E = \frac{\xi_{6p}}{2}\left[\frac{15}{4} - \frac{3}{4}\right] = \frac{3}{2}\xi_{6p}$$

and then

$$\xi_{6p} = \frac{2}{3}\Delta E = 0.045 \, \text{eV}.$$

The field (operator, Eq. (1.33)) is

$$\mathbf{H} = \frac{\hbar}{2emc}\frac{1}{r}\frac{dV}{dr}\mathbf{l}$$

with the spin-orbit Hamiltonian

$$\mathcal{H}_{spin-orbit} = -\boldsymbol{\mu}_s \cdot \mathbf{H}_{nl} = \xi_{6p}\mathbf{l} \cdot \mathbf{s}.$$

Thus

$$|\mathbf{H}_{6p}| = \frac{0.045 \, \text{eV} \, |\mathbf{l}|}{2\mu_B} \simeq 5.6 \cdot 10^6 \, \text{Oe} = 560 \, \text{Tesla}$$

Problem 2.3 In a maser ^{85}Rb atoms in the $63 \, ^2P_{3/2}$ state are driven to the transition at the $61 \, ^2D_{5/2}$ state. The quantum defects $\delta_{n,l}$ for the states are 2.64 and 1.34 respectively. Evaluate the transition frequency and compare it to the one deduced from the classical analogy for Rydberg atoms (Sect. 1.5). Estimate the isotopic shift for ^{87}Rb.

Solution:

From

$$E_{nl} = -R^* hc \frac{1}{n^{*2}}$$

where R^* is the Rydberg constant and $n^* = n - \delta_{n,l}$, the transition frequency turns out

$$\nu = -R^* c \left[\frac{1}{[n_i - \delta(n_i, l_i)]^2} - \frac{1}{[n_f - \delta(n_f, l_f)]^2} \right] \simeq 21.3 \, \text{GHz}$$

The classical analogy (see Problem 1.12) yields

$$\nu \approx -R^* c \frac{2 \Delta n^*}{(\bar{n}^*)^3} = 27.6 \, \text{GHz}.$$

The wavelengths are inversely proportional to the Rydberg constant:

$$\frac{\lambda_{87}}{\lambda_{85}} = \frac{R^*_{85}}{R^*_{87}} \approx 1 - 1.47 \cdot 10^{-7}.$$

Therefore the isotopic shift is $\Delta \nu \approx 3.16$ kHz or $\Delta \lambda \approx -20.6$ Å.

Problem 2.4 By considering Li as a Hydrogenic atom estimate the ionization energy. Discuss the result in the light of the real value (5.39 eV) in terms of percent of penetration of the optical electron in the $(1s)^2$ core.

Solution: By neglecting the core charge one would have $E_{2s} = -13.56 \, Z^2/n^2 = -30.6$ eV, while for total screening (i.e. zero penetration and $Z = 1$) $E_{2s} = -13.56 \, \text{eV}/4 = -3.4$ eV.

Then the effective charge experimented by the $2s$ electron can be considered $Z_{eff} \sim 1.27$, corresponding to about 15 % of penetration.

2.2 Helium Atom

2.2.1 Generalities and Ground State

The Helium atom represents a fruitful prototype to enlighten the effects due to the inter-electron interaction and then the arise of the central field potential $V(r)$, (see Sect. 1.1), the effects related to the exchange symmetry for indistinguishable electrons and to discuss the role of the spins and the *antisymmetry* of the total wavefunction.

First we shall start with the phenomenological examination of the energy levels diagram *vis-a-vis* to the one pertaining to Hydrogen atom (Fig. 2.7). A variety of comments is in order. It is noted that in He the state corresponding to the electronic

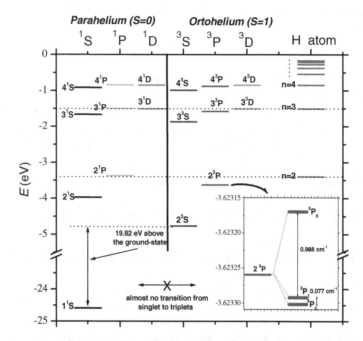

Fig. 2.7 Diagram of some energy levels for Helium atom and comparison with the correspondent levels for Hydrogen. The electron configuration of the states is $(1s)(nl)$. $E = 0$ corresponds to the first ionization threshold. The double-excited states (at weak transition probabilities and called *autoionizing states*) are unstable with respect to self-ionization (*Auger effect*) being at $E > 0$, within the continuum (Problem 2.14). In the inset the fine structure of the $2\,^3P$ state is reported, to be compared with the separation, about $9000\,\text{cm}^{-1}$, between the $2\,^3S$ and the $2\,^3P$ states. Note that this fine structure does not follow the multiplet rules described at Sect. 3.2

configuration $(1s)(nl)$ when compared to the n state in Hydrogen shows the removal of the accidental degeneracy in l. This could be expected, being the analogous of the effect for the optical electron in alkali atoms (Sect. 2.1). A somewhat unexpected result is the occurrence of a *double* series of levels, in correspondence to the same electronic configuration $(1s)(nl)$. The first series includes the ground state, with first ionization energy 24.58 eV. It is labelled as the group of *parahelium* states and all the levels are *singlets* (classification $^1S,^1P$, etc., see Sect. 1.7). The second series has the lowest energy state at 19.82 eV above the ground-state and identifies the *orthohelium* states. These states are all *triplets*, namely characterized by a fine structure (detailed in the inset of the Figure for the $2\,^3P$ state). Each level has to be thought as the superposition of almost degenerate levels, the degeneracy being removed by the spin-orbit interaction (Sect. 1.6). The orthohelium states are classified $^3S,^3P$, etc.. Among the levels of a given series the transition yielding the spectral lines correspond to the rule $\Delta L = \pm 1$, with an *almost* complete inhibition of the transitions from parahelium to orthohelium (i.e. almost no singlets$\leftrightarrow$ triplets transitions). Finally it can be remarked that while $(1s)^2$, at $S = 0$, is the ground state, the corresponding $(1s)^2$ triplet state is *absent* (as well as other states to be mentioned in the following).

In the assumption of infinite nuclear mass and by taking into account the Coulomb interactions only, the Schrodinger equation is

$$\left[-\frac{\hbar^2}{2m}(\nabla_1^2 + \nabla_2^2) - \frac{Ze^2}{r_1} - \frac{Ze^2}{r_2} + \frac{e^2}{r_{12}} \right] \phi(\mathbf{r}_1, \mathbf{r}_2) = E\phi(\mathbf{r}_1, \mathbf{r}_2) \qquad (2.4)$$

and it can be the starting point to explain the energy diagram. In Eq. (2.4) $Z = 2$ for the neutral atom.

Let us first assume that the inter-electron term e^2/r_{12} can be consider a perturbation of the hydrogenic-like Hamiltonian for two independent electrons (*independent electron approximation*). Then the unperturbed eigenfunction is

$$\phi_{n'l',n''l''}(\mathbf{r}_1, \mathbf{r}_2) = \phi_{n'l'}(\mathbf{r}_1)\phi_{n''l''}(\mathbf{r}_2) \qquad (2.5)$$

and

$$(E^0)_{n'l',n''l''} = Z^2 E^H_{n'l'} + Z^2 E^H_{n''l''} \qquad (2.6)$$

E^H_{nl} being the eigenvalues for Hydrogen (degenerate in l).

For the ground state $(1s)^2$ one has

$$\phi_{1s,1s}(\mathbf{r}_1, \mathbf{r}_2) = \frac{Z^3}{\pi a_0^3} e^{-\frac{Z(r_1+r_2)}{a_0}} \qquad (2.7)$$

and

$$E^0_{1s,1s} = 2Z^2 E^H_{1s} = -8\frac{e^2}{2a_0} \simeq -108.80 \text{ eV} \qquad (2.8)$$

In this oversimplified picture the first ionization energy would be 54.4 eV, evidently far from the experimental datum (see Fig. 2.7). This discrepancy had to be expected since the effect of the electron-electron repulsion had not yet been evaluated.

At the first order in the perturbative approach the repulsion reads

$$E^{(1)}_{1s,1s} = \int\int \phi^*_{1s,1s}(\mathbf{r}_1, \mathbf{r}_2)\frac{e^2}{r_{12}}\phi_{1s,1s}(\mathbf{r}_1, \mathbf{r}_2)d\tau_1 d\tau_2 \equiv$$

$$\equiv < 1s, 1s| \frac{e^2}{r_{12}} |1s, 1s > \equiv I_{1s,1s} \qquad (2.9)$$

$I_{1s,1s}$ is called *Coulomb integral* in view of its classical counterpart, depicted in Fig. 2.8.

The estimate of the Coulomb integral can be carried out by expanding r_{12}^{-1} in term of the associated Legendre polynomials (see Problem 2.5). For the particular case of $1s$ electrons, the Coulomb integral $I_{1s,1s}$ can be worked out in a straightforward way on the basis of the classical analogy for the electrostatic repulsion. The result is

Fig. 2.8 Illustrative plot sketching the classical analogy of the first order perturbation term $< e^2/r_{12} >$ for the ground-state, in terms of electrostatic repulsion of two electronic clouds

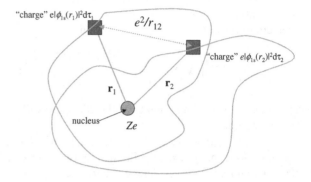

$$I_{1s,1s} = \frac{5}{4}Z(-E_{1s}^H) = \frac{5}{8}\frac{e^2}{a_0}Z \qquad (2.10)$$

The ground state energy corrected to the first order turns out

$$E_{1s,1s} = E_{1s,1s}^{(0)} + I_{1s,1s} = \left(2Z^2 - \frac{5}{4}Z\right)E_{1s}^H \simeq -74.8 \text{ eV} \qquad (2.11)$$

to be compared with the experimental value -78.62 eV.

The energy required to remove one electron is

$$\left[\left(2Z^2 - \frac{5}{4}Z\right) - Z^2\right] 13.6 \text{eV} \simeq 20.4 \text{ eV}$$

This estimate is not far from the value indicated in Fig. 2.7, in spite of the crudeness of the assumption for the unperturbed one-electron wavefunctions. An immediate refinement could be achieved by adjusting the hydrogen-like wave functions: in this way a good agreement with the experimental ionization energy would be obtained.

Another way to improve the description is to derive variationally an *effective nuclear charge* Z^*, which in indirect way takes into account the mutual screening of one electron by the other and the related correction in the wavefunctions. As shown in Problem 2.6, this procedure yields $Z^* = Z - (5/16)$, implying for the ground state

$$E_{1s,1s} = 2\left(Z - \frac{5}{16}\right)^2 \left(\frac{-e^2}{2a_0}\right) = -77.5 \text{ eV}$$

One can remark how the perturbative approach, without modification of the eigenfunctions, is rather satisfactory, in spite of the relatively large value of the first order energy correction.

The ground state energy for He turns out about 94.6 % of the "exact" one (numerically obtained via elaborate trial functions, see Sect. 3.4) with the first-order perturbative correction and 98 % with the variationally-derived effective charge. The agreement is even better for atoms with $Z \geq 3$, as Li^+ or Be^{2+}. At variance the analogous procedure fails for H^- (see Problem 2.8).

2.2.2 Excited States and the Exchange Interaction

The perturbative approach used for the ground state could be naively attempted for the excited states with an electron on a given nl state. For a trial wavefunction of the form

$$\phi(\mathbf{r}_1, \mathbf{r}_2) = \phi_{1s}(\mathbf{r}_1)\phi_{nl}(\mathbf{r}_2) \qquad (2.12)$$

the energy

$$E_{1s,nl} = E^0_{1s,nl} + \, < 1s, nl | \frac{e^2}{r_{12}} | 1s, nl > \qquad$$

would not account for the experimental data, numerically falling approximately in the middle of the singlet and triplet $(1s, nl)$ energy levels. The striking discrepancy is evidently the impossibility to infer two energy levels in correspondence to the same electronic configuration from the wavefunction in Eq. (2.12). The obvious inadequacy of the tentative wavefunction is that it disregards the *exchange symmetry* (discussed at Sect. 1.3). At variance with Eq. (2.12) one has to write the functions

$$\phi^{sym}_{1s,nl}(\mathbf{r}_1, \mathbf{r}_2) = \frac{1}{\sqrt{2}}\left[\phi_{1s}(\mathbf{r}_1)\phi_{nl}(\mathbf{r}_2) + \phi_{nl}(\mathbf{r}_1)\phi_{1s}(\mathbf{r}_2)\right] \qquad (2.13)$$

$$\phi^{ant}_{1s,nl}(\mathbf{r}_1, \mathbf{r}_2) = \frac{1}{\sqrt{2}}\left[\phi_{1s}(\mathbf{r}_1)\phi_{nl}(\mathbf{r}_2) - \phi_{nl}(\mathbf{r}_1)\phi_{1s}(\mathbf{r}_2)\right] \qquad (2.14)$$

granting indistinguishable electrons, the same weights being attributed to the configurations $1s(1)nl(2)$ and $1s(2)nl(1)$. The labels *sym* and *ant* correspond to the *symmetrical* and *antisymmetrical character* of the wavefunctions upon exchange of the electrons.

On the basis of the functions (2.13) and (2.14), along the same perturbative procedure used for the ground state, instead of Eq. (2.11) one obtains

$$E^{sym}_+ = Z^2 E^H_{1s} + Z^2 E^H_{nl} + I_{1s,nl} + K_{1s,nl} \qquad (2.15)$$

and

$$E^{ant}_- = Z^2 E^H_{1s} + Z^2 E^H_{nl} + I_{1s,nl} - K_{1s,nl} \qquad (2.16)$$

where

$$K_{1s,nl} = \int\int \phi^*_{1s}(\mathbf{r}_1)\phi^*_{nl}(\mathbf{r}_2)\frac{e^2}{r_{12}}\phi_{1s}(\mathbf{r}_2)\phi_{nl}(\mathbf{r}_1)d\tau_1 d\tau_2 \qquad (2.17)$$

is the *exchange integral, essentially positive* and without any classical interpretation, at variance with the Coulomb integral $I_{1s,nl}$. Thus double series of levels is justified by the quantum effect of *exchange symmetry*.[1]

The wavefunctions (2.13) and (2.14) are not complete, spin variables having not yet been considered. In view of the weakness of the spin-orbit interaction, as already stated (Sect. 1.6), one can factorize the spatial and spin parts. Then, again by taking into account indistinguishable electrons, the spin functions are

$$\alpha(1)\alpha(2), \quad \beta(1)\beta(2), \quad \frac{1}{\sqrt{2}}[\alpha(1)\beta(2) + \alpha(2)\beta(1)] \quad \text{for} \quad S = 1$$

$$\frac{1}{\sqrt{2}}[\alpha(1)\beta(2) - \alpha(2)\beta(1)] \quad \text{for} \quad S = 0 \quad (2.18)$$

The first group can be labelled $\chi_{S=1}^{sym}$ and it includes the three eigenfunctions corresponding to $S = 1$. The fourth eigenfunction is the one pertaining to $S = 0$. $\chi_{S=0}^{ant}$ is antisymmetrical upon the exchange of the electrons, while $\chi_{S=1}^{sym}$ are symmetrical.

Therefore the complete eigenfunctions describing the excited states of the Helium atom are of the form $\phi_{tot} = \phi_{1s,nl} \chi_S$ and in principle in this way one would obtain 8 spin-orbitals. However, from the comparison with the experimental findings (such as the spectral lines from which the diagram in Fig. 2.7 is derived) one is lead to conclude that only four states are actually found in reality. These states are the ones for which the total (spatial and spin) wavefunctions are *antisymmetrical* upon the exchange of the two electrons.

This requirement of antisymmetry is also known as *Pauli principle* and we shall see that it corresponds to require that the electrons differ at least in one of the four quantum numbers n, l, m and m_s. For instance, the lack of the triplet $(1s)^2$ is evidently related to the fact that in this hypothetical state the two electrons would have the same quantum numbers, meantime having a wavefunction of symmetric character $\phi_{tot} = \phi_{1s}\phi_{1s}\chi_{S=1}^{sym}$. Thus $\phi^{sym}\chi_{S=0}^{ant}$ describes the singlet states, while $\phi^{ant}\chi_{S=1}^{sym}$ describes the triplet states. Accordingly, one can give the following pictorial description

When	S	χ	$\phi(\mathbf{r})$	ϕ_{tot}	Energy
↑↓	0	ant	sym	ant	E_+
↑↑	1	sym	ant	ant	E_-

In other words, because of the exchange symmetry a kind of relationship, arising from electron-electron repulsion, between the mutual "direction" of the spin momenta and the energy correction does occur. For "parallel spins" one has $E_- < E_+$, the repulsion is decreased as the electron should move, on the average, more apart.

---

[1] Order of magnitude estimates yield $I_{1s,2s} \simeq 9$ eV, $I_{1s,2p} \simeq 10$ eV, $K_{1s,2s} \simeq 0.4$ eV and $K_{1s,2p} \simeq 0.1$ eV (see Problem 2.9).

The dependence of the energy from the spin orientation can be related to an exchange *pseudo-spin interaction*, in other words to an Hamiltonian operator of the form[2]

$$H = -2K\mathbf{s}_1 \cdot \mathbf{s}_2 \qquad (2.19)$$

In fact if we extend the vectorial picture to spin operators (in a way analogous to the definition of the **j** angular momentum for the electron (see Sect. 1.6)) and write

$$\mathbf{S} = \mathbf{s}_1 + \mathbf{s}_2, \qquad (2.20)$$

by "*squaring*" this sum one deduces $\mathbf{s}_1.\mathbf{s}_2 = (1/2)[\mathbf{S}^2 - \mathbf{s}_1{}^2 - \mathbf{s}_2{}^2]$. Thus, from the Heisenberg Hamiltonian (2.19) the two energy values

$$E' = -2K(1/2)[S(S+1) - 2(1/2)(1+1/2)] = -K/2$$

for $S = 1$ and

$$E'' = 3K/2$$

for $S = 0$ are obtained. In other words, from the Hamiltonian (2.19), for a given $1snl$ configuration, the singlet and the triplet states with energy separation and classification consistent with Eqs. (2.15) and (2.16), are deduced.

Now it is possible to justify the weak singlet↔triplet transition probability indicated by the optical spectra. The electric dipole transition element connecting parahelium to orthohelium states can be written

$$\mathbf{R}_{S=0 \leftrightarrow S=1} \propto < \chi_{ant} | \chi_{sym} > \int \int \phi^*_{sym}[\mathbf{r}_1 + \mathbf{r}_2] \phi_{ant} d\tau_1 d\tau_2. \qquad (2.21)$$

This matrix element is zero, both for the orthogonality of the spin states and because the function in the integral changes sign upon exchange of the indexes 1 and 2, then requiring zero as physically acceptable result. Thus one understands why orthohelium cannot be converted to parahelium and *vice-versa*. This selection rule would seem to prevent any transitions (including the ones related to magnetic dipole or electric quadrupole mechanisms) and then do not admit any violation. The weak singlet-triplet transitions actually observed in the spectrum are related to the non-total validity of the factorization in the form $\phi_{tot} = \phi(\mathbf{r}_1, \mathbf{r}_2)\chi_{spin}$. The spin-orbit interaction, by coupling spin and positional variables, partially invalidates that form of the wavefunctions. This consideration is supported by looking at the transitions in an atom similar to Helium, with two electrons outside the core. Calcium has the ground state electronic configuration $(1s)^2...(4s)^2$ and the diagram of the energy levels is strictly similar to the one in Fig. 2.7. At variance with Helium, because of the

---

[2]This Hamiltonian, known as *Heisenberg Hamiltonian*, is often assumed as starting point for quantum magnetism in bulk matter. Below a given temperature, in a three-dimensional array of atoms, this Hamiltonian implies a spontaneous ordered state, with magnetic moments cooperatively aligned along a common direction (see Sect. 4.4 for comments and Chap. 17).

increased strength of the spin-orbit interaction, the lines related to $S = 0 \leftrightarrow S = 1$ transitions are rather strong. Analogous case is Hg atom (see Fig. 3.9).

Problems

Problem 2.5 Evaluate the Coulomb integral for the ground state of the Helium atom.

Solution:

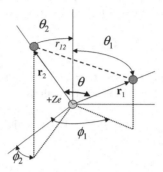

In the expectation value (for $e = a_0 = 1$)

$$< \frac{1}{r_{12}} > = \frac{Z^6}{\pi^2} \int e^{-2Z(r_1 + r_2)} \frac{1}{r_{12}} d\mathbf{r}_1 d\mathbf{r}_2.$$

$1/r_{12}$ is expanded in Legendre polynomials

$$\frac{1}{r_{12}} = \frac{1}{r_1} \sum_{l=0}^{\infty} \left(\frac{r_2}{r_1}\right)^l P_l(\cos\theta), \qquad r_1 > r_2$$

$$= \frac{1}{r_2} \sum_{l=0}^{\infty} \left(\frac{r_1}{r_2}\right)^l P_l(\cos\theta), \qquad r_1 < r_2$$

where θ is the angle between the vectors $\mathbf{r}_1$ and $\mathbf{r}_2$ and

$$\cos\theta = \cos\theta_1 \cos\theta_2 + \sin\theta_1 \sin\theta_2 \cos(\phi_1 - \phi_2)$$

In compact form

$$\frac{1}{r_{12}} = \sum_{l=0}^{\infty} \frac{(r_<)^l}{(r_>)^{l+1}} P_l(\cos\theta)$$

where $r_<$ is the smallest and $r_>$ the largest between r_1 and r_2. From the *addition theorem* one writes

$$\frac{1}{r_{12}} = \sum_{l=0}^{\infty} \sum_{m=-l}^{+l} \frac{4\pi}{(2l+1)} \frac{(r_<)^l}{(r_>)^{l+1}} Y_{lm}^*(\theta_1, \phi_1) Y_{lm}(\theta_2, \phi_2).$$

The function $exp[-2Z(r_1 + r_2)]$ is spherically symmetric and $Y_{00} = (4\pi)^{-\frac{1}{2}}$. By integrating over the polar angles one has

$$I'_{1s,1s} = \frac{Z^6}{\pi^2} \sum_{l=0}^{\infty} \sum_{m=-l}^{+l} \frac{(4\pi)^2}{(2l+1)} \int_0^{\infty} dr_1 r_1^2 \int_0^{\infty} dr_2 r_2^2 e^{-2Z(r_1+r_2)} \frac{(r_<)^l}{(r_>)^{l+1}}$$

$$\cdot \delta_{l,0} \delta_{m,0}.$$

All terms in the sum vanish, except the one for $l = m = 0$. Then

$$I'_{1s,1s} = 16Z^6 \int_0^{\infty} dr_1 r_1^2 \int_0^{\infty} dr_2 r_2^2 e^{-2Z(r_1+r_2)} \frac{1}{r_>}$$

$$= 16Z^6 \int_0^{\infty} dr_1 r_1^2 e^{-2Zr_1} \left[\frac{1}{r_1} \int_0^{r_1} dr_2 r_2^2 e^{-2Zr_2} + \int_{r_1}^{\infty} dr_2 r_2 e^{-2Zr_2} \right] = \frac{5}{8}Z$$

and properly including a_0 and e, $I_{1s,1s} = \frac{5}{4} Ze^2/2a_0$.

For spherically symmetric wavefunctions one can evaluate the Coulomb integral from the classical electrostatic energy:

$$I_{1s,1s} = \frac{Ze^2}{32\pi^2 a_0} \int \frac{e^{-\rho_1} e^{-\rho_2}}{\rho_{12}} d\tau_1 d\tau_2$$

where

$$\rho_{1,2} = \frac{2Zr_{1,2}}{a_0}, \qquad \rho_{12} = \frac{2Zr_{12}}{a_0}$$

and

$$d\tau_{1,2} = \rho_{1,2}^2 \sin\theta_{1,2} \, d\rho_{1,2} \, d\theta_{1,2} \, d\phi_{1,2}.$$

The electric potential from the shell $d\rho_1$ at ρ_1 is

$$d\Phi(r) = \ 4\pi\rho_1^2 e^{-\rho_1} d\rho_1 \frac{1}{\rho_1} \ \text{for } r < \rho_1,$$

$$4\pi\rho_1^2 e^{-\rho_1} d\rho_1 \frac{1}{r} \ \text{for } r > \rho_1.$$

Then the total potential turns out

$$\Phi(r) = \frac{4\pi}{r} \int_0^r e^{-\rho_1} \rho_1^2 d\rho_1 + 4\pi \int_r^{\infty} e^{-\rho_1} \rho_1 d\rho_1 = \frac{4\pi}{r} \{2 - e^{-r}(r+2)\}$$

and therefore

$$I_{1s,1s} = \frac{Ze^2}{32\pi^2 a_0} \int \Phi(\rho_2) e^{-\rho_2} d\tau_2 = \frac{Ze^2}{2a_0} \int_0^{\infty} [2 - e^{-\rho_2}(\rho_2 + 2)] e^{-\rho_2} \rho_2^2 d\rho_2 = \frac{Ze^2}{2a_0} \frac{5}{4}.$$

Problem 2.6 By resorting to the variational principle, evaluate the effective nuclear charge Z^* for the ground state of the Helium atom.

Solution: The energy functional is

$$E[\phi] = \frac{<\phi|H|\phi>}{<\phi|\phi>}$$

where

$$\phi(r_1, r_2) = \frac{Z^{*3}}{\pi} e^{-Z^*(r_1+r_2)}$$

with Z^* variational parameter ($e = a_0 = 1$).

Then

$$E[\phi] = \left\langle \phi \left| T_1 + T_2 - \frac{Z}{r_1} - \frac{Z}{r_2} + \frac{1}{r_{12}} \right| \phi \right\rangle$$

and

$$\langle\phi|T_1|\phi\rangle \equiv \langle\psi_{1s}^{Z^*}|T_1|\psi_{1s}^{Z^*}\rangle = \frac{1}{2}Z^{*2}, \qquad \langle\phi|T_2|\phi\rangle = \langle\phi|T_1|\phi\rangle,$$

while

$$\left\langle \phi \left| \frac{1}{r_1} \right| \phi \right\rangle = \left\langle \psi_{1s}^{Z^*} \left| \frac{1}{r_1} \right| \psi_{1s}^{Z^*} \right\rangle = Z^* = \left\langle \phi \left| \frac{1}{r_2} \right| \phi \right\rangle$$

Since

$$\left\langle \phi \left| \frac{1}{r_{12}} \right| \phi \right\rangle = \frac{5}{8}Z^* \quad \text{(see Eq. (2.10))}$$

one has

$$E[\phi] \equiv E(Z^*) = Z^{*2} - 2ZZ^* + \frac{5}{8}Z^*.$$

From

$$\frac{\partial E(Z^*)}{\partial Z^*} = 0, \quad Z^* = Z - 5/16.$$

Problem 2.7 In the light of the interpretation of the Coulomb integral in terms of repulsion between two spherically symmetric charge distributions, evaluate the effective potential energy for a given electron in the ground state of He atom and the effective charge $Z_{eff}(r)$.

Solution: The electric potential due to a spherical shell of radius R (thickness dR and density $-e\rho(R)$) at distance r from the center of the sphere is

$$-\frac{1}{4\pi e}d\phi(r) = R^2\rho(R)\frac{dR}{R} \quad \text{for } r \leq R,$$

$$R^2\rho(R)\frac{dR}{r} \quad \text{for } r \geq R.$$

By integrating over R and taking into account that

$$\varrho(r) \equiv |\psi_{1s}(r)|^2 = \left(\frac{Z}{a_0}\right)^3 \frac{e^{-\frac{2Zr}{a_0}}}{\pi},$$

one has

$$
\begin{aligned}
-\frac{\phi(r)}{4\pi e} &= \frac{1}{\pi}\left(\frac{Z}{a_0}\right)^3 \left[\frac{1}{r}\int_0^r dR\,R^2 e^{-\frac{2ZR}{a_0}} + \int_r^\infty dR\,R e^{-\frac{2ZR}{a_0}}\right] \\
&= \frac{1}{4\pi}\left(\frac{Z}{a_0}\right)\left[\frac{1}{u}\int_0^u dx\,x^2 e^{-x} + \int_u^\infty dx\,x e^{-x}\right] \\
&= \frac{1}{4\pi}\left(\frac{Z}{a_0}\right)\frac{1}{u}[2 - e^{-u}(u + 2)],
\end{aligned}
$$

where $u = \frac{2Zr}{a_0}$. Therefore

$$\phi(r) = -\frac{e}{r}\left[1 - e^{-\frac{2Zr}{a_0}}\left(\frac{Zr}{a_0} + 1\right)\right]$$

and from

$$-\frac{Z_{eff}(r)e^2}{r} = -\frac{Ze^2}{r} - e\phi(r)$$

for $Z = 2$ one finds

$$Z_{eff}(r) = 1 + e^{-\frac{4r}{a_0}}\left(1 + \frac{2r}{a_0}\right)$$

plotted below.

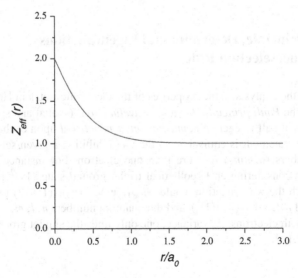

Problem 2.8 The *electron affinity* (energy gain when an electron is acquired) for Hydrogen atom is 0.76 eV. Try to derive this result in the framework of a perturbative approach for the ground state of H^-, as well as by considering a reduced nuclear charge.

Comment the results in the light of the almost-exact value which, at variance, is obtained only by means of a variational procedure with elaborate trial wavefunctions.

Solution: For H^-, by resorting to the results for He and setting $Z = 1$, in the perturbative approach one would obtain

$$E'_{H^-} = -2Z^2 R_H hc + \frac{5}{4} Z R_H hc = -\frac{3}{4} R_H hc$$

to be compared with $-R_H hc$ for H. With the variational effective charge $Z_{eff} = (1 - \frac{5}{16})$

$$E''_{H^-} = -2Z^2_{eff} R_H hc = -0.945 R_H hc$$

again less bound than the ground-state for neutral Hydrogen.

Only more elaborate calculations yield the correct value, the reason being that for small Z the perturbation is too large with respect to the unperturbed energy. By repeating the estimate for $Z = 3$ (Li^+), for $Z = 4$ (Be^{2+}) and for $Z = 5$ (B^{3+}) a convergence is noted towards the "exact" values of the ground state energy (namely 198.1, 371.7 and 606.8 eV, respectively) obtained from the variational procedure with elaborate trial functions. It should be remarked that the real experimental eigenvalues cannot be derived simply on the basis of the Hamiltonian in Eq. (2.4) which does not include the finite nuclear mass, the relativistic and the radiative terms (see for the Hydrogen atom the recall in Appendix 5.1).

2.3 Pauli Principle, Determinantal Eigenfunctions and Superselection Rule

In the light of the analysis of the properties of the electronic states in Helium atom, one can state the *Pauli principle*: *the total wavefunction* (spatial and spin) of electrons, particles at half integer spin, *must be antisymmetrical* upon exchange of two particles. This statement is equivalent to the one inhibiting a given set of the four quantum numbers $(nlmm_s)$ to more than one electron. For instance, this could be realized by considering an hypothetical triplet ground state $(1s)^2$ for orthohelium, for which the wavefunction would be $\phi_{1s}(\mathbf{r}_1)\phi_{1s}(\mathbf{r}_2)\alpha(1)\alpha(2)$ (or $\beta(1)\beta(2)$ or $(1/\sqrt{2})[\alpha(1)\beta(2) + \alpha(2)\beta(1)]$), and the quantum numbers n, l, m, m_s would be the same for both electrons. At variance, one only finds the singlet ground state, for which $m_s = \pm 1/2$.

From the specific case of Helium now we go back to the general properties of multi-electron atoms (see Sects. 1.1 and 1.3). Because of the exchange degeneracy and of the requirement of antisymmetrical wavefunction the total eigenfunction, instead of Eq. (1.10), must be written

$$\varphi_{tot} = \frac{1}{\sqrt{N!}} \sum_P P(-1)^P \varphi_\alpha(1)\varphi_\beta(2)...\varphi_\nu(N) \qquad (2.22)$$

where α, β, ... here indicate the group of quantum numbers ($nlmm_s$) and the numbers $1, 2, 3, \ldots, N$ include spatial and spin variables. P is an operator exchanging the electron i with the electron j and the wavefunction changes (does not change) sign according to an odd (even) number of permutations. The sum includes all possible permutations.

A total eigenfunction complying with exchange degeneracy and antisymmetry is the *determinantal wavefunction* devised by Slater [3]

$$\varphi_{tot} = \frac{1}{\sqrt{N!}} \begin{pmatrix} \varphi_\alpha(1) & \varphi_\alpha(2) & \cdots & \varphi_\alpha(N) \\ \varphi_\beta(1) & \varphi_\beta(2) & \cdots & \varphi_\beta(N) \\ \cdots & \cdots & \cdots & \cdots \\ \varphi_\nu(1) & \varphi_\nu(2) & \cdots & \varphi_\nu(N) \end{pmatrix}$$

accounting for all the possible index permutations with change of sign when two columns are exchanged. On the other hand the determinant goes to zero when two groups of quantum numbers (and then two rows) are the same.

Now it can be proved that no transition, by any mechanism, is possible between globally antisymmetric and symmetric states (in the assumption that they exist), sometimes known as *superselection rule*. In fact such a transition would be controlled by matrix elements of the form

$$\mathbf{R}_{ANT \leftrightarrow SYM} \propto \int \phi^*_{SYM}[\mathbf{O}_1 + \mathbf{O}_2 + ...]\phi_{ANT} d\tau_{gen} \qquad (2.23)$$

that must be zero in order to avoid the unacceptable result of havsing a change of sign upon exchange of indexes, since the integrand is globally antisymmetric.

In the light of what has been learned from the analysis of alkali atoms and of Helium atom, now we can move to a useful description of multi-electrons atoms which allows us to derive the structure of the eigenvalues and their classification in terms of proper quantum numbers (The *vectorial model*, Chap. 3). Other typical atoms, such as N, C and transition metals (Fe, Co, *etc...*) shall be discussed in that framework.

---

[3]This form is the basis for the multiplet theory in the perturbation approach dealing with operators r_i^{-1} and r_{ij}^{-1} (see Sect. 3.4).

Problems

Problem 2.9 By means of the perturbation approach for independent electrons derive the energy levels for the first excited states of Helium atom, in terms of Coulomb and exchange integrals, writing the eigenfunctions and plotting the energy diagram.

Solution: The first excited $1s2l$ states are

$$
\begin{aligned}
u_1 &= 1s(1)2s(2) \quad u_5 = 1s(1)2p_y(2) \\
u_2 &= 1s(2)2s(1) \quad u_6 = 1s(2)2p_y(1) \\
u_3 &= 1s(1)2p_x(2) \quad u_7 = 1s(1)2p_z(2) \\
u_4 &= 1s(2)2p_x(1) \quad u_8 = 1s(2)2p_z(1)
\end{aligned}
$$

From the unperturbed Hamiltonian without the electron-electron interaction, by setting $\hbar = 2m = e = 1$, one finds

$$
\mathcal{H}^° u_i(1,2) = -4\left(1 + \frac{1}{4}\right) u_i(1,2) = -5u_i(1,2)
$$

The secular equation involves the integrals

$$
I_s = \left\langle 1s(1)2s(2) \left| \frac{1}{r_{12}} \right| 1s(1)2s(2) \right\rangle
$$

$$
I_p = \left\langle 1s(1)2p(2) \left| \frac{1}{r_{12}} \right| 1s(1)2p(2) \right\rangle
$$

$$
K_s = \left\langle 1s(1)2s(2) \left| \frac{1}{r_{12}} \right| 1s(2)2s(1) \right\rangle
$$

$$
K_p = \left\langle 1s(1)2p(2) \left| \frac{1}{r_{12}} \right| 1s(2)2p(1) \right\rangle
$$

(p here represents p_x, p_y or p_z) and it reads

$$
\begin{vmatrix}
I_s - E' & K_s & 0 & 0 & 0 \\
K_s & I_s - E' & & & \\
0 & & I_p - E' & K_p & 0 & 0 \\
& & K_p & I_p - E' & & \\
0 & & 0 & & I_p - E' & K_p & 0 \\
& & & & K_p & I_p - E' & \\
0 & & 0 & & 0 & & I_p - E' & K_p \\
& & & & & & K_p & I_p - E'
\end{vmatrix} = 0
$$

From the first block $E' = I_s \pm K_s$, with the associated eigenfunctions

$$\phi_{1,2} = \frac{1}{\sqrt{2}}[1s(1)2s(2) \pm 1s(2)2s(1)].$$

From the second block $E'' = I_p \pm K_p$, with eigenfunctions

$$\phi_{3,4} = \frac{1}{\sqrt{2}}[1s(1)2p_x(2) \pm 1s(2)2p_x(1)]$$

and the analogous for y and z. Thus the following diagram (not in scale, see Sect. 2.2.2) is derived (I and $K > 0$ and $I_p > I_s$).

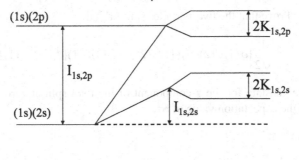

Problem 2.10 For the optical electron in Li atom consider the *hybrid orbital*

$$\Phi = (1 + \lambda^2)^{-\frac{1}{2}}[\phi_{2s} + \lambda\phi_{2p_z}]$$

ϕ_{2s} and ϕ_{2p_z} being normalized hydrogen-like wavefunctions, with effective nuclear charge Z. Find the pseudo-dipole moment $\mu = e\langle z \rangle$ and the value of λ yielding the maximum of μ (relevant connections for situations where hybrid orbitals are actually induced are to be found at Sects. 4.2 and 9.2).

Solution: The pseudo-dipole moment turns out

$$\mu = e \int \Phi^* z \Phi d\tau =$$

$$= \frac{e}{(1 + \lambda^2)} \left[\int \phi_{2s}^2(r)z d\tau + \lambda^2 \int \phi_{2p_z}^2(r)z d\tau + 2\lambda \int \phi_{2s}(r)\phi_{2p_z}(r)z d\tau \right],$$

where the first two integrals are 0. From Table 1.3 (with $e = a_0 = 1$)

$$\mu = \frac{2\lambda}{1+\lambda^2}\left(\frac{Z}{2}\right)^3\frac{1}{4\pi}\int_0^{2\pi}d\phi\int_0^\pi\cos^2\theta\sin\theta d\theta\int_0^\infty Zr^4(Zr-2)e^{-Zr}dr =$$

$$= \frac{\lambda}{1+\lambda^2}\frac{Z^3}{12}\left[\frac{Z^25!}{Z^6}-\frac{2Z4!}{Z^5}\right] = \frac{\lambda}{1+\lambda^2}\frac{6}{Z} ,$$

i.e. $\mu = (6ea_0/Z)\lambda/(1+\lambda^2)$ in complete form.

From

$$\frac{d\mu}{d\lambda} = \frac{6}{Z}\frac{(1+\lambda^2-2\lambda^2)}{(1+\lambda^2)^2} = \frac{6}{Z}\frac{(1-\lambda^2)}{(1+\lambda^2)^2} = 0$$

the maximum is found for $\lambda = 1$, as it could be expected.

Problem 2.11 Prove that the two-particles spin-orbital

$$\psi_{ANT} = \frac{1}{\sqrt{2}}\{\alpha(1)\beta(2)[\phi_a(1)\phi_b(2)] - \alpha(2)\beta(1)[\phi_a(2)\phi_b(1)]\}.$$

represents an eigenstate for the z-component of the total spin at zero eigenvalue. Then evaluate the expectation value of $\mathbf{S}^2$.

Solution: From

$$S_1^z\psi_{ANT} = \frac{1}{2}\frac{1}{\sqrt{2}}\{\alpha(1)\beta(2)[\phi_a(1)\phi_b(2)] + \alpha(2)\beta(1)[\phi_a(2)\phi_b(1)]\} = \frac{1}{2}\psi_{SYM}.$$

and

$$S_2^z\psi_{ANT} = -\frac{1}{2}\psi_{SYM}$$

Thus

$$S^z\psi_{ANT} = (S_1^z + S_2^z)\psi_{ANT} = 0$$

Since

$$S_1^z S_2^z\psi_{ANT} = -\frac{1}{4}\psi_{ANT}$$

while

$$S_1^x S_2^x\psi_{ANT} = \frac{1}{4}\frac{1}{\sqrt{2}}\{\beta(1)\alpha(2)[\phi_a(1)\phi_b(2)] - \alpha(1)\beta(2)[\phi_a(1)\phi_b(2)]\} \equiv \frac{1}{4}\psi'_{ANT}$$

and

$$< \psi_{ANT}|\psi'_{ANT} > \equiv -|\int \phi_a^*(\mathbf{r})\phi_b(\mathbf{r})d\tau|^2 \equiv -\mathcal{A}$$

(with the same result for the y component). By taking into account that $(\mathbf{S})^2 = (\mathbf{S}_1)^2 + (\mathbf{S}_2)^2 + 2\mathbf{S}_1.\mathbf{S}_2$, then

$$< \psi_{ANT}|(S_1^{x,y} + S_2^{x,y})^2|\psi_{ANT} >= \frac{1}{2}\{1 - \mathcal{A}\}$$

and

$$< \psi_{ANT}|(\mathbf{S})^2|\psi_{ANT} >= 1 - \mathcal{A}.$$

Problem 2.12 At Chap. 5 it will be shown that between one electron and one proton an hyperfine interaction of the form $A\mathbf{I} \cdot \mathbf{S}\delta(\mathbf{r})$ occurs, where $\mathbf{I}$ is the nuclear spin (*Fermi contact interaction*). An analogous term, i.e. $\mathcal{H}_p = A\mathbf{s}_1 \cdot \mathbf{s}_2\delta(\mathbf{r}_{12})$ (with $\mathbf{r}_{12} \equiv \mathbf{r}_1 - \mathbf{r}_2$) describes a relativistic interaction between the two electrons in the Helium atom. In this case A turns out $A = -(8\pi/3)(e\hbar/mc)^2$. Discuss the first-order perturbation effect of $\mathcal{H}_p$ on the lowest energy states of orthohelium and parahelium, showing that only a small shift of the ground-state level of the latter occurs (return to Problem 1.38 for similarities).

Solution: For orthohelium the lowest energy states is described by the spin-orbital

$$\phi_{tot}(1, 2) = \chi_{sym}^{S=1}\left[\phi_{1s}(\mathbf{r}_1)\phi_{2s}(\mathbf{r}_2) - \phi_{2s}(\mathbf{r}_1)\phi_{1s}(\mathbf{r}_2)\right].$$

The expectation value of $\mathcal{H}_p$ yields zero since two electrons at parallel spin cannot have the same spatial coordinates. For the ground state of parahelium since $\mathbf{s}_1 \cdot \mathbf{s}_2 = -3/4$ (see Eq. (2.20)), by using hydrogenic wave functions $\phi_{1s}(\mathbf{r}_1)$ and $\phi_{1s}(\mathbf{r}_2)$ one estimates

$$< 1s, 1s|\mathcal{H}_p|1s, 1s >= -\frac{3A}{4}\frac{Z^6}{\pi^2 a_0^6}\int\int e^{-2\frac{Z(r_1+r_2)}{a_0}}\delta(\mathbf{r}_{12})d\tau_1 d\tau_2 =$$

$$= -\frac{3A}{4}\frac{Z^6}{\pi^2 a_0^6}4\pi\int_0^\infty e^{-4Z\frac{r}{a_0}}r^2 dr = \frac{3}{32}\left(\frac{e\hbar}{mc}\right)^2\frac{Z^3}{a_0^3} \simeq 10^{-3}\text{eV},$$

a small shift compared to -78.62 eV.

Problem 2.13 The spin-orbit constant ξ_{2p} for the $2p$ electron in Lithium turns out $\xi_{2p} = 0.34$ cm^{-1}. Evaluate the magnetic field causing the first crossing between $P_{3/2}$ and $P_{1/2}$ levels, in the assumption that the field does not affect the structure of the doublet (return to Problems 1.23 and 1.17).

Solution: In the assumption that the field linearly affects the two levels, i.e.

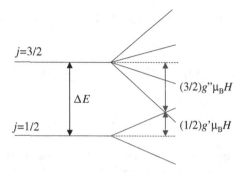

the first crossing takes place when

$$\left(\frac{3}{2}g'' + \frac{1}{2}g'\right)\mu_B H = \Delta E$$

Since $g' = 2/3$ and $g'' = 4/3$ the crossing occurs for $H = \Delta E/(7\mu_B/3)$.
The correction associated with the spin-orbit interaction is

$$\frac{\xi_{2p}}{2}[j(j+1) - l(l+1) - s(s+1)]$$

Then $\Delta E = (3/2)\xi_{2p}$ and $H \simeq 4370$ Oe.

When the *weak field* condition (corresponding to $\mu_B H \ll \xi_{2p}$) is released and the full Hamiltonian $\xi_{nl}\mathbf{l}\cdot\mathbf{s} + \mu_B\mathbf{H}\cdot(\mathbf{l}+2\mathbf{s})$ is diagonalized (as it would be more appropriate), the crossing is found at a slightly different field.

Try to estimate it after having read Chap. 4 (or see Problem 1.1.20,in the book by *Balzarotti, Cini* and *Fanfoni* or Problem 7.24 in the book by *Johnson* and *Pedersen*). A somewhat similar situation is the one discussed at Problem 5.13 with $\mathbf{l}$ substituted by the nuclear momentum $\mathbf{I}$

Problem 2.14 Refer to the double-excited electron state $2s4p$ of the Helium atom. In the assumption that the $2s$ electron in practice is not screened by the $4p$ electron, which in turn feels just the residual charge $Z(r) \simeq 1$ (see Sect. 2.1), evaluate the wavelength of the radiation required to promote the transition from the ground state to that double-excited state. After the *autoionization* of the atom, and decay to the ground-state of He$^+$, one electron is ejected. Estimate the velocity of this electron.

Solution:
   $E(2s, 4p) = -14.5$ eV, then $\lambda = c/\nu = 192$ Å and $v = 3.75 \times 10^8$ cm/s.

Specific References and Further Reading

1. A. Balzarotti, M. Cini, M. Fanfoni, *Atomi, Molecole e Solidi. Esercizi risolti*, (Springer Verlag, 2004).
2. C.S. Johnson and L.G. Pedersen, *Quantum Chemistry and Physics*, (Addison-Wesley, 1977).
3. B.H. Bransden and C.J. Joachain, *Physics of Atoms and Molecules*, (Prentice Hall, 2002).
4. B. Cagnac and J.C. Pebay - Peyroula, *Physique atomique, tome* 2, (Dunod Université, Paris, 1971).
5. J.A. Cronin, D.F. Greenberg, V.L. Telegdi, *University of Chicago Graduate Problems in Physics*, (Addison-Wesley, 1967).
6. W. Demtröder, *Atoms, Molecules and Photons*, (Springer Verlag, 2006).
7. R.N. Dixon, *Spectroscopy and Structure*, (Methuen and Co LTD, London, 1965).
8. R. Eisberg and R. Resnick, *Quantum Physics of Atoms, Molecules, Solids, Nuclei and Particles*, (J. Wiley and Sons, 1985).
9. H. Eyring, J. Walter and G.E. Kimball, *Quantum Chemistry*, (J. Wiley, New York, 1950).
10. A.P. French and E.F. Taylor, *An Introduction to Quantum Physics*, (The M.I.T. Introductory Physics Series, Van Nostrand Reinhold (UK), 1986).
11. H. Haken and H.C. Wolf, *Atomic and Quantum Physics*, (Springer Verlag, Berlin, 1987).
12. M.A. Morrison, T.L. Estle and N.F. Lane, *Quantum States of Atoms, Molecules and Solids*, (Prentice-Hall Inc., New Jersey, 1976).
13. J.C. Slater, *Quantum Theory of Matter*, (McGraw-Hill, New York, 1968).
14. S. Svanberg, *Atomic and Molecular Spectroscopy*, (Springer Verlag, Berlin, 2003).

Chapter 3
The Shell Vectorial Model

Topics

Electronic Structure: "aufbau" and Closed Shells
Coupling of Angular Momenta (LS and jj Schemes)
Rules for the Ground State
Low Energy States of C and N Atoms
Effective Magnetic Moments and Gyromagnetic Ratio
Approximate form of the Radial Wavefunctions
Hartree-Fock-Slater Theory for Multiplets
Selection Rules

3.1 Introductory Aspects

By resorting to the principles of quantum mechanics and after having dealt with specific atoms, one can now proceed to the description of the electronic structure in generic multi-electron atoms. We shall see that the sequence of electron states accounts for the microscopic origin of the periodic Table of the elements.

First the one-electron states, described by orbitals of the form $\phi_{nlm} = R_{nl}Y_{lm}$, have to be placed in the proper energy scale (*diagram*). Then the atom can be thought to result from the progressive accommodation of the electrons on the various levels, with the related eigenfunctions. This build-up principle (called *aufbau* from the German) has to be carried out by taking into account the *Pauli principle* (Sect. 2.3). Therefore a limited number of electrons can be accommodated on a given level and each electron has associated one (and only one) spin-orbital eigenfunction, differing in one or more of the quantum numbers n, l, m, m_s from the others.

The maximum number of electrons characterized by a given value of n is

$$\sum_{l=0}^{n-1} 2(2l+1) = 2n + 4\sum_{l=0}^{n-1} l = 2n + 4\frac{n(n-1)}{2} \tag{3.1}$$

© Springer International Publishing Switzerland 2015
A. Rigamonti and P. Carretta, *Structure of Matter*,
UNITEXT for Physics, DOI 10.1007/978-3-319-17897-4_3

When this maximum number is attained one has a *closed shell*. A closed sub-shell, often called *nl shell*, occurs when a given *nl* state (which defines the energy in the absence of spin-orbit and exchange interactions) accommodates $2(2l + 1)$ electrons, in correspondence to the degeneracy in the z-component of the orbital momentum and of the spin degeneracy.

A complete sub-shell (or shell) implies electron charge distribution at *spherical symmetry*[1] and the quantum numbers L (for the total orbital momentum) and S (total spin) are zero, obviously implying $J = 0$ and spectral notation (see Sect. 1.7) $^1S_0$.

For the electrons outside the closed *nl* shells one has to take into account the spin-orbit interaction yielding $\mathbf{j} = (\mathbf{l} + \mathbf{s})$ and the electron-electron interaction leading to the Coulomb and exchange integrals, as it has been discussed for alkali atoms (Sect. 2.1) and for Helium atom (Sect. 2.2). As a consequence, a variety of "couplings" is possible and a complex distribution of the energy levels occurs, the detailed structure depending on the relative strengths of the couplings. For instance, the sequence of levels seen for Helium (Sect. 2.2), with spin-orbit terms much weaker than the Coulomb and exchange integrals, can be considerably modified on increasing the atomic number, when the spin-orbit interaction is stronger than the inter-electron effects.

In order to take into account the various couplings and to derive the qualitative sequence of the eigenvalues (with the proper classification in terms of good quantum numbers corresponding to constants of motion) one can abide by the so-called *vectorial model*. Initiated by *Heisenberg* and by *Dirac*, this model leads to the structure of the energy levels and to their classification in agreement with more elaborate theories for the multiplets, although it does not provide the quantitative estimate of the energy separation of the levels.

In the vectorial model the angular momenta and the associated magnetic moments are thought as classical vectors, as seen in the *ad hoc* definition of $\mathbf{J}$ and of $\mathbf{L}$ and $\mathbf{S}$ at Sects. 1.6, 1.7 and 2.2.2. Furthermore, somewhat classical equations of motion are used (for instance the precessional motion is often recalled). Moreover constraints are taken into account in the couplings, so that the final results do have characteristics in agreement with the quantum conditions. For example, the angular momenta of two p electrons are coupled and pictorially sketched as shown in Fig. 3.1.

The interactions are written in the form

$$a) \quad a_{ik}\mathbf{l}_i \cdot \mathbf{s}_k \qquad b) \quad b_{ik}\mathbf{l}_i \cdot \mathbf{l}_k \qquad c) \quad c_{ik}\mathbf{s}_i \cdot \mathbf{s}_k \qquad (3.2)$$

where $a)$ can be considered a generalization of the spin-orbit interaction ($a_{ii} > 0$, as proved at Sect. 1.6); $b)$ is the analogous for the orbital couplings, while $c)$ is the extension of the exchange interaction discussed in Helium ($c_{jk} = -2K < 0$). In these coupling forms the constants a, b and c have usually the dimensions of energy, the angular momenta thus being in $\hbar$ units.

---

[1] The rule $\sum_{m=-l}^{+l} Y_{l,m}^*(\theta, \varphi)Y_{l,m}(\theta, \varphi) = (2l+1)/4\pi$ is known as *Unsold theorem* (See Problem 1.9 for a particular case).

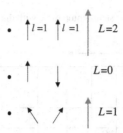

Fig. 3.1 Illustrative coupling of the angular momenta for two p electrons to yield the $L = 0$, $L = 1$ and $L = 2$ states. It is noted that the effective "lengths" of the "vectors" must be considered $\sqrt{l(l+1)}$ and $\sqrt{L(L+1)}$

On the basis of Eq. (3.2) the energy levels are derived by coupling the electrons outside the closed shells and the states are classified in terms of good quantum numbers. The values of a, b and c are left to be estimated on the basis of the experimental findings, for instance from the levels resulting from optical spectra.

In spite of these simplifying assumptions the many-body character of the problem prevents suitable solutions when a, b and c are of the same order of magnitude. Two limiting cases have to be considered:

(i) "small" atoms (nuclear charge Z not too large) so that the spin-orbit interaction is smaller than other coupling terms and the condition $a \ll c$ can be assumed. This assumption leads to the so-called *LS scheme*;

(ii) "heavy" atoms at large Z, where the strong spin-orbit interaction implies $a \gg c$ (*jj scheme*).

3.2 Coupling of Angular Momenta

3.2.1 LS Coupling Model

Within this scheme one couples s_i to obtain $\mathbf{S}$ and l_i for $\mathbf{L}$ (in a way to account for the quantum prescriptions). For

$$\mathbf{S} = \sum_i \mathbf{s}_i \quad \text{and} \quad \mathbf{L} = \sum_i \mathbf{l}_i \qquad (3.3)$$

the total spin number $S = 0, 1/2, 1, 3/2, \ldots$ and the total orbital momentum number $L = 0, 1, 2, \ldots$ are defined. Then the spin orbit interaction is taken into account with an Hamiltonian of the form

$$\mathcal{H}_{SO} = \xi_{LS} \mathbf{L} \cdot \mathbf{S}, \qquad (3.4)$$

an extension of the Hamiltonian derived at Sect. 1.6 (see Sect. 3.2.2 in order to understand that the precessions of l_i yield an average orbital momentum along $\mathbf{L}$, while the average spin momentum is along $\mathbf{S}$: then Eq. (3.4) follows).

Table 3.1 Derivation of the electronic states compatible with the Pauli principle for two equivalent p electrons

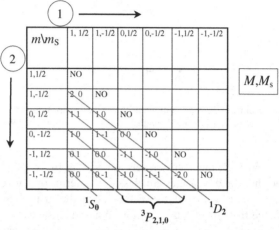

It is noted that the group ^3D cannot exists since states with $M = 2$ and $M_s = 1$ are not found. The values $M_s = 0$ and with M running from -2 to $+2$ are present and they correspond to ^1D states at $S = 0$, implying $J = 2$. The states at $M = 1$ and $M_s = 1$ are all found and then the multiplet $S = 1$ and $L = 1$ does exist, implying the values $J = 2, 1, 0$. Finally the last case corresponds to the singlet state at $S = 0$ and $L = 0$. The total number of original states is 36 (corresponding to $2 \times 2 \times (2l_1 + 1) \times (2l_2 + 1)$) and only 15 of them are allowed. Six states are eliminated because they violate Pauli principle. Of the remaining 30 states, only half are distinguishable

When the electrons to be coupled are *equivalent*, namely with the same quantum numbers n and l, one has to reject the coupling configurations that would invalidate the Pauli principle. In other words, one has to take into account the antisymmetry requirement for the total wavefunction and this corresponds to the problem of the *Clebsch-Gordan coefficients*. A simple method to rule out unacceptable states is shown in Table 3.1 for two np electrons. All the possible values for m and m_s are summed up to give M and M_s. Then the states along the diagonal are disregarded, since they correspond to four equal quantum numbers. The states above the diagonal are also to be left out, since they correspond to the exchange of equivalent electrons, the exchange degeneracy being taken into account by the spin-spin interaction. Finally the electronic states compatible with the values of M and M_s are found by inspection. This method corresponds to a brute-force counting of the states, as it is shown in the Problems for the low-energy electronic states in C and in N atoms (Problems 3.1 and 3.2).

When the electrons are *inequivalent* (differing in n or in l) no restrictions to the possible sums has to be considered (see Problem 3.3).

Once that L and S are found and the structure of the levels expected from the couplings 3.2 (*b* and *c*) is derived, then in the **LS** scheme one defines

$$\mathbf{J} = \mathbf{L} + \mathbf{S}, \tag{3.5}$$

characterized by the quantum number J. The spin-orbit interaction is taken into account according to Eq. (3.4) in order to derive the multiplets. Pictorially

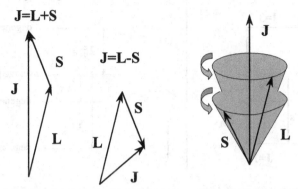

with coupling energy $E_{SO} = \xi_{LS}|\mathbf{L}|.|\mathbf{S}|cos\theta$ (θ angle between $\mathbf{L}$ and $\mathbf{S}$).

It is reminded that according to the classical equation of motion, a magnetic moment $\boldsymbol{\mu}_L \propto -\mathbf{L}$ in magnetic field precesses with angular frequency $\omega_L = \gamma H$, with γ the gyromagnetic ratio given by $\gamma = \mu_L/L$ (Problem 3.4). In terms of $\mathbf{L}$ and $\mathbf{S}$ and of the related torque of modulus $-\partial E_{SO}/\partial\theta$, a precession of each of them around the direction of $\mathbf{J}$ has to be expected. To show this one writes

$$\frac{d\mathbf{L}}{dt} = \xi_{LS}\mathbf{S} \times \mathbf{L} \tag{3.6}$$

$$\frac{d\mathbf{S}}{dt} = \xi_{LS}\mathbf{L} \times \mathbf{S}. \tag{3.7}$$

and since $\mathbf{S} \times \mathbf{S} = \mathbf{L} \times \mathbf{L} = 0$

$$\frac{d\mathbf{L}}{dt} = \xi_{LS}\mathbf{J} \times \mathbf{L}$$

$$\frac{d\mathbf{S}}{dt} = \xi_{LS}\mathbf{J} \times \mathbf{S},$$

implying the precessional motions of $\mathbf{L}$ and $\mathbf{S}$ around an effective magnetic field along the direction of $\mathbf{J}$, the angular frequency being proportional to ξ_{LS} (see Problem 3.4).

Therefore J and M_J are good quantum numbers while L_z and S_z are no longer constant of motion (z is here an arbitrary direction). Then the energies of the multiplet are derived by adding the corrections due to the spin-orbit Hamiltonian (in the form 3.4) to the energy $E^0(L, S)$ resulting from the couplings between $\mathbf{s}_i$ and between $\mathbf{l}_i$ (see examples in subsequent figures). From the definition of J (Eq. 3.5), again by the usual "*squaring rule*", one obtains

$$E(L, S, J) = E^0(L, S) + \frac{1}{2}\xi_{LS}\left[J(J+1) - L(L+1) - S(S+1)\right] \tag{3.8}$$

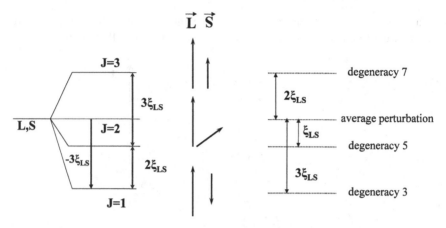

Fig. 3.2 Illustration of the interval rule for the multiplet arising from the $L = 2$ and $S = 1$ state

An empirical rule for ξ_{LS} is $\xi_{LS} \simeq \pm\xi_{nl}/2S$, with the sign $+$ when the number of the electrons in the sub-shell in less then half of the maximum number that can be accommodated and $-$ in the opposite case (according to Sect. 1.6 $\xi_{nl} = a_{11}$ in Eq. (3.1)). For sign $+$ the multiplet is called *regular*, namely the state at lowest energy is the one corresponding to J *minimum* (pictorially with **L** and **S** antiparallel). For sign $-$ the multiplet is *inverted*, the state at lowest energy being the one with maximum value for J (i.e. **L** and **S** parallel).

For regular multiplets one immediately derives the *interval rule*, giving the energy separation between the states at J and $(J + 1)$. From Eq. (3.8)

$$\Delta_{J,J+1} = (J + 1)\xi_{LS} \tag{3.9}$$

implying, for example for $L = 2$ and $S = 1$, the structure of the levels shown in Fig. 3.2. This rule can be used as a test to check the validity of the **LS** coupling scheme. It is noted that the "center of gravity" of the levels, namely the mean perturbation of all the states of a given term, is not affected by the spin-orbit interaction. In fact

$$< \Delta(E - E^0) > = \sum_{J=|L-S|}^{L+S} \frac{\xi_{LS}}{2}(2J + 1)\left[J(J + 1) - L(L + 1) - S(S + 1) \right] = 0 \tag{3.10}$$

(see Figs. 3.2, 3.4 and Problem 3.12).[2]

When more than two electrons are involved in the coupling, the procedure outlined above has to be applied by combining the third electron with the results of coupling the first two and so on. Examples (Problem 3.2) will clarify how to deal with more than two electrons.

---

[2]In fact $\sum_{J=0}^{N} J = N(N+1)/2$, $\sum_{J=0}^{N} J^2 = N(N+1)(2N+1)/6$ and $\sum_{J=0}^{N} J^3 = N^2(N+1)^2/4$.

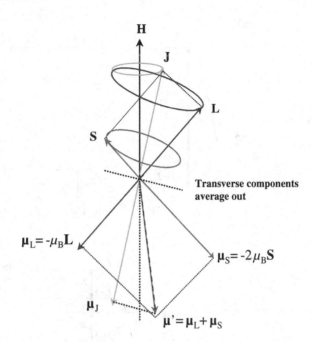

Fig. 3.3 Vectorial description of angular and magnetic moments in magnetic field, within the **LS** model. The interaction with the field is weak in comparison to the spin-orbit interaction and fast precessions of **L** and **S** around **J** occur, controlled by ξ_{LS}. Only the "result" of the precessional motion can effectively interact with the field: the precession of **J** at the Larmor frequency $\omega_L = \gamma H$ is induced. ω_L is much smaller than the precessional frequency of L and S around J (see Problem 3.4)

Transverse components average out

$\boldsymbol{\mu}_L = -\mu_B \mathbf{L}$

$\boldsymbol{\mu}_S = -2\mu_B \mathbf{S}$

$\boldsymbol{\mu}_J$

$\boldsymbol{\mu}' = \boldsymbol{\mu}_L + \boldsymbol{\mu}_S$

3.2.2 The Effective Magnetic Moment

At Sect. 1.6 the effect of an external magnetic field on one single electron has been considered. The quantum description for multi-electrons atom shall be given at Chap. 4. Here we derive the atomic magnetic moment that effectively interacts with the external field in the framework of the vectorial model and of the **LS** scheme.

The magnetic field, acting on $\boldsymbol{\mu}_L$ and $\boldsymbol{\mu}_S$, induces torques on **L** and on **S** while they are coupled by the spin-orbit interaction. A general solution for the motions of the momenta and for the energy corrections in the presence of the field can hardly be obtained. Rigorous results are derived in the limiting cases of *strong* and of *weak* magnetic field, namely for situations such that $\boldsymbol{\mu}_{L,S}.\mathbf{H} \gg \xi_{LS}$ and $\boldsymbol{\mu}_{L,S}.\mathbf{H} \ll \xi_{LS}$, respectively. Let us first discuss the case of weak magnetic field (Fig. 3.3).

In view of the meaning of **L.J** and of **S.J**, the angles between **L** and **J** and **S** and **J** can be written

$$cos\widehat{LJ} = \frac{L(L+1) + J(J+1) - S(S+1)}{2\sqrt{L(L+1)}\sqrt{J(J+1)}} \qquad (3.11)$$

$$cos\widehat{SJ} = \frac{S(S+1) + J(J+1) - L(L+1)}{2\sqrt{S(S+1)}\sqrt{J(J+1)}} \qquad (3.12)$$

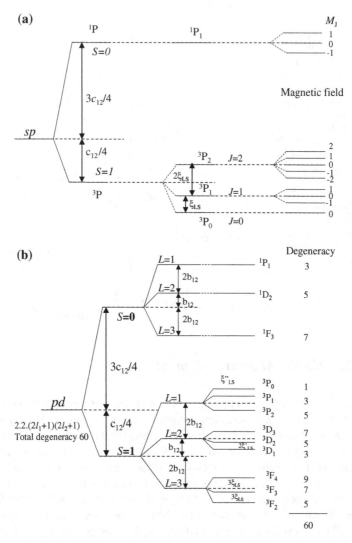

Fig. 3.4 Diagram of the energy levels and labeling of the electronic states within the **LS** scheme for: **a** one s and one p electron; **b** one p and one d electron (outside closed shells). For case **a** it is shown how a magnetic field removes all the degeneracies, while in case **b** the number of degenerate states are indicated on the right (ξ''_{LS} is negative)

Then the magnetic moment along the $\mathbf{J}$ direction, after averaging out the transverse components of $\mathbf{L}$ and $\mathbf{S}$ (due to fast precession induced by spin orbit interaction) is[3]

$$\boldsymbol{\mu}_J = -2\mu_B \mathbf{S} cos\widehat{SJ} - \mu_B \mathbf{L} cos\widehat{LJ}.$$

Therefore the effective magnetic moment turns out

$$\boldsymbol{\mu}_J = -\mu_B g \mathbf{J} \tag{3.13}$$

where g, called the *Landé factor*, is

$$g = 1 + \frac{J(J+1) + S(S+1) - L(L+1)}{2J(J+1)}. \tag{3.14}$$

Hence the energy corrections associated with the magnetic Hamiltonian are $\Delta E = -\boldsymbol{\mu}_J.\mathbf{H} = -\mu_J^z H = g\mu_B H M_J$. Thus the magnetic field removes the degeneracy in M_J and the energy levels, in weak magnetic field, turn out

$$E(L, S, J, M_J) = E^0(L, S, J) + \mu_B g H M_J. \tag{3.15}$$

In the opposite limit when the magnetic field is strong enough that the Hamiltonians $\boldsymbol{\mu}_L.\mathbf{H}$ and $\boldsymbol{\mu}_S.\mathbf{H}$ prevail over the spin-orbit interaction, one can first disregard this latter and the energy levels are derived in terms of the quantum magnetic numbers M and M_S. Vectorially this corresponds to the decoupling of the orbital and spin momenta and to their independent quantization along the axis of the magnetic field, around which they precess at high angular frequency. The magnetic moment is the sum of the independent components and therefore the energy correction is written

$$\Delta E = -[\mu_L^z H + \mu_S^z H] = \mu_B M H + 2\mu_B M_s H. \tag{3.16}$$

The spin-orbit interaction can be taken into account subsequently, as perturbation of the states labelled by the quantum magnetic numbers M and M_S. This will be described at Chap. 4 as the so-called *Paschen-Back regime*.

3.2.3 Illustrative Examples and the Hund Rules for the Ground State

In the framework of the **LS** scheme, by taking into account the signs of the coupling constants for the spin-orbit interaction (Sect. 1.6) and for the spin-spin interaction (Sect. 2.2.2), one can figure out simple rules to predict the configuration pertaining

---

[3]The formal proof is based on the *Wigner-Eckart theorem* (see Sect. 4.3).

to the ground state of the atom. This is an important step for the description of the magnetic properties of matter. The rules, first empirically devised by *Hund*, are the following:

(i) *maximize* the quantum number **S**. The reason for this is related to the sign of the exchange integral, since in the spin-spin coupling c_{12} plays the role of $-2K$, as already observed;

(ii) *maximize L*, in a way compatible with Pauli principle;

(iii) *minimize J for regular multiplets* while *maximize* it for *inverted multiplets*. This rule follows from the sign of ξ_{nl} and then of ξ_{LS} (see Eq. (3.8)).

As illustrative examples let us consider one atom of the transition elements, with incomplete $3d$ shell (Fe) and one of the rare earth group, with incomplete $4f$ shell (Sm). The electronic configuration of iron is $(1s)^2(2s)^2(2p)^6(3s)^2(3p)^6(3d)^6(4s)^2$. Maximization of S implies the spin vectorial coupling in the form $\uparrow\uparrow\uparrow\uparrow\uparrow\downarrow$ yielding $S = 2$. The coupling of five of the six orbital momenta must be zero, since the m numbers must be all different (from -2 to $+2$) in order to preserve Pauli principle. Then for the sixth electron we take the maximum, namely $L = 2$. The multiplet is inverted, because the maximum number of electrons that can be accommodated in the $3d$ sub-shell is 10. Then $J = 4$. Thus the ground state for iron is $^5D_4$. According to Eqs. (3.13) and (3.14) the magnetic moment would be $|\mu_J| = 4.9\mu_B$ while experimentally it turns out $|\mu_J| = 5.4\mu_B$ (for this discrepancy see Caption to Table 3.2).

Samarium has the electronic configuration ending with $(4f)^6(6s)^2$. Maximizing S yields $S = 3$. To complete half of the shell (that would give $L = 0$) one electron is

Table 3.2 Ground state of some magnetic ions of the $4f$ sub-shell, according to Hund's rules, and correspondent values of the effective magnetic moments

Ion	Shell	S	L	J	Atomic Configuration	$|\mu|$ (in Bohr magneton)
Ce^{3+}	$4f^1$	1/2	3	5/2	$^2F_{5/2}$	2.54
Pr^{3+}	$4f^2$	1	5	4	$^3H_4$	3.58
Nd^{3+}	$4f^3$	3/2	6	9/2	$^4I_{9/2}$	3.62
Pm^{3+}	$4f^4$	2	6	4	$^5I_4$	2.68
Sm^{3+}	$4f^5$	5/2	5	5/2	$^6H_{5/2}$	0.85
Eu^{3+}	$4f^6$	3	3	0	$^7F_0$	0
Gd^{3+}	$4f^7$	7/2	0	7/2	$^8S_{7/2}$	7.94
Tb^{3+}	$4f^8$	3	3	6	$^7F_6$	9.72
Dy^{3+}	$4f^9$	5/2	5	15/2	$^6H_{15/2}$	10.65
Ho^{3+}	$4f^{10}$	2	6	8	$^5I_8$	10.61
Er^{3+}	$4f^{11}$	3/2	6	15/2	$^4I_{15/2}$	9.58
Tm^{3+}	$4f^{12}$	1	5	6	$^3H_6$	7.56
Yb^{3+}	$4f^{13}$	1/2	3	7/2	$^2F_{7/2}$	4.54
Lu^{3+}	$4f^{14}$	0	0	0	$^1S_0$	0

It should be remarked that these data refer to free ions, while the magnetic properties can change when the crystalline electric field is acting (see Problems 4.11, 4.12 and 13.5)

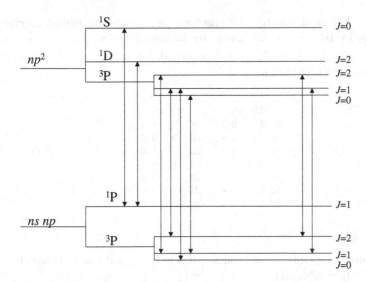

Fig. 3.5 Multiplet structure in the LS scheme for the $nsnp$ and the np^2 configurations and transitions allowed by the electric dipole mechanism (see Sect. 3.5)

missing. Then by taking the maximum possible value one has $L = 3$. The multiplet is regular and therefore the ground state is the one with $J = 0$, namely $^7F_0$. Other ground states are derived in Problem 3.5.

In Table 3.2 the ground state of some $4f$ magnetic ions often involved in paramagnetic crystals, with their effective magnetic moment $|\mu| = g\sqrt{J(J+1)}$ are reported.[4]

As illustrative examples of the structure and classification of the energy levels in the **LS** scheme according to the prescriptions described above, in Fig. 3.4 the cases of atoms with one s and one p electron and with one p and one d electron outside the closed shells are shown.

In Fig. 3.5 the energy levels of the p^2 configuration are reported and the transitions to the sp configuration (see Fig. 3.4a), driven by electric dipole mechanism, are indicated.

Problems

Problem 3.1 Derive and label the low-energy states of the *carbon* atom (ground state configuration $(1s)^2 (2s)^2 (2p)^2$) by taking into account the inter-electronic interactions, first disregarding the spin-orbit coupling.

---

[4]For Sm^{3+} and Eu^{3+} the agreement with the experimental estimates (1.5 μ_B and 3.4μ_B, respectively) is poor. If one takes into account higher order energy levels good agreement is found (see Eq. (4.38)).

Solution: The method to rule out unacceptable states for *equivalent* $2p$ electrons is shown in Table 3.1. Equivalently, by indicating $|m = 1, m_s = \frac{1}{2} >\equiv a$, $|0, -\frac{1}{2} >\equiv d, |1, -\frac{1}{2} >\equiv b, |-1, \frac{1}{2} >\equiv e, |0, \frac{1}{2} >\equiv c, |-1, -\frac{1}{2} >\equiv f$ one has the possibilities listed below:

	M_L	M_S			M_L	M_S	
ab	2	0	♯	bf	0	−1	•
ac	1	1	•	cd	0	0	♯
ad	1	0	♯	ce	−1	1	•
ae	0	1	•	cf	−1	0	♯
af	0	0	◇	de	−1	0	•
bc	1	0	•	df	−1	−1	•
bd	1	−1	•	ef	−2	0	♯
be	0	0	•	−			

♯ terms correspond to $L = 2$ and $S = 0$, • to $L = 1$ and $S = 1$, while ◇ to $L = 0$ and $S = 0$ (see Table 3.1).

The first low-energy states are $^1S_0$, $^1D_2$ and $^3P_{0,1,2}$, according to the vectorial picture and to the Hund rules:

$$\uparrow\downarrow \quad \uparrow\downarrow \;\; ^1S_0 \quad L = 0 \; S = 0$$
$$\uparrow\uparrow \quad \uparrow\downarrow \;\; ^1D_2 \quad L = 2 \; S = 0$$
$$\nwarrow\nearrow \uparrow\uparrow \;\; ^3P_{0,1,2} \; L = 1 \; S = 1$$

The correspondent energy diagram, including the experimentally detected splitting of the lowest energy ^3P state due to spin-orbit interaction, is

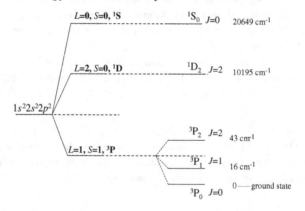

An extended energy diagram of the atom (with spin-orbit splitting not detailed) is

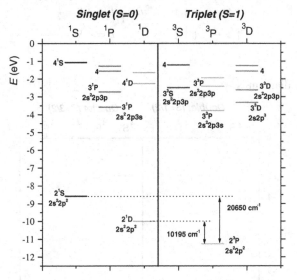

Problem 3.2 Derive and label the low-energy states for the N atom (electronic configuration $(1s)^2 (2s)^2 (2p)^3$) by taking into account the inter-electron couplings. By assuming a spin-spin interaction of the form $\sum'_{i,j} A\mathbf{s}_i \cdot \mathbf{s}_j$ evaluate the shift of the ground state.

Solution: According to the notation used in Problem 3.1, the possible one-electron states are $a\ b\ c\ d\ e\ f$.

The complete states, in agreement with the Pauli principle, are

	M_L	M_S		M_L	M_S
♯ abc	2	1/2	bcd	1	−1/2
♯ abd	2	−1/2	bce	0	1/2
♯ abe	1	1/2	bcf	0	−1/2
♯ abf	1	−1/2	bde	0	−1/2
acd	1	1/2	♯ cde	−1	1/2
ace	0	3/2	♯ cdf	−1	−1/2
acf	0	1/2	♯ def	−2	−1/2
♯ ade	0	1/2	bef	−1	−1/2
♯ adf	0	−1/2	♯ cef	−2	1/2
aef	−1	1/2	bdf	0	−3/2

(♯ terms corresponding to $L = 2$, $S = 1/2$, i.e. $^2D_{5/2,3/2}$, etc.).

Thus the three low-energy states are

$$^2P_{\frac{3}{2},\frac{1}{2}} \qquad ^2D_{\frac{5}{2},\frac{3}{2}} \qquad ^4S_{\frac{3}{2}}$$

correspondent to the vectorial picture

$$\nwarrow \nearrow \rightarrow \uparrow\downarrow\uparrow \qquad {}^2P_{\frac{3}{2},\frac{1}{2}} \quad L = 1 \; S = \tfrac{1}{2}$$

$$\uparrow\uparrow \rightarrow \;\; \uparrow\downarrow\uparrow \qquad {}^2D_{\frac{5}{2},\frac{3}{2}} \quad L = 2 \; S = \tfrac{1}{2}$$

$$\uparrow\downarrow \leftarrow \;\; \uparrow\uparrow\uparrow \qquad {}^4S_{\frac{3}{2}} \qquad L = 0 \; S = \tfrac{3}{2}$$

The energy diagram is

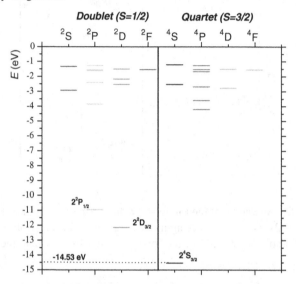

The shift of the ground state due to the spin-spin interaction is $3A/4$. In fact

$$\mathbf{S}^2 = \mathbf{s}_1^2 + \mathbf{s}_2^2 + \mathbf{s}_3^2 + 2[\mathbf{s}_1 \cdot \mathbf{s}_2 + \mathbf{s}_2 \cdot \mathbf{s}_3 + \mathbf{s}_1 \cdot \mathbf{s}_3]$$

and then

$$[\mathbf{s}_1 \cdot \mathbf{s}_2 + \mathbf{s}_2 \cdot \mathbf{s}_3 + \mathbf{s}_1 \cdot \mathbf{s}_3] = \frac{\mathbf{S}^2 - \mathbf{s}_1^2 - \mathbf{s}_2^2 - \mathbf{s}_3^2}{2} = \frac{3}{4}$$

The same structure and classification of the electronic states hold for *phosphorous* atom, in view of the same configuration $s^2 p^3$ outside the closed shells. On increasing the atomic number along the V group of the periodic Table, the increase in the spin-orbit interaction can be expected to invalidate the **LS** (see Sect. 3.3). However, for three electrons in the p sub-shell, since ξ_{LS} is almost zero (see Eq. (3.8)), the ${}^2P_{1/2}$ and ${}^2P_{3/2}$ states, for instance, have approximately the same energy ($\Delta E_{SO} \simeq$ 3.1 meV).

Problem 3.3 Reformulate the vectorial coupling for two inequivalent p electrons in the **LS** scheme, indicating the states that would not occur for equivalent electrons.

Solution:

$$\Sigma_i \, l_i \longrightarrow \qquad L \qquad\qquad \text{states}$$

↑↑	2	D
⋁↑	1	P
↑↓ •	0	S

$$\Sigma_i \, s_i \longrightarrow \qquad S \qquad\qquad \text{multiplets}$$

↑↑ ↑	1	3
↑↓ •	0	1

$$\mathbf{J} = \mathbf{L} + \mathbf{S} \qquad D \qquad J = 3, 2, 1 \;\; \text{for} \;\; S = 1$$

$$J = 2 \;\; \text{for} \;\; S = 0$$

$$P \qquad J = 2, 1, 0 \;\; \text{for} \;\; S = 1$$

$$J = 1 \;\; \text{for} \;\; S = 0$$

$$S \qquad J \equiv S = 1, 0.$$

For equivalent electrons only $^3P_{2,1,0}$, $^1D_2$ and $^1S_0$ are present (see Table 3.1).

Problem 3.4 By referring to a magnetic moment μ_L in magnetic field **H**, derive the precessional motion of **L** with the *Larmor frequency* $\omega_L = \gamma H$, where γ is the *gyromagnetic ratio* (see Problem 1.21).

Solution: The equation of motion is

$$\frac{d\mathbf{L}}{dt} = \mu_L \times \mathbf{H} = -\gamma \mathbf{L} \times \mathbf{H}$$

i.e.

$$\frac{dL_z}{dt} = 0 \quad \Rightarrow \quad L_z \equiv L \cos \theta = const$$

$$\dot{L_x} = -\gamma L_y H \qquad \dot{L_y} = \gamma L_x H$$

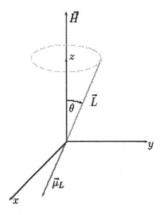

Then

$$\frac{d^2 L_x}{dt^2} = -\gamma^2 H^2 L_x$$

(and analogous for L_y), implying coherent rotation of the components in the (xy) plane with $\omega_L = eH/2mc$.

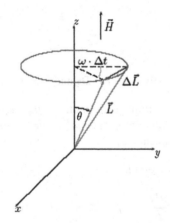

The frequency of the precessional motion can be obtained by writing (see Figure)

$$|\mathbf{\Delta L}| = L \sin\theta \, \omega_L \, \Delta t$$

so that

$$\omega_L = \frac{|\mathbf{\Delta L}|}{\Delta t} \frac{1}{L \sin\theta} = \frac{\mu_L H \sin\theta}{L \sin\theta} = \gamma H$$

Problem 3.5 Derive the ground states for Fe^{++}, V^{+++}, Co, As, La, Yb^{+++} and Eu^{++}, in the framework of the **LS** coupling scheme (a similar Problem is 3.10).

Solution: The ion Fe^{++} has six $3d$ electrons. According to the Pauli principle and the Hund rules m spin

2	↑ ↓
1	↑
0	↑
−1	↑
−2	↑

Then $S = 2$, $L = 2$, $J = L + S = 4 \implies$ state $^5D_4$;

V^{+++} has incomplete $3d$ shell (2 electrons):

m spin

2	↑
1	↑
0	
−1	
−2	

Then $S = 1$ $L = 3$ $J = L - S = 2 \implies$ state $^3F_2$.
Similarly

Co $(3d)^7 (4s)^2$ $S = \frac{3}{2}$ $L = 3$ $J = \frac{9}{2} \implies$ state $^4F_{\frac{9}{2}}$;

As $(3d)^{10} (4s)^2 (4p)^3$ $S = \frac{3}{2}$ $L = 0$ $J = \frac{3}{2} \implies$ state $^4S_{\frac{3}{2}}$;

La $(5d)^1 (6s)^2$ $S = \frac{1}{2}$ $L = 2$ $J = \frac{3}{2} \implies$ state $^2D_{\frac{3}{2}}$;

Yb^{+++} $(4f)^{13}$ $S = \frac{1}{2}$ $L = 3$ $J = L + S = \frac{7}{2} \implies$ state $^2F_{\frac{7}{2}}$;

Eu^{++} $(4f)^7$ $S = \frac{7}{2}$ $L = 0$ $J = S = \frac{7}{2} \implies$ state $^8S_{\frac{7}{2}}$.

(see Table 3.2).

3.3 jj Coupling Scheme

The experimental findings indicate that the interval rule (Eq. 3.9), characteristic of
the **LS** scheme, no longer holds for heavy atoms. This can be expected in view of
the increase of the spin-orbit interaction upon increasing Z, thus invalidating the
condition $a \ll c$ at the basis of the **LS** coupling. In the opposite limit of $a \gg c$ one
first has to couple the single-electron orbital and spin momenta to define **j** and then
construct the total momentum **J**:

$$\mathbf{j}_i = \mathbf{l}_i + \mathbf{s}_i \quad \text{with good quantum numbers} \quad j_i \text{ and } (m_j)_i \quad (3.17)$$

and

$$\mathbf{J} = \sum_i \mathbf{j}_i \quad \text{with good quantum numbers } J \text{ and } M_J \quad (3.18)$$

Fig. 3.6 Vectorial sketch of
the **jj** coupling and of the
precessional motions for two
electrons, leading to the total
J and μ_J precessing around
the external magnetic field

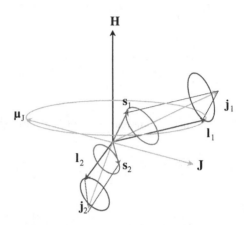

The final state is characterized by l, s, j of each electron and by J and M_J of the
whole atom. The vectorial picture is shown in Fig. 3.6. j_1 and j_2 are half integer
while J is always integer. To label the states, the individual j_i's are usually written
between parentheses while J is written as subscript.

In a way analogous to the couplings in Eqs. (3.3) and (3.5), by the "squaring rule"
$\mathbf{j}_1 \cdot \mathbf{j}_2$ leads to

$$\mathbf{j}_1 \cdot \mathbf{j}_2 = \frac{J(J+1) - j_1(j_1+1) - j_2(j_2+1)}{2} \tag{3.19}$$

The structure of the levels and their labelling is evidently different from the one
derived within the **LS** scheme, as it appears from the example for one s and one p
electron in Fig. 3.7 (to be compared with Fig. 3.4a). In Fig. 3.8 the comparison of the
LS and **jj** schemes for two equivalent p electrons is shown.

The jj coupling for two inequivalent p electrons is indicated below

j_1	j_2	J	Notation	Degeneracy	
3/2	3/2	3,2,1,0	$(3/2,3/2)_{3,2,1,0}$	16	
3/2	1/2	2,1	$(3/2,1/2)_{2,1}$	8	Total number of states 36
1/2	1/2	1,0	$(1/2,1/2)_{1,0}$	4	
1/2	3/2	2,1	$(1/2,3/2)_{2,1}$	8	

For equivalent p electrons the following cases are excluded

j_1	j_2	J	
3/2	3/2	3	
3/2	3/2	1	number of states excluded 13
1/2	1/2	1	

The first case implies parallel orbital momenta as well as parallel spins. The third
case corresponds to $l_1 = l_2 = 0$ and parallel spins. The middle term is not pictorially
evident (it is the analogous of the ^1P states at Table 3.1) and corresponds to a level
for which no additional distinguishable states are available.

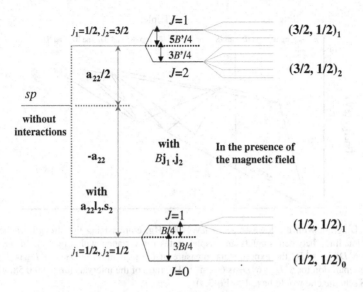

Fig. 3.7 jj coupling for s and p electrons. It is noted that for the state $j_1 = 1/2$ and $j_2 = 3/2$ the energy constant B' describing the coupling is equal and of opposite sign of the one (B) for the $j_1 = 1/2$ and $j_2 = 1/2$ state (this is proved in Problem 3.6)

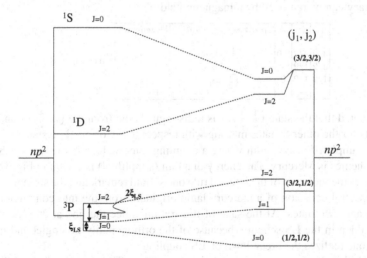

Fig. 3.8 Comparison of the structure and classification of the levels for two equivalent p-electrons in the **LS** (*left*) and **jj** (*right*) coupling schemes

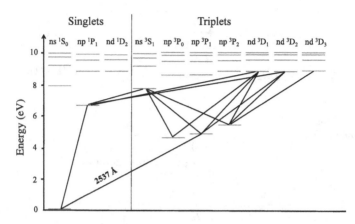

Fig. 3.9 Diagram with the lowest energy levels for Hg, emphasizing the strength of the inter-combination lines between singlets and triplet states (at variance with Fig. 2.7). In the triplet $6^3D_3 - 6^3D_2 - 6^3D_1$ the experimental measure of the separation $6^3D_3 - 6^3D_2$ is 35 cm^{-1}, while the separation for $6^3D_2 - 6^3D_1'$ is 60 cm^{-1}. The ratio of the intervals turns out 0.58, whereas in the **LS** scheme one would have 1.5 (Eq. 3.9)

The states allowed for equivalent p electrons are listed below, where the M_J degeneracy can be removed by a magnetic field:

j_1	j_2	J	Spectroscopic notation	Degeneracy	
3/2	3/2	2,0	$(\frac{3}{2},\frac{3}{2})_{2,0}$	6	
3/2	1/2	2,1	$(\frac{3}{2},\frac{1}{2})_{2,1}$	8	see Fig. 3.8
1/2	1/2	0	$(\frac{1}{2},\frac{1}{2})_0$	1	
-	-	-	-	Total 15	

It is noted that the state $(\frac{3}{2},\frac{1}{2})_{2,1}$ is indistinguishable from the $(\frac{1}{2},\frac{3}{2})_{2,1}$ and this accounts for the other 8 states missing with respect to the original 36 states.

An example of heavy atom where a coupling intermediate between the **LS** and the **jj** schemes is Mercury. The energy diagram (simplified) is shown in Fig. 3.9.

Besides the violation of the interval rule one should remark that the strongest lines in the spectral emission of a mercury lamp originate from the intercombination of the $^1S_0$ and $^3P_1$ states. At the sake of illustration, since the line at 2537 Å would be forbidden in the **LS** scheme (because of the orthogonality of singlet and triplet states), one realizes the breakdown of **LS** coupling.

In very heavy atoms pure **jj** coupling does occur. The tendency from **LS** to **jj** coupling scheme is shown schematically in Fig. 3.10 for the sequence C, Si, Ge, Sn, Pb, in terms of the (sp) outer electrons configuration.

Problems

Problem 3.6 Prove that for the **jj** coupling of one s and one p electrons in the state at $j_1 = 1/2$ and $j_2 = 3/2$ the fine structure constant B' is equal to $-B$ (see Fig. 3.7).

Solution: The couplings are

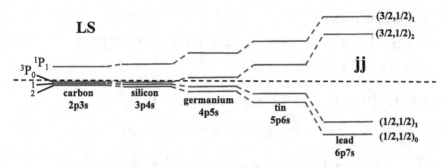

Fig. 3.10 Schematic view of the progressive changeover from **LS** scheme towards **jj** scheme on increasing the atomic number for the two electrons energy levels. It should be remarked that the **LS** scheme is often used to label the states eventhough their structure is rather close to the one pertaining to the **jj** coupling scheme

$$a_{11}\mathbf{l}_1 \cdot \mathbf{s}_1 + a_{22}\mathbf{l}_2 \cdot \mathbf{s}_2 + B(j_1, j_2)\mathbf{j}_1 \cdot \mathbf{j}_2 \quad \text{with} \quad a_{11} \text{ and } a_{22} > 0.$$

For s electron $\quad l_1 = 0 \quad s_1 = \frac{1}{2} \quad j_1 = \frac{1}{2}$

For p electron $\quad l_2 = 1 \quad s_2 = \frac{1}{2} \quad j_2 = \frac{3}{2}\,\frac{1}{2}$
corresponding to the configuration

$$
\begin{array}{ccc}
j_1 & j_2 & J \\
\frac{1}{2} & \frac{3}{2} & 2,1 \\
\frac{1}{2} & \frac{1}{2} & 1,0
\end{array}
$$

with $B\left(\frac{1}{2}, \frac{3}{2}\right) \equiv B'$ (a) and $B\left(\frac{1}{2}, \frac{1}{2}\right) \equiv B$ (b).
For case (a)

$$B\mathbf{j}_1 \cdot \mathbf{j}_2 = B'\mathbf{s}_1 \cdot (\mathbf{l}_2 + \mathbf{s}_2) = \underbrace{B'\mathbf{l}_2 \cdot \mathbf{s}_1}_{\text{negligible}} + B'\mathbf{s}_1 \cdot \mathbf{s}_2$$

while for case (b)

$$B\mathbf{j}_1 \cdot \mathbf{j}_2 = B\mathbf{s}_1 \cdot (\mathbf{l}_2 - \mathbf{s}_2) = \underbrace{B\mathbf{l}_2 \cdot \mathbf{s}_1}_{\text{negligible}} - B\mathbf{s}_1 \cdot \mathbf{s}_2$$

Thus $\quad B \equiv -c_{12} > 0 \quad$ and $\quad B' = -B$.

Problem 3.7 For an electron in the $lsjm_j$ state, express the expectation values of s_z, l_z, l_z^2 and l_x^2 (z is an arbitrary direction and x is perpendicular to z).

Solution: By using arguments strictly similar to the ones at Sect. 2.2.2 (see Eq. (3.12)) and taking into account that because of the spin-orbit precession **s** must be projected along **j**:

$\mathbf{s}_j = |\mathbf{s}|cos(\widehat{\mathbf{sj}})$ with

$cos(\widehat{\mathbf{sj}}) = \left[s(s+1) + j(j+1) - l(l+1)\right]/2\sqrt{s(s+1)}\sqrt{j(j+1)}.$

Then $< s_z > = \mathbf{s}_j\, m_j/|j| = m_j A,$

with $A = [s(s+1) + j(j+1) - l(l+1)]/2j(j+1)$

$$< l_z > = < j_z > - < s_z > = m_j(1 - A)$$

$$< l_z^2 > = < (j_z - s_z)^2 > = < j_z^2 > + < s_z^2 > -2 < j_z s_z > =$$

$$= m_j^2 + \frac{1}{4} - 2m_j < s_z > = m_j^2 + \frac{1}{4} - 2m_j^2 A$$

Since $< l^2 > = < l_z^2 > +2 < l_x^2 > (< l_x^2 > = < l_y^2 >)$ then

$$< l_x^2 > = \frac{1}{2}\left[l(l+1) - \frac{1}{4} - m_j^2(1 - 2A)\right]$$

The same result for $< s_z >$ is obtained from the Wigner-Eckart theorem (Eq. 4.25):
$< l, s, j, m_j|\mathbf{s_z}|l, s, j, m_j > = < |(\mathbf{s}\cdot\mathbf{j})j_z| > /j(j+1) =$
$= m_j < |\mathbf{s}\cdot\mathbf{j}| > /j(j+1) = m_j < |\mathbf{j}^2 - \mathbf{l}^2 + \mathbf{s}^2| > /2j(j+1).$

3.4 Quantum Theory for Multiplets. Slater Radial Wavefunctions

From the perturbative Hamiltonian reported in Eq. (1.11) and on the basis of the Slater determinantal eigenfunctions $D(1, 2, 3, \ldots)$ described at Sect. 2.3, one can develop a quantum treatment at the aim of deriving the multiplet structure discussed in the framework of the vectorial model. The perturbation theory for degenerate states has to be used. A particular form of this approach is described in Problem 2.9 for the $1s2l$ states of Helium. At Sect. 2.2.2 a similar treatment was practically given, without involving *a priori* the degenerate eigenfunctions corresponding to a specific electronic configuration.

In general the direct solution of the secular equation is complicated and the matrix elements include operators of the form r_i^{-1} and r_{ij}^{-1} and the spin-orbit term. Again two limiting cases of predominance of the spin-spin or of the spin-orbit interaction have to be used in order to fix the quantum numbers labelling the unperturbed states associated with the zero-order degenerate eigenfunctions. The eigenvalues are obtained in terms of generalized Coulomb and exchange integrals. First we shall limit ourselves to a schematic illustration of the results of the Slater theory for the electronic configuration $(np)^2$, to be compared with the results obtained at Sect. 3.2.3 in the framework of the vectorial model.

For two non-equivalent p electrons (say $2p$ and $3p$) the Slater multiplet theory yields the following eigenvalues in the **LS** scheme, $I_{0,2}$ and $K_{0,2}$ being Coulomb and exchange integrals for different one-electron states:

(a) $E(^3D) = E_0 + I_0 + \frac{I_2}{25} - K_0 - \frac{K_2}{25}$

(b) $E(^3P) = E_0 + I_0 - \frac{5I_2}{25} + K_0 - \frac{5K_2}{25}$

(c) $E(^3S) = E_0 + I_0 + \frac{10I_2}{25} - K_0 - \frac{10K_2}{25}$

(d) $E(^1D) = E_0 + I_0 + \frac{I_2}{25} + K_0 + \frac{K_2}{25}$

(e) $E(^1P) = E_0 + I_0 - \frac{5I_2}{25} - K_0 + \frac{5K_2}{25}$

(f) $E(^1S) = E_0 + I_0 + \frac{10I_2}{25} + K_0 + \frac{10K_2}{25}$

(the indexes 0 and 2 result from the expansion of $1/r_{12}$ in terms of *Legendre polynomials*). For equivalent $2p$ electrons only states (b), (d) and (f) occur, with energies (Fig. 3.11).

$$E(^3P) = E_0 + I_0 - \frac{5I_2}{25}$$

$$E(^1D) = E_0 + I_0 + \frac{I_2}{25}$$

$$E(^1S) = E_0 + I_0 + \frac{10I_2}{25}$$

(the exchange integral formally coincides with the Coulomb integral here).

The quantitative estimate of the energy levels cannot be given unless numerical computation of I and K in terms of one-electron eigenfunctions is carried out.

Approximate analytical expressions for the radial parts of the one-electron eigenfunction can be obtained as follows.

An effective potential energy of the form

$$V(r) = \frac{-(Z - \sigma)e^2}{r} + \frac{n^*(n^* - 1)\hbar^2}{2mr^2} \tag{3.20}$$

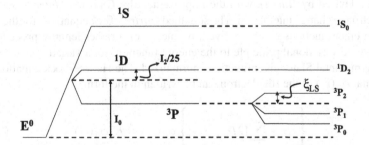

Fig. 3.11 Schematic diagram for equivalent p^2 electron configuration as derived in the Slater theory, in terms of Coulomb and exchange generalized integrals. The comparison with the results of the vectorial model (see Problem 3.1) clarifies that the same structure and classification of the levels is obtained. Quantitative estimates require the knowledge of the radial parts of the one-electron eigenfunctions

Table 3.3 The *Clementi-Raimondi* values for $Z - \sigma$ (ground states)

	H							He
1s	1							1.6875
	Li	Be	B	C	N	O	F	Ne
1s	2.6906	3.6848	4.6795	5.6727	6.6651	7.6579	8.6501	9.6421
2s	1.2762	1.9120	2.5762	3.2166	3.8474	4.4916	5.1276	5.7584
2p			2.4214	3.1358	3.8340	4.4532	5.1000	5.7584
	Na	Mg	Al	Si	P	S	Cl	Ar
1s	10.6259	11.6089	12.5910	13.5754	14.5578	15.5409	16.5239	17.5075
2s	6.5714	7.3920	8.2136	9.0200	9.8250	10.6288	11.4304	12.2304
2p	6.8018	7.8258	8.9634	9.9450	10.9612	11.9770	12.9932	14.0082
3s	2.5074	3.3075	4.1172	4.9032	5.6418	6.3669	7.0683	7.7568
3p			4.0656	4.2852	4.8864	5.4819	6.1161	6.7641

It can be noted that for He atom, since $n^* = n = 1$ the value of $Z - \sigma$ must coincide with Z^* variationally derived at Problem 2.6

is assumed, with σ and n^* parameters to be determined. This form is strictly similar to the one for Hydrogenic atoms, with a screened Coulomb term and a centrifugal term (see Sect. 1.4). Thus the associated eigenfunctions are

$$\phi_{nlm}(r, \theta, \varphi) = N Y_{l,m}(\theta, \varphi) r^{n^* - 1} e^{-\frac{(Z-\sigma)r}{n^* a_0}} \qquad (3.21)$$

with N normalization factor.

The eigenvalues are similar to the ones at Sect. 1.4 and depend on σ and n^*. Then $E(\sigma, n^*)$ is minimized to find the best approximate values for σ and n^* and the radial part of the eigenfunctions is derived.

Empirical rules to assign the proper values to σ and n^* are the following. For quantum number n one has the correspondence

$n = 1, 2, 3, 4, 5,$ and 6

$n^* = 1, 2, 3, 3.7, 4,$ and 4.2

while Table 3.3 gives the rules to derive $(Z - \sigma)$.

The best atomic orbitals are actually obtained by the numerical solutions along the lines devised by *Hartree* with the improvement by *Fock* and *Slater* to include the electron exchange interaction. The so-called *Hartree-Fock* equations for the one-electron eigenfunctions can be derived, by means of a rather lengthy procedure,[5] applying the variational principle to the energy function, for a variation that leaves the determinantal Slater eigenfunctions normalized. The Hartree-Fock equation for the orbital $\phi_\alpha(\mathbf{r}_i)$ of the ith electron can be written in the form

$$\left\{ \mathcal{H}_i + \sum_\beta [2I_\beta - K_\beta] \right\} \phi_\alpha(\mathbf{r}_i) = E_\alpha^i \phi_\alpha(\mathbf{r}_i) \qquad (3.22)$$

---

[5] See, for instance Sect. 16-3 in the book by *Slater* or Chap. 9 in the book by *Atkins* and *Friedman*.

$\mathcal{H}_i$ is the one-electron *core Hamiltonian* $(T_i - Z^* e^2 / r_i$, with $Z^* \equiv Z$ if no screening effects are considered), while I_β and K_β are the Coulomb and exchange operators that generalize the correspondent terms derived at Sect. 2.3 for He:

$$I_\beta \phi_\alpha(\mathbf{r}_i) = \left[\int \phi_\beta^*(\mathbf{r}_j) \frac{e^2}{r_{ij}} \phi_\beta(\mathbf{r}_j) d\tau_j \right] \phi_\alpha(\mathbf{r}_i) \qquad (3.23)$$

$$K_\beta \phi_\alpha(\mathbf{r}_i) = \left[\int \phi_\beta^*(\mathbf{r}_j) \frac{e^2}{r_{ij}} \phi_\alpha(\mathbf{r}_j) d\tau_j \right] \phi_\beta(\mathbf{r}_i).$$

E_α in Eq. (3.22) is the one-electron energy. After an iterative numerical procedure, once the best self-consistent ϕ's are obtained, by multiplying both sides of Eq. (3.22) by $\phi_\alpha^*(\mathbf{r}_i)$ and integrating, one obtains for the ith electron

$$E_\alpha = E_\alpha^o + \sum_\beta (2I_{\alpha\beta} - K_{\alpha\beta}) \qquad (3.24)$$

with $E_\alpha^o \equiv < \alpha | \mathcal{H}_i | \alpha >$ and $I_{\alpha\beta}$ and $K_{\alpha\beta}$ are the Coulomb and exchange integrals, respectively (with $I_{\beta\beta} \equiv K_{\beta\beta}$). A sum over all the energies E_α would count all the interelectron interactions twice. Thus, by taking into account that each orbital in a closed shell configuration is double occupied, the total energy of the atom is written

$$E_T = 2 \sum_\alpha E_\alpha - \sum_{\alpha, \beta} (2I_{\alpha\beta} - K_{\alpha\beta}) \qquad (3.25)$$

Although the eigenvalues obtained along the procedure outlined above are generally very close to the experimental data for the ground-state (for light atoms within 0.1 percent) still one could remark that any approach based on the model of independent electrons necessarily does not entirely account for the *correlation effects*.

Suppose that an electron is removed and that the other electrons do not readjust their configurations. Then the one-electron energy E_α corresponds to the energy required to remove a given electron from its orbital. This is the physical content of the *Koopmans theorem*, which identifies $|E_\alpha|$ with the ionization energy. Its validity rests on the assumption that the orbitals of the ion do not differ sizeably from the ones of the atom from which the electron has been removed.

The Hartree-Fock procedure outlined here for multi-electron atoms is widely used also for molecules and crystals, by taking advantage of the fast computers available nowadays which allow one to manipulate the Hartree-Fock equations. When the spherical symmetry of the central field approximation has to be abandoned numerical solutions along Hartree-Fock approach are anyway hard to be carried out. Thus particular manipulations of the equations have been devised, as the widely used *Roothaan's* one. Alternative methods are based on the *density functional theory* (DFT), implemented by the *local density approximation* (LDA). Correlation and relativistic effects are to be taken into account when detailed calculations are aimed, particularly for heavy atoms. Chapter 9 of the book by *Atkins* and *Friedman* adequately deals with the basic aspects of the computational derivation of the electronic structure.

Finally we mention that for atoms with a rather high number of electrons and when dealing in particular with the radial distribution function of the electron charge in the ground-state (and therefore to the expectation values), the semiclassical method devised by *Thomas* and *Fermi* can be used. This approach is based on the statistical properties of the so-called Fermi gas of independent non-interacting particles obeying to Pauli principle, that we shall encounter in a model of solid suited to describe the metals (Sect. 12.7.1). The Thomas-Fermi approach is often used as a first step in the self-consistent numerical procedure that leads to the Hartree-Fock equations.

3.5 Selection Rules

Here the selection rules that control the transitions among the electronic levels in the **LS** and in the **jj** coupling schemes are recalled. Their formal derivation (the extension of the treatment in Appendix 1.3) requires the use of the *Wigner-Eckart theorem* and of the properties of the *Clebsch-Gordan coefficients*. We will give the rules for electric dipole, magnetic dipole and electric quadrupole transition mechanisms, again in the assumption that one electron at a time makes the transition. This is the process having the strongest probability with respect to the one involving two electrons at the same time, that would imply the breakdown of the factorization of the total wavefunction, at variance to what has been assumed, for instance, at Sect. 2.1.

(A) **Electric dipole transition**

LS *coupling*
$\Delta L = 0, \pm 1$ and $\Delta S = 0$, non rigorous ($L = 0 \rightarrow L' = 0$ forbidden)
$\Delta J = 0, \pm 1$, transition $0 \leftrightarrow 0$ forbidden[6];
$\Delta M_J = 0, \pm 1$ for $\Delta J = 0$ the transition $M_J = 0 \rightarrow M'_J = 0$ forbidden.[6]
     For the electron making the transition one has $\Delta l = \pm 1$, according to parity arguments (see Appendix 1.3).

jj *coupling*
For the atom as a whole
$\Delta J = 0, \pm 1$, transition $0 \leftrightarrow 0$ forbidden[6];
$\Delta M_J = 0, \pm 1$ for $\Delta J = 0$ the transition $M_J = 0 \rightarrow M'_J = 0$ is forbidden.[6]
For the electron making the transition $\Delta l = \pm 1$, $\Delta j = 0, \pm 1$.

(B) **Magnetic dipole transitions**

$\Delta J = 0, \pm 1$ and $\Delta M_J = 0, \pm 1$ (general validity)

LS *scheme*
$\Delta S = 0, \Delta L = 0, \Delta M = \pm 1$

---

[6]Rules of general validity in both schemes.

(C) Electric quadrupole mechanism

$\Delta J = 0, \pm 1, \pm 2$ (general validity),

LS *scheme*
$\Delta L = 0, \pm 1, \pm 2$
$\Delta S = 0$

Problems

Problem 3.8 A beam of Ag atoms (in the ground state $5\,^2S_{1/2}$) flows with speed $v = 10^4$ cm/s, for a length $l_1 = 5$ cm, in a region of inhomogeneous magnetic field, with $dH/dz = 1$ T/cm. After the exit from this region the beam is propagating freely for a length $l_2 = 10$ cm and then collected on a screen, where a separation of about 0.6 cm between the split beam is observed (*Stern-Gerlach* experiment). From these data obtain the magnetic moment of Ag atom.

Solution: In the first path l_1 the acceleration is $a = F/M_{Ag} = (\mu_z/M_{Ag})(dH/dz)$ and the divergence of the atomic beam along z turns out $d' = (a/2)\,(l_1/v)^2$. In the second path l_2, with $v_z = al_1/v$ and then $d'' = al_1l_2/v^2$.

The splitting of the two beams with different z-component of the magnetic moment $(S = J = 1/2)$ turns out $d = 2(d' + d'') = (a/v^2)(l_1^2 + 2l_1l_2)$. Then

$$\mu_z = \frac{M_{Ag}v^2 d}{\frac{dH}{dz}(l_1^2 + 2l_1l_2)} \simeq 0.93 \cdot 10^{-20} \frac{\text{erg}}{\text{Gauss}}.$$

Problem 3.9 In the **LS** coupling scheme, derive the electronic states for the configurations $(ns, n's)$ (i), $(ns, n'p)$(ii), $(nd)^2$(iii) and $(np)^3$(iv). Then schematize the correlation diagram to the correspondent states in the **jj** scheme, for the nd^2 and for the np^3 configurations.

Solution:
(i) $S = 1$ $L = 0$ $^3S_1$; $S = 0$ $L = 0$ $^1S_0$
(ii) $^1P_1, ^3P_{0,1,2}$
(iii)

2		2	1	1	0	0	-1	-1	-2	-2 m_l
$\frac{1}{2}$		$-\frac{1}{2}$	$\frac{1}{2}$	$-\frac{1}{2}$	$\frac{1}{2}$	$-\frac{1}{2}$	$\frac{1}{2}$	$-\frac{1}{2}$	$\frac{1}{2}$	$-\frac{1}{2}$ m_s

2	$\frac{1}{2}$									
2	$-\frac{1}{2}$	4,0								
1	$\frac{1}{2}$	3,1 3,0								
1	$-\frac{1}{2}$	3,0 3,−1 2,0								
0	$\frac{1}{2}$	2,1 2,0 1,1 1,0								
0	$-\frac{1}{2}$	2,0 2,−1 1,0 1,−1 0,0								
-1	$\frac{1}{2}$	1,1 1,0 0,1 0,0 −1,1 −1,0								
-1	$-\frac{1}{2}$	1,0 1,−1 0,0 0,−1 −1,0 −1,−1 −2,0								
-2	$\frac{1}{2}$	0,1 0,0 −1,1 −1,0 −2,1 −2,0 −3,1 −3,0								
-2	$-\frac{1}{2}$	0,0 0,−1 −1,0 −1,−1 −2,0 −2,−1 −3,0 −3,−1 −4,0								
m_l	m_s									

then $^1S_0, ^3P_{2,1,0}, ^1D_2, ^3F_{4,3,2}, ^1G_4$.

The total number of states is $\begin{pmatrix} 10 \\ 2 \end{pmatrix} = 45$

(iv) $^4S_{\frac{3}{2}}, ^2P_{\frac{1}{2},\frac{3}{2}}, ^2D_{\frac{3}{2},\frac{5}{2}}$

The total number of states is $\begin{pmatrix} 6 \\ 3 \end{pmatrix} = 20$

The correlation between the two schemes is given below for the p^3

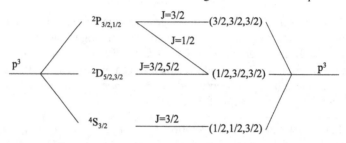

and d^2 configurations

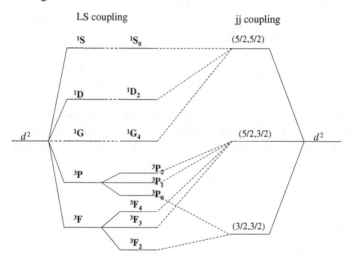

Problem 3.10 By resorting to the Hund rules derive the effective magnetic moments for Dy^{+++}, Cr^{+++} and Fe^{+++} (See Table 3.2).

Solution: The ion Dy^{+++} has incomplete $4f$ shell (9 electrons).

m spin

$$
\begin{array}{ll}
3 & \uparrow \ \downarrow \\
2 & \uparrow \ \downarrow \\
1 & \uparrow \quad S = \tfrac{5}{2} \quad L = 5 \quad J = L + S = \tfrac{15}{2} \quad \Longrightarrow \quad \text{state} \quad {}^{6}H_{\frac{15}{2}} \\
0 & \uparrow \\
-1 & \uparrow \\
-2 & \uparrow \\
-3 & \uparrow
\end{array}
$$

according to the Pauli principle and the Hund rules.

The Landé factor is

$$
g = 1 + \frac{J(J+1) + S(S+1) - L(L+1)}{2J(J+1)} = 1.33
$$

thus

$$
p = \frac{\mu}{\mu_B} = g\sqrt{J(J+1)} = 10.65
$$

In similar way

$$
\text{Cr}^{+++} \quad (3d)^3 \quad S = \tfrac{3}{2} \quad L = 3 \quad J = |L - S| = \tfrac{3}{2} \quad \Longrightarrow \quad \text{state} \quad {}^{4}F_{\frac{3}{2}};
$$

$$
g = 0.4 \text{ and } p = 0.77
$$

$$
\text{Fe}^{+++} \quad (3d)^5 \quad S = \tfrac{5}{2} \quad L = 0 \quad J \equiv S = \tfrac{5}{2} \quad \Longrightarrow \quad \text{state} \quad {}^{6}S_{\frac{5}{2}};
$$

$$
g = 2 \text{ and } p = 5.92
$$

Problem 3.11 When accelerated protons collide on ^{19}F nuclei an excited state of ^{20}Ne is induced and transition to the ground state yields γ *emission*. The emission spectrum, as a function of the energy of colliding protons, displays a line centered at 873.5 keV, with full width at half intensity of 4.8 keV. Derive the life time of the excited state of ^{20} Ne. Comment about the difference with the emission spectrum of ^{57}Fe, where the transition to the ground state from the first excited state yields a γ-photon at 14.4 keV, with life time 10^{-7} s.

By referring to ^{57}Fe, considering that the transition is due to a proton and assuming as radius of the nucleus 10^{-12} cm, by means of order of magnitude estimates discuss the transition mechanism (*electric dipole, electric quadrupole, magnetic dipole*) driving the γ transition at 14.4 keV in ^{57}Fe.

Solution: From

$$
\tau \simeq \frac{\hbar}{\Delta E} \simeq \frac{1.05 \cdot 10^{-27} \text{ erg s}}{4.8 \cdot 10^3 \cdot 1.6 \cdot 10^{-12} \text{ erg}} = 1.37 \cdot 10^{-19} \text{ s}.
$$

for ^{20}Ne, while for ^{57}Fe

$$\Delta E \simeq \frac{\hbar}{\tau} = 1.05 \cdot 10^{-20} \text{ erg} \simeq 6.6 \cdot 10^{-12} \text{ keV}.$$

The transition mechanism driving the γ transition at 14.4 keV in ^{57}Fe is discussed as follows:

(a) for electric dipole transition the spontaneous emission probability (see Appendix 1.3) is

$$A_{21}^E = \frac{32\pi^3(E_2 - E_1)^3}{3c^3\hbar\,h^3}| < 2|e\mathbf{R}|1 > |^2 \simeq 10^{11}\,\text{s}^{-1}$$

for $| < 2|e\mathbf{R}|1 > |^2 \simeq (e \cdot R_N)^2$. Then one would expect

$$\tau^E \sim (A_{21}^E)^{-1} \sim 10^{-11}\,\text{s};$$

(b) for electric quadrupole mechanism

$$\frac{A_{21}^E}{A_{21}^Q} \simeq \left(\frac{\lambda}{R_N}\right)^2 \frac{1}{4\pi^2}$$

and then

$$\tau^Q \sim \tau^E \cdot 1.9 \cdot 10^6 \simeq 1.9 \cdot 10^{-5}\,\text{s};$$

(c) for magnetic dipole mechanism

$$\frac{A_{21}^E}{A_{21}^M} \sim \left[\frac{eR_N}{\mu_N}\right]^2 \simeq 4100$$

$$\tau_M \sim \tau^E \cdot 4100 \sim 4 \cdot 10^{-7}\,\text{s},$$

having used for the magnetic moment $\mu_N \simeq 7.5 \times 10^{-24}$ erg/Gauss.

From the experimental value it may be concluded that the transition is due to magnetic dipole mechanism.

Problem 3.12 Derive the multiplets for the ^3F and the ^3D states and sketch the transitions allowed by the electric dipole mechanism.

Solution:

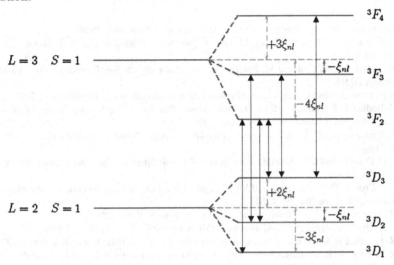

Problem 3.13 Estimate the order of magnitude of the ionization energy of ^{92}U in the case that Pauli principle should not operate (assume that the screened charge is $Z/2$) and compare it with the actual ionization energy (4 eV).

Solution: From

$$E = -\frac{\mu Z^2 e^4}{2\hbar^2 n^2} = -\frac{Z^2}{n^2} \cdot 13.6\,\text{eV}$$

and for $n = 1$ and $Z = 46$, the ionization energy would be

$$|E| = (46)^2 \cdot 13.6\,\text{eV} \simeq 2.9 \cdot 10^4\,\text{eV}.$$

Problem 3.14 The structure of the electronic states in the Oxygen atom can be derived in a way similar to the one for Carbon (Problem 3.1) since the electronic configuration $(1s)^2\,(2s)^2\,(2p)^4$ has two "holes" in the $2p$ shell, somewhat equivalent to the $2p$ two electrons. Discuss the electronic term structure for oxygen along these lines.

Solution: From Table 3.1 taking into account that for $(2p)^6$ one would have $M_L = 0$ and $M_S = 0$ the term ^3P, ^1D, ^1S are found. Since one has four electrons the spin-orbit constant changes sign, the multiplet is inverted and the ground state is $^3P_2$ instead of $^3P_0$ (see Problem 3.1).

Specific References and Further Reading

1. J.C. Slater, *Quantum Theory of Matter*, (Mc Graw-Hill, New York, 1968).
2. P.W. Atkins and R.S. Friedman, *Molecular Quantum Mechanics*, (Oxford University Press, Oxford, 1997).
3. A. Balzarotti, M. Cini and M. Fanfoni, *Atomi, Molecole e Solidi. Esercizi risolti*, (Springer Verlag, 2004).
4. B.H. Bransden and C.J. Joachain, *Physics of atoms and molecules*, (Prentice Hall, 2002).
5. D. Budker, D.F. Kimball and D.P. De Mille, *Atomic Physics - An Exploration Through Problems and Solutions*, (Oxford University Press, 2004).
6. B. Cagnac and J.C. Pebay - Peyroula, *Physique atomique, tome 2*, (Dunod Université, Paris, 1971).
7. E.U. Condon and G.H. Shortley, *The Theory of Atomic Spectra*, (Cambridge University Press, London, 1959).
8. J.A. Cronin, D.F. Greenberg and V.L. Telegdi, *University of Chicago Graduate Problems in Physics*, (Addison-Wesley, 1967).
9. W. Demtröder, *Atoms, Molecules and Photons*, (Springer Verlag, 2006).
10. R.N. Dixon, *Spectroscopy and Structure*, (Methuen and Co LTD, London, 1965).
11. R. Fieschi e R. De Renzi, *Struttura della Materia*, (La Nuova Italia Scientifica, Roma, 1995).
12. R. Eisberg and R. Resnick, *Quantum Physics of Atoms, Molecules, Solids, Nuclei and Particles*, (J. Wiley and Sons, 1985).
13. H. Eyring, J. Walter and G.E. Kimball, *Quantum Chemistry*, (J. Wiley, New York, 1950).
14. H. Haken and H.C. Wolf, *Atomic and Quantum Physics*, (Springer Verlag, Berlin, 1987).
15. C.S. Johnson and L.G. Pedersen, *Quantum Chemistry and Physics*, (Addison-Wesley, 1977).
16. M.A. Morrison, T.L. Estle and N.F. Lane, *Quantum States of Atoms, Molecules and Solids*, (Prentice -Hall Inc., New Jersey, 1976).
17. S. Svanberg, *Atomic and Molecular Spectroscopy*, (Springer Verlag, Berlin, 2003).

Chapter 4
Atoms in Electric and Magnetic Fields

Topics

Electric Polarizability of the Atom
Linear and Quadratic Field Dependences of the Energies
Energy Levels in Strong and in Weak Magnetic Fields
Atomic Paramagnetism and Diamagnetism
Paramagnetism in the Presence of Mean Field Interactions

4.1 Introductory Aspects

The analysis of the effects of magnetic or electric fields on atoms favors a deep understanding of the quantum properties of matter. Furthermore, electric or magnetic fields are tools currently used in several experimental studies.

In classical physics the prototype atom is often considered as an electron rotating on circular orbit around the fixed nucleus. In the presence of electric and magnetic fields (see Fig. 4.1), the equation of motion for the electron becomes

$$m\frac{d^2\mathbf{r}}{dt^2} = -\frac{e^2\mathbf{r}}{r^3} - e\mathcal{E} - \frac{e}{c}\left(\frac{d\mathbf{r}}{dt} \times \mathbf{H}\right) \tag{4.1}$$

For a static magnetic field $\mathbf{H}$ only (then the external electric field $\mathcal{E} = 0$) from Eq. (4.1) it is found that the Lorentz force induces a precessional motion of the charge around z, with angular frequency (see Problem 4.1)

© Springer International Publishing Switzerland 2015
A. Rigamonti and P. Carretta, *Structure of Matter*,
UNITEXT for Physics, DOI 10.1007/978-3-319-17897-4_4

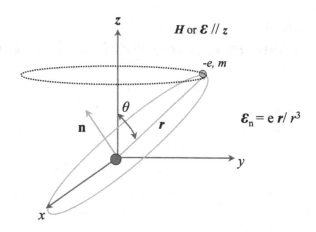

Fig. 4.1 Variables used to account for the effects of electric or magnetic field in the classical atom (Eq. (4.1))

$$\omega = \sqrt{\left(\frac{eH}{2mc}\right)^2 + \frac{e^2}{mr^3}} \pm \frac{eH}{2mc} \simeq \sqrt{\frac{e^2}{mr^3}} \pm \omega_L \qquad (4.2)$$

To give orders of magnitude, the orbital frequency in the plane of motion is $\omega_0 = e/\sqrt{mr^3} \sim 10^{16}$ rad s^{-1} while the Larmor frequency $\omega_L = eH/2mc$ is around 10^{11} rad s^{-1}, for field $H = 10^4$ Oe (1 T).

The current related to the orbital motion corresponds to the magnetic moment $\mu' = \mu_B \mathbf{n}$ (see Problem 1.18): its alignment along the field, contrasted by thermal excitation, implies the temperature dependent *paramagnetism*. The effective z component of the magnetic moment is expected of the order of $(\mu')_z \sim \mu_B(\mu_B H/k_B T)$ (formal description will be given at Sect. 4.4). Therefore the paramagnetic susceptibility $\chi_{para} = N(\mu'_z)/H$, for a number $N = 10^{22}$ of atoms per cubic cm, is of the order of $\chi_{para} \simeq N\mu_B^2/k_B T \sim 6 \times 10^{-5}$ (for $T \simeq 100$ K).

The current related to the precessional motion of the orbit is $i = (-e\omega_L/2\pi) = -e^2 H/4\pi mc$, along a ring of area $A = \pi(r\sin\theta)^2$ (see Fig. 4.1). The associated magnetic moment is

$$(\mu'')_z = \frac{iA}{c} = -\frac{e^2 H}{4\pi mc^2} \pi r^2 \sin^2\theta,$$

yielding a *diamagnetic susceptibility* $\chi_{dia} = N\mu''/H \simeq -e^2 N r^2/4mc^2$, as order of magnitude around -10^{-6} (again for $N = 10^{22}$ atoms per unit volume).

On the ground of qualitative arguments the effect of an electric field $\mathcal{E} \parallel \hat{z}$ can be understood by referring to the displacement δz of the orbit along the field direction: the component of the Coulomb force $e^2 \delta z / r^3$ equilibrates the force $e\mathcal{E}$ (see the sketch below).

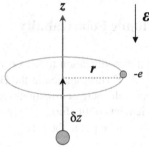

Then the dipole moment turns out $e\delta z = \mathcal{E}r^3$ and an *atomic polarizability* given by $\alpha \sim e\delta z / \mathcal{E} \sim r^3 \sim 10^{-24}\, \text{cm}^3$ can be predicted.

In the quantum mechanical description the electric and magnetic forces imply the one-electron Hamiltonian (see Eq. (1.26))

$$\mathcal{H} = \frac{1}{2m}\left(\mathbf{p} + \frac{e}{c}\mathbf{A}\right)^2 + V(r) + 2\mu_B \mathbf{s}.rot\mathbf{A} - e\varphi =$$

$$= \frac{p^2}{2m} + V(r) + 2\mu_B \mathbf{s}.rot\mathbf{A} - e\varphi + \underbrace{\frac{e}{2mc}\left[(\mathbf{p}.\mathbf{A}) + (\mathbf{A}.\mathbf{p})\right] + \frac{e^2 A^2}{2mc^2}}_{\mathcal{H}_P} \equiv \mathcal{H}_0 + \mathcal{H}_P$$

(4.3)

Here the magnetic term related to the spin moment (Eq. (1.32)) has been added, while $V(r)$ in $\mathcal{H}_0$ is the central field potential energy. φ and $\mathbf{A}$ in Eq. (4.3) are the scalar and vector potentials describing the perturbation applied to the atom.

For homogeneous electric field along the z direction

$$\mathbf{A} = 0, \qquad \text{and} \qquad \varphi = -\int_0^z \mathcal{E}dz = -z\mathcal{E} \qquad (4.4)$$

while for homogeneous magnetic field $\mathbf{H} = \hat{z}H_0$[1]

$$\varphi = 0, \qquad \text{and} \qquad \mathbf{A} = \frac{1}{2}\mathbf{H} \times \mathbf{r}. \qquad (4.5)$$

The corrections to the energy levels can be evaluated on the basis of the eigenfunctions of the zero-field Hamiltonian $\mathcal{H}_0$. In multi-electrons atoms this perturbative approach is generally hard to carry out, in view of the inter-electron couplings (as it can be realized by recalling the description in the framework of the vectorial

---

[1] In fact $(1/2)rot(\mathbf{H} \times \mathbf{r}) = \hat{z}H_0$.

model (Chap. 3)). In the following we shall describe the basic aspects of the effects
due to the fields by deriving the corrections to the atomic energy levels in some
simplifying conditions.

4.2 Stark Effect and Atomic Polarizability

Stark effect is usually called the modification to the energy levels in the presence
of the Hamiltonian $\mathcal{H}_P = \sum_i ez_i \mathcal{E}$ (first studied in the Hydrogen atom also by *Lo
Surdo*). In the perturbative approach energy corrections linear in the field in general
are not expected, the matrix elements of the form $\int \phi^*(\mathbf{r}_i)z_i\phi(\mathbf{r}_i)d\tau_i$ being zero.

The second order correction can be put in the form

$$\Delta E^{(2)} = -\frac{\alpha}{2}\mathcal{E}^2 \tag{4.6}$$

where, in the light of the classical analogy for the electric dipole[2]

$$\mu_e = -\frac{\partial \Delta E}{\partial \mathcal{E}}, \tag{4.7}$$

α defines the *atomic polarizability*. In fact, one can attribute to the atom an *induced
electric dipole moment* $\mu_e = \alpha\mathcal{E}$. The polarizability depends in a complicated way
from the atomic state, in terms of the quantum numbers J and M_J: $\alpha = \alpha(J, M_J)$.

Let us first evaluate the atomic polarizability α_{1s} for Hydrogen in the ground
state. Instead of carrying out the awkward sum of the second order matrix elements
we shall rather estimate the limits within which α_{1s} falls. From Eqs. (4.3) and (4.4)
one has

$$\Delta E^{(2)} = -\sum_{n>1} \frac{|<1s|\mathcal{H}_P|nlm>|^2}{E_n - E_1} = -\frac{1}{2}\alpha_{1s}\mathcal{E}^2. \tag{4.8}$$

$|<1s|\mathcal{H}_P|nlm>|^2$ is always positive and E_n increases on increasing n. Therefore
one can set the limits of variability of $\alpha_{1s}/2$:

$$-\frac{e^2}{E_1}\sum|<1s|z|nlm>|^2 < \frac{\alpha_{1s}}{2} < \frac{e^2}{(E_2 - E_1)}\sum|<1s|z|nlm>|^2. \tag{4.9}$$

(note that the state $n = 1$ can be included in the sum, since $<1s|z|1s> = 0$). On
the other hand

$$\sum<1s|z|nlm><nlm|z|1s> = <1s|z^2|1s> = \frac{1}{\pi a_0^3}\int \frac{4\pi}{3}r^4 e^{-2r/a_0}dr = a_0^2.$$

---

[2]Note that the field-related energy is $\Delta E = -\int_0^{\mathcal{E}} \mu_e d\mathcal{E}'$, so that for $\mu_e = \alpha\mathcal{E}'$ Eq. (4.6) follows.

From $E_1 = -e^2/2a_0$ while $(E_2 - E_1) = 3e^2/8a_0$, one deduces

$$4a_0^3 < \alpha_{1s} < \frac{16}{3}a_0^3.$$

It is recalled that the "brute-force" second order perturbative calculation yields $\alpha_{1s} = 4.66a_0^3$. Thus the electric polarizability turns out of the order of magnitude of the "size" of the atom to the third power, as expected from the qualitative argument at Sect. 4.1.

An approximate estimate of the polarizability of the ground state of the Hydrogen atom can also be obtained by means of variational procedures, on the basis of a trial function involving the mixture of the $1s$ and the $2p_z$ states:

$$\phi_{var} = c_1\phi_{1s} + c_2\phi_{2p_z}. \tag{4.10}$$

This form could be expected on the ground of physical arguments, as sketched below in terms of atomic orbitals:

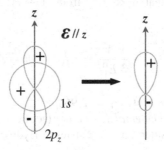

(see Problem 2.10).

The energy function is

$$E(c_1, c_2) = \frac{\int \phi_{var}^* \mathcal{H}\phi_{var}d\tau}{\int \phi_{var}^* \phi_{var}d\tau} \tag{4.11}$$

where $\mathcal{H}$ is the total Hamiltonian, while

$$\mathcal{H}_{11} \equiv< 1s|\mathcal{H}|1s >, \mathcal{H}_{22} \equiv< 2p_z|\mathcal{H}|2p_z >, \mathcal{H}_{12} \equiv< 1s|\mathcal{H}|2p_z >,$$
$$S_{12} \equiv< 1s|2p_z >= 0, S_{11} = S_{22} = 1 \tag{4.12}$$

From $\partial E/\partial c_{1,2} = 0$

$$c_1(\mathcal{H}_{11} - E) + c_2\mathcal{H}_{12} = 0$$
$$c_1\mathcal{H}_{12} + c_2(\mathcal{H}_{22} - E) = 0, \tag{4.13}$$

with secular equation

$$\begin{pmatrix} \mathcal{H}_{11} - E & \mathcal{H}_{12} \\ \mathcal{H}_{21} & \mathcal{H}_{22} - E \end{pmatrix} = 0 \qquad (4.14)$$

Since $\mathcal{H}_{11} = E^0_{1s}$, $\mathcal{H}_{22} = E^0_{1s}/4$, while from Table 1.3 for $Z = 1$
$\mathcal{H}_{12} = <1s|\mathcal{H}_0|2p_z> + <1s|z|2p_z> e\mathcal{E} = e\mathcal{E}2^8 a_0/3^5\sqrt{2} \equiv A$, Eq. (4.14) becomes

$$\begin{pmatrix} E^0_{1s} - E & A \\ A & \frac{E^0_{1s}}{4} - E \end{pmatrix} = 0, \qquad (4.15)$$

of roots

$$E_\pm = \frac{5}{8}E^0_{1s} \pm \frac{1}{2}\sqrt{\frac{9(E^0_{1s})^2}{16}\left(1 + \frac{64A^2}{9(E^0_{1s})^2}\right)} \qquad (4.16)$$

By taking into account that $A \ll E^0_{1s}$, from $(1 + x)^{1/2} \simeq 1 + x/2$ the lowest energy level turns out

$$E = E^0_{1s} + \frac{4}{3}\frac{A^2}{E^0_{1s}} \equiv E^0_{1s} - 2.96\frac{a_0^3}{2}\mathcal{E}^2,$$

corresponding to the polarizability $\alpha_{1s} = 2.96a_0^3$.

In the particular case of *accidental degeneracy* (see Sect. 1.4) Stark effect *linear in the field* occurs. Let us consider the $n = 2$ states of Hydrogen atom. The zero-order wavefunction is

$$\phi_l = c_1^{(l)}\phi_{2s} + c_2^{(l)}\phi_{2p_1} + c_3^{(l)}\phi_{2p_0} + c_4^{(l)}\phi_{2p_{-1}} \qquad (4.17)$$

and the corrected eigenvalues are obtained from

$$\begin{pmatrix} <2s| - ez\mathcal{E}|2s> - E & \cdots & \cdots \\ \cdots & <2p_1| - ez\mathcal{E}|2p_1> - E \cdots & \cdots \\ \cdots & \cdots & \cdots \\ \cdots & \cdots & \cdots \end{pmatrix} = 0 \qquad (4.18)$$

Again recalling the selection rules for the z-component of the electric dipole (Appendix 1.3), this determinant is reduced to

$$\begin{pmatrix} -E & 0 & B & 0 \\ 0 & -E & 0 & 0 \\ B & 0 & -E & 0 \\ 0 & 0 & 0 & -E \end{pmatrix} = 0 \qquad (4.19)$$

where $B = -3a_0 e\mathcal{E}$.

Fig. 4.2 Effect of the electric field on the $n = 2$ states of Hydrogen atom, illustrating how in the presence of accidental degeneracy a kind of *pseudo-orientational polarizability* arises, with energy correction linear in the field $\mathcal{E}$

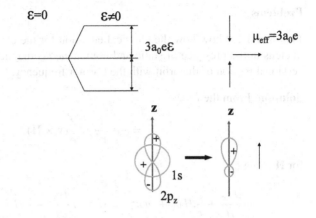

From the roots $R_{1,2} = 0$ and $R_{3,4} = \pm B$ the structure of the $n = 2$ levels in the presence of the field is deduced in the form depicted in Fig. 4.2.

The first-order Stark effect is observed in Hydrogen and in *F-centers* in crystals (where a vacancy of positive ion traps an electron and causes an effective potential of Coulombic character which yields the accidental degeneracy).

Finally in Fig. 4.3 the experimental observation of the Stark effect on the $D_{1,2}$ doublet of Na atom (see Fig. 2.2) is depicted. It is noted that the degeneracy in $\pm M_J \equiv \pm m_j$ is not removed, the energy correction being independent from the versus of the field.

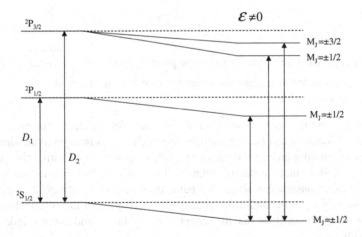

Fig. 4.3 Ground state and first excited states of Na atom upon application of electric field and modification of the D doublet. The energy shift of the ground state is $40.56\,\text{kHz/(kV/cm)}^2$, corresponding to an electric polarizability $\alpha = 24.11 \cdot 10^{-24}\,\text{cm}^3$. The shifts of the P states are about twice larger

Problems

Problem 4.1 Show how the classical equation for the electron in orbit around the nucleus in the presence of static and homogeneous magnetic field implies the precessional motion of the orbit with the Larmor frequency.

Solution: From the force

$$\mathbf{F} = -e^2 \frac{\mathbf{r}}{r^3} - \frac{e}{c}(\mathbf{v} \times \mathbf{H}),$$

for $\mathbf{H}$ along z

$$m\frac{dv_x}{dt} + \frac{e}{c}Hv_y + m\omega_0^2 x = 0, \quad m\frac{dv_y}{dt} - \frac{e}{c}Hv_x + m\omega_0^2 y = 0,$$

where $\omega_0 = \sqrt{e^2/mr^3}$ is the angular frequency of rotation in the orbit. The motion along z is unaffected. By transforming to

$$x(t) = r \cos \omega t, \quad y(t) = r \sin \omega t$$

one writes

$$\frac{dv_x}{dt} = -r\omega^2 \cos \omega t; \quad \frac{dv_y}{dt} = -r\omega^2 \sin \omega t.$$

From the equations of motion the equation for ω

$$\omega^2 - 2\omega_L \omega - \omega_0^2 = 0,$$

is found ($\omega_L = eH/2mc$), yielding (for the positive root) $\omega = \sqrt{\omega_0^2 + \omega_L^2} + \omega_L$.

Since, from order of magnitude estimates (see Sect. 4.1) $\omega_L^2 \ll \omega_0^2$, Eq. (4.2) follows. See also Problem 4.3.

Problem 4.2 In the classical model for the atom and for the electromagnetic radiation source (*Thomson* and *Lorentz* models) the electron was thought as an harmonic oscillator, oscillating around the center of a sphere of uniform positive charge (see Problem 1.5). Show that the electric polarizability $\alpha = e^2/k$ has to be expected, with effective elastic constant $k = 4\pi\rho e/3$, ρ being the (uniform) positive charge density.

By resorting to the second-order perturbative derivation of the polarizability for the quantum oscillator show that the same result is obtained and that it is indeed the exact result.

Solution: The restoring force is $F = -(4\pi x^3 \rho/3)e/x^2$ and then $k = 4\pi\rho e/3$, corresponding for the electron to an oscillating frequency $\nu_0 = (1/2\pi)\sqrt{k/m} \simeq 2.53 \times 10^{15} \, \text{s}^{-1}$. From $e\mathcal{E} = F = kx$ and dipole moment $ex = e^2\mathcal{E}/k$, $\alpha = e^2/k$ follows.

From Eq. (4.8), with perturbation Hamiltonian $\mathcal{H}_p = -e\mathcal{E}z$ and quantum oscillator ground and excited states

$$\alpha = 2e^2 \sum_{exc \neq f} \frac{|<exc|z|f>|^2}{E^0_{exc} - E^0_f}.$$

From the matrix elements $<v|z|v-1> = \sqrt{v\hbar/2m\omega}$ (according to the properties of *Hermite polynomials*, see Sect. 10.3.1) only the first excited state $|exc>$ has to be taken into account. Then

$$\alpha = \frac{2e^2}{\hbar} \frac{|<f+1|z|f>|^2}{\omega_0} = \frac{2e^2}{\hbar} \frac{\hbar}{2m\omega_0^2} = \frac{e^2}{k}$$

The proof that this is the *exact result* is achieved by rewriting the Hamiltonian of the linear oscillator to include the electric energy $ez\mathcal{E}$ and observing that a shift of the eigenvalues by $-(e\mathcal{E})^2/2k$ occurs (see the analogous Problem 10.16 for the vibrational motion of molecules, where it is also shown that α does not depend from the state $|v>$ of the oscillator).

4.3 Hamiltonian in Magnetic Field

From Eqs. (4.3) and (4.5), by including now the spin-orbit interaction, the perturbation of the central field Hamiltonian for multi-electron atoms is written

$$\mathcal{H}_P^{(1)} = \mu_B H \sum_i l_z^i + 2\mu_B H \sum_i s_z^i + \sum_i \xi_{nl}^i \mathbf{l}_i \cdot \mathbf{s}_i. \qquad (4.20)$$

The term

$$\mathcal{H}_P^{(2)} = \sum_i \frac{e^2 A_i^2}{2mc^2} \qquad (4.21)$$

has been left out: it shall be taken into account in discussing the diamagnetism (Sect. 4.5). In writing Eq. (4.20) we have used the interaction in the form $-\boldsymbol{\mu}_{l,s} \cdot \mathbf{H}$, as it has been proved possible at Sect. 1.6. The magnetic field is considered static, homogeneous and applied along the z-direction.

One could emphasize that in the hypothetical absence of the spin Eq. (4.20) would reduce to

$$\mathcal{H}_P^{(1)} = \mu_B H L_z \qquad (4.22)$$

implying corrections to the energy levels in the form $\Delta E = \mu_B M H$. Therefore, in the light of the selection rule $\Delta M = 0, \pm 1$ (see Sect. 3.5), one realizes that for a given emission line the magnetic field should induce a triplet, characteristic of the so-called *normal Zeeman effect* (this terminology being due to the fact that for such a triplet an explanation in terms of classical Lorentz oscillators appeared possible, see Problem 4.3). The experimental observation that the effect of the magnetic field on the spectral lines is more complex, as shown in the following, can be considered stringent evidence for the existence of the spin. The real *Zeeman effect* (at first erroneously considered as "*anomalous*") in general does not consists in a triplet (see the case of the Na doublet in the following). The triplet actually can occur, in principle, in the presence of very strong field (*Paschen-Back effect*), as we shall see at Sect. 4.3.2.

4.3.1 Zeeman Regime

In order to derive the energy of the atom from the Hamiltonian 4.20 one has to consider the relative magnitude of the terms $\mu_B H$ (magnetic field energy) and ξ_{nl} (spin-orbit energy). In the *weak field regime*, for $\mu_B H \ll \xi_{nl}$ and in the **LS** coupling scheme, the Hamiltonian is considered in the form

$$\mathcal{H}_P^{(1)} = \mu_B \mathbf{H} \cdot (\mathbf{L} + 2\mathbf{S}) \tag{4.23}$$

and acting as a perturbation on the states $|E_0, J, M_J >$ resulting from the central field Hamiltonian, with the coupling $\sum_i \mathbf{l}_i$ and $\sum_i \mathbf{s}_i$ and the spin-orbit interaction in the form $\xi_{LS} \mathbf{L} \cdot \mathbf{S}$.

The operator $(\mathbf{L} + 2\mathbf{S})$ has to be projected along $\mathbf{J}$ by using *Wigner-Eckart* theorem

$$< E_0, J, M'_J | L_z + 2S_z | E_0, J, M_J > = g < E_0, J, M'_J | J_z | E_0, J, M_J > = g M_J \delta_{M'_J, M_J}, \tag{4.24}$$

the constant g being obtained from the component of $(\mathbf{L} + 2\mathbf{S})$ along $\mathbf{J}$:

$$\begin{aligned}
g &= < E_0, J, L, S \left| \frac{(\mathbf{L} + 2\mathbf{S}) \cdot \mathbf{J}}{J^2} \right| E_0, J, L, S > = \\
&= < E_0, J, L, S \left| \frac{(\mathbf{L} + \mathbf{S}) \cdot \mathbf{J} + \mathbf{S} \cdot \mathbf{J}}{J^2} \right| E_0, J, L, S > = \\
&= 1 + \frac{J(J+1) + S(S+1) - L(L+1)}{2J(J+1)}
\end{aligned} \tag{4.25}$$

This result is in close agreement with the deduction of the *Lande' factor* within the vectorial coupling model (Sect. 3.2.2). Then the energy corrections are given by

$$\Delta E = \mu_B H g M_J, \tag{4.26}$$

the result that one would anticipate by assigning to the atom a magnetic moment $\mu_J = -\mu_B g \mathbf{J}$ and by writing the perturbation Hamiltonian as $\mathcal{H}_P = -\mu_J.\mathbf{H}$.

As a consequence of Eqs. (4.25) and (4.26), in general the structure of the atomic levels in the magnetic field, in the Zeeman regime, is more complicated than the one for $S = 0$. The spectral lines are modified in a form considerably different from a triplet. At the sake of illustration, the case of the Na doublet D_1 and D_2 is schematically reported in Fig. 4.4. By taking into account the selection rules $\Delta M_J = 0, \pm 1$, for Na coinciding with the ones for single electron (see Sect. 2.1), also the polarization of the emission lines is justified.

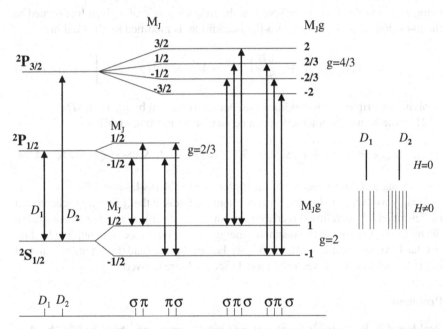

Fig. 4.4 Structure of the $^2S_{1/2}$ ground-state and of the ^2P doublet of Na atom in a magnetic field and transitions allowed by the electric dipole selection rules $\Delta S = 0$, $\Delta J = 0, \pm 1$ and $\Delta M_J = 0, \pm 1$. The D_1 line splits into four components, the D_2 line into six. Similar structure of the levels hold for the other alkali atoms. On increasing the magnetic field strength the structure of the lines, here shown for the *weak field regime*, progressively changes towards a central π line and two σ^+ and σ^- doublets (see Problem 4.7). π lines correspond to $\Delta M_J = 0$, while σ lines to $\Delta M_J = \pm 1$

4.3.2 Paschen-Back Regime

When the strength of the magnetic field is increased the structure of the spectral lines predicted within the **LS** coupling model and weak field condition is progressively altered and in the limit of very strong field the condition of a triplet (as one would expect for $S = 0$) is restored. This crossover is related to the fact that for $\mu_B H \gg \xi_{LS}$ the effect of the magnetic perturbation has to be evaluated for unperturbed states characterized by quantum numbers M and M_S pertaining to L_z and S_z, while the spin-orbit interaction can be taken into account only as a subsequent perturbation. This is the so-called *Paschen-Back*, or *strong field*, *regime*.

From the field-related Hamiltonian in Eq. (4.23), in a way similar to the derivation within the vectorial model (see Sect. 3.2), the energy correction turns out

$$\Delta E = \mu_B H (M + 2 M_S). \qquad (4.27)$$

From the selection rules $\Delta M = 0, \pm 1$ and $\Delta M_S = 0$ (the spin-orbit interaction being absent at this point) one sees that the frequency $\nu_{12}^{(0)}$ of a given line related to the transition $|2> \to |1>$ in zero-field condition, is modified by the field in

$$\nu_{12}^{(H)} = \nu_{12}^{(0)} + \frac{\mu_B H}{h} \left[(M^2 - M^1) + 2(M_S^2 - M_S^1) \right] \qquad (4.28)$$

implying the triplet, with two lines symmetrically shifted by $(e/4\pi mc)H$.

Then the spin-orbit interaction can be taken into account, yielding

$$\Delta E' = \xi_{LS} < |L_x S_x + L_y S_y + L_z S_z| > = \xi_{LS} < |L_z S_z| > = \xi_{LS} M M_S \qquad (4.29)$$

and causing a certain structure of the triplet (see Problems 4.4 and 4.7).

Finally we mention that the effect of magnetic fields in the **jj** coupling scheme can be described by operating directly on the single-electron **j** moment and considering the relationship between the magnetic energy and the inter-electron coupling leading to total **J**. Again one has to use the Wigner-Eckart theorem and the results anticipated in the framework of the vectorial model (Sect. 3.3) are derived.

Problems

Problem 4.3 By taking into account the Larmor precession (Problem 4.1), the classical picture of the Lorentz radiation in magnetic field implies a triplet for observation perpendicular to the field and a doublet for longitudinal observation.

Discuss the polarization of the radiation in terms of the selection rules for the quantum magnetic number.

Solution: The sketch of the experimental observation for classical oscillator in a magnetic field is given below:

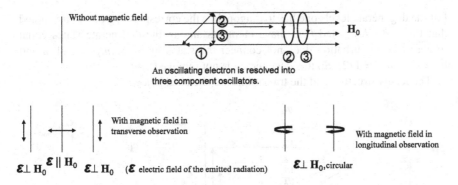

An oscillating electron is resolved into three component oscillators.

With magnetic field in transverse observation

With magnetic field in longitudinal observation

$\mathcal{E}\perp \mathbf{H}_0$ $\mathcal{E}\parallel \mathbf{H}_0$ $\mathcal{E}\perp \mathbf{H}_0$ ($\mathcal{E}$ electric field of the emitted radiation) $\mathcal{E}\perp \mathbf{H}_{0,\text{circular}}$

The frequency shift $\delta\omega$ can be calculated as shown in Problem 4.1. The frequency of the electron oscillating in the z direction (see sketch above) remains unchanged. The equations for x and y can be written in terms of $u = x + iy$ and $v = x - iy$, to find for $\omega_0 \gg \omega_L$

$$u = u_0 e^{i\left(\omega_0 + \frac{eH_0}{2mc}\right)t} \quad \text{and} \quad v = v_0 e^{i\left(\omega_0 - \frac{eH_0}{2mc}\right)t},$$

namely the equations for left-hand and right-hand circular motions at frequencies $\omega_0 \pm \delta\omega$, with $\delta\omega = eH_0/2mc$. The oscillators 2 and 3 in the sketch above have to emit or absorb radiation at frequency $(\omega_0 \pm \delta\omega)$, *circularly polarized* when detected along $\mathbf{H}_0$.

Oscillator 1 is along the field and therefore the intensity of the radiation is zero along that direction. If the radiation from the oscillators 2 and 3 is observed along the perpendicular direction is *linearly polarized*.

The polarizations of the Zeeman components have their quantum correspondence in the $\Delta M_J = 0$ and $\Delta M_J = \pm 1$ transitions. These rules are used in the so-called *optical pumping*: the exciting light is polarized in a way to allow one to populate selectively individual Zeeman levels, thus inducing a given orientation of the magnetic moments (somewhat equivalent to the magnetic resonance, see Chap. 6).

Problem 4.4 Illustrate the Paschen-Back regime for the $2P \longleftrightarrow 2S$ transition in Lithium atom, by taking into account *a posteriori* the spin-orbit interaction. Sketch the levels structure and the resulting transitions, with the correspondent polarizations.

Solution: The degenerate block

$$\mu_B H < nlm'm_s'|l_z + 2s_z|nlmm_s> = \mu_B H(m + 2m_s)$$

is diagonal. The degeneracy is not completely removed. For the non-degenerate levels the spin-orbit interaction yields the correction

$$\xi_{nl} < mm_s|l_z s_z|mm_s> = \hbar^2 \xi_{nl} m m_s.$$

For the degenerate levels one has to diagonalize the corresponding block. It is noted that the terms l_+s_- and l_-s_+ have elements among the degenerate states equal to zero (the perturbation does not connect the states $m = 1$, $m_s = -1/2$ and $m = -1$, $m_s = 1/2$). So this degeneracy is not removed.

The levels structure and the transitions are sketched below:

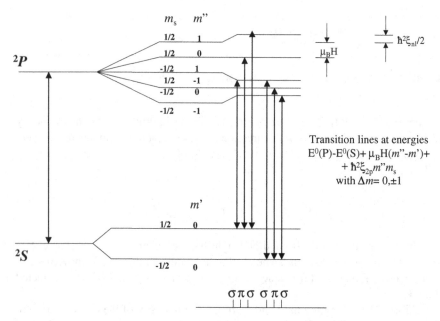

Transition lines at energies
$E^0(P)-E^0(S)+\mu_B H(m''-m')+$
$+\hbar^2\xi_{2p}m''m_s$
with $\Delta m= 0,\pm 1$

Problem 4.5 Evaluate the shift of a spectral line at $\lambda = 1894.6\,\text{Å}$ due to the transition from the $^1P_1$ to the $^1S_0$ state when a magnetic field of 1 T is applied.

Solution: The magnetic field gives rise to the triplet ($\Delta M_J = 0,\pm 1$) and the separation between the components is

$$\Delta\nu = \frac{E_1 - E_0}{h} = \frac{g\mu_B H}{h} \simeq 1.4 \cdot 10^{10}\,\text{Hz}$$

From $\Delta\lambda \simeq -(\lambda_0/\nu_0)\Delta\nu = -(\lambda_0^2/c)\Delta\nu$ one has $\Delta\lambda \simeq 1.67 \times 10^{-2}\,\text{Å}$.

Problem 4.6 Show that in the low-energy state of the positronium atom ($^1S_0$ and $^3S_1$) no Zeeman effect occurs (the magnetic moment of the positron is $\mu_p = \mu_B g S^p$).

Solution: The Hamiltonian is

$$\mathcal{H} = -(\mu_e + \mu_p) \cdot \mathbf{H} = a(S_z^e - S_z^p),$$

with $a = \mu_B g H$. From the energy corrections

$$E = a < \phi|S_z^e - S_z^p|\phi >$$

since in the singlet state the spin eigenfunction is antisymmetric and the operator $(S_z^e - S_z^p)$ is antisymmetric, the matrix element must be zero. The same is true also for the triplet state $^3S_1$.

A more formal proof can be obtained by applying the operator $(S_z^e - S_z^p)$ on the four spin eigenfunctions $\alpha_p \alpha_e$, $\beta_p \beta_e$, etc. for the two particles.

Problem 4.7 From the Paschen-Back structure of the $D_{1,2}$ doublet of Sodium atom imagine to decrease the magnetic field until the Zeeman weak field regime is reached. Classify the states and connect the levels in the two regimes.

Solution:

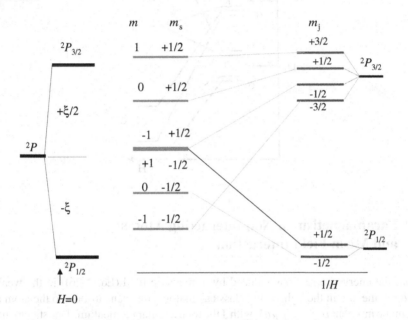

Paschen-Back regime

$$\Delta E = \mu_B H (m + 2m_s)$$

with $E_0(n, l, m, m_s)$

Zeeman regime

$$\Delta E = g \mu_B H m_j$$

with $E_0(n, l, j, m_j)$

Upon increasing the field (from right to left in the figure) the D_1 and D_2 lines (Fig. 4.4) modify their structures as schematically shown below:

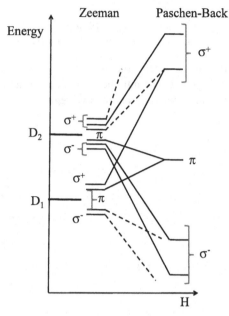

4.4 Paramagnetism of Non-interacting Atoms and Mean Field Interaction

From the energy corrections induced by a magnetic field (Eq. (4.26)) in the weak field regime and in the light of the classical analogy, one can attribute to the atom a magnetic moment $\mu_J = -\mu_B g\mathbf{J}$, with $\mathbf{J}$ the total angular momentum. This statement, already used in the vectorial description at Sect. 3.2, is at the basis of the theory for the magnetic properties of matter.

As illustrative example we shall show how the magnetic properties of an assembly of atoms can be derived by referring to the statistical distribution on the levels, when the thermal equilibrium at a given temperature T is achieved. The atoms will first be considered as non-interacting (the only weak interactions occurring with the other degrees of freedom of the thermal reservoir, so that statistical equilibrium can actually be attained).

In the absence of field, degeneracy in the magnetic quantum number M_J occurs, pictorially corresponding to equiprobable orientations of the magnetic moments with respect to a given z-direction, as sketched in Fig. 4.5. When the field is switched on, in a characteristic time usually called *spin-lattice relaxation time* T_1 (for some detail on this process see Chap. 6), statistical equilibrium is achieved, with the populations

Fig. 4.5 Pictorial sketch of non-interacting atomic magnetic moments in the absence (**a**) and in the presence (**b**) of the field. The field removes the degeneracy in M_J and after some time (of the order of T_1) the statistical distribution yields an excess population on the low energy levels so that an effective component of the magnetic moment along the field is induced

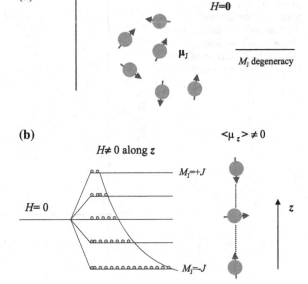

on the magnetic levels as depicted in Fig. 4.5b and with an average (statistical) expectation value of the magnetic moment along the field $< \mu_z > \neq 0$.

$< \mu_z >$ is written

$$< \mu_z > = -g\mu_B \frac{\sum_{M_J} M_J e^{-xM_J}}{\sum_{M_J} e^{-xM_J}} \qquad (4.30)$$

where $x = g\mu_B H / k_B T$. For $x \ll 1$ one has

$$< \mu_z > \simeq -g\mu_B \frac{\sum_{M_J} M_J (1 - xM_J)}{\sum_{M_J} (1 - xM_J)}$$

and since $\sum_{M_J} M_J^2 = J(J+1)(2J+1)/3$,

$$< \mu_z > = g\mu_B x \frac{J(J+1)(2J+1)}{3(2J+1)} = \frac{\mu_J^2 H}{3k_B T} \qquad (4.31)$$

with

$$|\mu_J| = g\mu_B \sqrt{J(J+1)}.$$

Fig. 4.6 Normalized value
of the effective magnetic
moment along the field
direction as a function of the
dimensionless variable
$(g\mu_B H/k_B T)$, according to
Eq. (4.32), for different J's

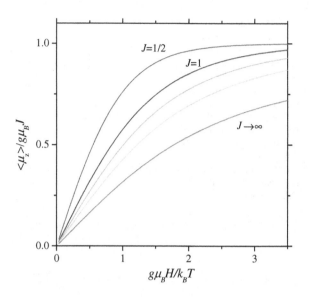

The volume paramagnetic susceptibility is $\chi = N\chi_a$, with N number of atoms
per unit volume and χ_a atomic susceptibility, given by $\chi_a = \mu_J^2/3k_B T$, according
to Eq. (4.31). Thus the quantum derivation of the Curie law has been obtained.
Without the approximation of low field (or high temperature), Eq. (4.30) gives

$$< \mu_z > = g\mu_B J\left[\frac{2J+1}{2J}coth\frac{(2J+1)x}{2} - \frac{1}{2J}coth\frac{x}{2}\right], \qquad (4.32)$$

the function depicted in Fig. 4.6 and known as *Brillouin function*. For $J \to \infty$, the
Brillouin function becomes the *Langevin function*, while for $J = 1/2$ it reduces to
$tanh(x/2)$.

The saturation magnetization $M_{sat} = N < \mu_z >_{T \to 0}$ corresponds to the situation where all the atoms are found on the lowest energy level of Fig. 4.5b and
$(< \mu_z >)_{T \to 0} = g\mu_B J$.

According to Eq. (4.31) on decreasing temperature the paramagnetic susceptibility
(in evanescent field) diverges as $1/T$. However, when the temperature is approaching
zero so that the condition $x \ll 1$ no longer holds, partial saturation is achieved and χ
reaches a maximum and then decreases on cooling. In practice this can happen only
in strong fields (of the order of several Tesla) and at low temperature.

The reference to an ideal paramagnet in practice corresponds to the assumption
that the local magnetic field is the one externally applied (apart from the diamagnetic
correction, see Sect. 4.5). This condition does not hold when some type of interaction
among the atomic magnetic moments is active, as it is common in crystals with
magnetic ions. In this case the susceptibility can diverge at finite temperature, as
sketched in Fig. 4.7.

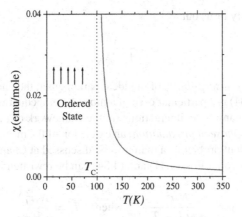

Fig. 4.7 Sketchy behavior of the temperature dependence of the paramagnetic susceptibility in presence of interactions among the magnetic moments. The state below T_c corresponds to spontaneous ordering of the magnetic moments along a given direction as a consequence of a cooperative process, typical of *phase transitions* in *many-body systems*, driven by the interaction among the components (see Chaps. 15 and 17)

A simple method to deal with the interactions is the *mean field approximation*, namely to assume that the local field is the external one $\mathbf{H}_{ext}$ plus a second contribution, related to the interactions, proportional to the magnetization:

$$\mathbf{H} = \mathbf{H}_{ext} + \lambda \mathbf{M}$$

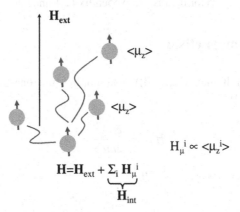

Then the magnetization reads

$$M = N \left[\chi_0 (\mathbf{H}_{ext} + \mathbf{H}_{int}) \right] \tag{4.33}$$

and the susceptibility turns out

$$\chi = \frac{\chi_0}{1 - \lambda\chi_0} \qquad (4.34)$$

where χ_0 is the *bare susceptibility* of the ideal paramagnet, the one without interactions. Equation (4.34) is a particular case of a more general equation, for any system in the presence of many-body interactions (in the framework of the *linear response theory* and *random phase approximation*, an extension of the mean field approximation to time-dependent problems, a matter to be discussed at Chap. 17).

By taking into account Eqs. (4.31) and (4.34) can be rewritten in the form

$$\chi = \frac{N\mu_J^2}{3k_B(T - T_c)}, \quad \text{where} \quad T_c = \frac{N\mu_J^2\lambda}{3k_B} \qquad (4.35)$$

For $T \to T_c^+$ one has the divergence of the magnetic response and a *phase transition* to an ordered state, with spontaneous magnetization in zero field, is induced. Typical transition is the one from the paramagnetic to the ferromagnetic state and it can be expected to occur when the thermal energy k_BT is of the order of the interaction energy.

It is noted that the values of T_c's in most ferromagnets (as high as $T_c = 1044$ K, for instance for Fe bcc), indicate that the transition is driven by interactions much stronger than the dipolar one. This latter, in fact, for an interatomic distance d of the order of 1 Å, would imply $T_c \sim \mu_J^2/d^3 k_B$, of the order of a few degrees K. Instead the interaction leading to the ordered states (ferromagnetic or antiferromagnetic, depending on the sign of λ in Eq. (4.34)) is the one related to the *exchange integral*, as mentioned at Sect. 2.2 (for details see Appendix 13.1 and Chap. 17).

4.5 Atomic Diamagnetism

The magnetic Hamiltonian (Eq. (4.3)) also implies the one-electron term (see Eqs. (4.21) and (4.5))

$$\mathcal{H}_P^{(2)} = \frac{e^2 A^2}{2mc^2}$$

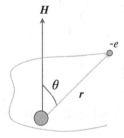

with $\mathbf{A} = (1/2)\mathbf{H} \times \mathbf{r} = (1/2)Hr\sin\theta$, the term usually neglected in comparison with the one linear in the field and leading to paramagnetism. Instead $\mathcal{H}_P^{(2)}$ is
responsible of the atomic *diamagnetism*.

Let us refer to atoms in the ground state where $\mu_L = \mu_S = 0$. The effect of
$\mathcal{H}_P^{(2)}$ can be evaluated in the form of perturbation for states having L, S, M and
M_S as good quantum numbers, the spin-orbit interaction being absent. Thus, from
first-order perturbation theory the energy correction due to $\mathcal{H}_P^{(2)}$ is

$$\Delta E = \frac{e^2 H^2}{8mc^2} \sum_i \; < |r_i^2 \sin^2\theta_i| >$$ (4.36)

where the sum is over all the electrons. By resorting to $\mu = -(\partial E / \partial H)$, Eq. (4.36)
implies an atomic magnetic moment *linear* in the field and in the *opposite direction*.
Therefore the diamagnetic susceptibility is written

$$\chi_{dia} = -N\frac{e^2}{4mc^2} \sum_i \frac{2}{3} < r_i^2 >$$ (4.37)

(N number of atoms per unit volume), the assumption of isotropy having been made,
so that $< x^2 > = < y^2 > = 1/3 < r^2 >$. In the Table below the molar diamagnetic
susceptibilities for inert-gas atoms (to a good approximation the same values apply
in condensed matter) are reported:

	He	Ne	Ar	Kr	Xe
$\chi_{dia}(cm^3/mole)(\times 10^{-6})$	−2.36	−8.47	−24.6	−36.2	−55.2
Z	2	10	18	36	54

When the perturbation effects from the magnetic Hamiltonian are extended up to
the second order, a mixture of states is induced and a further energy correction is
obtained, *quadratic in the field* and causing a *decrease* of the energy. Thus, even
in atoms where in the ground state no paramagnetic moment is present, a *positive*
paramagnetic-like susceptibility (*Van Vleck paramagnetism*) of the form

$$\chi_{vv} = 2N\mu_B^2 \sum_{n\neq 0} \frac{|< \phi_0|(L_z + 2S_z)|\phi_n >|^2}{E_n^0 - E_0^0}$$ (4.38)

is found. For a quantitative estimate the electronic wavefunctions ϕ_0 of the ground
and of the excited states ϕ_n are required. The Van-Vleck susceptibility is usually
temperature-independent and small with respect to Curie susceptibility.

Problems

Problem 4.8 Evaluate the molar diamagnetic susceptibility of Helium in the ground state, by assuming Hydrogen-like wavefunctions with the effective nuclear charge derived in the variational procedure (Problem 2.6). Estimate the variation of the atom energy when a magnetic field of 1 T is applied.

Solution: From Eq. (4.37)

$$\chi_{dia} = -\frac{Ne^2}{6mc^2}[< r_1^2 > + < r_2^2 >]$$

and in hydrogenic atoms (Table 1.4) $< r^2 > = 3\left(\frac{a_0}{Z}\right)^2$. For effective charge $Z^* = Z - \frac{5}{16} = \frac{27}{16}$

$$\chi_{dia} \simeq -1.46 \cdot 10^{-6} \frac{emu}{mole}.$$

The energy variation is $\Delta E = (e^2 H^2/12mc^2)[< r_1^2 > + < r_2^2 >] \simeq 10^{-10}\,eV$, very small compared to the ground state energy.

Problem 4.9 In a diamagnetic crystal Fe^{3+} paramagnetic ions are included, with density $d = 10^{21}$ ions/cm^3. By neglecting interactions among the ions and the diamagnetic contribution, derive the magnetization at $T = 300\,K$, in a magnetic field $H = 1000\,Oe$. Then estimate the magnetic contribution to the specific heat (per unit volume).

Solution: From Problem 3.10 for Fe^{3+} in the ground state the effective magnetic moment is $\mu = p\mu_B$ with $p = g\sqrt{J(J+1)} = \sqrt{35}$.
From Eq. (4.31)

$$M = d\frac{\mu^2 H}{3k_B T} = 0.0242 \text{ erg cm}^{-3}\,Oe^{-1}.$$

The energy density is $E = -\mathbf{M} \cdot \mathbf{H}$ and for $\mu_B H \ll k_B T$ the specific heat is

$$C_V = \left(\frac{\partial E}{\partial T}\right)_V = d\frac{\mu^2 H^2}{3k_B T^2} = 0.081 \text{ erg K}^{-1}\,cm^{-3}.$$

Problem 4.10 For non-interacting spins in external magnetic field, in the assumption of high temperature, derive the Curie susceptibility from the density matrix for the expectation value of the effective magnetic moment.

Solution: The density matrix is

$$\rho = \frac{1}{Z}exp\left(-\frac{\mathcal{H}_{Zeeman}}{k_B T}\right),$$

where Z is the partition function, namely the sum over all the n−states of the statistical factors $exp(-E_n/k_BT)$, or $Z = Tr\{exp(-\frac{\mathcal{H}}{k_BT})\}$. Then the quantum and statistical average (see Sect. 6.1) is written

$$\overline{<\mu_z>} = \frac{1}{Z}Tr\left\{\mu_z exp\left(-\frac{\mathcal{H}_{Zeeman}}{k_BT}\right)\right\} \simeq \frac{1}{Z}Tr\left\{\mu_z\left(1-\frac{\mathcal{H}_{Zeeman}}{k_BT}\right)\right\}$$

with

$$\mu_z = -g\mu_B S_z \qquad \mathcal{H}_{Zeeman} = S_z g\mu_B H.$$

Since

$$Tr\, S_z^2 = \frac{1}{3}(S+1)S(2S+1) \quad \text{and} \quad Z \simeq 2S+1$$

one obtains for the single particle susceptibility

$$\chi = \frac{S(S+1)g^2\mu_B^2}{3k_BT}$$

(as in Eq. (4.31) for $S \equiv J$).

Appendix 4.1 Electromagnetic Units and Gauss System

Throughout this book we are using the CGS system of units that when involving the electromagnetic quantities is known as the *Gauss system*. This system corresponds to have assumed for the dielectric constant ε_0 and for the magnetic permeability μ_0 of the vacuum the dimensionless values $\varepsilon_0 = \mu_0 = 1$, while the velocity of light in the vacuum is necessarily given by $c = 3 \times 10^{10}$ cm/s.

As it is known, the most common units in practical procedures (such as Volt, Ampere, Coulomb, Ohm and Faraday) are better incorporated in the MKS system of units (and in the international SI). These systems of units are derived when in the Coulomb equation instead of assuming as arbitrary constant $k = 1$, one sets $k = 1/4\pi\varepsilon_0$, with $\varepsilon_0 = 8.85 \times 10^{-12}$ C^2/Nm^2, as electrical permeability of the vacuum. In the SI system the magnetic field **B**, defined through the Lorentz force

$$\mathbf{F} = q\mathbf{E} + q\mathbf{v} \times \mathbf{B}$$

is measured in Weber/m^2 or *Tesla*.

The auxiliary field **H** is related to the current due to the free charges by the equation $H = nI$, corresponding to the field in a long solenoid with n turns per meter, for a current of I Amperes. The unit of H is evidently Ampere/m. Thus in the vacuum one has $\mathbf{B} = \mu_0\mathbf{H}$, with $\mu_0 = 4\pi 10^{-7}$ N/A$^2 \equiv 4\pi 10^{-7}$ H/m.

In the matter the magnetic field is given by

$$\mathbf{B} = \mu_0(\mathbf{H} + \mathbf{M})$$

where $\mathbf{M}$ is the magnetic moment per unit volume.

The SI system is possibly more convenient in engineering and for some technical aspects but it is not suited in physics of matter. In fact, the Maxwell equations in the vacuum are symmetric in the magnetic and electric fields only when $\mathbf{H}$ is used, while $\mathbf{B}$ and not $\mathbf{H}$ is the field involved in the matter. The SI system does not display in a straightforward way the electromagnetic symmetry. In condensed matter physics the Gauss system should be preferred.

Thus within this system the electric and magnetic fields have the same dimensions (another appealing feature), the Lorentz force is

$$\mathbf{F} = q\mathbf{E} + \frac{q}{c}\mathbf{v} \times \mathbf{B}, \tag{A.4.1.1}$$

$\mathbf{B}$ is related to $\mathbf{H}$ by

$$\mathbf{B} = \mathbf{H} + 4\pi\mathbf{M} = \mu\mathbf{H} \qquad \text{with} \qquad \mu = 1 + 4\pi\chi. \tag{A.4.1.2}$$

$\mathbf{M} = \chi\mathbf{H}$ defines the dimensionless *magnetic susceptibility* χ. For $\mu_B H \ll k_B T$, often called *evanescent field condition*, χ is field independent. As already mentioned μ_0 and ε_0 are equal to unit, dimensionless.

The practical units can still be used, just by resorting to the appropriate conversion factors, such as

1 volt →	$\frac{1}{299.8}$ statvolt or erg/esu (esu electrostatic unit)
1 ampere	2.998×10^9 esu/sec
1 Amp/m	$4\pi \times 10^{-3}$ Oersted (see below)
1 ohm	1.139×10^{-12} sec/cm
1 farad	0.899×10^{12} cm
1 henry	1.113×10^{-12} sec²/cm
1 Tesla	10^4 Gauss
1 Weber	10^8 Gauss/cm²

The Bohr magneton, which is *not an SI unit*, is often indicated as $\mu_B = 9.274 \times 10^{-24}$ J/T, equivalent to our definition $\mu_B = 0.9274 \times 10^{-20}$ erg/Gauss. The gyromagnetic ratio is measured in the Gauss system in (rad/s.Oe) and in the SI system in (rad.m/Amp.s).

Unfortunately, some source of confusion is still present when using the Gauss system. According to Eq. (A.4.1.2), $\mathbf{B}$ and $\mathbf{H}$ have the same dimensions and are related to the currents in the very same way. In spite of that, while $\mathbf{B}$ is measured in *Gauss*, without serious reason the unit of $\mathbf{H}$ is called *Oersted*. Furthermore, there are two ways to describe electromagnetism in the framework of the CGS system. One with electrostatic units *(esu)* and the other with electromagnetic units *(emu)*. The latter is usually preferred in magnetism. Thus the magnetic moment is measured in the *emu unit*, which is nothing else than a volume and therefore cm^3. The magnetic susceptibility (per unit volume) is dimensionless and often indicated as emu/cm^3. The symmetric Gauss-Hertz-Lorentz system (commonly known as Gauss system) corresponds to a mixing of the *esu* and of the *emu* systems, having assumed both $\varepsilon_0 = 1$ and $\mu_0 = 1$.

Here we do not have the aim to set the final word on the *vexata quaestio* of the most convenient system of units. Further details can be achieved from the books by *Purcell* and by *Blundell*.

A Table is given below for the magnetic quantities in the Gauss system and in the SI system, with the conversion factors.

Quantity	Symbol	Gauss	SI	Conversion factor*
Magnetic Induction	B	G $\equiv$ *Gauss*	T	10^{-4}
Magnetic field intensity	H	Oe	A m^{-1}	$10^3/4\pi$
Magnetization	M	erg/(G cm^3)	A m^{-1}	10^3
Magnetic moment	μ	erg/G($\equiv$ emu)	J/T($\equiv$ Am^2)	10^{-3}
Specific magnetization	σ	emu/g	A m^2/kg	1
Magnetic flux	ϕ	Mx (maxwell)	Wb (Weber)	10^{-8}
Magnetic energy density	E	erg/cm^3	J/m^3	10^{-1}
Demagnetizing factor	N_d	–	–	$1/4\pi$
Susceptibility(unit volume)	χ	–	–	4π
Mass susceptibility	χ_g	erg/(G g Oe)	m^3/kg	$4\pi \times 10^{-3}$
Molar susceptibility	χ_{mol}	emu/(mol Oe)	m^3/mol	$4\pi \times 10^{-6}$
Magnetic permeability	μ	G/Oe	H/m	$4\pi \times 10^{-7}$
Vacuum permeability	μ_0	G/Oe	H/m	$4\pi \times 10^{-7}$
Anisotropy constant	K	erg/cm^3	J/m^3	10^{-1}
Gyromagnetic ratio	γ	rad/(s Oe)	rad m/(A s)	$4\pi \times 10^{-3}$

*To obtain the values of the quantities in the SI, the corresponding Gauss values should be multiplied by the conversion factor

Finally a mention to the *atomic units* (a.u.), frequently used, is in order. In this system of units (derived from the SI system) one sets $e = \hbar = m = 1$ and $4\pi\varepsilon_0 = 1$. Thus the Bohr radius for atomic Hydrogen (infinite nuclear mass) becomes $a_0 = 1$, the ground state energy becomes $E_{n=1} = -1/2$ a.u., the a.u. for velocity is $v_0 = \alpha c$ with $\alpha \simeq 1/137$ the *fine structure constant*, so that the speed of light is $c \simeq 137$ a.u.. The Bohr magneton is $1/2$ a.u. and the *flux quantum* is $\Phi_0 = 2\pi$ (see Appendix 13.1). Less practical are the a.u.'s for other quantities. For instance the a.u. for the magnetic field corresponds to 2.35×10^5 T and the one for the electric field to 5.13×10^9 V/cm.

Problems

Problem 4.11 The magnetization curves for crystals containing paramagnetic ions Gd^{3+}, Fe^{3+} and Cr^{3+} display the saturation (for about $H/T = 20\,kGauss/K$) about at the values 7, 5, and $3\mu_B$ (per ion), respectively. From the susceptibility measurements at $T = 300$ K for evanescent magnetic field one evaluates the magnetic moments 7.9, 5.9, and $3.8\mu_B$, respectively. Comment on the differences. Then obtain the theoretical values of the magnetic moments for those ions and prove that *quenching of the orbital momenta* occurs (see Problem 4.12).

Solution: The susceptibility $\chi = Ng^2 J(J+1)\mu_B^2/3k_B T$ involves an effective magnetic moment $\mu_{eff} = g\mu_B\sqrt{J(J+1)}$ different from $<\mu_z>_{max} = g\mu_B J$ obtained from the saturation magnetization, related to the component of $\mathbf{J}$ along the direction of the field.

For Gd^{3+}, electronic configuration $(4f)^7$, one has $S = \frac{7}{2}$, $L = 0$, $J = \frac{7}{2}$ and $g = 2$.

Then $\mu_{eff} = g\mu_B\sqrt{S(S+1)} \simeq 7.9\,\mu_B$, while $<\mu_z>_{max} \simeq 2\mu_B 7/2 = 7\mu_B$, in satisfactory agreement with the data.

For Fe^{3+} (see Problem 4.9) $J = 5/2$ and $g = 2$ and then $\mu_{eff} = 5.92\,\mu_B$ and $<\mu_z>_{max} \simeq 5\,\mu_B$.

For Cr^{3+}, electronic configuration $(3d)^3$, $S = 3/2$, $L = 3$, $J = 3/2$ and $g = 2/5 = 0.4$. For unquenched $\mathbf{L}$ one would have $\mu_{eff} = (2/5)\mu_B\sqrt{15/4} = 0.77\,\mu_B$, while for $L = 0$, $\mu_{eff} = 2\mu_B\sqrt{15/4} \simeq 3.87\,\mu_B$.

Problem 4.12 By referring to the expectation value of l_z in $2p$ and $3d$ atomic states, in the assumption that the degeneracy is removed by crystal field, justify the quenching of the orbital momenta.

Solution: When the degeneracy is removed the wavefunction $\phi_{2px,y,z}$ *are real.* Then, since

$$< l_z > = -i\hbar \int \phi^* \frac{\partial}{\partial\varphi}\phi d\tau$$

cannot be imaginary, one must have $< l_z > = 0$. Analogous consideration holds for $3d$ states and for any non-degenerate state. Details on the role of the crystal field in quenching the expectation values of the components of the angular momenta are given at Sect. 13.3 and at Problem 13.5.

Problem 4.13 In Hydrogen, the lines resulting from the transitions $^2P_{3/2} \longrightarrow$ $^2S_{1/2}$ and $^2P_{1/2} \longrightarrow$ $^2S_{1/2}$ (see Appendix 5.1) occur at $(1210 - 3.54 \cdot 10^{-3})$ Åand $(1210 + 1.77 \cdot 10^{-3})$ Å, respectively. Evaluate the effect of a magnetic field of 500 Gauss, by estimating the shifts in the wavelengths of these lines, in the *weak field* regime.

Solution: The relationship between the splitting of lines and the applied field is found from

$$dE_2 - dE_1 = -\frac{hc}{\lambda^2} d\lambda,$$

namely

$$d\lambda = -\frac{\lambda^2}{hc}(dE_2 - dE_1) = \left(-118\frac{\text{Å}}{\text{eV}}\right)(dE_2 - dE_1)$$

The values for dE_2 and dE_1 are given in Table below. There are 10 transitions that satisfy the electric dipole selection rule $\Delta M_J = \pm 1, 0$. The deviation of each of these lines from $\lambda_0 = 1210\,\text{Å}$ is also given.

$d\lambda_0$ Å$\cdot 10^{-3}$	dE_2 eV$\cdot 10^{-5}$	dE_1 eV$\cdot 10^{-5}$	$d\lambda$ Å$\cdot 10^{-3}$	$d\lambda_T = d\lambda_0 + d\lambda$ Å$\cdot 10^{-3}$
-3.54	$+0.579$	$+0.289$	-0.342	-3.88
-3.54	$+0.193$	$+0.289$	$+0.114$	-3.43
-3.54	$+0.193$	-0.289	-0.570	-4.11
-3.54	-0.193	$+0.289$	$+0.570$	-2.97
-3.54	-0.193	-0.289	-0.114	-3.65
-3.54	-0.579	-0.289	$+0.342$	-3.20
1.77	$+0.096$	$+0.289$	$+0.228$	$+2.00$
1.77	$+0.096$	-0.289	-0.456	$+1.31$
1.77	-0.096	$+0.289$	$+0.456$	$+2.23$
1.77	-0.096	-0.289	-0.228	$+1.54$

Problem 4.14 Refer to the H_α line in Hydrogen (see Problem 1.4). From the splitting of the s and p levels when a magnetic field of 4.5 T is applied, by taking into account that the separation between two adjacent lines is $6.29 \cdot 10^{10}$ Hz and by ignoring the fine structure, evaluate the specific electronic charge (e/m). Compare the estimate with the one obtained from the observation that a field of 3 T induces the splitting of the spectral line in Ca atom at 4226 Å in a triplet with separation 0.25 Å (do not consider in this case the detailed structure of the energy levels).

Solution: In the Paschen-Back regime the energy correction is

$$\Delta E_{m,m_s} = \mu_B H_0(m + 2m_s),$$

with electric dipole selection rules. One observes three lines with splitting $\Delta\bar{\nu} = 2.098\,\text{cm}^{-1}$. Then from

$$\Delta E = \mu_B H_0 = \frac{e\hbar}{2mc}H_0 = h\Delta\nu$$

one has

$$\frac{|e|}{m} = \frac{4\pi \Delta \nu c}{H_0} = 5.28 \times 10^{17} \frac{\text{u.e.s.}}{g}.$$

For Ca, from $\Delta E \simeq e\hbar H/2mc = -hc\Delta\lambda/\lambda^2$ again

$$\frac{|e|}{m} = \frac{4\pi c^2}{H} \frac{\Delta\lambda}{\lambda^2} \simeq 5.2 \cdot 10^{17} \frac{\text{u.e.s.}}{g}.$$

Problem 4.15 Two particles at spin $s = 1/2$ and magnetic moments $a\boldsymbol{\sigma}_1$ and $b\boldsymbol{\sigma}_2$, with $\boldsymbol{\sigma}_{1,2}$ spin operators, interact through the Heisenberg exchange Hamiltonian (see Sect. 2.2). Derive eigenstates and eigenvalues in the presence of magnetic field.

Solution: From the Hamiltonian

$$\mathcal{H} = -\frac{(a+b)}{2}(\sigma_{1z} + \sigma_{2z})H - \frac{(a-b)}{2}(\sigma_{1z} - \sigma_{2z})H + K\boldsymbol{\sigma}_1 \cdot \boldsymbol{\sigma}_2.$$

one writes

$$\mathcal{H} = -(a+b)HS_z + \frac{K}{2}(4S(S+1) - 6) - \frac{(a-b)}{2}(\sigma_{1z} - \sigma_{2z})H.$$

The first two terms are diagonal (in the representation in which the total spin is diagonal). Both the triplet and the singlet states have definite parity for exchange of particles (even and odd respectively). Thus the only non-zero matrix element of the last term (which is odd for exchange) is the one connecting singlet and triplet states. One finds

$$(\sigma_{1z} - \sigma_{2z})\left(\frac{a(1)b(2) - a(2)b(1)}{\sqrt{2}}\right) = 2\left(\frac{a(1)b(2) + a(2)b(1)}{\sqrt{2}}\right).$$

the only non-zero matrix element being

$$< 10|\mathcal{H}|00 > = -(a-b)H.$$

Therefore $S = 1$, $S_z = \pm 1$ classify the eigenstates, with eigenvalues

$$E_\pm = \mp(a+b)H + K,$$

For the states with $S_z = 0$, the Hamiltonian can be represented by the matrix

$$\mathcal{H}(S_z = 0) = \begin{pmatrix} K & -(a-b)H \\ -(a-b)H & -3K \end{pmatrix},$$

where the out of diagonal elements involve the triplet-singlet mixture. From the secular equation the eigenvalues turn out

$$E_\pm(S_z = 0) = -K \pm \sqrt{4K^2 + (a-b)^2 H^2}.$$

Problem 4.16 The saturation magnetization (per unit volume) of Iron (Fe^{2+}) is often reported to be $1.7 \cdot 10^6$ A/m. Derive the magnetic moment per atom and compare it to the theoretical estimate (density of iron $7.87\,g/cm^3$).

Solution: From

$$M_{sat} = 1.7 \cdot 10^3 \text{ erg Gauss}^{-1} \text{ cm}^{-3}$$

and $n_{atom} = 0.85 \cdot 10^{23} \text{ cm}^{-3}$, one derives

$$\mu_a \simeq 2 \cdot 10^{-20} \text{ erg Gauss}^{-1}$$

or equivalently $\mu_a = 2.2\mu_B$. For Fe^{2+} ($S = 2$, $L = 2$, $J = 4$ and $g = \frac{3}{2}$) one would expect $\mu = g\mu_B J = 6\mu_B$. For quenched orbital momentum $\mu = 2S\mu_B = 4\mu_B$ (see Problem 4.12 and Sect. 13.3).

Problem 4.17 A bulb containing Hg vapor is irradiated by radiation propagating along the x axis and linearly polarized along z, along which a constant magnetic field is applied. When the wavelength of the radiation is $2537\,\text{Å}$, absorption and meantime re-emission of light along the y direction, with the same polarization, is detected. When a RF coil winding the bulb along the y direction is excited at the frequency $200\,MHz$ one notes re-emission of light also along the z direction, light having about the same wavelength and circular polarization. Explain such a phenomenology and estimate the strength of the field.

Solution: Since spin-orbit interaction is very strong the weak-field regime holds (see Sect. 3.3 and Fig. 3.9). The electric dipole selection rule $\Delta M_J = 0$ requires linearly polarized radiation. In the absence of radio frequency excitation, π radiation is re-emitted again, observed along y. Along the z direction the radiation is not observed.

The radio-frequency induces magnetic dipole transitions at $\Delta M_J = \pm 1$ among the Zeeman levels. The re-emission of light in such a way is about at the same wavelength λ. In fact

$$\Delta\lambda \simeq \frac{\lambda^2}{c} \Delta\nu = 4.29 \cdot 10^{-4}\,\text{Å}.$$

On the other hand, since among the levels involved in the emission $\Delta M_J = \pm 1$, one has circular σ polarization and so the radiation along z can be observed.

From the resonance condition

$$\nu = \frac{\Delta E}{h} = \frac{g\mu_B H}{h}$$

with $g = 3/2$, one deduces

$$H = \frac{h\nu}{g\mu_B} = 95.26\,\text{Oe.}$$

The levels (in the **LS** classification and in the weak field regime) for the Hg $^1S_0$ and $^3P_1$ states (see Fig. 3.9) and the transitions are sketched below:

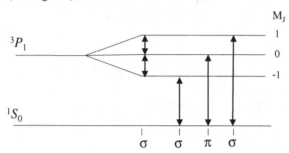

Problem 4.18 Consider a paramagnetic crystal, with non-interacting magnetic ions at $J = 1/2$. Evaluate the fluctuations $< \Delta M^2 >$ of the magnetization and show that it is related to the susceptibility $\chi = \partial < M >/\partial H$ by the relation $\chi = < \Delta M^2 >/k_BT$ (particular case of the *fluctuation-dissipation theorem*).

Solution: The density matrix is $\rho = (1/Z)e^{\beta H M_z}$ ($\beta \equiv 1/k_BT$) and the partition function $Z = Tr\left[e^{\beta H M_z}\right]$. The magnetization can be written (see also Sect. 6.1)[3]

$$< M_z > = \frac{1}{H}\frac{\partial}{\partial\beta}\ln Z.$$

Then

$$\chi \equiv \frac{\partial < M_z >}{\partial H} = \beta\left[\frac{Tr(e^{\beta H M_z}M_z^2)}{Tr(e^{\beta H M_z})} - \left[\frac{Tr(e^{\beta H M_z}M_z)}{Tr(e^{\beta H M_z})}\right]^2\right] = \beta < \Delta M_z^2 >.$$

Without involving the density matrix, from the single-ion fluctuations

$$< \Delta\mu_z^2 > = < \mu_z^2 > -(< \mu_z >)^2$$

---

[3]From

$$< M_z > = -\frac{\partial}{\partial H}(-\frac{1}{\beta}\ln Z)_T,$$

with

$$\frac{\partial}{\partial H} = \frac{\beta}{H}\frac{\partial}{\partial\beta}.$$

with $< \mu_z >$ statistical average of $\mu = -g\mu_B \mathbf{J}$, since

$$< \mu_z^2 > = \frac{1}{4}g^2\mu_B^2$$

and (see Sect. 4.4)

$$(< \mu_z >)^2 = \frac{1}{4}g^2\mu_B^2 \tanh^2\left(\frac{1}{2}\frac{g\mu_B H}{k_B T}\right),$$

by taking into account that

$$\chi = \frac{\partial M}{\partial H} = \frac{1}{4}g^2\mu_B^2 \frac{N}{k_B T}\cosh^{-2}\left(\frac{1}{2}\frac{g\mu_B H}{k_B T}\right)$$

one finds[4]

$$< \Delta M^2 > = N < \Delta\mu_z^2 > = \frac{1}{4}g^2\mu_B^2 N \cosh^{-2}\left(\frac{1}{2}\frac{g\mu_B H}{k_B T}\right)$$

and then

$$\chi = \frac{1}{k_B T} < \Delta M^2 > = \beta < \Delta M^2 > .$$

Problem 4.19 Consider an ensemble of $N/2$ pairs of atoms at $S=1/2$ interacting through an Heisenberg-like coupling $\mathcal{H} = K\mathbf{S}_1 \cdot \mathbf{S}_2$ with $K > 0$. By neglecting the interactions among different pairs, derive the magnetic susceptibility. Express the density matrix and the operator S_z on the basis of the singlet and triplet states. Finally derive the time-dependence of the statistical ensemble average $< S_1^z(0) \cdot S_1^z(t) >$, known as *auto-correlation function*.

Solution: The eigenvalues are $E_s = (K/2)S(S+1)$, with $S = 0$ and $S = 1$. The susceptibility is

$$\chi = \left(\frac{N}{2}\right)(p_0\chi_0 + p_1\chi_1),$$

where

$$p_S = \frac{(2S+1)e^{-\frac{E_S}{k_B T}}}{e^{-\frac{E_0}{k_B T}} + 3e^{-\frac{E_1}{k_B T}}}$$

and

$$\chi_S = \frac{g^2\mu_B^2 S(S+1)}{3k_B T}.$$

---

[4]The single μ's are uncorrelated i.e. $< \Delta\mu_n \Delta\mu_m > = < \Delta\mu_n >< \Delta\mu_m >$, for $n \neq m$.

Then

$$\chi = \frac{Ng^2\mu_B^2}{3k_BT} \frac{3e^{-\frac{K}{k_BT}}}{1+3e^{-\frac{K}{k_BT}}} .$$

On the basis given by the states

$$|1> = |++>, \quad |2> = |-->, \quad |3> = \frac{1}{\sqrt{2}}(|+->+|-+>) \quad \text{and}$$

$$|4> = \frac{1}{\sqrt{2}}(|+->-|-+>)$$

omitting irrelevant constants, one has

$$<i|\mathcal{H}|j> = \begin{pmatrix} K & 0 & 0 & 0 \\ 0 & K & 0 & 0 \\ 0 & 0 & K & 0 \\ 0 & 0 & 0 & 0 \end{pmatrix}$$

Then the density matrix is

$$<i|\rho|j> = <i|e^{-\beta\mathcal{H}}|j> = \begin{pmatrix} e^{-\beta K} & 0 & 0 & 0 \\ 0 & e^{-\beta K} & 0 & 0 \\ 0 & 0 & e^{-\beta K} & 0 \\ 0 & 0 & 0 & 1 \end{pmatrix}$$

By letting S_1^z act on the singlet and triplet states, one has

$$<i|S_1^z|j> = \frac{1}{2}\begin{pmatrix} 1 & 0 & 0 & 0 \\ 0 & -1 & 0 & 0 \\ 0 & 0 & 0 & 1 \\ 0 & 0 & 1 & 0 \end{pmatrix}$$

The autocorrelation-function is

$$g(t) = <\{S_1^z(t) \cdot S_1^z(0)\}> = Re[< S_1^z(t) \cdot S_1^z(0) >] \quad \text{where} \quad \{A, B\} = \frac{1}{2}(AB + BA),$$

$$<S_1^z(t) \cdot S_1^z(0)> = Tr\left[\frac{\rho}{Z}e^{-\frac{iHt}{\hbar}}S_1^z e^{\frac{iHt}{\hbar}}S_1^z\right]$$

By setting $\omega_e = \frac{K}{\hbar}$, (Heisenberg exchange frequency), one writes

$$<S_1^z(t) \cdot S_1^z(0)> = \frac{1}{4(1+3e^{-\beta K})} \times$$

$$Tr\left\{\begin{pmatrix} e^{-\beta K} & 0 & 0 & 0 \\ 0 & e^{-\beta K} & 0 & 0 \\ 0 & 0 & e^{-\beta K} & 0 \\ 0 & 0 & 0 & 1 \end{pmatrix} \times \right.$$

$$\times \begin{pmatrix} e^{-i\omega_e t} & 0 & 0 & 0 \\ 0 & e^{-i\omega_e t} & 0 & 0 \\ 0 & 0 & e^{-i\omega_e t} & 0 \\ 0 & 0 & 0 & 1 \end{pmatrix} \begin{pmatrix} 1 & 0 & 0 & 0 \\ 0 & -1 & 0 & 0 \\ 0 & 0 & 0 & 1 \\ 0 & 0 & 1 & 0 \end{pmatrix} \times$$

$$\left. \begin{pmatrix} e^{i\omega_e t} & 0 & 0 & 0 \\ 0 & e^{i\omega_e t} & 0 & 0 \\ 0 & 0 & e^{i\omega_e t} & 0 \\ 0 & 0 & 0 & 1 \end{pmatrix} \begin{pmatrix} 1 & 0 & 0 & 0 \\ 0 & -1 & 0 & 0 \\ 0 & 0 & 0 & 1 \\ 0 & 0 & 1 & 0 \end{pmatrix} \right\},$$

and then

$$g(t) = Re[< S_1^z(t) \cdot S_1^z(0) >]$$

$$= \frac{1}{4(1 + 3e^{-\beta K})}[2e^{-\beta K} + e^{-\beta K}\cos(\omega_e t) + \cos(\omega_e t)].$$

For $k_B T \gg K$

$$g(t) = \frac{1}{8}[1 + \cos(\omega_e t)].$$

$1/\omega_e$ can be defined as the *correlation time*, in the infinite temperature limit.

Problem 4.20 By resorting to the *Bohr-Sommerfeld quantization* rule (Problem 1.4) for the canonical moment, derive the *cyclotron frequency* and the energy levels for a free electron (without spin) moving in the xy plane, in the presence of a constant homogeneous magnetic field along the z axis.

Solution: The canonical moment (see Eq. (1.26)) is $\mathbf{p} = m\mathbf{v} - e\mathbf{A}/c$. From the quantization along the circular orbit

$$\oint \mathbf{p}.d\mathbf{q} = \oint \left(m\mathbf{v} - \frac{e}{c}\mathbf{A}\right) d\mathbf{q} = mv2\pi R - \frac{e}{c}\pi R^2 H = \frac{\pi e R^2 H}{c}$$

(R radius of the orbit). The equilibrium condition along the trajectory implies $v = eHR/mc$ and then the quantization rule yields

$$\frac{\pi R_n^2 e H}{c} = nh, \quad (\text{with } n = 1, 2, \ldots).$$

The energy becomes

$$E_n = \frac{mv_n^2}{2} = \frac{e\hbar H}{mc}n \equiv \hbar\omega_c n \equiv 2\mu_B H$$

with

$$\omega_c = \frac{eH}{mc}$$

cyclotron frequency. For the quantum description, which includes $n = 0$ and the zero-point energy $\hbar\omega_c/2$, see Appendix 13.1.

Problem 4.21 By referring to a Rydberg atom (Sect. 1.5) and considering that the diamagnetic correction to a given n-level increases with n, discuss the limit of applicability of the perturbative approach, giving an estimate of the breakdown value of n for magnetic field of 1 T (see Eq. (4.36)). Then discuss why the Rydberg atoms are highly polarizable and ionized by a relatively small electric field.

Solution: In

$$\Delta E_n = \frac{e^2 H^2}{8mc^2}\frac{2}{3} <r_n^2>$$

consider (see Table 1.4)

$$<r^2>_{nlm} = \frac{n^2}{2}(\frac{a_0}{Z})^2[5n^2 + 1 - 3l(l+1)] \simeq a_0^2 n^4$$

for large n and l and $Z = 1$, as for ideal total screen. Then, by assuming that the perturbation approach can be safely used up to a diamagnetic correction $\Delta E_n(H) \sim 0.2 E_n^0$, one obtains

$$\frac{e^2 H^2}{12mc^2}a_0^2 n_{lim}^4 \sim 0.2\frac{e^2}{2a_0}\frac{1}{n_{lim}^2}$$

from which a limiting value of the quantum number n turns out around $n_{lim} \sim 65$.

As regards the electric polarizability, by considering that in Eq. (4.8) the relevant matrix elements increase with n^2 while the difference in energy at the denominator varies as $1/n^3$ (remember the correspondence principle, Problem 1.12) one can deduce that the electric polarizability must increase as n^7.

Specific References and Further Reading

1. E.M. Purcell, *Electricity and Magnetism, Berkeley Physics Course Vol.2*, (McGraw-Hill, 1965).
2. S.J. Blundell, *Magnetism in Condensed Matter*, (Oxford Master Series in Condensed Matter Physics, Oxford U.P., 2001).
3. A. Balzarotti, M. Cini, M. Fanfoni, *Atomi, Molecole e Solidi. Esercizi risolti*, (Springer Verlag, 2004).
4. B.H. Bransden and C.J. Joachain, *Physics of atoms and molecules*, (Prentice Hall, 2002).
5. D. Budker, D.F. Kimball and D.P. De Mille, *Atomic Physics - An Exploration Through Problems and Solutions*, (Oxford University Press, 2004).
6. B. Cagnac and J.C. Pebay - Peyroula, *Physique atomique, tome 2*, (Dunod Université, Paris, 1971).
7. E.U. Condon and G.H. Shortley, *The Theory of Atomic Spectra*, (Cambridge University Press, London, 1959).
8. J.A. Cronin, D.F. Greenberg, V.L. Telegdi, *University of Chicago Graduate Problems in Physics*, (Addison-Wesley, 1967).
9. W. Demtröder, *Atoms, Molecules and Photons*, (Springer Verlag, 2006).
10. R.N. Dixon, *Spectroscopy and Structure*, (Methuen and Co LTD, London, 1965).
11. R. Eisberg and R. Resnick, *Quantum Physics of Atoms, Molecules, Solids, Nuclei and Particles*, (J. Wiley and Sons, 1985).
12. H. Eyring, J. Walter and G.E. Kimball, *Quantum Chemistry*, (J. Wiley, New York, 1950).
13. H. Haken and H.C. Wolf, *Atomic and Quantum Physics*, (Springer Verlag, Berlin, 1987).
14. C.S. Johnson and L.G. Pedersen, *Quantum Chemistry and Physics*, (Addison-Wesley, 1977).
15. M.A. Morrison, T.L. Estle and N.F. Lane, *Quantum States of Atoms, Molecules and Solids*, (Prentice-Hall Inc., New Jersey, 1976).
16. J.C. Slater, *Quantum Theory of Matter*, (McGraw-Hill, New York, 1968).
17. S. Svanberg, *Atomic and Molecular Spectroscopy*, (Springer Verlag, Berlin, 2003).

Chapter 5
Nuclear Moments and Hyperfine Interactions

Topics

Angular, Magnetic and Quadrupole Moments of the Nucleus
Magnetic Electron-Nucleus Interaction
Quadrupolar Electron-Nucleus Interaction
Hyperfine Structure and Quantum Number F
Hydrogen Atom Re-Examined: Fine and Hyperfine Structure

5.1 Introductory Generalities

Until now the nucleus has been often considered as a point charge with infinite mass, when compared to the electron mass. The *hyperfine structure* in high resolution optical spectra and a variety of experiments that we shall mention at a later stage, point out that the nuclear charge is actually distributed over a finite volume. Several phenomena related to such a charge distribution occur in the atom and can be described as due to *nuclear moments*. One can state the following:

(i) most nuclei have an *angular momentum*, usually called *nuclear spin*. Accordingly one introduces a nuclear spin operator $\mathbf{I}\hbar$, with related quantum numbers I and M_I, of physical meaning analogous to the one of J and M_J for electrons.

Nuclei having *even* A and *odd* N have integer quantum spin number I (hereafter *spin*) while nuclei at *odd* A have semi-integer spin $I \leq 9/2$; nuclei with both A and N *even* have $I = 0$.

(ii) associated with the angular momentum one has a *dipole magnetic moment*, formally described by the operator

$$\boldsymbol{\mu}_I = \gamma_I \mathbf{I}\hbar = g_n M_n \mathbf{I} \tag{5.1}$$

© Springer International Publishing Switzerland 2015
A. Rigamonti and P. Carretta, *Structure of Matter*,
UNITEXT for Physics, DOI 10.1007/978-3-319-17897-4_5

where g_n is the *nuclear Landé factor* and M_n the *nuclear magneton*, given by $M_n = \mu_B/1836.15 = e\hbar/2M_p c$, with M_p proton mass. γ_I is the *gyromagnetic ratio* (see Problems 1.21 and 3.4). g_n (which depends on the intrinsic nuclear properties) in general is different from the values seen to characterize the electron Landé factor. For instance, the proton has $I = 1/2$ and $\mu_I = 2.796 M_n$ and then $g_n = 5.59$. For deuteron one has $I = 1$ and $\mu_I = 0.86 M_n$. Since the angular momentum of the neutron is $I = 1/2$, from the comparison of the moments for proton and deuteron one can figure out a "vectorial" composition with the neutron and proton magnetic moments pointing along opposite directions.

At variance with most nuclei, for which g_n is *positive*, neutron as well as the nuclei ^{3}He, ^{15}N and ^{17}O have magnetic moment *opposite* to the angular momentum. Thus for them g_n is *negative*, similarly to electron. The pictorial composition indicated for deuteron does not account for a discrepancy of about 0.023 M_n, which is attributed to the fact that the ground state of the deuteron involves also the D excited state, with a little weight (about 4 percent). Properties of some nuclei are listed in Table 5.1.

(iii) nuclei with $I \geq 1$ are characterized by a charge distribution lacking spherical symmetry. Therefore, in analogy with classical concepts, they possess a *quadrupole electric moment*. For charge rotationally symmetric along the z axis the quadrupole moment is defined

$$Q = \frac{1}{Ze} \int \left[3z^2 - r^2 \right] \rho(\mathbf{r}) d\tau \tag{5.2}$$

For uniform charge density $\rho(\mathbf{r})$, one has $Q = (2/5)(b^2 - a^2)$, a and b being the axes of the ellipsoid (see Problem 5.8). Since the average radius of the nucleus can be written $R_n = (a^2 b)^{1/3}$, by indicating with δR_n the departure from the sphere (i.e. $b = R_n + \delta R_n$) one has $Q = (6/5) R_n^2 [\delta R_n / R_n]$.

Table 5.1 Properties of some nuclei

Nucleus	Z	N	I	μ/M_n	g_n
neutron	0	1	1/2	-1.913	-3.826
^{1}H	1	0	1/2	2.793	5.586
^{2}H	1	1	1	0.857	0.857
^{3}He	2	1	1/2	-2.12	-4.25
^{4}He	2	2	0	-	-
^{12}C	6	6	0	-	-
^{13}C	6	7	1/2	0.702	1.404
^{14}N	7	7	1	0.404	0.404
^{16}O	8	8	0	-	-
^{17}O	8	9	5/2	-1.893	-0.757
^{19}F	9	10	1/2	2.628	5.257
^{31}P	15	16	1/2	1.132	2.263
^{133}Cs	55	78	7/2	2.579	0.737

(iv) since proton and neutron ($I = 1/2$) are *fermions*, a nucleus with mass number A *odd* is a fermion while for A *even* the nuclei must obey to the *Bose-Einstein statistics*; in fact the exchange of two nuclei corresponds to the exchange of A pairs of fermions.

5.2 Magnetic Hyperfine Interaction—F States

The nuclear magnetic moment induces an interaction with the electrons of magnetic character, that can be thought to arise from the coupling between μ_J and μ_I. In the framework of the vectorial model one can extend the usual assumptions (see Sect. 3.2) to yield a coupling Hamiltonian of the form

$$\mathcal{H}_{hyp}^{mag} = a_J \mathbf{I} \cdot \mathbf{J} \tag{5.3}$$

In the quantum mechanical description the magnetic hyperfine interaction is obtained by considering the one-electron magnetic Hamiltonian (see Eq. (4.3))

$$\mathcal{H}_m = \frac{e}{2mc}\left(\mathbf{p}\cdot\mathbf{A} + \mathbf{A}\cdot\mathbf{p}\right) + 2\mu_B \mathbf{s}\cdot\nabla\times\mathbf{A} \tag{5.4}$$

with avector potential $\mathbf{A}$ due to the dipole moment μ_I at the origin (see the sketchy description in Fig. 5.1).

By means of some vector algebra (see Problem 5.21) and by singling out the terms with a singularity at the origin, as it could be expected on physical grounds the magnetic hyperfine Hamiltonian can be written in the form

$$\mathcal{H}_{hyp}^{mag} = -\mu_I \cdot \mathbf{h}_{eff}, \tag{5.5}$$

namely the one describing the magnetic moment μ_I in an effective field given by

$$\mathbf{h}_{eff} = \mathbf{h}_1(\mathbf{l}) + \mathbf{h}_2(\mathbf{s}) + \mathbf{h}_3(sing.). \tag{5.6}$$

Fig. 5.1 Pictorial view of the nucleus-electron interaction of magnetic origin

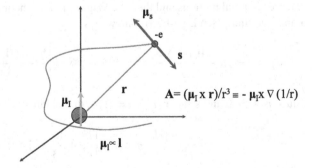

$$\mathbf{A} = (\mu_I \times \mathbf{r})/r^3 \equiv -\mu_I \times \nabla(1/r)$$

The orbital term

$$\mathbf{h}_1 = -\frac{2\mu_B\mathbf{l}}{r^3} \qquad (5.7)$$

is derived by considering the magnetic field at the nucleus due to the electronic current, in the a way similar to the deduction of the spin-orbit interaction (see Sect. 1.6), from $\mathbf{h}_1 = \mathbf{E} \times \mathbf{v}/c = -el\hbar/mcr^3$.

The field

$$\mathbf{h}_2 = \frac{2\mu_B}{r^3}\left(\mathbf{s} - 3\frac{(\mathbf{s}\cdot\mathbf{r})\mathbf{r}}{r^2}\right) \qquad (5.8)$$

is the classical field at the origin from a dipole at $\mathbf{r}$:

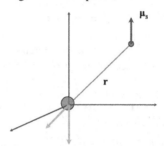

Finally

$$\mathbf{h}_3 = -\frac{2\mu_B 8\pi}{3}\mathbf{s}\delta(\mathbf{r}) \qquad (5.9)$$

is a term that includes all the singularities at the origin related to the expectation values of operator of the form r^{-3} (see Table 1.4), for s states. The *contact term* $\mathcal{H}_{cont} \propto (\mathbf{I}\cdot\mathbf{s})\delta(\mathbf{r})$ can be derived from a classical model where the nucleus is treated as a sphere uniformly magnetized (see Problem 5.21).

It is remarked that an analogous contact term of the form $A\mathbf{s}_1 \cdot \mathbf{s}_2\delta(\mathbf{r}_{12})$, with $A = -(8\pi/3)(e\hbar/mc)^2$, is involved in the electron-electron *magnetic* interaction, as already recalled at Problem 2.12.

The three fields in Eq. (5.6) are along different directions. However, by recalling the precessional motions and then the Wigner-Eckart theorem and the precession of $\mathbf{l}$ and $\mathbf{s}$ around $\mathbf{j}$ (see Eq. (4.25)) one writes

$$a_j = -\gamma_I\hbar\frac{<\mathbf{h}_{eff}\cdot\mathbf{j}>}{<j^2>} = -\gamma_I\hbar < l,s,j\left|\frac{\mathbf{h}_{eff}\cdot(\mathbf{l}+\mathbf{s})}{j^2}\right|l,s,j> \qquad (5.10)$$

Since $\mathbf{r}\cdot\mathbf{l} = 0$ and $< |\mathbf{s}\cdot\mathbf{r}/r|^2 > = 1/4$, one obtains

$$a_j = \frac{2\mu_B\gamma_I\hbar}{j(j+1)} < l,s,j\left|\frac{l^2}{r^3} + \frac{8\pi}{3}\mathbf{s}\cdot\mathbf{j}\delta(\mathbf{r})\right|l,s,j > . \qquad (5.11)$$

Then

$$a_j = \frac{16\pi}{3} \mu_B \gamma_I \hbar |\phi(0)|^2_{l=0} \qquad \text{for } s \text{ electrons} \qquad (5.12)$$

and

$$a_j = 2\mu_B \gamma_I \hbar \frac{l(l+1)}{j(j+1)} <r^{-3}>_{l\neq0} \qquad \text{for } l \neq 0, \text{ with } j = l \pm \frac{1}{2}. \quad (5.13)$$

For $l = 0$ the angular average of $\mathbf{h}_2$ and $\mathbf{h}_1$ (Eqs. (5.7) and (5.8)) yields zero: only the contact term related to $\mathbf{h}_3$ contributes to a_j, once that the expectation values are evaluated (see Problem 5.5).

From the Hydrogenic wavefunctions (Sect. 1.4) one evaluates $|\phi(0)|^2_{l=0} = Z^3/\pi a_0^3 n^3$ and

$$<r^{-3}>_{l\neq0} = Z^3/a_0^3 n^3 l(l + 1/2)(l + 1), \text{ in Eq. (5.13)}.$$

Therefore the effective field turns out of the order of $8 \times 10^4 (Z^3/n^3)$ G for s electrons and of the order of $3 \times 10^4 (Z^3/n^5)$ G for $l \neq 0$.

Values of the hyperfine field at the nucleus due to the optical electron, for the lowest energy states in alkali atoms, are reported in Table 5.2.

For two or more electrons outside the closed shells, in the **LS** coupling scheme one has to extend Eqs. (5.7)–(5.10) to total **L**, **S** and **J**, thus specifying a_J in Eq. (5.3).

The energy corrections related to the magnetic hyperfine interaction can be expressed by introducing the total angular momentum **F**

$$\mathbf{F} = \mathbf{I} + \mathbf{J} \qquad (5.14)$$

with the related quantum numbers F (integer or half integer) and M_F taking the values from $-F$ to $+F$.

The structure of the *hyperfine energy levels* turns out

$$<L, S, J, F|a_J\mathbf{I}.\mathbf{J}|L, S, J, F> = \frac{a_J}{2}\left(F(F+1) - I(I+1) - J(J+1)\right) \quad (5.15)$$

Table 5.2 Magnetic field (in Tesla) at the nucleus in alkali atoms, as experimentally obtained by direct magnetic dipole transitions between hyperfine levels (see Chap. 6) or by high-resolution irradiation in beams (see Fig. 5.3)

	$^2S_{1/2}$	$^2P_{1/2}$	$^2P_{3/2}$
Na	45	4.2	2.5
K	63	7.9	4.6
Rb	130	16	28.6
Cs	210	28	13

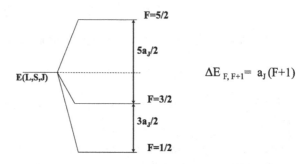

Fig. 5.2 Magnetic hyperfine structure for $J = 3/2$ and $I = 1$

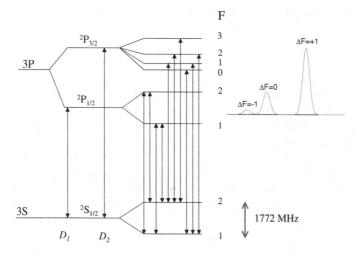

Fig. 5.3 Hyperfine magnetic structure of the low-energy states in Na atom, with the schematic illustration of three lines detected by means of resonance irradiation for the D_2 component in atomic beams (for details see Chap. 6) by using a narrow band variable-frequency dye laser (Problems 5.9 and 5.20)

The hyperfine structure for the electronic state $J = 3/2$ and nuclear spin $I = 1$ is illustrated in Fig. 5.2, showing the *interval rule* $\Delta_{F,F+1} = a_J(F+1)$. In Fig. 5.3 the hyperfine structure of the $D_{1,2}$ doublet in Na atom is reported.

It is reminded that the definition of the second as time unit and its metrological measure is obtained through the magnetic dipole $F = 4 \Leftrightarrow F = 3$ transition, at 9172.63 MHz, in ^{133}Cs atom ($I = 7/2$) in the ground state $^2S_{1/2}$.

The hyperfine energy levels for the ground state of Hydrogen are sketched in Fig. 5.4, with the indication of the spontaneous emission line at 21 cm, largely used in the astrophysical studies of galaxies.

A complete description of the fine and hyperfine structure of the energy levels in Hydrogen, including the results by *Dirac* and from the *Lamb* electrodynamics, is given in Appendix 5.1.

Fig. 5.4 Magnetic hyperfine structure of the ground state in Hydrogen and line at 21 cm resulting from the spontaneous emission from the $F = 1$ state, the transition being driven by the magnetic dipole mechanism

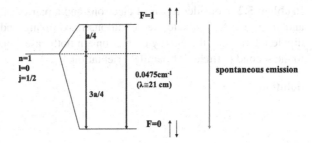

Finally we mention that the effect of an external magnetic field on the hyperfine states of the atom can be studied in a way strictly similar to what has been discussed at Chap. 4 in regards of the fine structure levels. *Zeeman* as well as *Paschen-Back* regimes are currently observed (see Problem 5.10).

Problems

Problem 5.1 Evaluate the dipolar and the contact hyperfine splitting for the ground state of positronium. Estimate the effective magnetic field experimented by the electron.

Solution: The hyperfine dipolar Hamiltonian is

$$\mathcal{H}_d = \frac{(2\mu_B)^2}{r^3} \left[\frac{3(\mathbf{r} \cdot \mathbf{s}_1)(\mathbf{r} \cdot \mathbf{s}_2)}{r^2} - \mathbf{s}_1 \cdot \mathbf{s}_2 \right],$$

while the Fermi contact term is

$$\mathcal{H}_F = \frac{8\pi}{3} (2\mu_B)^2 |\psi(0)|^2 \mathbf{s}_1 \cdot \mathbf{s}_2$$

yielding

$$E_F(S) = \frac{8\pi}{3} (2\mu_B)^2 |\psi(0)|^2 \frac{1}{2} S(S+1) + const.,$$

where $|\psi(0)|^2$ represents the probability of finding the electron and the positron in the same position, in the $1s$ state. Being zero the contribution from $\mathcal{H}_d$, the separation between the singlet and triplet levels is given by

$$E_F(S = 0) - E_F(S = 1) = \frac{8\pi}{3} \frac{(2\mu_B)^2}{\pi a_p^3} = \frac{4\mu_B^2}{3a_o^3} = 5 \cdot 10^{-4}\, \text{eV},$$

with a_p Bohr radius for positronium. The magnetic field experimented by the electron is about 4×10^4 G.

Problem 5.2 Consider a pair of electrons and a pair of protons at the distance e–e and p–p of 2 Å. Evaluate the conditions maximizing and minimizing the dipole-dipole interaction, the energy corrections in both cases, and the magnetic field due to the second particle, for parallel orientation.

Solution:

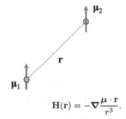

$$H(r) = -\nabla \frac{\mu \cdot r}{r^3}.$$

From the interaction energy

$$E_{int} = -\mu_1 \cdot H_2(r) = \frac{\mu_1 \cdot \mu_2}{r^3} - \frac{3(\mu_1 \cdot r)(\mu_2 \cdot r)}{r^5}$$

$E_{int} = 0$ for $\mu_1 \perp H_2(r)$.

For parallel orientation of the μ's $E_{int} = (\mu_1 \mu_2 / r^3)(1 - 3\cos^2\theta)$ and for $|r|$ fixed the extreme values are

$$E' = -\frac{2\mu_1\mu_2}{r^3} \text{ for } \theta = 0, \pi$$

$$E'' = \frac{\mu_1\mu_2}{r^3} \text{ for } \theta = \frac{\pi}{2}.$$

For two electrons at $|r| = 2$ Å, for $\mu_s = 2\mu_B\sqrt{s(s+1)}$

$$E' = -6.45 \cdot 10^{-17} \text{erg and } |H'| = 4016 \text{ Oe}$$

$$E'' = +3.22 \cdot 10^{-17} \text{erg and } |H''| = 2008 \text{ Oe}.$$

For two protons

$$\mu_p = g_p \mu_N \sqrt{I(I+1)}, \quad I = 1/2, \quad \mu_N = \mu_B/1836, \quad g_I = 5.6$$

and then

$$E' = -1.5 \cdot 10^{-22} \text{erg and } |H'| = 6 \text{ Oe}$$

$$E'' = 7.5 \cdot 10^{-23} \text{erg and } |H''| = 3 \text{ Oe}.$$

For the derivation of the eigenstates and eigenvalues see Problem 5.15.

Problem 5.3 Evaluate the magnetic field at the nuclear site in the Hydrogen atom for the electron in the states $1s$, $2s$ and $3s$. Estimate the energy difference between the states for parallel and antiparallel nuclear and electronic spins.

Solution: From Eq. (5.15) with

$$a_j = \frac{g_I \mu_N}{\sqrt{j(j+1)}} h_J = \frac{8\pi}{3} g_e \mu_B \, g_I \, \mu_N \, |\phi(0)|^2$$

and $|\phi(0)|^2 = 1/\pi a_0^3 n^3$ the field can be written

$$h_j = \frac{8\pi}{3} g_e \, \mu_B \, |\phi(0)|^2 \, [j(j+1)]^{1/2}.$$

The energy separation a for $j \equiv s = 1/2$ (Fig. 5.4) and the field turn out

n	$|\phi(0)|^2 (\mathrm{cm}^{-3})$	$a\ (\mathrm{cm}^{-1})$	h_J (kGauss)
1	$2.15 \cdot 10^{24}$	0.0474	289
2	$2.69 \cdot 10^{23}$	0.00593	36.1
3	$7.96 \cdot 10^{22}$	0.00176	10.7

Problem 5.4 Consider a *muonic atom* (negative muon) and Hydrogen, both in the $2p$ state. Compare in the two atoms the following quantities:

 (i) expectation values of the distance, kinetic and potential energies;
 (ii) the spin-orbit constant and the separation between doublet due to $2p - 1s$ transition (see Sect. 1.6);
(iii) the magnetic hyperfine constant and the line at 21 cm (note that the magnetic moment of the muon is about 10^{-2} Bohr magneton).

Solution:

 (i) $\langle r \rangle \propto a_0^*$ with $a_0^* = a_0/186$; $E_n^\mu = 186 E_n^H$, with $E_n^H = -e^2/2a_0 n^2$.

By resorting to the *virial theorem*, $\langle V \rangle = 2 \cdot 186 E_n^H$ and $\langle T \rangle = -\frac{\langle V \rangle}{2}$.

(ii) From

$$\mathcal{H}_{s.o.} = \frac{e^2 \hbar^2}{2m_\mu^2 c^2} \frac{1}{r^3} \mathbf{l} \cdot \mathbf{s}$$

by taking into account the scale factors for m_μ and for r, one finds $\xi_{2p}^\mu = 186\, \xi_{2p}^H$, with doublet separation $\frac{3}{2} \xi_{2p}^\mu$.

Alternatively, by considering the spin-orbit Hamiltonian in the form $\mu_l^e \, \mu_s^e \, \langle r^{-3} \rangle$, since $\mu_l^\mu \sim \mu_l^e / 186$ and $\mu_s^\mu \sim 10^{-2}$, the order of magnitude of the correcting factor can be written $(186)^3 / 186 \cdot 100$.

(iii) From $\mathcal{H}_{hyp} = -\boldsymbol{\mu}_n \cdot \mathbf{h}_{eff}$ and $|h_{eff}| \propto \mu_\mu |\psi(0)|^2$, since $\mu_\mu \sim 10^{-2}\mu_B$ and $|\phi(0)|^2 \sim (186)^3 / a_0^3$, one has $a_{1s}^\mu \approx a_{1s}^H \cdot 6.5 \cdot 10^4$ and $\lambda_{21}^\mu = \lambda_{21}^H / 6.5 \cdot 10^4$.

Problem 5.5 Estimate the dipolar magnetic field that the electron in $2p_{1,0}$ states and spin eigenfunction α creates at the nucleus in the Hydrogen atom.

Solution: From Eq. (5.8), since $\boldsymbol{\mu}_s = -2\mu_B \mathbf{s}$ and $-2\mu_B s_z = -\mu_B$, by taking into account that for symmetry reasons only the z-component is effective (the terms of the form $z \cdot x$ and $z \cdot y$ being averaged out) one has

$$h_z^{dip} = -\mu_B \left[-\frac{1}{r^3} + \frac{3z^2}{r^5} \right] = \frac{\mu_B}{r^3} \left[1 - 3cos^2\theta \right].$$

The expectation value of $3cos^2\theta / r^3$ on $R_{21}(r) Y_{11}(\theta, \varphi)$ reads

$$< \frac{1}{r^3} > \frac{9}{4} \int_{-1}^{1} d(cos\theta) sin^2\theta cos^2\theta = \frac{3}{2} < \frac{1}{r^3} >$$

Thus $< h_z^{dip} > = -(\mu_B / 2) < 1/r^3 >$. By taking $< 1/r^3 >_{211}$ from Table 1.4, $|h_z^{dip}| \simeq 1.5 \times 10^3$ G, to be compared to Eq. (5.13).

For the electron in the $2p_0$ state one obtains $< (1 - 3cos^2\theta)/r^3 > = -(4/5) < 1/r^3 >$ (again considering as effective only the z component).

This condition is the one usually occurring in strong magnetic fields where only the z components of $\mathbf{s}$ and of $\mathbf{I}$ are of interest.

The vanishing of $< \mathbf{h}_2 >$ (Eq. (5.8)) in s states arises from $\int (1 - 3cos^2\theta) sin\theta d\theta = 0$.

5.3 Electric Quadrupole Interaction

Since the first studies of the hyperfine structure by means of high resolution spectroscopy, it was found that in some cases the interval rule $\Delta_{F,F+1} = a_J(F + 1)$ was not obeyed. The breakdown of the interval rule was ascribed to the presence of a further electron-nucleus interaction of *electrical character*, related to the electric quadrupole moment of the nucleus. As we shall see, this second hyperfine interaction is described by an Hamiltonian different from the form $a_J \mathbf{I} \cdot \mathbf{J}$ which is at the basis of the interval rule.

To derive the electric quadrupole Hamiltonian one can start from the classical energy of a charge distribution in a site dependent electric potential:

$$E = \int \rho_n(\mathbf{r}) V_P(\mathbf{r}) d\tau_n \qquad (5.16)$$

By expanding the potential V_P due to electrons around the center of charge, one writes

$$E = \int \rho_n(\mathbf{r}) V_P(0) d\tau_n + \sum_\alpha \left(\frac{\partial V_p}{\partial x_\alpha}\right)_0 \int \rho_n(\mathbf{r}) x_\alpha d\tau_n +$$
$$\frac{1}{2} \sum_{\alpha,\beta} \left(\frac{\partial^2 V_p}{\partial x_\alpha \partial x_\beta}\right)_0 \int \rho_n(\mathbf{r}) x_\alpha x_\beta d\tau_n + \cdots \qquad (5.17)$$

where one notices the *monopole interaction* (already taken into account as potential energy in the electron core Hamiltonian), a *dipole term* which is *zero* (the nuclei do not have electric dipole moment) and the *quadrupole term* of the form

$$E_Q = \frac{1}{2} \sum_{i,j} Q'_{i,j} V_{i,j}, \quad \text{with } V_{i,j} = \int \rho_{elec.}(\mathbf{r}) \frac{3x_i x_j - \delta_{i,j} r^2}{r^5} d\tau. \qquad (5.18)$$

In the quantum description

$$Q'_{i,j} = e \sum_{nucleons} \left(3x_i^n x_j^n - \delta_{i,j} r_n^2\right) \qquad (5.19)$$

is the quadrupole moment operator, while $V_{i,j}$ is the electric field gradient operator, a sum of terms of the form $-e(3x_i x_j - \delta_{i,j} r^2)/r^5$.

Without formal derivation (for details see Problem 5.22), we specify the correspondent Hamiltonian in the form

$$\mathcal{H}^Q_{hyp} = b_J \left[3(\mathbf{I} \cdot \mathbf{J})^2 + \frac{3}{2} \mathbf{I} \cdot \mathbf{J} - I(I+1)J(J+1)\right] \qquad (5.20)$$

where $b_J = eQV_{zz}/2I(2I-1)J(2J-1)$, with eQV_{zz} the *quadrupole coupling constant*. The z-component of the electric field gradient is

$$V_{zz} = < J, M_J = J \left| -e \sum_{elec.} \frac{(3z_e^2 - r_e^2)}{r_e^5} \right| J, M_J = J > \qquad (5.21)$$

while

$$eQ = < I, M_I = I |eQ_z| I, M_I = I > = < I, I \left| e \sum_n (3z_n^2 - r_n^2) \right| I, I > \qquad (5.22)$$

Fig. 5.5 Hyperfine magnetic and electric quadrupole energy levels for an atom with $I = 3/2$ and $J = 1$; a_J is the hyperfine constant while here $b = eQV_{zz}$ (see Eqs. (5.20)–(5.22)). Q has been assumed positive. The value of a_J/b is arbitrary

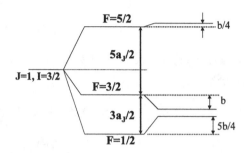

is the quantum equivalent of the classical quadrupole moment. Q is measured in cm^2 and a practical unit is 10^{-24} cm^2, called *barn*. Q positive means elongation of the nuclear charge along the spin direction while for negative Q the nuclear ellipsoid has its major axis perpendicular to **I**. For $I = 0$ or $I = 1/2$, $Q = 0$.

It is remarked that in the Hamiltonian (5.20) the first term is the one implying the breakdown of the interval rule.

With the usual procedure to evaluate the coupling operators in terms of the correspondent squares of the angular momentum operators, one can derive the energy corrections associated with the Hamiltonian (5.20). In Fig. 5.5 the structure of the hyperfine (magnetic and electric) levels for $I = 3/2$ and $J = 1$ is shown.

The one-electron *electric field gradient* (Eq. (5.21)) is written $< j, j|(3cos^2\theta - 1)/r^3|j, j >$.

For a wavefunction of the form $\varphi_{j,j} = R_{n,l}Y_{l,l}\chi^{spin}$, since

$$\int Y_{l,l}^* \left(3cos^2\theta - 1\right) Y_{l,l}sin\theta d\theta = -\frac{2l}{(2l+3)}$$

for $\chi^{spin} \equiv \alpha$, one has

$$q \equiv -\frac{V_{zz}}{e} = \frac{2l}{2l+3} < r^{-3} > . \qquad (5.23)$$

In terms of j one can write

$$q_j = \frac{(2j-1)}{(2j+2)} < r^{-3} > \qquad (5.24)$$

valid for $m_s = 1/2$ as well as for $m_s = -1/2$. For s states the spherical symmetry of the charge distribution implies $q = 0$.

For Hydrogenic wavefunctions the order of magnitude of the quadrupole coupling constant is

$$e^2qQ \sim 10^6 Q\frac{Z^3}{n^6} \simeq 10^{-6}\frac{Z^3}{n^6}eV \qquad (5.25)$$

for $Q \sim 10^{-24}$ cm^2.

In the condensed matter the operators V_{jk} can be substituted by the correspondent expectation values. The electric field gradient tensor has a principal axes frame of reference in which $V_{XY} = V_{XZ} = V_{YZ} = 0$, while $\sum_\alpha V_{\alpha\alpha} = 0$, with $|V_{ZZ}| > |V_{YY}| > |V_{XX}| \cdot \eta = (V_{XX} - V_{YY})/V_{ZZ}$ is defined the *asymmetry parameter* (see Problems 5.6 and 5.7).

Problems

Problem 5.6 Find eigenvalues, eigenstates and transition probabilities for a nucleus at $I = 1$ in the presence of an electric field gradient at cylindrical symmetry (expectation values $V_{ZZ} = eq$, $V_{XX} = V_{YY} = -eq/2$).

Repeat for an electric field gradient lacking of the cylindrical symmetry $(V_{XX} \neq V_{YY})$.

Solution: From Eq. (5.20), along the lines of Problem 5.22, and by referring to the expectation values for the electric field gradient, the quadrupole Hamiltonian is

$$\mathcal{H}_Q = A\left\{3\hat{I}_z^2 - \hat{I}^2 + \frac{1}{2}\eta(\hat{I}_+^2 + \hat{I}_-^2)\right\}$$

where

$$A = \frac{e^2qQ}{4I(2I-1)} = \frac{1}{4}e^2qQ, \quad eq = V_{ZZ}, \quad \eta = \frac{V_{XX} - V_{YY}}{V_{ZZ}}.$$

For cylindrical symmetry $\eta = 0$ and $\mathcal{H}_Q$ commutes with I_z and I^2. The eigenstates are

$$|1> = \begin{pmatrix} 1 \\ 0 \\ 0 \end{pmatrix}, |0> = \begin{pmatrix} 0 \\ 1 \\ 0 \end{pmatrix}, |-1> = \begin{pmatrix} 0 \\ 0 \\ 1 \end{pmatrix}.$$

In matrix form

$$\mathcal{H}_Q = A\begin{pmatrix} 1 & 0 & 0 \\ 0 & -2 & 0 \\ 0 & 0 & 1 \end{pmatrix}$$

and

$$\mathcal{H}_Q|\pm 1> = A|\pm 1> \quad \mathcal{H}_Q|0> = -2A|0>.$$

It can be noticed that *magnetic dipole transitions*, with $\Delta M_I = \pm 1$ and circular polarized radiation, are allowed (read Sect. 6.2).

For $\eta \neq 0$ the Hamiltonian in matrix form can be written[1]

---

[1] The matrices of the angular momentum operators for $I = 1$ in a basis which diagonalizes I_z and I^2 are

$$\mathcal{H}_Q = A \begin{pmatrix} 1 & 0 & \eta \\ 0 & -2 & 0 \\ \eta & 0 & 1 \end{pmatrix}.$$

From

$$\mathrm{Det}(A^{-1}\mathcal{H}_Q - \epsilon I) = 0 \implies (2+\epsilon)(1+\epsilon^2 - 2\epsilon - \eta^2) = 0$$

the eigenvalues turn out

$$E = A\epsilon = -2A, \quad (1 \pm \eta)A$$

with corresponding eigenvectors

$$|-2A> = \begin{pmatrix} 0 \\ 1 \\ 0 \end{pmatrix} \quad |(1 \pm \eta)A> = \frac{1}{\sqrt{2}} \begin{pmatrix} 1 \\ 0 \\ \pm 1 \end{pmatrix}.$$

The unitary transformation that diagonalizes $\mathcal{H}_Q$ is

$$\mathcal{H}'_Q = U\mathcal{H}_Q U^+$$

with

$$U = \frac{1}{\sqrt{2}} \begin{pmatrix} 1 & 0 & 1 \\ 0 & \sqrt{2} & 0 \\ 1 & 0 & -1 \end{pmatrix}.$$

Then

$$\mathcal{H}'_Q = A \begin{pmatrix} 1+\eta & 0 & 0 \\ 0 & -2 & 0 \\ 0 & 0 & 1-\eta \end{pmatrix}.$$

(Footnote 1 continued)

$$I_x = \frac{1}{\sqrt{2}} \begin{pmatrix} 0 & 1 & 0 \\ 1 & 0 & 1 \\ 0 & 1 & 0 \end{pmatrix} \quad I_y = \frac{1}{\sqrt{2}} \begin{pmatrix} 0 & -i & 0 \\ i & 0 & -i \\ 0 & i & 0 \end{pmatrix}$$

$$I_z = \begin{pmatrix} 1 & 0 & 0 \\ 0 & 0 & 0 \\ 0 & 0 & -1 \end{pmatrix} \quad I^2 = 2 \begin{pmatrix} 1 & 0 & 0 \\ 0 & 1 & 0 \\ 0 & 0 & 1 \end{pmatrix}$$

$$I_z^2 = \begin{pmatrix} 1 & 0 & 0 \\ 0 & 0 & 0 \\ 0 & 0 & 1 \end{pmatrix} \quad I_+ = \frac{1}{\sqrt{2}} \begin{pmatrix} 0 & 2 & 0 \\ 0 & 0 & 2 \\ 0 & 0 & 0 \end{pmatrix}$$

$$I_+^2 = \begin{pmatrix} 0 & 0 & 2 \\ 0 & 0 & 0 \\ 0 & 0 & 0 \end{pmatrix}$$

The structure of the energy levels is

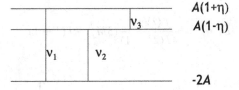

with transition frequencies

$$\nu_1 = \frac{3A}{h}\left(1+\frac{\eta}{3}\right), \quad \nu_2 = \frac{3A}{h}\left(1-\frac{\eta}{3}\right)$$

and

$$\nu_3 = 2A\eta/h.$$

From the interaction Hamiltonian with a radio frequency field $\mathbf{H}^{RF}$ (see Problem 5.13 and for details Chap. 6)

$$\mathcal{H}_I = -\gamma\hbar\mathbf{H}^{RF} \cdot \mathbf{I}$$

and by taking into account that

$$\mathcal{H}'_I = U\mathcal{H}_I U^+$$

one finds

$$\mathcal{H}'_I = -\gamma\hbar\begin{pmatrix} 0 & H_x^{RF} & H_z^{RF} \\ H_x^{RF} & 0 & iH_y^{RF} \\ H_z^{RF} & -iH_y^{RF} & 0 \end{pmatrix}.$$

The transition amplitudes are

$$\langle A(1+\eta)|\mathcal{H}'_I| - 2A\rangle = -\gamma\hbar H_x^{RF}$$

$$\langle -2A|\mathcal{H}'_I|A(1-\eta)\rangle = i\gamma\hbar H_y^{RF}$$

$$\langle A(1+\eta)|\mathcal{H}'_I|A(1-\eta)\rangle = -\gamma\hbar H_z^{RF}.$$

All the transitions are allowed, with intensity depending on the orientation of the radio frequency field with respect to the electric field gradient.

Problem 5.7 Consider a ^{23}Na nucleus at distance 1 Å from a fixed charge $-e$. Estimate the eigenvalues of the electric quadrupole interaction and the frequency of the radiation which induces transitions driven by the magnetic dipole mechanism(the electric quadrupole moment of ^{23}Na is $Q = +0.1 \cdot 10^{-24}$ cm^2).

Solution: From the eigenvalues of the electric quadrupole Hamiltonian (see Problem 5.6, for $\eta = 0$)

$$E_{I,M_I} = \frac{eQV_{ZZ}}{4I(2I-1)}(3M_I^2 - I(I+1)).$$

For $I = \frac{3}{2}$

$$E_{\pm 3/2} = \frac{1}{4}eQV_{ZZ} \qquad E_{\pm 1/2} = -\frac{1}{4}eQV_{ZZ}$$

The transition probabilities related to a perturbation Hamiltonian of the form $\mathcal{H}_P \propto \mathbf{H_x}^{RF} \cdot \mathbf{I} \propto I_\pm$ (see Sect. 6.2) involve the matrix elements

$$\left\langle \frac{3}{2}, \frac{3}{2} \left| \hat{I}_+ \right| \frac{3}{2}, \frac{1}{2} \right\rangle = \sqrt{3}\left\langle \frac{3}{2}, \frac{1}{2} \left| \hat{I}_+ \right| \frac{3}{2}, -\frac{1}{2} \right\rangle = 2$$

$$\left\langle \frac{3}{2}, -\frac{1}{2} \left| \hat{I}_+ \right| \frac{3}{2}, -\frac{3}{2} \right\rangle = \sqrt{3}$$

Then $W_{\frac{3}{2},\frac{1}{2}} \propto 3$, $W_{\frac{1}{2},-\frac{1}{2}} \propto 4$, $W_{-\frac{1}{2},-\frac{3}{2}} \propto 3$. The transition frequencies turn out

$$\nu_{\frac{3}{2},\frac{1}{2}} = \frac{1}{h}\left(E_{\frac{3}{2}} - E_{\frac{1}{2}}\right) = \frac{1}{2}\frac{eQV_{ZZ}}{h}, \quad \nu_{-\frac{3}{2},-\frac{1}{2}} = \frac{1}{h}\left(E_{-\frac{3}{2}} - E_{-\frac{1}{2}}\right) = \frac{1}{2}\frac{eQV_{ZZ}}{h}$$

From

$$V_{ZZ} = \frac{3z^2 - r^2}{r^5}e = \frac{2e}{r^3} \quad \text{and} \quad V_{XX} = V_{YY} = -\frac{e^2}{r^3}$$

(note that the Laplace equation holds) one estimates

$$\nu_{\frac{3}{2},\frac{1}{2}} = \nu_{-\frac{3}{2},-\frac{1}{2}} = \frac{e^2 Q}{hr^3} \simeq 3.5\,\text{MHz}.$$

Appendix 5.1 Fine and Hyperfine Structure in Hydrogen

Having introduced the various interaction terms (spin-orbit, relativistic corrections and hyperfine interaction) to be taken into account for one-electron states in atoms, it is instructive to reconsider the Hydrogen atom and to look at the detailed energy diagram (Fig. 5.6).

The solution of the non-relativistic Schrodinger equation (Sect. 1.4) provided the eigenvalues $E_{n,l} = -R_H hc/n^2$. Then the spin-orbit Hamiltonian $\mathcal{H}_{so} = (e^2/2m^2c^2r^3)\mathbf{l}\cdot\mathbf{s}$ was introduced (Sect. 1.6). However we did not really discuss at that point the case of Hydrogen (where other relativistic effects are of comparable strength) dealing instead with heavier atoms (Sect. 2.2 and Chap. 3) where the most

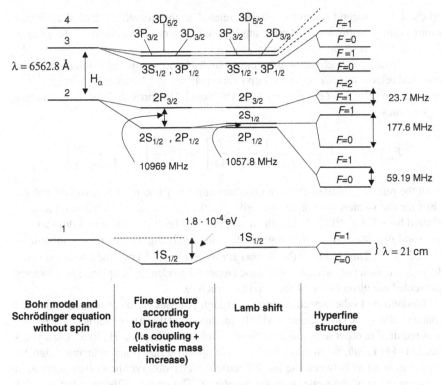

Fig. 5.6 Energy levels in Hydrogen including the effects contributing to its detailed structure. The scale is increased from *left* to *right* and some energy splittings are numerically reported to give an idea of the energy separations. The fine structure of the $n=2$ level is detailed in Fig. 5.7

relevant contribution to the fine structure arises from $\mathcal{H}_{so}$. At Sect. 2.2 and Problem 1.38 it was pointed out that a more refined relativistic description would imply a shift of the s-states (where $l=0$ while at the same time a divergent behaviour for $r \to 0$ is related to the positional part of $\mathcal{H}_{so}$). Finally the hyperfine magnetic interaction was introduced (Sect. 5.2) where $\mathcal{H}_{hyp.} = a_j \mathbf{I} \cdot \mathbf{j}$, with $I = 1/2$, $j = l \pm 1/2$ and a_j given by Eqs. (5.12) and (5.13).

A simple relativistic correction which could remove the accidental degeneracy in l was already deduced in the old quantum theory. As a consequence of the relativistic mass $m = m(v)$, for elliptical orbits in the Bohr model, *Sommerfeld* derived for the energy levels

$$E_{n,k} = -\frac{R_H hc}{n^2}\left[1 + \frac{\alpha^2}{n^2}\left(\frac{n}{k} - \frac{3}{4}\right) + \dots\right]$$

where k is a second quantum number related to the quantization of the angular momentum $\int p_\theta d\theta = kh$ (θ polar angle)(see Problem 1.4), while $\alpha = (e^2/\hbar c) \simeq 1/137$ is the *fine-structure constant*.

The *Dirac electrodynamical theory*, which includes spin-orbit interaction and classical relativistic effects (the relativistic kinetic energy being $c(p^2 + m^2c^2)^{1/2} - mc^2 \simeq (p^2/2m) - (p^4/8m^3c^2)$ (see Problem 1.38), provided the fine-structure eigenvalues

$$E_{n,j}^{fs} = -\frac{R_H hc\alpha^2}{n^3}\left[\frac{1}{j+1/2} - \frac{3}{4n}\right] = E_n^0\frac{\alpha^2}{n^2}\left[\frac{n}{j+1/2} - \frac{3}{4}\right],$$

with the relevant findings that the quantum number j and not l is involved and the shift for the s-states is explicit. Accordingly, the ground state of Hydrogen atom is shifted by -1.8×10^{-4} eV and the $n = 2$ energy level is splitted in a doublet, the $p_{3/2}$ and $p_{1/2}$ states (this latter degenerate with $s_{1/2}$) being separated by an amount of $0.3652\,\mathrm{cm}^{-1}$. The H_α line of the *Balmer series* (at 6562.8 Å) was then detected in the form of a doublet of two lines, since the Doppler broadening in optical spectroscopy prevented the observation of the detailed structure.

Giulotto and other spectroscopists, through painstaking measurements, noticed that the relativistic Dirac theory had to be modified and that a more refined description was required in order to account for the detailed structure of the H_α line. A few years later (1947) *Lamb*, by means of microwave spectroscopy (thus inducing magnetic dipole transitions between the levels) could directly observe the energy separation between terms at the same quantum number j. The energy difference between $2S$ and $2P$ states turned out $0.03528\,\mathrm{cm}^{-1}$ and the line had a fine structure of five lines, some of them broadened. Later on, by Doppler-free spectroscopy using dye lasers (Hansch et al., see Problem 5.20 for an example) the seven components of the H_α line consistent with the Lamb theory could be inferred. It was also realized that this result had to be generalized and the states with the same n and j quantum numbers, but different l, have different energy.

The *Lamb shift* (reported in detail in Fig. 5.7 for the $n = 2$ states) triggered the development of the *quantum electrodynamical theory*, which fully account for the fine structure of the levels on the basis of physical grounds that electrons are continuously emitting and adsorbing photons by transitions to virtual states. These states are poorly defined in energy due to their very short lifetimes. Qualitatively the Lamb shift can be considered the result of zero-point fluctuations of the set of harmonic oscillators describing the electromagnetic radiation field. These fluctuations induce analogous effects on the motion of the electron. Since the electric field in the atom is not uniform, the effective potential becomes different from the one probed by the electron in the average position.

The shift of the ground state due to the Lamb correction is about six times larger than the magnetic hyperfine splitting.

As regards the hyperfine splitting in the Hydrogen atom, at Sect. 5.2 it has been shown how the structure depicted in Fig. 5.6 is originated.

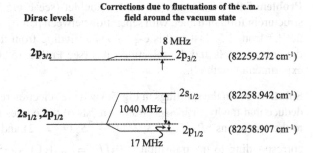

Fig. 5.7 Lamb shift for the $n = 2$ levels in Hydrogen

Problems

Problem 5.8 The electric quadrupole moment of the deuteron is $Q = 2.8 \cdot 10^{-3}$ barn. By referring to an ellipsoid of uniform charge, evaluate the extent of departure of the nuclear charge distribution from the sphere. Assume for average nuclear radius $R_n \simeq 1.89 \cdot 10^{-13}$ cm.

Solution: From

$$Q = \frac{1}{Ze} \rho \int_V (3z^2 - r^2) d\tau,$$

for the ellipsoid, defined by the equation $(x^2 + y^2)/a^2 + (z^2/b^2) = 1$, one obtains $Q = (2/5)(b^2 - a^2)$. If the average nuclear radius is taken to be $R_n^3 = a^2 b$ (the volume of the ellipsoid is $\frac{4}{3}\pi a^2 b$), with $R_n + \delta R_n = b$, then, for $\delta R_n \ll R_n$

$$a^2 = \frac{R_n^3}{R_n + \delta R_n} = \frac{R_n^2}{1 + \frac{\delta R_n}{R_n}} \approx R_n^2 \left(1 - \frac{\delta R_n}{R_n}\right)$$

and

$$b^2 - a^2 \approx R_n^2 \left[1 + 2\left(\frac{\delta R_n}{R_n}\right) + \left(\frac{\delta R_n}{R_n}\right)^2\right] - R_n^2 \left(1 - \frac{\delta R_n}{R_n}\right)$$

$$= R_n^2 \left[3\left(\frac{\delta R_n}{R_n}\right) + \left(\frac{\delta R_n}{R_n}\right)^2\right] \approx 3R_n^2 \left(\frac{\delta R_n}{R_n}\right).$$

Hence

$$Q = \frac{6}{5} R_n^2 \left(\frac{\delta R_n}{R_n}\right)$$

corresponding to $\left(\frac{\delta R_n}{R_n}\right) \approx 6.5 \cdot 10^{-2}$.

Problem 5.9 The D_2 line of the Na doublet (see Fig. 5.3) displays an hyperfine structure in form of triplet, with separation between pairs of adjacent lines in the ratio not far from 1.5. Justify this experimental finding from the hyperfine structure of the energy levels and the selection rules (see Problem 5.20 for some detail on the experimental method).

Solution: From the splittings in Fig. 5.3 and the selection rule $\Delta F = 0, \pm 1$ one can deduce that the hyperfine spectrum consists of three lines $\nu_{\{3,2,1\}\leftrightarrow 2}$ corresponding to the transitions $^2P_{\frac{3}{2}}(F = 3, 2, 1) \leftrightarrow ^2S_{\frac{1}{2}}(F = 2)$ and of three lines $\nu_{\{2,1,0\}\leftrightarrow 1}$ corresponding to the transitions $^2P_{\frac{3}{2}}(F = 2, 1, 0) \leftrightarrow ^2S_{\frac{1}{2}}(F = 1)$. From the interval rule

$$\frac{\nu_{3,2} - \nu_{2,2}}{\nu_{2,2} - \nu_{1,2}} = \frac{3}{2} \quad \text{and} \quad \frac{\nu_{2,1} - \nu_{1,1}}{\nu_{1,1} - \nu_{0,1}} = 2.$$

The lines in Fig. 5.3 correspond to the transitions $^2P_{\frac{3}{2}}(F = 3, 2, 1) \leftrightarrow ^2S_{\frac{1}{2}}(F = 2)$ (see Problem 5.20).

Problem 5.10 Plot the magnetic hyperfine levels for an atom in the electronic state $^2S_{1/2}$ and nuclear spin $I = 1$. Then derive the corrections due to a magnetic field, in the *weak* and *strong field regimes* (with respect to the hyperfine energy). Classify the states in the two cases and draw a qualitative correlation between them.

Solution: In the weak-field regime the effective magnetic moment is along **F**. By neglecting the contribution from the nuclear magnetic moment one writes

$$\boldsymbol{\mu}_F = -g_F \mu_B \mathbf{F}$$

and the hyperfine correction is

$$\Delta E = g_F \mu_B H_0 m_F.$$

g_F is calculated by projecting $\boldsymbol{\mu}_J$ along **F**: $\mu_F = -g_J |\mathbf{J}| \cos\widehat{FJ}.\mathbf{F}/|\mathbf{F}|$, with $\cos\widehat{FJ} = [J(J + 1) + F(F + 1) - I(I + 1)]/2\sqrt{J(J + 1)}\sqrt{F(F + 1)}$.
Thus

$$g_F = g_J \frac{F(F + 1) + J(J + 1) - I(I + 1)}{2F(F + 1)}.$$

A relatively small field breaks up the $\mathbf{I} \cdot \mathbf{J}$ coupling and the hyperfine Zeeman effect is replaced by the hyperfine Paschen-Back effect. The oscillating components in the x and y directions average to zero and the final result is that the nuclear angular momentum vector **I** is oriented along H_0. The quantum number F is no longer defined while the quantum numbers m_I and m_J describe **I** and **J**. The splitting involves three terms, one being $g_J \mu_B H_0 m_J$, already considered in the Zeeman effect (Sect. 4.3.2), the other is $am_I m_J$ and the third one, $-\mu_N g_I m_I H_0$, is negligible. See Fig. 5.8 for a pictorial view of the angular momenta and of the correspondent magnetic moments.

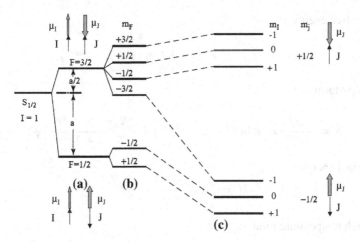

Fig. 5.8 Hyperfine structure of the $S_{1/2}$ state with $I = 1$: **a** *in zero field*; **b** *in weak field*, Zeeman regime; **c** *in strong field*, Paschen-Back regime

Problem 5.11 In the Na atom the hyperfine interaction for the P state is much smaller than the one in the S ground state. In poor resolution the hyperfine structure is observed in the form of a doublet, with relative intensities 5 and 3. From this observation derive the nuclear spin (see also Problem 5.20).

Solution: From

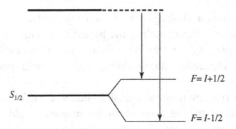

The intensity being $\propto (2F + 1)$ and the ratio $(I + 1)/I = 5/3$, then $I = 3/2$.

Problem 5.12 For a solid ideally formed by a mole of non-interacting deuterons in an electric field gradient, derive the contributions to the entropy and to the specific heat, in the high temperature limit (see Problem 5.6).

Solution: The quadrupolar interaction $e^2 q Q [3M^2 - I(I + 1)]/4$ yields two energy levels, one doubly degenerate ($M = 0, \pm 1$).

By indicating with ϵ the separation between the levels, the partition function is

$$Z(\beta) = \left(1 + 2e^{-\beta\epsilon}\right)^{N_A}, \text{ with } \beta = 1/k_B T.$$

From the free energy

$$F(T) = -k_B T \ln Z = -RT \ln \left(1 + 2e^{-\epsilon/k_B T}\right),$$

the entropy turns out

$$S = -\frac{\partial F}{\partial T} = R \ln \left(1 + 2e^{-\epsilon/k_B T}\right) + \frac{2N_A \epsilon}{T} \frac{e^{-\epsilon/k_B T}}{1 + 2e^{-\epsilon/k_B T}}.$$

The internal energy is

$$U = 2N_A \epsilon \frac{1}{e^{\epsilon/k_B T} + 2}.$$

In the high temperature limit

$$U \simeq \frac{2}{3} N_A \epsilon \left(1 - \frac{\epsilon}{3k_B T}\right),$$

so that

$$C = \frac{\partial U}{\partial T} = \frac{2}{9} R \left(\frac{\epsilon}{k_B T}\right)^2 \propto T^{-2},$$

namely the high-temperature tail of a *Schottky anomaly* (a "bump" in the specific heat versus temperature), typical of two-levels systems.

Problem 5.13 Consider the Hydrogen atom, in the ground state, in a magnetic field H_0 and write the Hamiltonian including the hyperfine interaction. First derive the eigenvalues and the spin eigenvectors in the limit $H_0 \rightarrow 0$ and estimate the frequencies of the transitions induced by an oscillating magnetic field (perpendicular to the quantization axis).

Then derive the correction to the eigenvalues due to a weak magnetic field.

Finally consider the opposite limit of strong magnetic field. Draw the energy levels with the appropriate quantum numbers, again indicating the possibility of inducing magnetic dipole transitions between the hyperfine levels (this is essentially the *EPR* experiment, see for details Chap. 6) and from the resulting lines show how the hyperfine constant can be extracted.

Figure out a schematic correlation diagram connecting the eigenvalues for variable external field.

Solution: From the Hamiltonian

$$\mathcal{H}_s = 2\mu_B \mathbf{S} \cdot \mathbf{H}_0 - \gamma \hbar \mathbf{I} \cdot \mathbf{H}_0 + a \mathbf{I} \cdot \mathbf{S}$$

(γ nuclear gyromagnetic ratio, a hyperfine interaction constant, with $a = hc/\lambda$ and $\lambda = 21$ cm).

For $H_0 \to 0$ the eigenstates are classified by S, I, F and M_F and for $I = S = 1/2$ two magnetic hyperfine states, $F = 0$ and $F = 1$, occur. Then $E_{\frac{1}{2},\frac{1}{2},1} = a/4$ and $E_{\frac{1}{2},\frac{1}{2},0} = -3/4a$.

The spin eigenvectors are the same of any two spins system, i.e.

$$\left.\begin{array}{l} |\alpha_e \alpha_p > \\ |\beta_e \beta_p > \\ \frac{1}{\sqrt{2}} |\alpha_e \beta_p + \alpha_p \beta_e > \end{array}\right\} \text{ defining the triplet } T_{1,1} \ T_{1,0} \ T_{1,-1}$$

and

$$\frac{1}{\sqrt{2}} |\alpha_e \beta_p - \alpha_p \beta_e > \text{ defining the singlet } S_{0,0}.$$

The oscillating magnetic field acts as a perturbation involving the operator $\mu_x = 2\mu_B S_x - \gamma \hbar I_x$ (for details see Sect. 6.2). The matrix elements for the triplet and singlet states turn out

$$< 1, 1 |\mu_x| 1, 0 > = \frac{1}{2} \frac{1}{\sqrt{2}} (g\mu_B - \gamma \hbar)$$

$$< 1, 1 |\mu_x| 0, 0 > = \frac{1}{2} \frac{1}{\sqrt{2}} (-g\mu_B - \gamma \hbar)$$

$$< 1, 0 |\mu_x| 1, -1 > = \frac{1}{2} \frac{1}{\sqrt{2}} (g\mu_B - \gamma \hbar)$$

$$< 0, 0 |\mu_x| 1, -1 > = \frac{1}{2} \frac{1}{\sqrt{2}} (g\mu_B + \gamma \hbar)$$

$$< 1, 0 |\mu_x| 0, 0 > = < 1, 1 |\mu_x| 1, -1 > = 0.$$

Therefore the allowed transitions are $S \to T_1$ and $S \to T_{-1}$ corresponding to the transition frequency $\nu = \frac{a}{h} = 1420$ MHz (and formally $T_{-1} \to T_0$, $T_0 \to T_{+1}$ at $\nu = 0$).

For weak field H_0, neglecting the interaction with the proton magnetic moment and considering that the perturbation acts on the basis where F^2, F_z, I^2 and S^2 are diagonal, the matrix for $\mathcal{H} = a\mathbf{I} \cdot \mathbf{S} + 2\mu_B \mathbf{S} \cdot \mathbf{H}_0$ is

$$\begin{pmatrix} a/4 + \mu_B H_0 & 0 & 0 & 0 \\ 0 & a/4 & 0 & -\mu_B H_0 \\ 0 & 0 & a/4 - \mu_B H_0 & 0 \\ 0 & -\mu_B H_0 & 0 & -3a/4 \end{pmatrix}$$

From the secular equation the eigenvalues are found by solving

$$\frac{a}{4} + \mu_B H_0 - E = 0 \qquad \frac{a}{4} - \mu_B H_0 - E = 0$$

$$\left(\frac{a}{4} - E\right)\left(-\frac{3a}{4} - E\right) - \mu_B^2 H_0^2 = 0$$

yielding $E_{1,2} = a/4 \pm \mu_B H_0$ and $E_{3,4} = -a/4 \pm (a/2)[1 + 4\mu_B^2 H_0^2/a^2]^{1/2}$ (the states $F = 1, M_F = 0$ and $F = 0, M_F = 0$ being little affected by a weak magnetic field).

The *Breit-Rabi diagram*, as reported below, holds

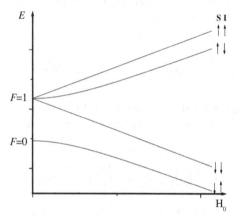

In the strong field regime the eigenvalues are the ones for S_z, $I_z S_z$ and I_z:

$$E = 2\mu_B H_0 m_S + a m_S M_I - \gamma \hbar M_I H_0$$

The first term is dominant and the diagram is

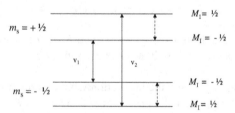

with the electronic transitions $\Delta m_S = \pm 1$ (and $\Delta M_I = 0$) at the frequencies

$$\nu_{1,2} = \frac{2\mu_B H_0 \pm a/2}{h}.$$

The nuclear transitions $\Delta M_I = \pm 1$ (and $\Delta m_S = 0$) occur at $a/2h$.

Since the internal field due to the electron is usually much larger than H_0 the third term can be neglected (see the figure below).

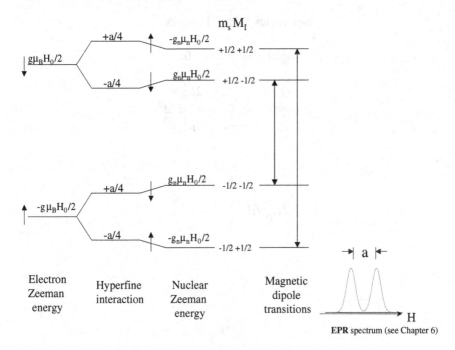

EPR spectrum (see Chapter 6)

Problem 5.14 The ^{209}Bi atom has an excited $^2D_{5/2}$ state, with 6 sublevels due to hyperfine interaction. The separations between the hyperfine levels are 0.23, 0.31, 0.39, 0.47 and 0.55 cm^{-1}. Evaluate the nuclear spin and the hyperfine constant.

Solution: From $E(F, I, J) = (a/2)[F(F+1) - I(I+1) - J(J+1)]$ and $E_{F+1} - E_F = a(F+1)$, one finds $a = 0.08$ cm^{-1} and $F_{max} = 7$.
Therefore $F = 2, 3, 4, 5, 6, 7$ and since $J = 5/2$ the nuclear spin must be $I = \frac{9}{2}$.

Problem 5.15 A proton and an anti-proton, at a given distance d, interact through the magnetic dipole-dipole interaction. Derive the total spin eigenstates and eigenvalues in term of the proton magnetic moment(it is reminded that the magnetic moment of the antiproton is the same of the proton, with *negative* gyromagnetic ratio).

Solution: From

$$\mathcal{H} = \frac{\mu_1 \cdot \mu_2}{r^3} - 3\frac{(\mu_1 \cdot \mathbf{r})(\mu_2 \cdot \mathbf{r})}{r^5}.$$

with $\mu_1 = 2\mu_p \mathbf{s}_1$ and $\mu_2 = -2\mu_p \mathbf{s}_2$, by choosing the z axis along $\mathbf{r}$

$$\mathcal{H} = -4\frac{\mu_p^2}{d^3} \mathbf{s}_1 \cdot \mathbf{s}_2 + 12\frac{\mu_p^2}{d^3} s_1^z s_2^z.$$

Since $\mathbf{s}_1 \cdot \mathbf{s}_2 = S(S+1)/2 - 3/4$ and $s_1^z s_2^z = (1/2)M_S^2 - (1/2) \cdot (1/2)$, one finds

Eigenstates S M_S Energies

singlet   0 0   0

		1	$2\mu_p^2/d^3$
triplet	1	0	$-4\mu_p^2/d^3$
		-1	$2\mu_p^2/d^3$

i.e.

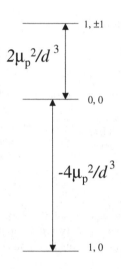

Problem 5.16 Two electrons interact through the dipolar Hamiltonian. A strong magnetic field is applied along the z-direction, at an angle θ with the line connecting the two electrons. Find the eigenvalues and the corresponding eigenfunctions for the two spins system, in terms of the basis functions $\alpha_{1,2}$ and $\beta_{1,2}$.

Solution: In the light of Eq. (5.8) for the dipolar field (see also Problem 5.2) the total Hamiltonian is

$$\mathcal{H} = \mathcal{H}_0 + \mathcal{H}_d$$
$$= 2\mu_B H_0(s_z^{(1)} + s_z^{(2)}) + \frac{4\mu_B^2}{r^3}\left\{ \mathbf{s}_1 \cdot \mathbf{s}_2 - \frac{3}{r^2}[(\mathbf{s}_1 \cdot \mathbf{r})(\mathbf{s}_2 \cdot \mathbf{r})] \right\}.$$

In order to evaluate the matrix elements it is convenient to write the perturbation Hamiltonian in the form (called *dipolar alphabet*)

$$\mathcal{H}_d = \frac{4\mu_B^2}{r^3}[A + B + C + D + E + F]$$

where

$$A = s_z^{(1)} s_z^{(2)} \left[1 - 3\cos^2\theta\right], \quad B = -\frac{1}{4}\left[s_+^{(1)} s_-^{(2)} + s_-^{(1)} s_+^{(2)}\right]\left(1 - 3\cos^2\theta\right),$$

θ angle between H_0 and r. The terms C, D, E and F involve operators of the form $s_+^{(1)} s_z^{(2)}$, $s_-^{(1)} s_z^{(2)}$, $s_+^{(1)} s_+^{(2)}$, $s_-^{(1)} s_-^{(2)}$ and can be neglected. In fact these terms are off-diagonal and produce admixtures of the zero-order states to an amount of the order of $\left(\mu_B/r^3\right)/H_0$ (i.e. $\sim 10^{-4}$ for $H_0 = 1\,\text{T}$).

Thus the dipolar Hamiltonian is written in the form

$$\mathcal{H}_d = \underbrace{\frac{4\mu_B^2}{r^3}\left(1 - 3\cos^2\theta\right)}_{\mathcal{A}}\left[s_z^{(1)} s_z^{(2)} - \frac{1}{4}\left(s_+^{(1)} s_-^{(2)} + s_-^{(1)} s_+^{(2)}\right)\right],$$

most commonly used.

The complete set of the basis functions is $\alpha_1\alpha_2$, $\alpha_1\beta_2$, $\alpha_2\beta_1$ and $\beta_1\beta_2$ and the matrix elements are

$$< \alpha\alpha|\mathcal{H}_T|\alpha\alpha > = 2\,\mu_B\,H_0 + \frac{1}{4}\mathcal{A}$$

$$< \alpha\beta|\mathcal{H}_T|\alpha\beta > = < \beta\alpha|\mathcal{H}_T|\beta\alpha > = -\frac{\mathcal{A}}{4}$$

$$< \alpha\beta|\mathcal{H}_T|\beta\alpha > = < \beta\alpha|\mathcal{H}_T|\alpha\beta > = -\frac{\mathcal{A}}{4}$$

$$< \beta\beta|\mathcal{H}_T|\beta\beta > = -2\mu_B H_0 + \mathcal{A}/4$$

It is noted that while the term A is completely diagonal, the term B only connects $|m_s^{(1)} m_s^{(2)} >$ to states $< m_s^{(1)}+1, m_s^{(2)}-1|$ or $< m_s^{(1)}-1, m_s^{(2)}+1|$. B simultaneously flips one spin up and the other down.

The secular equation is

$$\begin{vmatrix} (+2\mu_B H_0 + \frac{\mathcal{A}}{4}) - E & 0 & 0 & 0 \\ 0 & -\frac{\mathcal{A}}{4} - E & -\frac{\mathcal{A}}{4} & 0 \\ 0 & -\frac{\mathcal{A}}{4} & -\frac{\mathcal{A}}{4} - E & 0 \\ 0 & 0 & 0 & \left(-2\mu_B H_0 + \frac{\mathcal{A}}{4}\right) - E \end{vmatrix} = 0.$$

and the eigenvalues turn out

$$E_1 = -2\mu_B\left(H_0 - \frac{\mu_B}{2r^3}(1 - 3\cos^2\theta)\right)$$

$$E_2 = 0$$

$$E_3 = -\frac{2\mu_B^2}{r^3}(1 - 3\cos^2\theta)$$

$$E_4 = +2\mu_B \left(H_0 - \frac{\mu_B}{2r^3}(1 - 3\cos^2\theta)\right).$$

The correspondent eigenfunctions being $\alpha_1\alpha_2$, $\beta_1\beta_2$ and $\frac{1}{\sqrt{2}}[\alpha_1\beta_2 \pm \alpha_2\beta_1]$, as expected.

Problem 5.17 In the Ba atom the line due to the transition from the $6s\,6p\;J = 1$ to the $(6s)^2$ ground state in high resolution is evidenced as a triplet, with line intensities in the ratio 1, 2 and 3. Evaluate the nuclear spin.

Solution: Since $\mathbf{F} = \mathbf{I} + \mathbf{J}$

$$\text{for } J = 0 \text{ one has } I = F \implies \text{ no splitting}$$

$$\text{for } J = 1 \implies \text{ splitting in } (2I + 1) \text{ or in } (2J + 1) \text{ terms.}$$

$$I = 0 \implies \text{ no splitting,}$$

$$\text{for } I = \frac{1}{2} \text{ and } J = 1 \; \Delta F = 0, \pm 1 \implies \text{ two lines}$$

$$I = 1 \text{ or } I > 1 \implies \text{ three lines.}$$

Looking at the intensities, proportional to $e^{-E/k_B T}(2F + 1)$, where the energy E is about the same

$$\text{for } I = 1 \; F = 0, 1, 2 \implies \text{ intensities: } 1, 3, 5$$

$$\text{for } I = \frac{3}{2} \; F = \frac{1}{2}, \frac{3}{2}, \frac{5}{2} \implies \text{ intensities: } 2, 4, 6.$$

Therefore $I = \frac{3}{2}$.

Problem 5.18 In the assumption that in a metal the magnetic field on the electron due to the hyperfine interaction with $I = 1/2$ nuclei is $H_z = (a/N)\Sigma_n I_n^z$ (a constant and same population on the two states) prove that the odd moments of the distribution are zero and evaluate $< H_z^2 >$. Then evaluate $< H_z^4 >$ and show that for large N the distribution tends to be Gaussian, the width going to zero for $N \to \infty$.

Solution: $\langle \hat{H}_z^{2n+1} \rangle = 0$ for symmetry. Since $(I_n^z)^2 = \frac{1}{4}$

$$\left\langle \left(\sum_n I_n^z\right)^2 \right\rangle = \sum_n \langle (I_n^z)^2 \rangle + \sum_{n \neq m} \langle I_n^z I_m^z \rangle = \frac{1}{4}N,$$

and then $\langle H_z^2 \rangle = (\frac{a}{2N})^2 N$.

$$\langle H_z^4 \rangle = \left(\frac{a}{N}\right)^4 \sum_{i,j,k,l} \langle I_i^z I_j^z I_k^z I_l^z \rangle = \left(\frac{a}{N}\right)^4 3 \sum_{i,j} \left\langle (I_i^z)^2 (I_j^z)^2 \right\rangle - \left(\frac{a}{N}\right)^4 \sum_i \left\langle (I_i^z)^4 \right\rangle$$

$$= \left(\frac{a}{2N}\right)^4 (3N^2 - N).$$

In the thermodynamical limit one has $\langle H_z^4 \rangle \simeq 3 \left(\frac{a}{2N}\right)^4 N^2$: the first two even moments correspond to the Gaussian moments.

Problem 5.19 Evaluate the transition probability from the state $M = -1/2$ to $M = +1/2$ by spontaneous emission, for a proton in a magnetic field of 7500 Oe.

Solution: From the expression for A_{21} derived in Appendix 1.3 and extending it to magnetic dipole transitions, one can write

$$A_{21} = \frac{4\omega_L^3}{3c^3\hbar} | < 2|\mu|1 > |^2$$

$$= \frac{4\omega_L^3}{3c^3\hbar} \left\{ \left| < \frac{1}{2} \left| \mu_x \right| -\frac{1}{2} > \right|^2 + \left| < \frac{1}{2} \left| \mu_y \right| -\frac{1}{2} > \right|^2 \right\}$$

with $\mu = \gamma\hbar\mathbf{I}$. From $I_\pm = I_x \pm iI_y$ one derives $A_{21} = (2/3)(\gamma^2\hbar/c^3)\omega_L^3$ and for $\gamma = 42.576 \cdot 2\pi \cdot 10^2\,\text{Hz/G}$, $\omega_L = \gamma H_0 = 2\pi \cdot 31.9\,\text{MHz}$, yielding

$$A_{21} \simeq 1.5 \times 10^{-25}\,\text{s}^{-1}.$$

Problem 5.20 High-resolution laser spectroscopy allows one to evidence the hyperfine structure in the optical lines with almost total elimination of the *Doppler broadening*.

The figure below

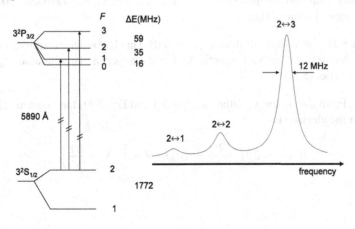

shows the hyperfine structure of the $^2S_{1/2} - ^2P_{3/2}$ D_2 line of Na at 5890 Å (transitions $\Delta F = 0, \pm 1$ from the $F = 2$ level of the electronic ground state). (This spectrum is obtained by irradiating a collimated beam of sodium atoms at right angles by means of a narrow-band single-mode laser and detecting the fluorescent light after the excitation. This and other high resolution spectroscopic techniques are described in the book by *Svanberg*).

From the figure, discuss how the magnetic and electric hyperfine constants could be derived and estimate the life-time of the $^2P_{3/2}$ state (in the assumption that is the only source of broadening).

Then compare the estimated value of the life time with the one known (from other experiments), $\tau = 1.6$ ns. In the assumption that the extra-broadening is due to Doppler *second-order relativistic shift*, quadratic in (v/c), estimate the temperature of the oven from which the thermal atomic beam is emerging, discussing the expected order of magnitude of the broadening (see Problem 1.30).

Solution: For the ground-state $^2S_{1/2}$, $I = 3/2$ and $J = S = 1/2$, the quadrupole interaction being zero, from Eq. (5.15) the separation between the $F = 2$ and $F = 1$ states yields the magnetic hyperfine constant $a = \Delta_{1,2}/(F+1) = 886$ MHz, corresponding to an effective magnetic field of about 45 T.

The sequence of the hyperfine levels for the $^2P_{3/2}$ state does not follow exactly the interval rule. In the light of Eq. (5.20) an estimate of the quadrupole coupling constant b can be derived (approximate, the correction being of the order of the intrinsic line-widths).

In the assumption that the broadening (12 MHz) is due only to the life-time one would have $\tau = 1/2\pi\Delta\nu \simeq 13.3 \times 10^{-9}$ s, a value close to the one ($\tau \simeq 16 \times 10^{-9}$s) pertaining to the $3^2P_{3/2}$ state ($\Delta\nu \simeq 10$ MHz).

The most probable velocity of the beam emerging from the oven is $v = \sqrt{3k_BT/M_{Na}}$, that for $T \simeq 500$ K corresponds to about 7×10^4 cm/s.

While the first-order Doppler broadening is in the range of a few GHz, scaling by a term of the order of v/c leads to an estimate of the *second-order Doppler broadening* in the kHz range. Thus the extra-broadening of a few MHz is likely to be due to the *residual first-order broadening* (for a collimator ratio of the beam around 100 being typically around some MHz).

Problem 5.21 From the perturbation generated by nuclear magnetic moment on the electron, derive the effective magnetic field in the hyperfine Hamiltonian $\mathcal{H}_{hyp} = -\mu_I \cdot \mathbf{h}_{eff}$ (Eq. (5.6)).

Solution: From the vector potential (see Fig. 5.1 and Eq. (5.4)) the magnetic Hamiltonian for the electron is

$$\mathcal{H}_{hyp} = 2\mu_B \frac{\mathbf{l} \cdot \mu_I}{r^3} + 2\mu_B \mathbf{s} \cdot \nabla \times \left[-\nabla \times \frac{\mu_I}{r} \right]$$

Since

$$\nabla \times \left[-\nabla \times \frac{\mu_I}{r} \right] = -g_n M_n \left\{ \frac{\mathbf{I}}{r^3} - \frac{3(\mathbf{I} \cdot \mathbf{r})\mathbf{r}}{r^5} \right\} + g_n M_n \mathbf{I} div \left(\frac{\mathbf{r}}{r^3} \right),$$

while $div(\mathbf{r}/r^3) = 4\pi\delta(\mathbf{r})$, one writes

$$\mathcal{H}_{hyp} = 2\mu_B g_n M_n \frac{\mathbf{I} \cdot \mathbf{l}}{r^3} - 2\mu_B g_n M_n \left\{ \frac{\mathbf{s} \cdot \mathbf{I}}{r^3} - \frac{3(\mathbf{I} \cdot \mathbf{r})(\mathbf{s} \cdot \mathbf{r})}{r^5} \right\} + 2\mu_B g_n M_n (\mathbf{s} \cdot \mathbf{I}) 4\pi\delta(\mathbf{r}) \equiv$$

$$\equiv A + B + C,$$

To deal with the singularities at the origin involved in B and C, let us define with V_ε a little sphere of radius ε centered at $r = 0$. Then in the integral for the expectation values

$$I = \int_{V_\varepsilon} B\phi^*(\mathbf{r})\phi(\mathbf{r})d\tau \equiv \int_{V_\varepsilon} Bf(\mathbf{r})d\tau$$

one can expand $f(\mathbf{r})$ in Taylor series, within the volume V_ε

$$f(\mathbf{r}) = f(0) + \mathbf{r} \cdot \nabla f(\mathbf{r}) + \frac{1}{2}(\mathbf{r} \cdot \nabla)(\mathbf{r} \cdot \nabla)f(\mathbf{r})$$

In I there are two types of terms, one of the form

$$s_x I_x \frac{\partial^2}{\partial x^2} \left(\frac{1}{r} \right) \quad (a)$$

the other of the form

$$\left(s_x I_y + s_y I_x \right) \frac{\partial^2}{\partial x \partial y} \left(\frac{1}{r} \right) \quad (b)$$

In the expansion $(\mathbf{r} \cdot \nabla f)$ is odd while (a) terms are even, thus yielding zero in I. The product of (a) terms with the third term in $f(\mathbf{r})$ when even, contributes with a term quadratic in ε.

The terms of type (b) are odd in the two variables, while $\mathbf{r} \cdot \nabla f$ includes odd terms in a single variable. In the same way are odd (and do not give contribution) the terms $(b)f(0)$. Finally the terms (b) times the third term in the expression again contribute to I only to the second order in ε. Therefore, one can limit I to

$$I = 2g_n M_n \mu_B f(0) \frac{1}{3} \int_{V_\varepsilon} (\mathbf{s} \cdot \mathbf{I}) \nabla^2 \left(\frac{1}{r} \right) d\tau.$$

Since $\nabla^2(1/r) = -4\pi\delta(\mathbf{r})$ the magnetic hyperfine hamiltonian can be rewritten[2]

$$\mathcal{H}_{hyp} = 2\mu_B g_n M_n \frac{\mathbf{I}\cdot\mathbf{l}}{r^3} - 2\mu_B g_n M_n \left[\frac{\mathbf{s}\cdot\mathbf{I}}{r^3} - \frac{3(\mathbf{s}\cdot\mathbf{r})(\mathbf{I}\cdot\mathbf{r})}{r^5}\right]^* + \frac{16\pi}{3}\mu_B g_n M_n \mathbf{s}\cdot\mathbf{I}\delta(\mathbf{r})$$

Thus the effective field $\mathbf{h}_{eff}$ in the form given in Eq. (5.6) is justified.

A model which allows one to derive similar results for the dipolar and the contact terms is to consider the nucleus as a small sphere with a uniform magnetization $\mathbf{M}$, namely a magnetic moment $\boldsymbol{\mu}_n = (4\pi R^3/3)\mathbf{M}$. For $r > R$ the magnetic field is the one of a point magnetic dipole. Inside the sphere $\mathbf{H}_{int} = (8\pi/3)\mathbf{M}$. By taking the limit $R \to 0$, keeping μ_n constant and then assuming that $\mathbf{M} \to \infty$, so that $\int_{r<R} \mathbf{H}_{int}d\mathbf{r}_n = 8\pi\mu_n/3$, the complete expression of the field turns out

$$\mathbf{H} = -\frac{\boldsymbol{\mu}_n}{r^3} + 3\frac{(\boldsymbol{\mu}_n\cdot\mathbf{r})\mathbf{r}}{r^5} + \frac{8\pi}{3}\boldsymbol{\mu}_n\delta(\mathbf{r}).$$

Problem 5.22 From the energy of the nuclear charge distribution in the electric potential due to the electron (Eq. (5.16)) derive the hyperfine quadrupole Hamiltonian (Eq. (5.20)).

Solution: By starting from Eq. (5.18) a new tensor Q_{ij} so that $\sum_l Q_{ll} = 0$ is defined

$$Q_{ij} = 3Q'_{ij} - \delta_{ij}\sum_l Q'_{ll}$$

and in terms of Q'_{ij} one has

$$E_Q = \frac{1}{6}\sum_{ij} Q_{ij}V_{ij} + \frac{1}{6}\sum_l Q'_{ll}\sum_j V_{jj}$$

The second term can be neglected since $\sum_j V_{jj} \simeq 0$. Thus

$$\mathcal{H}^Q_{hyp} = \sum_{ij} \hat{Q}_{ij}\frac{\hat{V}_{ij}}{6}$$

where the operators are

$$\hat{Q}_{ij} = e\sum_n \left(3x_i^n x_j^n - \delta_{ij}r_n^2\right)$$

---

[2]The star in the following equation means that in the expectation value a small sphere at the origin can be excluded in the integration and then ε set to zero. All singularities are included in the contact term.

$$\hat{V}_{ij} = -e \sum_e \frac{(3x_i x_j - \delta_{ij} r_e^2)}{r_e^5}$$

This Hamiltonian can be simplified by expressing the five independent components of Q_{ij} in terms of one. Semiclassically this simplification originates from the precession of the nuclear charges around $\mathbf{I}$, yielding a charge distribution with cylindrical symmetry around the z direction of the nuclear spin.

Then $Q_{ij} = 0$ for $i \neq j$ and being $\sum_l Q_{ll} = 0$, one has $\hat{Q}_{11} = Q_{22} = -Q_{33}/2$ with $Q_{33} = \int \rho_n(\mathbf{r})(3z^2 - r^2) d\tau_n$.

In the quantum description the reduction of $\mathcal{H}_{hyp}^Q$ is obtained by considering that only the dependence from the orientation is relevant. Thus, for the matrix elements $< I, M_I' |\hat{Q}_{ij}| I, M_I >$ (other quantum numbers for the nuclear state being irrelevant), by using *Wigner-Eckart theorem* one writes

$$< I, M_I' |\hat{Q}_{ij}| I, M_I > = C < I, M_I' \left| \frac{3}{2}(I_i I_j + I_j I_i) - \delta_{ij} I^2 \right| I, M_I >.$$

By defining, in analogy to the classical description, the quadrupole moment Q in proton charge units as

$$Q = < II \left| \frac{\hat{Q}_{zz}}{e} \right| II > \equiv < II \left| \sum_n (3z_n^2 - r_n^2) \right| II >$$

the constant C is obtained:

$$C < II \left| 3I_z^2 - I^2 \right| II > = C \left[3I^2 - I(I+1) \right] = eQ$$

Therefore all the components Q_{ij} are expressed in terms of Q, which has the classical physical meaning (see Eqs. (5.2) and (5.22)). Then the quadrupole operator is

$$\hat{Q}_{ij} = \frac{eQ}{I(2I-1)} \left\{ \frac{3}{2}(I_i I_j + I_j I_i) - \delta_{ij} I^2 \right\}$$

Analogous procedure can be carried out for the electric field gradient operator:

$$\hat{V}_{ij} = \frac{eq_J}{J(2J-1)} \left\{ \frac{3}{2}(J_i J_j + J_j J_i) - \delta_{ij} J^2 \right\}$$

where

$$q_J = < JJ \left| \frac{\hat{V}_{zz}}{e} \right| JJ > = < JJ \left| -\frac{\sum_e (3z_e^2 - r_e^2)}{r_e^5} \right| JJ >$$

Finally, since

$$\sum_{ij} I_i I_j J_i J_j = \left(\sum_i I_i J_i \right)^2 = (\mathbf{I} \cdot \mathbf{J})^2$$

$$\sum_{ij} I_i I_j \delta_{ij} J^2 = \left(\sum_i I_i \right)^2 J^2 = I^2 J^2$$

$$\sum_{ij} I_i I_j J_j J_i = (\mathbf{I} \cdot \mathbf{J})^2 + (\mathbf{I} \cdot \mathbf{J})$$

the quadrupole hyperfine Hamiltonian is written

$$\mathcal{H}_{hyp}^{Q} = \frac{eq_J Q}{2I(2I-1)J(2J-1)} \left\{ 3(\mathbf{I} \cdot \mathbf{J})^2 + \frac{3}{2}(\mathbf{I} \cdot \mathbf{J}) - I^2 J^2 \right\},$$

as in Eq. (5.20) (see also Eq. (5.24)).

Problem 5.23 At Sect. 1.5 the isotope effect due to the *reduced mass* correction has been mentioned. Since two isotopes may differ in the nuclear radius R by an amount δR, once that a finite nuclear volume is taken into account a further shift of the atomic energy levels has to be expected. In the assumption of nuclear charge Ze uniformly distributed in a sphere of radius $R = r_F A^{1/3}$ (with Fermi radius $r_F = 1.2 \times 10^{-13}$ cm) estimate the *volume shift* in an hydrogenic atom and in a muonic atom. Finally discuss the effect that can be expected in muonic atoms with respect to ordinary atoms in regards of the hyperfine terms.

Solution: The potential energy of the electron is $V(r) = -Ze^2/r$ for $r \geq R$, while (see Problem 1.6)

$$V(r) = -3 \frac{Ze^2}{2R} \left(1 - \frac{r^2}{3R^2} \right) \quad \text{for } r \leq R$$

The first-order correction, with respect to the nuclear point charge hydrogenic Hamiltonian, turns out

$$\Delta E = \frac{Ze^2}{2R} \int_0^R |R_{nl}(r)|^2 \left(-3 + \frac{r^2}{R^2} + \frac{2R}{r} \right) r^2 dr \simeq \frac{Ze^2}{10} R^2 |R_{nl}(0)|^2$$

The correction is negligible for non-s states, where $R_{nl}(0) \simeq 0$, while for s states one has

$$\Delta E = \frac{2}{5} e^2 R^2 \frac{Z^4}{a_0^3 n^3}$$

In terms of the difference δR in the radii (to the first order) the shift turns out

$$\delta E \simeq \frac{4}{5} e^2 R^2 \frac{Z^4}{a_0^3 n^3} \frac{\delta R}{R}$$

In muonic atoms (see Sect. 1.5) because of the change in the reduced mass and in the Bohr radius a_0, the volume isotope effect is dramatically increased with respect to ordinary hydrogenic atoms.

As regards the hyperfine terms one has to consider the decrease in the Bohr radius and in the Bohr magneton ($\mu_B \propto 1/m$) (see Problem 5.4). For the hyperfine quadrupole correction small effects have to be expected, since only states with $l \neq 0$ are involved.

Finally it is mentioned that an *isomeric shift*, analogous to the volume isotope shift, occurs when a radiative decay (e.g. from ^{57}Co to ^{57}Fe) changes the radius of the nucleus. The isomeric shift is experimentally detected in the *Mössbauer resonant absorption spectrum* (see Sect. 14.6).

Specific References and Further Reading

1. S. Svanberg, *Atomic and Molecular Spectroscopy*, Springer Verlag, Berlin (2003).
2. A. Abragam, *L'effet Mossbauer et ses applications a l'etude des champs internes*, (Gordon and Breach, 1964).
3. A. Balzarotti, M. Cini, M. Fanfoni, *Atomi, Molecole e Solidi. Esercizi risolti*, (Springer Verlag, 2004).
4. B.H. Bransden and C.J. Joachain, *Physics of atoms and molecules*, (Prentice Hall, 2002).
5. D. Budker, D.F. Kimball and D.P. De Mille, *Atomic Physics - An Exploration Through Problems and Solutions*, (Oxford University Press, 2004).
6. B. Cagnac and J.C. Pebay - Peyroula, *Physique atomique, tome 2*, (Dunod Université, Paris, 1971).
7. E.U. Condon and G.H. Shortley, *The Theory of Atomic Spectra*, (Cambridge University Press, London, 1959).
8. R. Eisberg and R. Resnick, *Quantum Physics of Atoms, Molecules, Solids, Nuclei and Particles*, (J. Wiley and Sons, 1985).
9. H. Haken and H.C. Wolf, *Atomic and Quantum Physics*, (Springer Verlag, Berlin, 1987).
10. C.S. Johnson and L.G. Pedersen, *Quantum Chemistry and Physics*, (Addison-Wesley, 1977).
11. M.A. Morrison, T.L. Estle and N.F. Lane, *Quantum States of Atoms, Molecules and Solids*, (Prentice -Hall Inc., New Jersey, 1976).
12. N.F. Ramsey, *Nuclear Moments*, (J. Wiley Inc., New York, 1953).
13. C.P. Slichter, *Principles of Magnetic Resonance*, (Springer Verlag, Berlin, 1990).

Chapter 6
Spin Statistics, Magnetic Resonance, Spin Motion and Echoes

Topics

Spin Temperature and Spin Thermodynamics
Magnetic Resonance and Magnetic Dipole Transitions
NMR and EPR
Spin Echo
Cooling at Extremely Low Temperatures

This chapter, dealing with nuclear and electronic angular momenta in magnetic fields, further develops topics already discussed in Chaps. 4 and 5. The new arguments involve some aspects of *spin statistics* and of *magnetic resonance* (namely how to drive the angular and magnetic moments and to change their components along a magnetic field). The magnetic resonance experiment in most cases is equivalent to induce magnetic dipole transitions among Zeeman-like levels.

6.1 Spin Statistics, Spin-Temperature and Fluctuations

Let us refer to a number N (of the order of the Avogadro number) weakly interacting spins $S = 1/2$, each carrying magnetic moment $\mu = -2\mu_B S$, in static and homogeneous magnetic field $\mathbf{H}$ along the z-axis. At the thermal equilibrium the statistical distribution depicted in Fig. 6.1 occurs. The number of spins (*statistical populations*) on the two energy levels are

$$N_- = N \frac{e^{\frac{\mu_B H}{k_B T}}}{e^{\frac{\mu_B H}{k_B T}} + e^{\frac{-\mu_B H}{k_B T}}} \equiv \frac{N}{Z} e^{\frac{\mu_B H}{k_B T}} \tag{6.1}$$

© Springer International Publishing Switzerland 2015
A. Rigamonti and P. Carretta, *Structure of Matter*,
UNITEXT for Physics, DOI 10.1007/978-3-319-17897-4_6

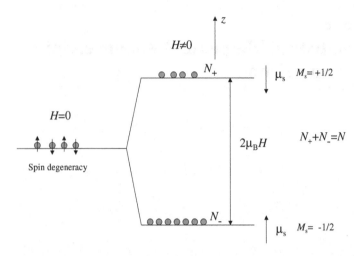

Fig. 6.1 Pictorial view of the statistical distribution of N spins $S = 1/2$ on the two "Zeeman levels" in a magnetic field, with $N_- > N_+$ at thermal equilibrium. In a field of 1 T the separation energy $2\mu_B H$ is 1.16×10^{-4} eV, corresponding to the magnetic temperature $2\varepsilon = 2\mu_B H/k_B = 1.343$ K. An equivalent description holds for protons, with $I = 1/2$, with the lowest energy level corresponding to quantum magnetic number $M_I = +1/2$, the gyromagnetic ratio being positive (Sect. 5.1). The energy separation between the two levels, for proton magnetic moments, is $2\mu_p H$, with $\mu_p = M_n g_n I$ and M_n the nuclear magneton, $g_n = 5.586$ the nuclear g-factor. In a field of 1 T, for protons the separation turns out 1.76×10^{-7} eV (or 20.4×10^{-4} K)

and

$$N_+ = \frac{N}{Z} e^{\frac{-\mu_B H}{k_B T}}, \qquad (6.2)$$

with Z the partition function (for reminds see Sect. 4.4, Problem 4.18). The contribution to the thermodynamical energy is

$$U = N_-(-\mu_B H) + N_+(\mu_B H) = \mu_B H(2N_+ - N) \equiv \left[\frac{2N}{Z} e^{-\varepsilon/T} - N\right]\mu_B H   \qquad (6.3)$$

with

$$\varepsilon = \frac{\mu_B H}{k_B} \qquad \text{the "magnetic temperature"}$$

(having assumed $U = 0$ in the absence of the magnetic field).

The statistical populations N_+ or N_- are modified when the temperature (or the field) is changed and after some time a new equilibrium condition is attained. $N_\pm$ can be varied, while keeping the temperature of the thermal reservoir and the magnetic field constant, by proper irradiation at the transitional frequency $\nu = 2\mu_B H/h$, by resorting to the *magnetic dipole transition mechanism*(the methodology is known, in general, as *magnetic resonance*, described in some detail at Sect. 6.2).

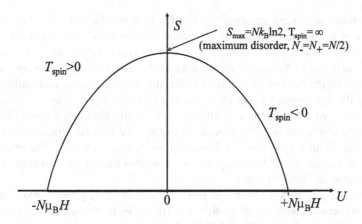

Fig. 6.2 Entropy S as a function of the energy U in a spin system. The statistical entropy is defined as the logarithm of the number of ways a given spin distribution can be attained (See Sect. 6.4). The zeroes at $U = \pm N\mu_B H$ correspond to all spins in a single state (see also Problems 6.2, 6.4 and 6.7)

When $N_\pm$ are modified, in principle the energy U can take any value in between $-N\mu_B H$ (corresponding to full occupation of the state at $M_S = -1/2$) and $+N\mu_B H$ (complete reversing of all the spins, with $N_+ = N$).

From thermodynamics, no volume variation being involved, the entropy of the spin system can be defined

$$S_{spin} \equiv S = \int \frac{1}{T}\left(\frac{\partial U}{\partial T}\right)_V dT \tag{6.4}$$

and therefore, from Eq. (6.3),

$$S = 2\mu_B H \int \frac{1}{T}\left(\frac{\partial N_+}{\partial T}\right)_V dT \tag{6.5}$$

When the statistical distribution on the levels is modified the entropy changes, in the way sketched in Fig. 6.2 in terms of the energy U.

Since the temperature can be expressed as

$$\frac{1}{T} = \frac{\partial S}{\partial U} \tag{6.6}$$

(in the partial differentiation keeping constant all the other thermodynamical variables), one can define a *spin temperature* T_{spin} in terms of N_+ and N_-. Thus a spin temperature is defined also for $U > 0$, eventhough there is not a correspondent thermal equilibrium temperature T of the reservoir. When, by means of magnetic resonance methods (or, for example, simply by suddenly reversing the magnetic field) the equilibrium distribution is altered, then $T_{spin} \neq T$. It should be remarked that this non-equilibrium situation can last for time intervals of experimental significance only when the probability of spontaneous emission (see Appendix 1.2) is not so strong to cause fast restoring. This is indeed the case for states of magnetic moments in magnetic fields (see the estimate in Problem 5.19). However, exchanges of energy with the thermal bath, related to the time-dependence of the Hamiltonians coupling the spin system to all other degrees of freedom (the *"lattice"*), usually occur. This is why a given non-equilibrium spin distribution rather fast attains the equilibrium condition, usually through an exponential process characterized by a time constant called *spin-lattice relaxation time* T_1 (see Sect. 6.2). The relaxation times T_1's, particularly at low temperatures, are often long enough to allow one to deal with non-equilibrium states.

Let us imagine to have prepared one spin system at $T_{spin} = -300\,\mathrm{K}$ and to bring it in thermal contact with another one, strictly equivalent but at thermal equilibrium, namely at $T_{spin} = T = 300\,\mathrm{K}$. The two systems reach a common equilibrium by means of *spin-spin transitions* in which two spins exchange their relative orientations (this process involves a *spin-spin relaxation time* T_2 usually much shorter than T_1). The total energy is constant while the temperatures of both the two sub-systems evolve, as well as the entropy. The final spin temperatures are $+\infty$ and $-\infty$ and the entropy takes its maximum value. The internal equilibrium, with $T_{spin} = \pm\infty$, is attained in very short times (for $T_2 \ll T_1$). Then the spin-lattice relaxation process drives the system towards the thermodynamical equilibrium condition, where $T_{spin} = T$.

Now we return to the field induced magnetization

$$M = N < \mu_z >_H, \tag{6.7}$$

$< \mu_z >_H$ being the statistical average of the component of the magnetic moment along the field (see Sect. 4.4).

From

$$\frac{-g\mu_B \sum_{M_J} M_J e^{-g\mu_B M_J H/k_B T}}{Z} = k_B T \frac{\partial \left(ln(\sum_{M_J} e^{-g\mu_B M_J H/k_B T}) \right)}{\partial H} \tag{6.8}$$

with Z the partition function (see Eq. (4.30)), the magnetization can be written

$$M = Nk_B T \left(\frac{\partial lnZ}{\partial H} \right)_T. \tag{6.9}$$

For $J = S = 1/2$

$$M = \frac{N}{2} 2\mu_B tanh \left(\frac{\mu_B H}{k_B T} \right) \tag{6.10}$$

Let us now evaluate the mean square deviation of the magnetization from this average equilibrium value, i.e. its *fluctuations* $< (M- < M >)^2 >$ (now we have added the symbol <> to M in Eqs. (6.9) or (6.10) to mean its average character). The magnetization has a Gaussian distribution around the average value $< M >$, zero for $H = 0$ (See Problem 6.1) and the one in Eq. (6.10) in the presence of the field.

For the fluctuations one has

$$< \Delta M^2 > = < (M - < M >)^2 > =$$
$$= < M^2 > - 2 < M < M >> + < M >^2 = < M^2 > - < M >^2 \quad (6.11)$$

The single $< \mu_z >$'s are *uncorrelated* and therefore $< \Delta M^2 > = N < \Delta \mu_z^2 >$ with $< \Delta \mu_z^2 > = < \mu_z^2 > -(< \mu_z >)^2$, yielding

$$< M^2 > = N < \mu_z^2 >_H = 4 N \mu_B^2 \frac{\sum_{M_s} M_s^2 e^{-x M_s}}{Z}$$

with $x = (2\mu_B H / k_B T)$ and $M_S^2 = 1/4$.
Then $< M^2 > = N \mu_B^2$ and finally, from Eqs. (6.10) and (6.11)

$$< \Delta M^2 > = N \mu_B^2 \left[1 - tanh^2 \left(\frac{\mu_B H}{k_B T} \right) \right] \quad (6.12)$$

Now we look for the relationship of the fluctuations to the *response function*, the magnetic susceptibility $\chi = \partial < M > / \partial H$. Again, from Eq. (6.10) one derives

$$\chi = N \mu_B \left[1 - tanh^2 \left(\frac{\mu_B H}{k_B T} \right) \right] \frac{\mu_B}{k_B T} \quad (6.13)$$

and therefore

$$< \Delta M^2 > = k_B T \chi. \quad (6.14)$$

This relationship is a particular case of the *fluctuation-dissipation theorem*, relating the spectrum of the fluctuations to the response functions (see Problem 4.18 for an equivalent derivation).

The considerations carried out in the present paragraph are a few illustrative examples of the topic that one could call *spin thermodynamics*. This field includes the method of *adiabatic demagnetization*, which allows one to reach the lowest temperatures (Sect. 6.4). An introduction to statistical physics with paramagnets, leading step by step the reader to the concepts suited for extending the arguments recalled in the present paragraph, can be found in Chaps. 4 and 5 of the book by *Amit* and *Verbin*.

Problems

Problem 6.1 Express the probability distribution of the total "magnetization" along a given direction in a system of N independent spin $S = 1/2$, in zero magnetic field.

Solution: Along the z-direction two values $\pm \mu_B$ are possible for the magnetic moment. The probability of a given sequence is $(1/2)^N$. A magnetization $M = n\mu_B$ implies $\frac{1}{2}(N+n)$ magnetic moments "up" and $\frac{1}{2}(N-n)$ magnetic moments "down" (see Fig. 6.1). The total number of independent sequences giving such a distribution is

$$W(n) = \frac{N!}{\left[\frac{1}{2}(N+n)\right]! \left[\frac{1}{2}(N-n)\right]!}.$$

The probability distribution for the magnetization is thus $W(M) = W(n)(1/2)^N$. From Stirling approximation and series expansion

$$ln\left(1 \pm \frac{n}{N}\right) \approx \pm\frac{n}{N} - \frac{n^2}{2N^2} \pm \ldots$$

one has

$$ln\,W(M) \approx -\frac{1}{2}ln\left(\frac{\pi N}{2}\right) - \frac{n^2}{2N}$$

so that

$$W(M) \approx \left(\frac{2}{\pi N}\right)^{1/2} exp\left[-\frac{n^2}{2N}\right]$$

namely a *Gaussian distribution* around the value $<M> = 0$, at width about $(N)^{1/2}$. It is noted that the *fractional width* goes as $N^{-1/2}$, rapidly decreasing for large N.

Problem 6.2 Express the entropy of an ensemble of $S = 1/2$ non-interacting spins in a magnetic field and discuss the spin temperature recalled in Fig. 6.2.

Solution: The number of available states is

$$W = \frac{N!}{(N_+)!(N_-)!}.$$

Resorting to the Stirling approximation (see Problem 6.1) the entropy is

$$S = k_B ln\,W = k_B[N ln N - N_+ ln N_+ - N_- ln N_-].$$

The energy U (in Fig. 6.2) can be written $U = N_+\alpha$, by setting the low-energy level at zero and $\alpha = 2\mu_B H$. Being $N_- = N - N_+$ the entropy becomes

$$S = k_B[N ln N - u ln u - (N - u)ln(N - u)],$$

with $u = U/\alpha \equiv N_+$. From Eq. (6.6), with $\partial S/\partial U = (1/\alpha)\partial S/\partial u$

$$\frac{1}{T} = \frac{k_B}{\alpha}[ln N_- - ln N_+]$$

or

$$T = \frac{\alpha}{k_B ln(\frac{N}{u} - 1)}$$

justifying the plot in Fig. 6.2. The maximum of S occurs for $u = N/2$, i.e. $N_+ = N/2$. From the free energy

$$F = -\frac{N}{\beta} ln[1 + exp(-2\beta \mu_B H)],$$

and $S = - (\partial F/\partial T)_H$ the same expression for the entropy in terms of u is obtained.

Problem 6.3 Two identical spin systems at $S = 1/2$, prepared at spin temperatures $T_a = E/2k_B$ and $T_b = -E/k_B$ are brought into interaction. Find the energy and the spin temperature of the final state.

Solution: By setting $E = 0$ for the low energy level, $U_x = U_a + U_b$ is written

$$U_x = 2NE\frac{exp(-E/k_B T_x)}{1 + exp(-E/k_B T_x)}.$$

Since

$$U_a = NE\frac{exp(-2)}{1 + exp(-2)}$$

and

$$U_b = NE\frac{exp(1)}{1 + exp(+1)},$$

one has

$$exp(E/k_B T_x) = \frac{e^2 + e^{-1} + 2e}{2 + e^2 + e^{-1}} \equiv z$$

and then

$$T_x = \frac{E}{k_B \, ln \, z} \approx 3.3 \frac{E}{k_B}.$$

Problem 6.4 Show that the entropy (per particle) of a system can be written

$$S = -k_B \sum_n p_n \, ln \, p_n$$

where p_n is the probability that the system is found in the state at energy E_n, namely for a canonical ensemble

$$p_n = \frac{exp(-E_n/k_B T)}{Z},$$

with Z partition function. This form of S is known as *Shannon-Von Neumann* entropy and it holds also for microcanonical and grand canonical ensembles.

Solution: In fact

$$S = -\frac{k_B}{Z} \sum_n exp\left(-\frac{E_n}{k_B T}\right)\left[-\frac{E_n}{k_B T} - ln Z\right]$$

$$= k_B \frac{ln Z}{Z} \sum_n exp\left(-\frac{E_n}{k_B T}\right) + \frac{1}{T} \sum_n exp\left(-\frac{E_n}{k_B T}\right) \frac{E_n}{Z}$$

$$= k_B ln Z + \frac{1}{T} \sum_n exp\left(-\frac{E_n}{k_B T}\right) \frac{E_n}{Z}.$$

On the other hand, from

$$F = -k_B T ln Z$$

one can write

$$S = \frac{U - F}{T} = k_B \frac{\partial (T ln Z)}{\partial T}$$

and

$$U = k_B T^2 \frac{\partial ln Z}{\partial T}.$$

Then

$$S = -\left[\frac{\partial F}{\partial T}\right]_{v,H} = k_B ln Z + k_B T \frac{\partial ln Z}{\partial T}.$$

Since

$$\frac{\partial ln Z}{\partial T} = \frac{1}{k_B T^2} \sum_n exp\left(-\frac{E_n}{k_B T}\right) \frac{E_n}{Z}$$

one has

$$S = k_B ln Z + \frac{1}{T} \sum_n exp\left(-\frac{E_n}{k_B T}\right) \frac{E_n}{Z}.$$

Problem 6.5 A model widely used in statistics and in magnetism is the *Ising* model, for which an Hamiltonian of the form $\mathcal{H} = -K \sum_{i,j} s_i s_j$ is assumed, with the spin variables s_i taking the values $+1$ and -1. K is the *exchange integral* (see Sect. 2.2.2).[1] Derive the partition function Z, the free energy F, the thermodynamical energy U and the specific heat C_V, for a system of N spins.

---

[1] It can be remarked that having assumed a site-independent interaction, this model corresponds to the mean-field description, or equivalently to an infinite range of the interactions (see Chap. 17).

Solution: By indicating with $N_{p,a}$ the number of parallel (p) and antiparallel (a) spins with $N_p + N_a = N - 1$ the number of interacting pairs, the energy of a given spin configuration is $E = -K(N_p - N_a) = -K(2N_p + 1 - N)$.

The number of permutations of the $(N-1)$ pairs is $(N-1)!$, of which $[(N-1)!/N_a! N_p!]$ are distinguishable. Therefore the sum over the states reads

$$Z = 2 \sum_{N_p=0}^{N-1} \left[\frac{(N-1)!}{N_a! N_p!} \right] exp \left[+ \frac{K(2N_p + 1 - N)}{k_B T} \right]$$

$$= 2 \, exp \left[+ \frac{K(1-N)}{k_B T} \right] \sum_{N_p} \left[\frac{(N-1)!}{(N-1-N_p)! N_p!} \right] exp \left[+ \frac{K(2N_p)}{k_B T} \right]$$

(the factor 2 accounts for the configurations arising under the reversing of all the spins without changing N_p or N_a). The sum is the expansion of $\{1 + exp[(2K/k_B T)]\}^{N-1}$ and therefore

$$Z = 2^N \left[\cosh \frac{K}{k_B T} \right]^{N-1} .$$

Then

$$F = -k_B T \, lnZ = -k_B T \left[N \, ln2 + (N-1) \, ln \left(\cosh \frac{K}{k_B T} \right) \right],$$

$$U = -\frac{\partial (lnZ)}{\partial \beta} = -(N-1) K \tanh(\beta K), \quad \text{with} \quad \beta = \frac{1}{k_B T}$$

and

$$C_V = \left(\frac{\partial U}{\partial T} \right)_N = (N-1) \left(\frac{K^2}{k_B T^2} \right) \left[\frac{1}{(\cosh \beta K)^2} \right].$$

6.2 The Principle of Magnetic Resonance and the Spin Motion

Transitions involving hyperfine states or nuclear and/or electronic Zeeman-states in magnetic fields are carried out by resorting to the magnetic dipole mechanism. These transitions are usually performed by exploiting the phenomenon elsewhere called *magnetic resonance*, which allows one to drive electronic or nuclear magnetic moments. This type of experiments are at the core of modern *microwave* and *radiofrequency spectroscopies*.

The first experiment of magnetic resonance, performed by *Rabi*, involved molecular beams (see Fig. 6.3).

The vectorial description, with classical equation of motion (Chap. 3) is the following (see Fig. 6.4). The motion of the angular momentum **L** in **H$_0$** is described by

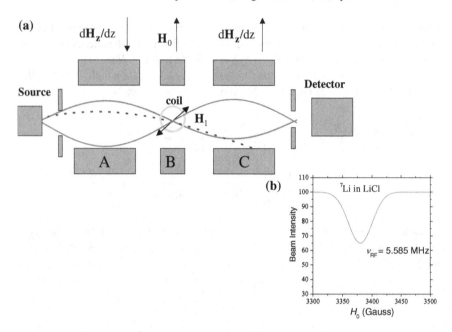

Fig. 6.3 **a** Sketch of the experimental setup for magnetic resonance in beams (*ABMR*). The magnetic fields A and C have gradients along opposite directions. In region B the magnetic field $H_0 \parallel z$ is homogeneous. The radiofrequency (or the microwave) field H_1 in region B is perpendicular to H_0. In part **b** of the figure the sketch of a typical magnetic resonance signal is shown, detected as a minimum in the arrival of the atoms when in region C the refocusing of the deviations is inhibited (*dotted line*) by the resonance driven by H_1 in region B (see text)

$$\frac{d\mathbf{L}}{dt} = \boldsymbol{\mu}_L \times \mathbf{H}_0 \quad \text{i.e.} \quad \frac{d\boldsymbol{\mu}_L}{dt} = -\gamma(\boldsymbol{\mu}_L \times \mathbf{H}_0) \tag{6.15}$$

implying the precession at the Larmor frequency $\omega_L = \gamma H_0$ (see Sect. 3.2 and Problem 3.4). In a frame of reference rotating at angular frequency ω, Eq. (6.15) becomes[2]

$$\frac{d\boldsymbol{\mu}_L}{dt} = \gamma\left(\mathbf{H}_0 + \frac{\omega}{\gamma}\right) \times \boldsymbol{\mu}_L. \tag{6.16}$$

Thus in the presence of the radiofrequency (or microwave) irradiation the effective field is

$$\mathbf{H}_{eff} = \left(H_0 + \frac{\omega_{RF}}{\gamma}\right)\hat{k} + H_1\hat{i} \tag{6.17}$$

---

[2]It is reminded that

$$\left(\frac{d\boldsymbol{\mu}}{dt}\right)_{lab.frame} = \left(\frac{\partial\boldsymbol{\mu}}{\partial t}\right)_{relative\ to\ rot.\ frame} + \left(\frac{d\boldsymbol{\mu}}{dt}\right)_{rot.frame}$$

the latter being $\omega \times \boldsymbol{\mu}$. For $H_x = H_1 cos\omega_{RF}t$, $H_y = H_1 sin\omega_{RF}t$ and $H_z = H_0$, in the rotating frame the magnetic field is constant: $H'_x = H_1$, $H'_y = 0$ and $H'_z = H_0$ (see Fig. 6.6).

Fig. 6.4 Precessional motion of the magnetic moment μ at the angular frequency $\omega_L = \gamma H_0$ and rotation of the field H_1 at ω_{RF}. For $\omega_L = \omega_{RF}$ the magnetic resonance occurs. The gyromagnetic ratio γ is $\mu_I/I\hbar$ for nuclear moment (see Sect. 5.1) or $\gamma = \mu_J/J\hbar$ for electron magnetic moment

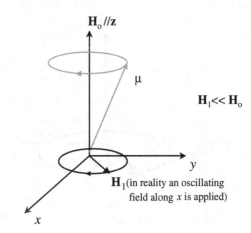

Fig. 6.5 Quantum magnetic levels for magnetic moment $\mu_I = gM_n\mathbf{I} = \gamma\mathbf{I}\hbar$, for $I = 1/2$ in a magnetic field. The resonance corresponds to transitions from $M_I = +1/2$ to $M_I = -1/2$, driven by the magnetic dipole mechanism

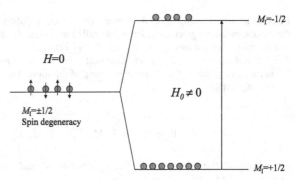

When $\omega_{RF} = -\gamma H_0$ (the sign minus refers to clockwise precession), in the rotating frame of reference only $\mathbf{H}_1$ is active and the magnetic moment precesses around it, thus changing its component with respect to $\mathbf{H}_0$. As a consequence of the change in the z-component of the magnetic moment in the region B of the Rabi experimental set up (Fig. 6.3) the compensation of the deviations due to $\mathbf{F} = \pm\mu_z(d\mathbf{H}/dz)$ in the regions A and C does no longer occur. Then a minimum in the number of atoms (or molecules) reaching the detector is observed.

The quantum description of the magnetic resonance corresponds to the situation depicted in Fig. 6.5 for nuclear spin $I = 1/2$.

The eigenvalues are $\pm M_I g_n M_n H_0$ and magnetic dipole transitions, with selection rule $\Delta M_I = \pm 1$, are possible when the condition $h\nu_{RF} = g_n M_n H_0 \equiv \hbar\omega_L$ is verified. The perturbation operator is

$$\mathcal{H}_P = -\mu_I.\mathbf{H}_1 \equiv -\gamma_N\hbar(H_x I_x + H_y I_y) = -\gamma_N\hbar\frac{H_1}{2}(I_- e^{i\omega_{RF}t} + I_+ e^{-i\omega_{RF}t}),$$

with only out-of-diagonal elements. By extending the description in Appendix 1.3, the transition probability has to be written

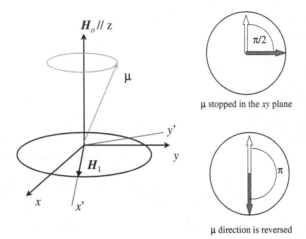

Fig. 6.6 Illustrative examples of the spin motions induced in pulse magnetic resonance, by stopping the irradiation after a given time. For the so-called $\pi/2$ *pulse* the time of irradiation turns out (see text for the precession around H_1) $\tau = (\pi/2)/\omega_1 = (\pi/2)/\gamma H_1 = (\pi/4)\hbar/H_1\mu_I$ for nuclear spin $I = 1/2$ and $\tau = (\pi/2)\hbar/H_1\mu_B$ for electron at $S = 1/2$. The π *pulse* requires an irradiation time 2τ and it corresponds to the complete reversing of the spins in the magnetic field H_0 (x', y', z' is the rotating frame)

$$W_{RF} \propto \ | < I, M_I' |I_+ + I_-|I, M_I > \ |^2. \qquad (6.18)$$

According to the properties of the $I_\pm$ operators[3] and to the orthogonality of states at different M_I, Eq. (6.18) leads to the selection rule $\Delta M_I = \pm 1$. The circular polarization required for $\Delta M_I = \pm 1$ transitions is the counterpart of the rotating field $\mathbf{H}_1$ perpendicular to the z-quantization axis.

A treatment of quantum character is possible (for free spins) by considering the time evolution of the expectation valuesfor the spin components (Problem 6.6).

The description in terms of spin motion is particularly suited for understanding the modern *pulse resonance techniques*, which allow one to drive the magnetic moments along a given direction by controlling the length of the radiofrequency irradiation. Examples are shown in Fig. 6.6.

Finally we mention that resonance experiments (*NMR* for nuclear, *EPR* for electron) nowadays are generally carried out in condensed matter, with a number of interesting applications.

In condensed matter the interactions with the other degrees of freedom (the "*lattice*") or among spins themselves, play a relevant role. Phenomenologically the interactions are taken into account by the *Bloch equations*, that for the expectation

---

[3] $< M|I_+|M-1 > = \sqrt{(I+M)(I-M+1)}, < M|I_-|M+1 > = \sqrt{(I-M)(I+M+1)},$

all other elements being zero.

values of the spin components, averaged over the statistical ensemble, in the rotating frame can be written

$$\frac{d < \overline{I_x} >}{dt} = -\frac{< \overline{I_x} >}{T_2} \tag{6.19a}$$

$$\frac{d < \overline{I_y} >}{dt} = -\frac{< \overline{I_y} >}{T_2}. \tag{6.19b}$$

These equations account for the decay of the transverse components of $< \overline{\mathbf{I}} >$, that at long time must vanish. For the longitudinal component the Bloch equation is

$$\frac{d < \overline{I_z} >}{dt} = \frac{(I_z^0 - < \overline{I_z} >)}{T_1} \tag{6.20}$$

(where I_z^0 is the expectation value of the z-component at the thermal equilibrium). This equation describes the relaxation process towards equilibrium, after a given alteration of the statistical populations (see Sect. 4.4 for a qualitative definition of the relaxation time T_1).

In order to have a complete description of the spin motions Eqs. (6.19) and (6.20) must be coupled to the equation

$$\frac{d < \overline{\mathbf{I}} >}{dt} = -\frac{g_n M_n}{\hbar} < \overline{\mathbf{I}} > \times \mathbf{H}_{eff} \tag{6.21}$$

where the effective field is defined in Eq. (6.17). Then one has a system of equations (6.19–6.21) for the expectation values of the spin components (often written in terms of the *nuclear magnetization* $\mathbf{M}_{nuclear} \propto \sum_i < \overline{\mathbf{I}_i} >$). These equations can be solved under certain approximations, to yield the time evolution of $< \overline{\mathbf{I}} >$ or of $\mathbf{M}_{nuclear}$.

The quantum description of the time evolution of the spin operators in magnetic resonance experiments, in the presence of the relaxation processes imbedded in Eqs. (6.19) and (6.20), is usually based on a variant of the time-dependent perturbation theory, the *density matrix method*. The textbook by *Slichter* deals with this matter to the due extent. We shall limit ourselves, in the next paragraph, to describe a very important phenomenon, the *spin echo*, that in simple circumstances can satisfactorily be treated on the basis of the semiclassical motions of the spin operators and of the Bloch equations.

Problems

Problem 6.6 Consider a single spin **s** in a constant and homogeneous magnetic field along the z-direction. From time-dependent Schrödinger equation derive the expectation values of the spin components and show that the precessional motion occurs.

Then consider a small oscillating magnetic field along the x-direction and prove that at the resonance one has reversing of $\mathbf{s}$ with respect to the static field. Discuss the cases of pulse application of the oscillating field for time intervals so that the rotation of $\mathbf{s}$ is by angles $\pi/2$ and π. Qualitatively figure out what happens if spin-spin and spin-lattice interactions are taken into account.

Solution: It can be noticed that the perturbation theory leading to Eq. (6.18) is valid only for short times, so that the probability of finding the spin in the original state is still close to unity. A solution valid for any time t can be given by means of a procedure based on the Rabi description of two-level systems (Appendix 1.2), for the case $S = 1/2$. From

$$\mu_B \mathbf{H} \begin{pmatrix} 1 & 0 \\ 0 & -1 \end{pmatrix} |\phi(t)> = i\hbar \frac{d|\phi(t)>}{dt}$$

where

$$|\phi(t)> = \alpha(t)|\uparrow> + \beta(t)|\downarrow>, \quad |\alpha(t)|^2 + |\beta(t)|^2 = 1,$$

one derives

$$\alpha(t) = a\, exp\,(-i\omega_L t/2) \qquad \beta(t) = b\, exp\,(i\omega_L t/2)$$

with ω_L Larmor frequency. The expectation values are

$$<\phi(t)|s_z|\phi(t)> = \frac{\hbar}{2}[|\alpha|^2 - |\beta|^2], \quad \text{time-independent,}$$

while

$$<\phi(t)|s_x|\phi(t)> = (a\,b\,\hbar)\cos(\omega_L t)$$

$$<\phi(t)|s_y|\phi(t)> = (a\,b\,\hbar)\sin(\omega_L t),$$

indicating the precession depicted in Figs. 6.4 and 6.6.
  In the presence of H_1 rotating in the (xy) plane

$$H_1(t) = H_1 exp[\pm i\omega t],$$

from the Schrödinger equation one derives

$$\frac{\hbar}{2}\omega_L\alpha + \mu_B H_1 exp[-i\omega t]\beta = i\hbar\frac{d\alpha}{dt}$$

$$\mu_B H_1 exp[+i\omega t]\alpha - \frac{\hbar}{2}\omega_L\beta = i\hbar\frac{d\beta}{dt}.$$

By writing the coefficients α and β in the form

$$\alpha = \Gamma(t)exp\left[-\frac{i\omega_L t}{2}\right] \qquad \beta = \Delta(t)exp\left[+\frac{i\omega_L t}{2}\right]$$

those equations are rewritten

$$\mu_B H_1 exp[-i(\omega - \omega_L)t]\Delta = i\hbar\frac{d\Gamma}{dt}$$

$$\mu_B H_1 exp[+i(\omega - \omega_L)t]\Gamma = i\hbar\frac{d\Delta}{dt}.$$

At the resonance

$$\mu_B H_1 \Delta = i\hbar\frac{d\Gamma}{dt} \quad and \quad \mu_B H_1 \Gamma = i\hbar\frac{d\Delta}{dt}.$$

From the derivative of the first, substituted in the second, one finds

$$\Gamma = \sin(\Omega t + \psi) \quad and \quad \Delta = i\cos(\Omega t + \psi), where \quad \Omega = \frac{\mu_B H_1}{\hbar}.$$

By setting $\psi = 0$, by repeating the derivation of the expectation values one has

$$< \phi(t)|s_z|\phi(t) > = -\frac{\hbar}{2}\cos(2\Omega t)$$

$$< \phi(t)|s_x|\phi(t) > = -\frac{\hbar}{2}\sin(2\Omega t)\sin(\omega_L t)$$

$$< \phi(t)|s_y|\phi(t) > = \frac{\hbar}{2}\sin(2\Omega t)\cos(\omega_L t).$$

These equations can be interpreted in terms of the motion of **s** as the superposition of the *precession around* z at the Larmor frequency and the *rotation around* H_1 at the angular frequency $2\mu_B H_1/\hbar$.

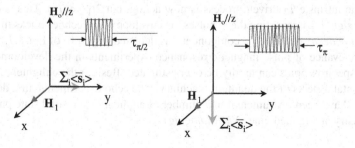

In the rotating frame (see Fig. 6.6) where $\mathbf{H}_1$ is fixed, one has the rotation of $\mathbf{s}$ by a given angle, depending on the duration of the irradiation. Thus, in principle, one can prepare the magnetization $\mathbf{M} \propto \sum_i < \overline{\mathbf{s}_i} >$ along any direction, as schematically illustrated above.

It is noted that the $(\pi/2)$ pulse corresponds to equalize of the statistical populations in the two Zeeman levels and magnetization in the xy plane. The π pulse corresponds to the inversion of $\mathbf{M}$ and therefore to *negative spin temperature* (see Sect. 6.1).

The spin-spin interaction implies the decay to zero, in a time of the order of T_2, of the transverse components of $\mathbf{M}$. The spin-lattice interaction, with transfer of energy to the reservoir, drives the relaxation process towards the thermal equilibrium distribution, with $\mathbf{M}$ along $\mathbf{H}_0$, attained in a time of the order of T_1 (see Problem 6.11).

6.3 Spin and Photon Echoes

Let us imagine that a system of electronic or nuclear spins has been brought in the xy plane (perpendicular to the z-axis along the field $\mathbf{H}_0$) by a $\pi/2$ pulse, by means of the experimental procedure described at Sect. 6.2 and Problem 6.6. Once in the plane, the transverse components have to decay towards zero according to Eq. (6.19), in a time of the order of T_2, yielding in a proper receiver a signal called *free induction decay (FID)*.

Now let us suppose that in times much shorter than T_2 another mechanism, different from the spin-spin interaction, causes a *distribution of precessional frequencies*. This mechanism could be due for instance to magnetic field inhomogeneities, to spatially varying diamagnetic or paramagnetic corrections to the external field H_0 or to a field gradient created by external coils. Because of the spread in the precessional frequencies, in a time usually called T_2^* and much shorter than T_2, the transverse components of the total magnetization $\mathbf{M}_{x,y} \propto \sum_i < \overline{\mathbf{I}}_{x,y}(i) >$ decay to value close to zero. After a time t_1 larger than T_2^* but shorter than T_2, a second pulse, of duration π, is applied (see Fig. 6.7). Since all the spins are flipped by 180° around the $x\prime$-axis, the ones precessing faster now are forced to return in phase with the ones precessing slower.

Thus after a further time interval t_1, refocusing of all the spins along a common direction occurs, yielding the "original" strength of the signal (only the reduction due to the intrinsic T_2-driven process is now acting, but $2t_1 \ll T_2$). This is called the *echo signal*. By repeating the π-pulses the envelope of the echoes tracks the real, *irreversible decay* of the $M_{x,y}$ components, as depicted in part b) of Fig. 6.7.

The relevance of pulse magnetic resonance experiments in the development of modern spectroscopies can hardly be over estimated. Besides the enlightenment of fundamental aspects of the quantum machinery, the echo experiments, first devised by *Hahn*, have been instrumental in a number of applications in solid state physics, in chemistry and in medicine (*NMR imaging*).

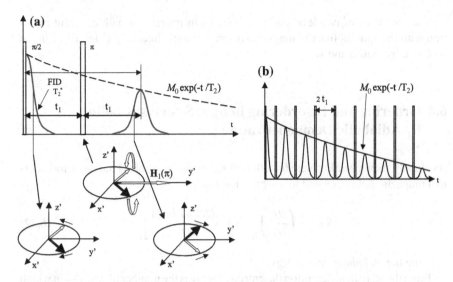

Fig. 6.7 Schematic representation of the spin motions generating the echo signal upon application of a sequence of $\pi/2$ and π pulses. Part **a** shows the FID signal following the $\pi/2$ pulse and how the echo signal is obtained at the time $2t_1$ owing to the *reversible decay* of the magnetization in a time shorter than T_2. The rotation of the spins in the (x, y) plane, as seen in the rotating frame of reference, evidences how refocusing generates the echo. It should be remarked that with pulse techniques, by switching the phase of the RF field it is possible to apply the second pulse (at time t_1) along a direction different from the one of the first pulse at $t = 0$ (e.g. from x' to y' in the rotating frame, see Fig. 6.6). Part **b** shows the effect of a sequence of π pulses (after the initial $\pi/2$), with a train of echoes, the envelope yielding the intrinsic *irreversible decay* of the transverse magnetization due to the T_2-controlled mechanism

Furthermore the pulse magnetic resonance methodology has been transferred in the field of the *optical spectroscopy*, by using lasers. In this case special techniques are required, because in the optical range the "dipoles" go very fast out of phase (the equivalent of T_2 is very short).

In this respect we only mention that the *pseudo-spin formalism* can be applied to any system where approximately only two energy levels, corresponding to the spin-up and spin-down states, can be considered relevant. For a pair of states in atoms, to a certain extent *coherent electric radiation* can be used to induce the analogous of the inversion of the magnetization described at Sect. 6.2 and Problem 6.6, in terms of the populations on the lower and on the upper atomic or molecular levels. The "oscillating" electric dipole moment $\mathbf{R}_{21}$ (see Appendix 1.3) plays the role analogous to the magnetic moment in the magnetic resonance phenomenon. After the "saturation of the line" corresponding to the equalization of the two levels (to a $\pi/2$-pulse), a second pulse π at a time t_1 later, can force the diverging phases of the oscillating electric dipoles to come back in phase: a "light pulse", the *photon echo*, is observed at the time $2t_1$.

The analogies of two-levels atomic systems in interaction with coherent radiation with the spin motions in magnetic resonance experiments, are described in the textbook by *Haken* and *Wolf.*

6.4 Ordering and Disordering in Spin Systems: Cooling by Adiabatic Demagnetization

As already shown (see Problem 6.4), the entropy of an ensemble of magnetic moments in a magnetic field is related to the partition function Z:

$$S = -\left(\frac{\partial F}{\partial T}\right)_H = \left[\frac{\partial (Nk_B T \ln Z)}{\partial T}\right]_H \tag{6.22}$$

F being the *Helmholtz free energy.*

From the statistical definition the entropy involves the number of ways W in which the magnetic moments can be arranged: $S = k_B \ln W$. For angular momenta $\mathbf{J}$, in the high temperature limit the M_J states are equally populated and $W = (2J + 1)^N$. The statistical entropy is

$$S = Nk_B \ln(2J + 1) \tag{6.23}$$

For $T \to 0$, in finite magnetic field, there is only one way to arrange the magnetic moments (see Sect. 6.1) and then the spin entropy tends to zero. In general, since the probability $p(M_J)$ that J_z takes the value M_J is given by

$$p(M_J) = \frac{e^{-M_J g \mu_B H / k_B T}}{Z}, \tag{6.24}$$

the statistical entropy has to be written (see Problems 6.4 and 6.7)

$$S = -Nk_B \sum_{M_J} p(M_J) \ln(p(M_J)) \tag{6.25}$$

By referring for simplicity to non interacting magnetic ions with $J = S = 1/2$, at finite temperature the entropy is

$$S(T) = Nk_B \left((\ln 2)\cosh\left(\frac{\epsilon}{T}\right) - \frac{\epsilon}{T}\tanh\left(\frac{\epsilon}{T}\right)\right), \tag{6.26}$$

where $\epsilon = \mu_B H / k_B$ is the *magnetic temperature* and $S(T \to \infty) = Nk_B \ln 2$ (Eq. (6.23)).

The temperature dependence of the entropy is plotted below:

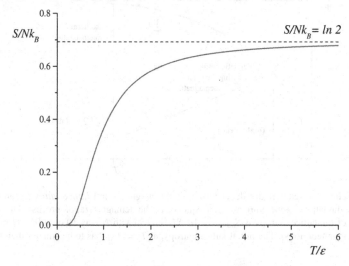

Now we describe the basic principle of the process called *adiabatic demagnetization*, used in order to achieve extremely low temperatures.

A crystal with magnetic ions, almost non-interacting (usually a paramagnetic salt) is in thermal contact by means of an exchange gas (typically low-pressure Helium) with a reservoir, generally a bath of liquid Helium at $T = 4.2$ K. (This temperature can be further reduced, down to about 1.6 K, by pumping over the liquid so that the pressure is decreased).

In zero external field the spin entropy is practically given by $Nk_B ln2$. Only at very low temperature the residual internal field (for instance the one due to dipolar interaction or to the nuclear dipole moments) would anyway induce a certain ordering. The schematic form of the temperature dependence of the magnetic entropy is the one given by curve 1 in Fig. 6.8. Then the external field is applied, in isothermal condition at $T = T_{init}$, up to a certain value H_m. In a time of the order of the spin-lattice relaxation time T_1, spin alignment is achieved, the magnetic temperature is increased and $(T/\epsilon) \ll 1$. Therefore the magnetic entropy is decreased down to S_{init} (curve 2 in the Figure), at the same temperature of the thermal bath and of the crystal. The value S_{init} in Figure corresponds to Eq. (6.26) for $T \ll \epsilon$ and implies a large difference in the populations N_+ and N_- (see Fig. 6.1, where now the point at the energy $E = -N\mu_B H$ is approached). The external bath (the liquid Helium) absorbs the heat generated in the process, while the magnetic energy is decreased. The "internal" reservoir of the sample (namely all the other degrees of freedom besides the spins, already defined *"lattice"*) has its own entropy $S_{lattice}$ related to the vibrational excitations of the ions (in number N', ten or hundred times the number N of the magnetic ions).

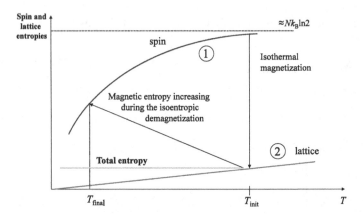

Fig. 6.8 Schematic temperature dependences of the *magnetic* and *lattice entropies* and of the decrease of the lattice temperature as a consequence of the demagnetization process. The order of magnitude of the lattice entropy is $S_{lattice} \sim 10^{-6} N' k_B T^3$ (with N' say 10 or $100N$, N being the number of magnetic ions). The initial lattice entropy, at $T = T_{init}$ has to be smaller than the spin entropy

Since in general the entropy is

$$S \propto \int \frac{\delta Q}{T} = \int \frac{C_V}{T} dT,$$

by considering that at low temperature the specific heat C_V of the lattice goes as T^3 (see the Debye contribution from acoustical vibrational modes at Sect. 14.5) one approximately has

$$S_{lattice} \sim 10^{-6} N' k_B T^3$$

(curve 2 in Fig. 6.8).

Now the exchange gas is pumped out and the sample remains in poor thermal contact with the external bath. The magnetic field is slowly decreased towards zero and the *demagnetization* proceed ideally in *isoentropic* condition. The total entropy stays constant while the magnetic entropy, step after step, each in time of the order of T_1, has to return to curve 1.

Therefore $S_{lattice}$ has to *decrease* of the same amount of the *increase* of the magnetic entropy S. Then the temperature of the "internal" thermal bath has to decrease to $T_{final} \ll T_{init}$.

The amount of cooling depends from the initial external field, from the lattice specific heat and particularly from the internal residual field H_{res} that limits the value of the magnetic entropy at low temperature. In fact, it prevents the total randomization of the magnetic moments. As an order of magnitude one has $T_{final} = T_{init}(H_{res}/H_{init})$.

The adiabatic demagnetization corresponds to the exchange of entropy between the spin system and the lattice excitations. In the picture of the spin temperature (Sect. 6.1) one has an increase of the spin temperature at the expenses of the lattice temperature. The final temperature usually is in the range of milliKelvin, when the electronic magnetic moments are involved in the process. Nuclear magnetic moments are smaller than the electronic ones by a factor 10^{-3}–10^{-4} and then sizeable ordering of the nuclear spins can require temperature as low as 10^{-6} K or very strong fields. In principle, by using the nuclear spins the adiabatic demagnetization could allow one to reach extremely low temperatures. However, one has to take into account that the relaxation times T_1 become very long at low temperatures (while the spin-spin relaxation time T_2 remains of the order of milliseconds). The experimental conditions are such that *negative spin temperature* can easily be attained, for instance by reversing the magnetic field.

From these qualitative considerations it can be guessed that a series of experiments of thermodynamical character based on *spin ordering* and *spin disordering* can be carried out, involving non equilibrium states when the characteristic times of the experimental steps are shorter than T_1 or T_2.

We shall limit ourself to mention that by means of adiabatic demagnetization temperature as low as 2.8×10^{-10} K have been obtained. The *nuclear* moments of Copper have been found to order antiferromagnetically at 5.8×10^{-8} K, while in Silver they order antiferromagnetically at $T_N = 5.6 \times 10^{-10}$ K and ferromagnetically at $T_c = -1.9 \times 10^{-9}$ K.

Problems

Problem 6.7 A magnetic field H of 10 T is applied to a solid of 1 cm^3 containing $N = 10^{20}$, $S = 1/2$ magnetic ions. Derive the magnetic contribution to the specific heat C_V and to the entropy S. Then estimate the order of magnitude of C_V and S at $T = 1$ K and $T = 300$ K.

Solution: The thermodynamical quantities can be derived from the partition function Z. From the single particle statistical average the energy is

$$< E >= \sum_i p_i E_i$$

and from Maxwell-Boltzmann distribution function the probability of occupation of the ith state is

$$p_i = \frac{exp(-E_i/k_B T)}{\sum_i exp(-E_i/k_B T)} \equiv \frac{exp(-E_i \beta)}{Z}.$$

$\sum_i p_i = 1$ and the partition function normalizes the probability p_i.

The total contribution from the magnetic ions to the thermodynamical energy U (per unit volume) is

$$U = N < E >$$

and

$$< E >= \frac{\sum_i E_i \, exp(-E_i/k_B T)}{Z} = -\frac{1}{Z}\frac{\partial Z}{\partial \beta} = -\frac{\partial \ln Z}{\partial \beta}$$

thus yielding

$$U = -N\frac{\partial \ln Z}{\partial \beta}.$$

For $\mu = -g\,\mu_B\,\mathbf{S}$ and $g = 2$ (see Sect. 4.4) one has

$$Z = exp\left(\frac{\mu_B H}{k_B T}\right) + exp\left(-\frac{\mu_B H}{k_B T}\right) \equiv 2\cosh x$$

with

$$x = \frac{\mu_B H}{k_B T} \equiv \beta\mu_B H.$$

For independent particles $Z_{total} = Z^N$. Then $U = -N\mu_B H \tanh x$ and

$$C_V = \left(\frac{\partial U}{\partial T}\right)_H = \left(\frac{\partial \beta}{\partial T}\right)\left(\frac{\partial E}{\partial \beta}\right)_H = -k_B\,\beta^2\left(\frac{\partial U}{\partial \beta}\right)_H,$$

i.e.

$$C_V = Nk_B\,x^2 \text{sech}^2 x \equiv \frac{N\,k_B\,x^2}{\cosh^2 x}$$

plotted below.

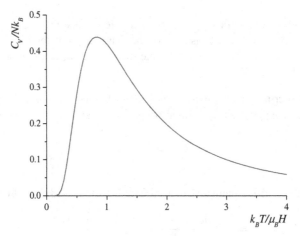

For the entropy, since (see Problem 6.4) $S = -k_B \sum_i p_i\,ln\,p_i$

$$S = -k_B \sum_i \left[\frac{exp(-\beta E_i)}{Z}\right](-\beta E_i - lnZ) = \frac{<E>}{T} + k_B\,lnZ,$$

so that

$$S = N\,k_B\,[ln\,(2\cosh x) - x\tanh x],$$

as it could also be obtained from $S = -(\partial F/\partial T)_H$ with $F = -Nk_B T lnZ$ (see the plot at Sect. 6.4).

Numerically, for $H = 10\,T$, $T = 1\,K$ corresponds to $x \gg 1$ and $T = 300\,K$ to $x \ll 1$, so that

$$T = 1\,K \quad C_V \approx 0, \quad S \approx 0$$
$$T = 300\,K \quad C_V \approx 0, \quad S \approx k_B N ln2$$

Problem 6.8 A spin system ($S = 1/2$) in a magnetic field of $10\,T$, is prepared at a temperature close to $0\,K$ and then put in contact with an identical spin system prepared in the condition of equipopulation of the two spin states. Find the spin temperature reached by the system after spin-spin exchanges, assuming that meantime no exchange of energy with the lattice occurs. Discuss the behavior of the entropy.

Solution: The thermodynamical energies are

$$U_1 = 0 \qquad U_2 = \frac{N}{2}\mathcal{E}, \quad \text{with } \mathcal{E} \text{ energy separation between the two spin states.}$$

From the final energy

$$U_{final} = U_1 + U_2 = \frac{N}{2}\mathcal{E}$$

the spin temperature is obtained by writing

$$U_{final} = \frac{2N\mathcal{E}}{Z} exp\left(-\frac{\mathcal{E}}{k_B T_{spin}}\right).$$

Thus

$$1 = \frac{4}{exp(\mathcal{E}/k_B T_{spin}) + 1}$$

and $T_{spin} \simeq \mathcal{E}/(k_B \cdot 1.1)$. For $\mathcal{E} = 2\mu_B H$ one has $T_{spin} = 12.2$ K.

For the entropy see Problem 6.7 and look at Fig. 6.1, by taking into account that no energy exchange with the reservoir is assumed to occur.

It is noted that the increase of the entropy can be related to the irreversibility of the process.

Problem 6.9 Prove that the mean square deviation of the energy of a system from its mean value (due to exchange of energy with the reservoir) is given by $k_B T^2 C_V$, C_V being the heat capacity.

Solution: The mean square deviation is

$$<(E - <E>)^2> = <E^2 - 2E<E> + <E>^2> = <E^2> - <E>^2$$

where

$$<E> = \sum_i \frac{E_i \, exp(-E_i/k_B T)}{Z} = -\frac{1}{Z}\frac{\partial Z}{\partial \beta}$$

while

$$<E^2> = \sum_i \frac{E_i^2\, exp(-E_i/k_B T)}{Z} = \frac{1}{Z}\frac{\partial^2 Z}{\partial \beta^2}.$$

Therefore

$$<E^2> - <E>^2 = \frac{\partial}{\partial\beta}\left[\frac{1}{Z}\frac{\partial Z}{\partial\beta}\right] = -\frac{\partial<E>}{\partial\beta}.$$

Since

$$\frac{\partial}{\partial\beta} = -k_B T^2 \frac{\partial}{\partial T} \quad and \quad \frac{\partial<E>}{\partial T} = C_V$$

one has

$$<(E - <E>)^2> = k_B T^2 C_V,$$

another example of fluctuation-dissipation relationships (see Eq. (6.14)).

The *fractional deviation* of the energy $\left[(<E^2> - <E>^2)/<E>^2\right]^{\frac{1}{2}}$ at high temperatures, where $<E> \approx N k_B T$ and $C_V \approx N k_B$ is of the order of $N^{-1/2}$, a very small number for N of the order of the Avogadro number (see Problem 6.1).

Problem 6.10 Compare the magnetic susceptibility of non-interacting magnetic moments $S = 1/2$ with the classical $S = \infty$ limit (where any orientation with respect to the magnetic field is possible).

Solution: For $S = \infty$

$$<\mu \cos\theta> = \mu \int \cos\theta\, exp\left(\frac{\mu H \cos\theta}{k_B T}\right)\sin\theta\, d\theta / \int exp\left(\frac{\mu H \cos\theta}{k_B T}\right)\sin\theta\, d\theta$$

$$= \mu\left[\coth\frac{\mu H}{k_B T} - \left(\frac{\mu H}{k_B T}\right)^{-1}\right] \approx \frac{\mu^2 H}{3 k_B T}$$

yielding the Langevin-like susceptibility. For $S = 1/2$ (see Sect. 4.4)

$$<\mu> = \mu\frac{\left[exp\left(\frac{\mu H}{k_B T}\right) - exp\left(-\frac{\mu H}{k_B T}\right)\right]}{\left[exp\left(\frac{\mu H}{k_B T}\right) + exp\left(-\frac{\mu H}{k_B T}\right)\right]} \approx \frac{\mu^2 H}{k_B T}$$

Problem 6.11 By taking inspiration from Fig. 6.7, devise an experimental procedure suitable to measure the spin-lattice relaxation time T_1.

Solution:

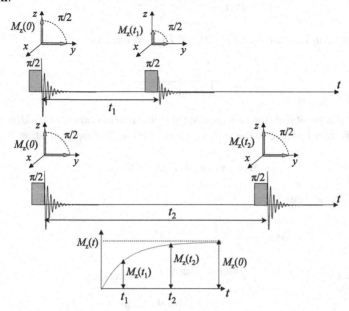

At $t = 0$ the ($\pi/2$) pulse brings the magnetization along y, saturating the populations of the two levels and yielding the *FID* signal (see Sect. 6.3). After a time t_1 a second ($\pi/2$) pulse measures the magnetization $M_z(t_1)$ during the recovery towards the equilibrium value. By applying pairs of pulses with different t_1's (e.g. t_2) the recovery plot towards the equilibrium is constructed and T_1 is extracted.

Problem 6.12 For an ensemble of particles with a ground state at spin $S = 0$ and the excited state at energy Δ and spin $S = 1$, derive the paramagnetic susceptibility.

Then, by resorting to the fluctuation-dissipation theorem (see Eq. (6.14) and Problem 4.18) show that the same result is obtained.

Solution: The energy levels are sketched below:

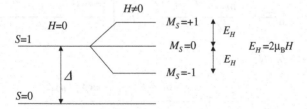

The direct expression for the single particle susceptibility is

$$\chi = \chi_0 p_0 + \chi_1 p_1$$

where $\chi_0 = 0$, $\chi_1 = \mu_B^2 g^2 S(S+1)/3k_B T = 8\mu_B^2/3k_B T$ and

$$p_{0,1} = \frac{(2S+1)e^{-\beta E_{0,1}}}{Z},$$

with Z partition function. For $E_0 = 0$ and $E_1 = \Delta$ one has

$$\chi = \frac{8\mu_B^2}{k_B T} \frac{e^{-\beta\Delta}}{(1 + 3e^{-\beta\Delta})}$$

It is noted that the above equation is obtained in the limit of *evanescent field*, condition that will be retained also in the subsequent derivation. The magnetization is

$$M = N_{-1}\mu_z + N_0.0 - N_{+1}\mu_z$$

with $\mu_z = 2\mu_B$. Then

$$M = \frac{N2\mu_B}{Z} \left\{ e^{-\beta(\Delta - E_H)} - e^{-\beta(\Delta + E_H)} \right\}$$

with $Z = 1 + e^{-\beta(\Delta - E_H)} + e^{-\beta\Delta} + e^{-\beta(\Delta + E_H)}$. Therefore,

$$M = N2\mu_B \frac{e^{-\beta\Delta}[e^{\beta E_H} - e^{-\beta E_H}]}{1 + e^{-\beta\Delta}[e^{\beta E_H} + 1 + e^{-\beta E_H}]}$$

and for $\beta E_H \ll 1$

$$M = 2\mu_B N e^{-\beta\Delta} \frac{2\beta E_H}{1 + 3e^{-\beta\Delta}} = \frac{8\mu_B^2 N}{k_B T} \frac{e^{-\beta\Delta}}{1 + 3e^{-\beta\Delta}} H,$$

yielding the susceptibility obtained from the direct expression.

From the fluctuation-dissipation relationship (see Eqs. (6.11)–(6.14)), being the fluctuations uncorrelated $< \Delta M^2 > = N < \Delta\mu_z^2 >$ with $< \Delta\mu_z^2 > = < \mu_z^2 > - < \mu_z >^2$, and

$$< \mu_z^2 > = 4\mu_B^2 \frac{\sum_{M_S,S} M_S^2 e^{-\beta E(M_s,S)}}{Z} = 4\mu_B^2 \left\{ \frac{e^{-\beta(\Delta - E_H)} + e^{-\beta(\Delta + E_H)}}{Z} \right\}.$$

From $< \mu_z > = M/N$

$$< \mu_z >^2 = 4\mu_B^2 \frac{[e^{-\beta(\Delta - E_H)} - e^{-\beta(\Delta + E_H)}]^2}{Z^2}$$

Thus

$$< \Delta M^2 > = 4N\mu_B^2 \frac{e^{-\beta\Delta}}{Z} \left\{ e^{\beta E_H} + e^{-\beta E_H} - \frac{e^{-2\beta\Delta}}{Z}(e^{\beta E_H} - e^{-\beta E_H})^2 \right\}$$

and again for $\beta E_H \ll 1$

$$< \Delta M^2 > = 4N\mu_B^2 \frac{e^{-\beta\Delta}}{Z} \left\{ 2 - \frac{e^{-2\beta\Delta}}{Z}(\beta E_H)^2 \right\} \simeq 8N\mu_B^2 \frac{e^{-\beta\Delta}}{1 + 3e^{-\beta\Delta}} = k_B T \chi.$$

Problem 6.13 Consider an ideal paramagnet, with $S = 1/2$ magnetic moments. Derive the expression for the relaxation time T_1 in terms of the transition probability W (due to the time-dependent spin-lattice interaction) driving the recovery of the magnetization to the equilibrium, after a perturbation leading to a spin temperature T_s, different from the temperature $T = 300$ K of the thermal reservoir. Find the time-evolution of the spin temperature starting from the initial condition $T_s = \infty$.

Solution: The instantaneous statistical populations are

$$N_- = \frac{N}{Z} e^{\mu_B H \beta_s} \simeq \frac{N}{Z}(1 + \beta_s E_-)$$

with $\beta_s = 1/k_B T_s$

$$N_+ = \frac{N}{Z} e^{-\mu_B H \beta_s} \simeq \frac{N}{Z}(1 - \beta_s E_+),$$

while at the thermal equilibrium

$$N_\mp^{eq} \simeq \frac{N}{Z}(1 \pm \beta E_\mp) \simeq \frac{N}{2}(1 \pm \beta \Delta E)$$

with $\beta = 1/k_B T$ and $\Delta E = 2\mu_B H$.

From the equilibrium condition $N_- W_{-+} = N_+ W_{+-}$ one deduces

$$W_{+-} = W_{-+} \frac{N_-}{N_+} \simeq W_{-+} \frac{1 + \beta E_-}{1 - \beta E_+}$$

and $W_{+-} \simeq W(1 + \beta\Delta E)$, with $W_{-+} \equiv W$.

Since

$$\frac{dN_-}{dt} = -N_- W + N_+ W(1 + \beta\Delta E) = -N_- W + (N - N_-)W(1 + \beta\Delta E) = -2N_- W + 2N_-^{eq} W,$$

then $N_-(t) = ce^{-2Wt} + N_-^{eq}$ and from the initial condition

$$N_-(t) = (N_-^{init} - N_-^{eq})e^{-2Wt} + N_-^{eq}$$

Evidently $dN_+/dt = -dN_-/dt$.

From the magnetization $M_z(t) \propto (N_- - N_+)$ one has $dM_z/dt \propto 2(dN_-/dt)$ and

$$M_z(t) = \left(M_z^{init} - M_z^{eq} \right) e^{-2Wt} + M_z^{eq}$$

implying $1/T_1 = 2W$.

M_z is also inversely proportional to T_s and then one approximately writes

$$\beta_s(t) = \left(\beta_s^{init} - \beta\right)e^{-2Wt} + \beta$$

and for $\beta_s^{init} = 0$, $\beta_s(t) = \beta(1 - e^{-2Wt})$

$$T_s(t) = \frac{T}{1 - e^{-2Wt}} \qquad (a)$$

For exact derivation, over all the temperature range, from Problem 6.2 $T_s = (2\mu_B H/k_B)/ln(u^{-1} - 1)$ with $u = N_+/N$. Then the expression of the spin temperature is

$$T_s = \frac{2\mu_B H}{k_B}\left[ln\left(\frac{(N/2 - N_-^{eq})e^{-2Wt} + N_-^{eq}}{N - (N/2 - N_-^{eq})e^{-2Wt} - N_-^{eq}}\right)\right]^{-1} \qquad (b)$$

See plots in Fig. 6.9.

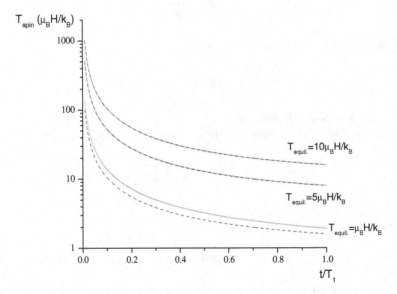

Fig. 6.9 Plot of Eq. a) (*solid line*) and Eq. b) (*dashed line*) showing the equivalence of the two procedures for $T > \mu_B H/k_B$. (In plotting Eq. b) keep at least three significant digits in the expansion.)

Problem 6.14 An hypothetical crystal has a mole of Na atoms, each at distance $d = 1\,\text{Å}$ from a point charge ion of charge $-e$ (and no magnetic moment). By taking into account the quadrupole interaction (Sect. 5.2) derive the energy, the entropy and the specific heat of the crystal around room temperature (Na nuclear spin $I = 3/2$ and nuclear quadrupole moment $Q = 0.14 \times 10^{-24}\ \text{cm}^2$).

Solution: The eigenvalues being $E_{\pm 1/2} = 0$ and $E_{\pm 3/2} = eQV_{zz}/2 = E$ (see Problem 5.7), the partition function is written

$$Z = \left[\sum_{M_I} e^{-\beta E_{M_I}} \right]^{N_A} = \left[2\left(1 + e^{-\beta E}\right) \right]^{N_A}.$$

Then the free energy is (return to Problem 5.12)

$$F = -k_B T \ln Z = -N_A k_B T \ln(1 + e^{-\beta E}) - N_A k_B T \ln 2$$

and

$$U = -\frac{\partial}{\partial \beta} \ln Z = \frac{N_A E e^{-\beta E}}{(1 + e^{-\beta E})}$$

and

$$S = -\frac{\partial F}{\partial T} = N_A k_B \ln(1 + e^{-\beta E}) + \frac{N_A E}{T} \frac{e^{-\beta E}}{(1 + e^{-\beta E})} + k_B N_A \ln 2$$

Since $E \sim 10^{-8}$ eV $\ll k_B T$, U and S can be written

$$U = \frac{N_A E}{2}\left(1 - \frac{1}{2}\frac{E}{k_B T}\right)$$

and

$$S \simeq R\left(2\ln 2 - \frac{1}{8}\frac{E^2}{k_B^2 T^2}\right)$$

(see Problem 6.7 for the analogous case).[4] From $C_V = \partial U/\partial T$, in the high temperature limit $C_V \simeq (1/4)R(E/k_B T)^2$, the high-temperature tail of the Schottky anomaly already recalled at Problem 5.12. For $T = 0$, $U = 0$ and $S = R\ln 2$.

---

[4] $\ln(1 + e^{-x}) \simeq \ln 2 + \ln\left(1 - \dfrac{x}{2} + \dfrac{x^2}{4}\right) \simeq \ln 2 - \dfrac{x}{2} + \dfrac{x^2}{8}.$

Specific References and Further Reading

1. D.J. Amit and Y. Verbin, *Statistical Physics - An Introductory course*, (World Scientific, 1999).
2. C.P. Slichter, *Principles of Magnetic Resonance*, (Springer Verlag, Berlin, 1990).
3. H. Haken and H.C. Wolf, *Atomic and Quantum Physics*, (Springer Verlag, Berlin, 1987).
4. B.H. Bransden and C.J. Joachain, *Physics of atoms and molecules*, (Prentice Hall, 2002).
5. D. Budker, D.F. Kimball and D.P. De Mille, *Atomic Physics - An Exploration Through Problems and Solutions*, (Oxford University Press, 2004).
6. B. Cagnac and J.C. Pebay - Peyroula, *Physique atomique, tome 2*, (Dunod Université, Paris, 1971).
7. R. Fieschi e R. De Renzi, *Struttura della Materia*, (La Nuova Italia Scientifica, Roma, 1995).
8. A.P. Guimaraes, *Magnetism and Magnetic Resonance in Solids*, (J. Wiley and Sons, 1998).
9. C. Kittel, *Elementary Statistical Physics*, (J. Wiley and Sons, 1958).
10. M.A. Morrison, T.L. Estle and N.F. Lane, *Quantum States of Atoms, Molecules and Solids*, (Prentice -Hall Inc., New Jersey, 1976).
11. H. Semat, *Introduction to Atomic and Nuclear Physics*, (Chapman and Hall LTD, 1962).

Chapter 7
Molecules: General Aspects

Topics

Separation of Electronic and Nuclear Motions
Symmetry Properties in Diatomic Molecules
Labels for Electronic States
One Electron in Axially Symmetrical Potential

In this chapter we shall discuss the general aspects of the first state of "bonded matter", the aggregation of a few atoms to form a molecule. The related issues are also relevant for biology, medicine, astronomy etc. The knowledge of the quantum properties of the electronic states in molecules is the basis in order to create new materials, as the ones belonging to the "artificial matter", often obtained by means of subtle manipulations of atoms by means of special techniques.

We shall understand why the molecules are formed, why the H_2 molecule exists while two He atoms do not form a stable system, why the law of definite proportions holds or why there are multiple valences, what controls the geometry of the molecules. These topics have to follow as extension of the atomic properties. Along this path new phenomena, typical of the realm of the molecular physics, will be emphasized.

In principle, the Schrödinger equation for nuclei and electrons contains all the information we wish to achieve. In practice, even the most simple molecule, the Hydrogen molecule-ion H_2^+, cannot be exactly described in the framework of such an approach: the Schrödinger equation is solved only when the nuclei are considered fixed. Therefore, in most cases we will have to deal with simplifying assumptions or approximations, which usually are not of mathematical character but rather based on the physical intuition and that must be supported by experimental findings.

The first basic assumption we will have to take into account is the *Born-Oppenheimer* approximation, essentially relying on the large ratio of the nuclear and electronic masses. It allows one to deal with a kind of separation between the

© Springer International Publishing Switzerland 2015
A. Rigamonti and P. Carretta, *Structure of Matter*,
UNITEXT for Physics, DOI 10.1007/978-3-319-17897-4_7

motions of the electrons and of the nuclei. Another approximation that often will be used involves tentative wavefunctions for the electronic states as linear combination of a set of basis functions, that can help in finding appropriate solutions. For instance, a set of wavefunctions *centered at the atomic sites* will allow one to arrive at the *secular equation* for the approximate eigenvalues.

Finally in this chapter we have to find how to label the electronic states in terms of *good quantum numbers*. This will be done in a way similar to the one in atoms, by relying on the *symmetry properties* of the potential energy (for example, the cylindrical symmetry) and by referring to the limit atomic-like situations of united-atoms or of separated-atoms.

7.1 Born-Oppenheimer Separation and the Adiabatic Approximation

For a system of nuclei and electrons the Hamiltonian is written (see Fig. 7.1)

$$\mathcal{H} = -\frac{\hbar^2}{2}\sum_\alpha \frac{\nabla_\alpha^2}{M_\alpha} - \frac{\hbar^2}{2m}\sum_i \nabla_i^2 + \sum_{i<j}\frac{e^2}{r_{ij}} + \sum_{\alpha<\beta}\frac{Z_\alpha Z_\beta e^2}{R_{\alpha\beta}} - \sum_{\alpha,i}\frac{Z_\alpha e^2}{r_{i\alpha}} \equiv$$

$$\equiv T_n + T_e + V_{ee} + V_{nn} + V_{ne} \tag{7.1}$$

The corresponding wave function $\phi(\mathbf{R}, \mathbf{r})$ involves both the group $\mathbf{R}$ of the nuclear coordinates and the group $\mathbf{r}$ for the electrons. In the Hamiltonian the spin-orbit interactions and the hyperfine interactions have not been included, since at a first stage they can be safely neglected.

In order to solve the Schrödinger equation for $\phi(\mathbf{R}, \mathbf{r})$ one observes the large difference in nuclear and electronic masses (and the related differences in the electronic and roto-vibrational energies, as it will appear in subsequent chapters). This difference suggests that in time intervals much shorter than the ones required for the nuclei to sizeably change their positions, the electrons have been able to take

Fig. 7.1 Nuclear and electronic coordinates used in Eq. (7.1)

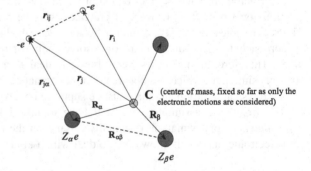

the quantum configuration pertaining to ideally fixed coordinates $\mathbf{R}$. Then one can attempt an eigenfunction of the form

$$\phi(\mathbf{R}, \mathbf{r}) = \phi_n(\mathbf{R})\phi_e(\mathbf{R}, \mathbf{r}), \tag{7.2}$$

where $\phi_n(\mathbf{R})$ pertains to the nuclei, while the electronic wavefunction ϕ_e involves only *parametrically* the nuclear coordinates, these latter ideally *frozen* in the configuration specified by $\mathbf{R}$. When such a function is included in the Schrödinger equation involving the Hamiltonian in Eq. (7.1) and the following equivalences are taken into account

$$T_e\phi_n\phi_e = \phi_n T_e\phi_e,$$

$$T_n\phi_n\phi_e \equiv -\hbar^2 \sum_\alpha \frac{\nabla_\alpha^2}{2M_\alpha} \phi_n\phi_e = -\hbar^2 \sum_\alpha \frac{1}{2M_\alpha} \nabla_\alpha \cdot \{\phi_e \nabla_\alpha \phi_n + \phi_n \nabla_\alpha \phi_e\} =$$

$$= \phi_e T_n\phi_n + \phi_n T_n\phi_e - 2\hbar^2 \sum_\alpha \frac{1}{2M_\alpha} \nabla_\alpha \phi_e \cdot \nabla_\alpha \phi_n,$$

then one has

$$\left[-\sum_\alpha \frac{\hbar^2}{2M_\alpha} (2\nabla_\alpha \phi_e \cdot \nabla_\alpha \phi_n) - \sum_\alpha \frac{\hbar^2}{2M_\alpha} \phi_n \nabla_\alpha^2 \phi_e\right] + \tag{7.3}$$

$$+ \phi_e T_n\phi_n + \phi_n T_e\phi_e + (V_{nn} + V_{ne} + V_{ee})\phi_e\phi_n = E\phi_e\phi_n.$$

Let us assume that the terms included in the square brackets can be neglected (the conditions for such an approximation, essentially corresponding to the so-called *adiabatic approximation*, shall be discussed subsequently). For the electronic wavefunction one can write

$$T_e\phi_e + (V_{ne} + V_{ee})\phi_e = E_e^{(g)}\phi_e, \tag{7.4}$$

where $E_e^{(g)}(\mathbf{R})$ is the eigenvalue for the electrons in a frozen nuclear configuration. Then, from Eq. (7.3), by neglecting the terms in square-brackets, after dividing by ϕ_e one obtains the equation for the nuclear motions:

$$\left[T_n + V_{eff}(\mathbf{R})\right]\phi_n = E\phi_n \tag{7.5}$$

with $V_{eff}(\mathbf{R}) = V_{nn}(\mathbf{R}) + E_e^{(g)}(\mathbf{R})$. In Eq. (7.5) the effective Hamiltonian includes the eigenvalue for the electrons $E_e^{(g)}$, for given $\mathbf{R}$'s, as *effective potential energy*.

Thus, by assigning to the nuclear and electronic states the appropriate set of *quantum numbers* ν and g, under the approximations discussed above the wavefunction solution for the Hamiltonian 7.1 is

$$\phi^{(g,\nu)}(\mathbf{r}, \mathbf{R}) = \phi_e^{(g)}(\mathbf{r}, \mathbf{R})\phi_n^{(\nu)}(\mathbf{R}), \tag{7.6}$$

with ϕ_e and ϕ_n eigenfunctions from Eqs. (7.4) and (7.5) respectively.

The electronic eigenvalue $E_e^{(g)}$, entering the effective potential energy in Eq. (7.5), is not a number as in atoms but *parametrically* depends from the nuclear coordinates. The total energy of the molecule can be written

$$E^{(g,\nu)} = E_e^{(g)}(\mathbf{R}_m) + V_{nn}(\mathbf{R}_m) + E_n^{(\nu)}, \tag{7.7}$$

where $\mathbf{R}_m$ means the nuclear configuration corresponding to the minimum for $V_{eff}(\mathbf{R})$ (see Eq. (7.5)).

The physical contents of such a framework are more easily grasped by referring to a diatomic molecule, where in practice the only parameter required to fix the nuclear configuration is the distance R_{AB} between the two nuclei, since as a first approximation the electronic states can be considered unaffected by the rotation of the molecule. A schematic view of the energy of the molecule for the electronic states as a function of R_{AB} and of the effect of the eigenvalue for the vibrational motion of the nuclei (corresponding to a variation of R_{AB}) is given in Fig. 7.2. Complete understanding of this illustration will be achieved after reading Sects. 8.1 and 10.3.1.

Let us briefly comment on the possibility to neglect the terms in square brackets in Eq. (7.3), corresponding to the validity of the adiabatic approximation. The order of magnitude of the contribution of those terms to the energy can be estimated by looking at the expectation values

$$< \phi_e \phi_n | \left[\ldots \right] | \phi_e \phi_n >,$$

Therefore a first term is

$$-\frac{\hbar^2}{M_\alpha} \int \phi_e^* \phi_n^* \nabla_n \phi_e \cdot \nabla_n \phi_n d\tau_n d\tau_e$$

that for $\phi_e(\mathbf{r}, \mathbf{R})$ in real form, becomes proportional to

$$\int \phi_e^* \nabla_n \phi_e d\tau_e \propto \nabla_n \int \phi_e^* \phi_e d\tau_e$$

which is zero for a given electronic state g_1. The second term is

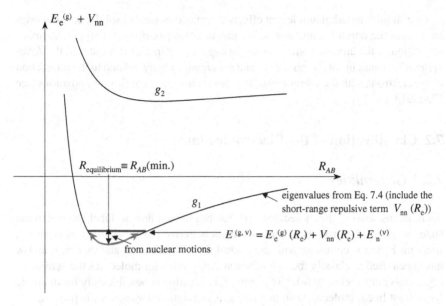

Fig. 7.2 Schematic view of the separation of the electronic and vibrational energies in a diatomic molecule and of the role of $E_e^{(g)}$ as effective potential energy for the nuclear motion, within the adiabatic approximation. The vibrational motion occurs in an effective potential energy, while the electrons follow adiabatically this motion

$$-\frac{\hbar^2}{M_\alpha} \int \phi_n^* \phi_n d\tau_n \int \phi_e^* \nabla_n^2 \phi_e d\tau_e,$$

and by taking into account that the electronic wavefunction depends on $(\mathbf{r} - \mathbf{R})$, one can write

$$-\frac{\hbar^2}{M_\alpha} \int \phi_e^* \nabla_n^2 \phi_e d\tau_e \simeq \frac{m}{M_\alpha} < |T_e| >,$$

which is of the order of the contribution to the energy from the electronic kinetic term scaled by the factor m/M_α and thus negligible.

Finally one would have to consider the non-diagonal terms involving the operator ∇_n, of the form

$$\int \phi_e^{(g_2)*} \nabla_n \phi_e^{(g_1)} d\tau_e. \tag{7.8}$$

These terms can be different from zero and in principle they drive transitions between electronic states associated with the nuclear motions, in other words to the *non-adiabatic* contributions. For large separation between the electronic states compared to the energy of the thermal motions, the transition probability is expected to be small.

One should remark that relevant effects in molecules (and in solids) actually originate from the non-adiabatic terms. We just mention *pre-dissociation* (spontaneous
separation of the atoms), some *removal of degeneracy* in electronic states, the *Jahn-
Teller* effect and, in solids, *resistivity* and *superconductivity*, related to the interaction
of the electrons with the vibrations of the ions around their equilibrium positions (see
Chaps. 13 and 14).

7.2 Classification of the Electronic States

7.2.1 Generalities

As in atoms, also in the molecules first one has to find how to label the electronic
states in terms of constants of motions, namely derive the *good quantum numbers*.
In atoms $\mathbf{l}^2$ and $\mathbf{l}_z$ commute with the central field Hamiltonian and then n, l, and m
have been used to classify the one-electron states. Also for molecules the symmetry
arguments play a relevant role: a rigorous classification is possible only for diatomic
(or at least linear molecules) so that one axis of rotational symmetry is present.

Let us refer to Fig. 7.3. When the z axis is aligned along the molecular axis the
potential energy V is a function of the cylindrical coordinates z and ρ, while it does
not involve the angle φ. Then the l_z operator $-i\hbar\frac{\partial}{\partial\varphi}$ commutes with the Hamiltonian:

$$\left[l_z, V\right] \propto \left(x\frac{\partial}{\partial y} - y\frac{\partial}{\partial x}\right) V\phi - V \left(x\frac{\partial}{\partial y} - y\frac{\partial}{\partial x}\right)\phi = (\mathbf{r} \times \boldsymbol{\nabla} V)_z,$$

which is zero when the z axis is along the molecular axis.

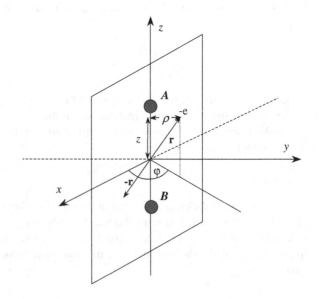

Fig. 7.3 Schematic view for
the discussion of the
symmetry arguments
involved in the classification
of one-electron states in a
diatomic molecule. A and B
are nuclei dressed by the
electrons uninvolved in the
bonding mechanism. When
$A = B$ the molecule is
homonuclear and it acquires
the *inversion symmetry* with
respect to the center. Then
$\phi_e(\mathbf{r}) = \pm\phi_e(-\mathbf{r})$ and the
classification *gerade* or
ungerade, according to the
sign of the wavefunction
upon inversion (parity),
becomes possible

For *homonuclear* molecules ($A = B$) in terms of the positional vector **r** one has

$$|\phi(\mathbf{r})|^2 = |\phi(-\mathbf{r})|^2 \quad i.e. \quad \phi(\mathbf{r}) = -\phi(-\mathbf{r}) \quad or \quad \phi(\mathbf{r}) = +\phi(-\mathbf{r})$$

and one can classify the states with a letter **g** (from *gerade*) or **u** (from *ungerade*) according to the *even* or *odd* parity under the inversion of **r** with respect to the center of the molecule (see Fig. 7.3). One should also remark that the reflection with respect to the yz plane, bringing x in $-x$, changes the sign of the z-component of the angular momentum while the Hamiltonian is invariant. It follows that the energy must depend on the square of the l_z-eigenvalue while this operator has to convert the eigenfunction in the one having eigenvalue of opposite sign. The electronic states with l_z-eigenvalue different from zero must be *double degenerate*, each of the two states corresponding to different direction of the projection of the orbital angular momentum along the z-axis.[1] On the other hand, for l_z-eigenvalue equal to zero a further $-$ or $+$ sign has to be used to describe the behavior of the wavefunction upon reflection with respect to the planes containing the molecular axis.

Finally, in these introductory remarks it is noted that the z-component of the total angular momentum, implying an algebric sum $L_z = \sum_i l_z^i$, is also a constant of motion, with associated a good quantum number M_L (see Sect. 7.2.3).

7.2.2 Schrödinger Equation in Cylindrical Symmetry

By referring again to Fig. 7.3 and in the framework of the Born-Oppenheimer separation, the Schrödinger equation for the one-electron wavefunction is

$$\frac{-\hbar^2}{2m} \nabla^2_{z,\rho,\varphi} \phi + V\phi = E(R_{AB})\phi, \tag{7.9}$$

where $V = V(z, \rho)$. One should remark that if A and B are protons, namely we are dealing with the Hydrogen molecule ion, then

$$V = -\frac{e^2}{r_A} - \frac{e^2}{r_B}. \tag{7.10}$$

---

[1] This two-fold degeneracy is removed when the interaction between the electronic and rotational motions is taken into account. Then the terms at $l_z \neq 0$ would split into two nearby levels. For a multi-electron molecule, in the **LS** scheme, where $\Lambda = \Lambda(\Pi, \Delta, ...)$ (see end of Sect. 7.2.3) characterizes the states at $L_z \neq 0$, the splitting is known as Λ-doubling.

By using ellipsoidal coordinates with the nuclei at the foci of the ellipse, then Eq. (7.9), with V as in Eq. (7.10), is exactly solvable in a way similar to

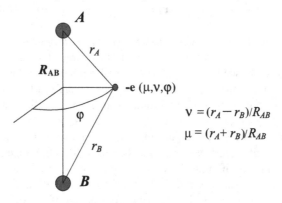

$$\nu = (r_A - r_B)/R_{AB}$$
$$\mu = (r_A + r_B)/R_{AB}$$

Hydrogen atom, with separation of the variables.

This solution would not be of much help, since when diatomic molecules with the nuclei dressed by the atomic (*core*) electrons have to be considered, the potential is no longer of the form in Eq. (7.10) and therefore relevant modifications can be expected. A similar modification in the atom is the removal of the accidental degeneracy upon abandoning the Coulomb potential. Thus we prefer to disregard the formal solution of Eq. (7.9) for strictly Coulomb-like potential and first give the general properties of electronic states just by referring to the cylindrical symmetry of V (again in a way analogous to atoms, where only the spherical symmetry of the Hamiltonian in the central field approximation was taken into account). Subsequently approximate methods will allow us to derive specific forms of the wavefunctions of more general use, rather than the exact expressions pertaining to the Hydrogen molecule ion.

The kinetic energy operator in cylindrical coordinates reads

$$\nabla^2_{z,\rho,\varphi} = \frac{\partial^2}{\partial\rho^2} + \frac{1}{\rho}\frac{\partial}{\partial\rho} + \frac{\partial^2}{\partial z^2} + \frac{1}{\rho^2}\frac{\partial^2}{\partial\varphi^2} \qquad (7.11)$$

and by factorizing ϕ in the form $\phi = \chi(z,\rho)\Phi(\varphi)$ Eq. (7.9) is rewritten

$$\frac{2m\rho^2}{\hbar^2}\left[E(R_{AB}) - V(z,\rho)\right] + \frac{\rho^2}{\chi}\left[\frac{\partial^2\chi}{\partial z^2} + \frac{\partial^2\chi}{\partial\rho^2}\right] + \frac{\rho}{\chi}\frac{\partial\chi}{\partial\rho} = -\frac{1}{\Phi}\frac{\partial^2\Phi}{\partial\varphi^2}, \qquad (7.12)$$

where at the first member one has only operators and functions of z and ρ while at the second member only of φ. As a consequence, Eq. (7.12) leads to solutions of the form $\phi = \chi\Phi$, where χ and Φ originate from the separate equations in which both members are equal to a constant independent on z, ρ and φ. We label that constant λ^2 and then

$$\frac{\partial^2 \Phi}{\partial \varphi^2} = -\lambda^2 \Phi$$

so that

$$\Phi = A e^{i\lambda\varphi} + B e^{-i\lambda\varphi}. \tag{7.13}$$

The boundary condition for Φ is

$$\Phi(\varphi + 2\pi n) = \Phi(\varphi)$$

and $exp(i\lambda n 2\pi) = 1$, thus yielding λ *integer*.

The meaning of the number λ can be directly grasped by looking for the eigenvalue of the z component of the angular momentum of the electron:

$$l_z \phi = a\phi \quad i.e. \quad -i\hbar\frac{\partial}{\partial\varphi}\chi\Phi = \chi(-i\hbar\frac{\partial e^{\pm i\lambda\varphi}}{\partial\varphi}) = \pm\lambda\hbar\chi\Phi = \pm\lambda\hbar\phi,$$

namely λ measures in $\hbar$ unit the component of $\mathbf{l}$ along the molecular axis, a constant of motion, as it was anticipated.

From Eq. (7.12) it is realized that the eigenvalue $E(R_{AB})$ depends on λ^2. Therefore we understand that from a given atomic-like state of angular momentum $\mathbf{l}$, the presence of the second atom at the distance R_{AB} generates $(l+1)$ states of different energy. These states correspond to $l_z = 0, \pm 1, \pm 2...$ and are in general double degenerate, in agreement with the fact that the energy cannot depend on the sign of l_z, as we have previously observed. These one-electron states are labelled by the letters $\sigma, \pi, \delta.....$ in correspondence to $0, 1, 2$, etc. similarly to the atomic states $s, p, d...$

7.2.3 Separated-Atoms and United-Atoms Schemes and Correlation Diagram

Other good quantum numbers for the electronic states to be associated with (z, ρ) in Eq. (7.12) can be introduced only when $\chi(z, \rho)$ can be factorized in two functions, involving separately z and ρ. This happens when one refers to the limit situations of united atoms (i.e. $R_{AB} \rightarrow 0$) or of separated atoms (i.e. $R_{AB} \rightarrow \infty$). In the *united-atoms* classification scheme (for example the Hydrogen molecule ion H_2^+ tends to become the He^+ atom) the two further quantum numbers are n and l, while λ tends to become m. Then the sequence of the states is

$$1s\sigma, 2s\sigma, 2p\sigma, 2p\pi...$$

and the parity **g** or **u** is fixed by the value of l, namely l even **g**, l odd **u**.

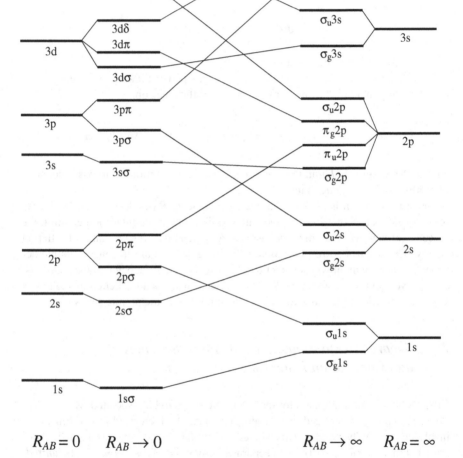

$$R_{AB} = 0 \qquad R_{AB} \rightarrow 0 \qquad\qquad\qquad R_{AB} \rightarrow \infty \qquad R_{AB} = \infty$$

Fig. 7.4 Classification schemes for diatomic homonuclear molecules and correlation lines yielding a sketchy behavior of the eigenvalues $E(R_{AB})$ as a function of the interatomic distance. Some of the correlations of the high-energy states are not straightforward and can involve the mixing of other states in the LCAO scheme

For $R_{AB} \rightarrow \infty$ the atoms are far away (H_2^+ becomes H with a proton at large distance) and for heteronuclear molecule one has

$$\sigma 1s_A, \sigma 1s_B, \sigma 2s_A, \sigma 2s_B, \sigma 2p_A \ldots$$

For $A = B$ (homonuclear molecule) $(nl)_A = (nl)_B$, the splitting of the level due to the perturbing effect of the other nucleus (e.g. H^+ in the Hydrogen molecule ion) removes the degeneracy and the character **g** or **u** can be assigned.

The two classification schemes are obviously correlated. For the lowest energy levels the correlation can be established by direct inspection, by taking into account that λ and the **g** or **u** character do not depend on the distance R_{AB} (see Fig. 7.4). A pictorial view of the correlation diagram in terms of transformation of the orbitals upon changing the distance R_{AB} is given in Fig. 7.5, having assumed the one-electron wavefunction in the form of linear combination of $1s$-atomic like wave functions and $2p_x$-wavefunctions, centered at the two sites A and B (for the proper description see Sect. 8.1).

The correlation diagram for heteronuclear diatomic molecules is shown in Fig. 7.6.

It should be observed that there is a rule that helps in establishing the correlation diagram, the so-called *non-intersection* or *non-crossing* rule (*Von Neumann-Wigner rule*). This rule states that two curves $E_1(R_{AB})$ and $E_2(R_{AB})$ cannot cross if the correspondent wavefunctions ϕ_1 and ϕ_2 belong to the same symmetry species. In other words they can cross if they have different values either of λ or of the parity (**g** and **u**) or different multiplicities.

Finally we mention that the electronic states in a multielectron molecule can be classified in a way similar to the one used in the **LS** scheme for the atom (Chap. 3). From the algebraic sum $S_z = \sum_i s_z^{(i)}$ we construct M_S, while to $L_z = \sum_i l_z^{(i)}$ M_L is associated. The symbols $\Sigma, \Pi, \Delta \ldots$ (generic Λ) are used for $M_L = 0, 1, 2$ etc. Then the state is labelled as

$$^{2S+1}\Lambda_{g,u},$$

g and **u** for homonuclear molecule.

For the state Σ, namely the one with zero component of the total angular momentum along the molecular axis, in view of the consideration on the property upon reflection with respect to a plane containing the axis, one adds the symbol $+$ or $-$ as right apex. Illustrative examples shall be given in dealing with particular diatomic molecules (Sect. 8.2). For the Λ-doubling see footnote 1.

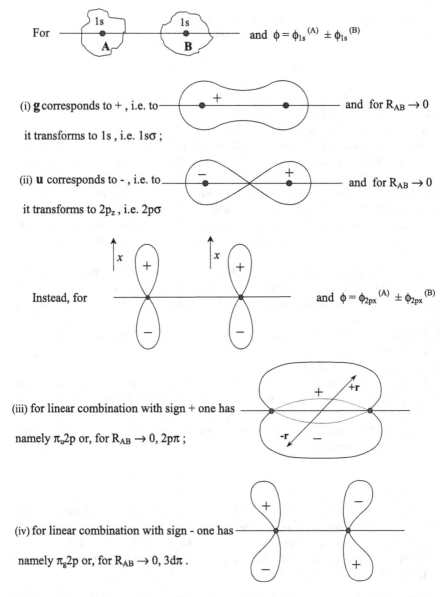

Fig. 7.5 Schematic view of the correlation diagram by referring to the transformation with the interatomic distance R_{AB} of the shape of the molecular orbitals generated by linear combination of atomic $1s$ (cases (i) and (ii)) and $2p_x$ orbitals (cases (iii) and (iv)) centered at the A and B sites (see Sect. 8.1 for details)

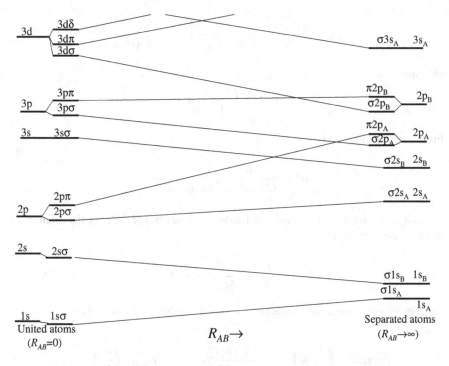

Fig. 7.6 Correlation diagram for heteronuclear diatomic molecules. For $R_{AB} \to \infty$ the assumption of effective nuclear charge $Z_A > Z_B$ has been made

Problems

Problem 7.1 From order of magnitude estimates of the frequencies to be associated with the motions of the electrons of mass m and of the nuclei of mass M in a molecule of "size" d, derive the correspondent velocities by resorting to the Heisenberg principle. By using analogous arguments derive the amplitude of the vibrational motion.

Solution: From Heisenberg principle $p \sim \hbar/d$. The electronic frequencies can be defined

$$\nu_{elect} \simeq \frac{E_{elect}}{h} \sim \frac{1}{h}\frac{p^2}{2m} \sim \frac{1}{h}\frac{\hbar^2}{2d^2 m} = \frac{\hbar}{4\pi m d^2}$$

For the vibrational motion, by assuming for the elastic constant K
$K\, d^2 \sim E_{elect}$ (a crude approximation, see Sect. 10.3 and Problem 8.3) and $\nu_{vib} = (1/2\pi)\sqrt{K/M}$ $E_{vib} \sim h\nu_{vib} \sim \left(\frac{m}{M}\right)^{\frac{1}{2}} E_{elect}$ and $\nu_{vib} \sim (\hbar/d^2 \sqrt{mM})(1/2\pi)$.
Approximate expressions for the correspondent velocities are

$$v_{elect} \sim \sqrt{\frac{E_{elect}}{m}} \sim \frac{\hbar}{md},$$

$$v_{vib} \sim \sqrt{\frac{E_{vib}}{M}} \sim \left[\frac{h\,\hbar}{\sqrt{mM^3d^2}}\right]^{\frac{1}{2}} \sim \frac{\hbar}{m^{\frac{1}{4}}M^{\frac{3}{4}}d}$$

yielding

$$\frac{v_{vib}}{v_{elect}} \sim \left(\frac{m}{M}\right)^{\frac{3}{4}} \ll 1.$$

From $a^2K \sim h\nu_{vib}$ and then

$$a^2 \sim \frac{h^2}{(Mm)^{1/2}d^2}\frac{1}{(E_{elect}/d^2)}$$

one can derive for the vibrational amplitude $a \simeq 2\pi d(2m/M)^{1/4}$. For the rotational motion (see Sect. 10.1)

$$E_{rot} \simeq \frac{P^2}{2I} \sim \frac{\hbar^2}{Md^2} \sim \frac{m}{M}E_{elect}$$

(P angular momentum and I moment of inertia, see Sect. 10.1) and then

$$v_{rot} \sim \left(\frac{E_{rot}}{M}\right)^{\frac{1}{2}} \sim \left[\frac{\left(\frac{m}{M}\right)m\,v_{elect}^2}{M}\right]^{\frac{1}{2}} \sim v_{elect}\cdot\left(\frac{m}{M}\right).$$

Further Reading

1. P.W. Atkins and R.S. Friedman, *Molecular Quantum Mechanics*, (Oxford University Press, Oxford, 1997).
2. B.H. Bransden and C.J. Joachain, *Physics of atoms and molecules*, (Prentice Hall, 2002).
3. C.A. Coulson, *Valence*, (Oxford Clarendon Press, 1953).
4. W. Demtröder, *Molecular Physics*, (Wiley-VCH, 2005).
5. H. Eyring, J. Walter and G.E. Kimball, *Quantum Chemistry*, (J. Wiley, New York, 1950).
6. H. Haken and H.C. Wolf, *Molecular Physics and Elements of Quantum Chemistry*, (Springer Verlag, Berlin, 2004).
7. C.S. Johnson and L.G. Pedersen, *Quantum Chemistry and Physics*, (Addison-Wesley, 1977).
8. M.A. Morrison, T.L. Estle and N.F. Lane, *Quantum States of Atoms, Molecules and Solids*, (Prentice-Hall Inc., New Jersey, 1976).
9. J.C. Slater, *Quantum Theory of Matter*, (McGraw-Hill, New York, 1968).
10. J.M. Ziman, *Principles of the Theory of Solids*, (Cambridge University Press, 1964).

Chapter 8
Electronic States in Diatomic Molecules

Topics

H_2^+ as Prototype of the Molecular Orbital Approach (MO)
H_2 as Prototype of the Valence Bond Approach (VB)
How MO and VB Become Equivalent
The Quantum Nature of the Bonding Mechanism
Some Multi-Electron Molecules (N_2, O_2)
The Electric Dipole Moment

In this chapter we specialize the concepts given in Chap. 7 for the electronic states by introducing specific forms for the wavefunctions in diatomic molecules. Two main lines of description can be envisaged. In the approach known as *molecular orbital (MO)* the molecule is built up in a way similar to the *aufbau* method in atoms, namely by ideally adding electrons to one-electron states. The prototype for this description is the Hydrogen molecule ion H_2^+. In the *valence bond (VB)* approach, instead, the molecule results from the interaction of atoms dressed by their electrons. The prototype in this case is the Hydrogen molecule H_2.

8.1 H_2^+ as Prototype of MO Approach

8.1.1 Eigenvalues and Energy Curves

In the Hydrogen molecule ion the Schrödinger equation for the electronic wavefunction $\phi(r, \theta, \varphi)$, or equivalently $\phi(z, \rho, \varphi)$ (see Fig. 8.1) is written

$$\mathcal{H}\phi = \left\{ \frac{-\hbar^2}{2m}\nabla^2 - \frac{e^2}{r_A} - \frac{e^2}{r_B} + \frac{e^2}{R_{AB}} \right\} \phi = E(R_{AB})\phi. \qquad (8.1)$$

© Springer International Publishing Switzerland 2015
A. Rigamonti and P. Carretta, *Structure of Matter*,
UNITEXT for Physics, DOI 10.1007/978-3-319-17897-4_8

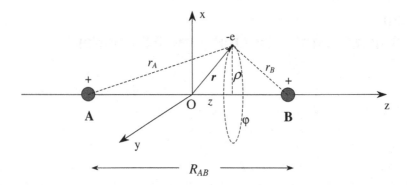

Fig. 8.1 Schematic view of the Hydrogen molecule ion H_2^+ and definition of the coordinates used in the MO description of the electronic states

As already mentioned the exact solution of this equation can be carried out in elliptic coordinates. Having in mind to describe H_2^+ as prototype for more general cases we shall not take that procedure.

It should be remarked that in the Hamiltonian in Eq. (8.1) the proton-proton repulsion e^2/R_{AB} (V_{nn} in Eq. (7.1); see Fig. 7.2) has been included, so that the total energy of the molecule, for a given inter-proton distance R_{AB}, will be found.

By taking into account that for $R_{AB} \rightarrow \infty$ the molecular orbital must transform into the atomic wavefunction ϕ_{1s} centered at the site A or at the site B, one can tentatively write

$$\phi = c_1 \phi_{1s}^{(A)} + c_2 \phi_{1s}^{(B)}. \tag{8.2}$$

This is a particular form of the *molecular orbital*, written as in the so-called *MO-LCAO method*, namely with the wavefunction as linear combination of atomic orbitals.[1]

From the variational procedure, by deriving with respect to c_i the energy function

$$E(c_1, c_2) = \frac{\int \phi^* \mathcal{H} \phi \, d\tau}{\int \phi^* \phi \, d\tau} \tag{8.3}$$

with the tentative wavefunction given by Eq. (8.2), the usual equations

$$c_1(H_{AA} - E) + c_2(H_{AB} - E S_{AB}) = 0 \tag{8.4}$$

$$c_1(H_{AB} - E S_{AB}) + c_2(H_{BB} - E) = 0$$

---

[1] A similar method is used also in more complex molecules, by writing $\phi = \sum_i c_i \phi_i$ and constructing the energy function $E = E(c_i)$ on the basis of the complete electronic Hamiltonian $\mathcal{H} = \sum_i (-\hbar^2/2m)\nabla_i^2 - e^2 \sum_{\alpha,i} Z_\alpha/R_{i\alpha} + e^2 \sum_{i,j}' 1/r_{ij}$, by iterative procedure evaluating the self-consistent coefficients c_i. This is the MO-LCAO-SCF method.

are obtained. Here

$$H_{AA} = H_{BB} = \int \phi_{1s}^{(A)*} \mathcal{H} \phi_{1s}^{(A)} d\tau$$

represents the energy of the H$^+$H or of the HH$^+$ configuration.

$$H_{AB} = H_{BA} = \int \phi_{1s}^{(B)*} \mathcal{H} \phi_{1s}^{(A)} d\tau = \int \phi_{1s}^{(A)*} \mathcal{H} \phi_{1s}^{(B)} d\tau \qquad (8.5)$$

called *resonance integral*, will be discussed at a later stage.

$$S_{AB} = \int \phi_{1s}^{(A)*} \phi_{1s}^{(B)} d\tau$$

is the *overlap integral*, a measure of the region where $\phi_{1s}^{(A)}$ and $\phi_{1s}^{(B)}$ are both different from zero:

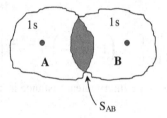

From Eq. (8.4) the secular equations yields

$$E_\pm = \frac{H_{AA} \pm H_{AB}}{1 \pm S_{AB}}, \qquad (8.6)$$

with $c_1 = c_2$ for the sign $+$ and $c_1 = -c_2$ for the sign $-$. Thus

$$\phi_+ = \frac{1}{\sqrt{2(1 + S_{AB})}} \left\{ \phi_{1s}^{(A)} + \phi_{1s}^{(B)} \right\} \qquad (8.7)$$

$$\phi_- = \frac{1}{\sqrt{2(1 - S_{AB})}} \left\{ \phi_{1s}^{(A)} - \phi_{1s}^{(B)} \right\}.$$

In order to discuss the dependence of the approximate eigenvalues $E_\pm$ on the inter-atomic distance R_{AB} one has to express H_{AA}, H_{AB} and S_{AB}. One writes

$$H_{AA} = \int \phi_{1s}^{(A)*} \{\mathcal{H}_{hydr.}\} \phi_{1s}^{(A)} d\tau + \int \phi_{1s}^{(A)*} \frac{e^2}{R_{AB}} \phi_{1s}^{(A)} d\tau - \overbrace{\int \phi_{1s}^{(A)*} \frac{e^2}{r_B} \phi_{1s}^{(A)} d\tau}^{\varepsilon_{AA}}$$
$$(8.8)$$

The first term is $-R_H hc$ (with R_H Rydberg constant), the second is e^2/R_{AB}. The third term, ε_{AA}, represents the somewhat classical interaction energy of an electron centered at A with the proton at B:

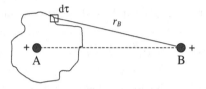

ε_{AA} can be evaluated by introducing confocal elliptic coordinates (see Sect. 7.2.2). Then

$$\frac{1}{\pi a_0^3} \int_0^{2\pi} \int_1^\infty \int_{-1}^{+1} \frac{R_{AB}^3(\mu^2 - \nu^2)e^{-(\mu+\nu)R_{AB}/2a_0}}{4R_{AB}(\mu-\nu)} \, d\phi \, d\mu \, d\nu$$

$$= \frac{1}{R_{AB}}\left[1 - \left(1 + \frac{R_{AB}}{a_0}\right)e^{\frac{-2R_{AB}}{a_0}}\right],$$

plotted below as a function of the internuclear distance in a_0 units:

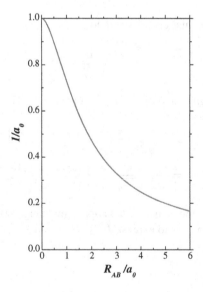

Therefore

$$\epsilon_{AA} = -\frac{e^2}{R_{AB}} \cdot \left[1 - e^{\frac{-2R_{AB}}{a_o}} \left(1 + \frac{R_{AB}}{a_o} \right) \right]$$

In analogous way the overlap integral S_{AB} and the resonance integral H_{AB} are evaluated.

$$S_{AB} = \frac{R_{AB}^3}{8\pi a_0^3} \int_0^{2\pi} \int_1^\infty \int_{-1}^{+1} (\mu^2 - \nu^2)\, e^{-\mu R_{AB}/a_0}\, d\mu\, d\nu\, d\varphi =$$

$$= \left[1 + \frac{R_{AB}}{a_o} + \frac{1}{3}\left(\frac{R_{AB}}{a_0} \right)^2 \right] e^{\frac{-R_{AB}}{a_o}} \qquad (8.9)$$

is plotted below

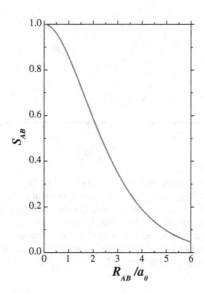

while

$$H_{AB} = \int \phi_{1s}^{(B)*} \left\{ \mathcal{H}_{hydr.} \right\} \phi_{1s}^{(A)} d\tau + \frac{e^2}{R_{AB}} S_{AB} - \int \phi_{1s}^{(B)*} \frac{e^2}{r_B} \phi_{1s}^{(A)} d\tau =$$

$$= S_{AB}\left(\frac{e^2}{R_{AB}} - R_H hc \right) + \varepsilon_{AB},$$

with

$$\varepsilon_{AB} = -\int \phi_{1s}^{(B)*} \frac{e^2}{r_B} \phi_{1s}^{(A)} d\tau = -\frac{e^2}{a_o} e^{\frac{-R_{AB}}{a_o}} \left(1 + \frac{R_{AB}}{a_o} \right), \qquad (8.10)$$

is approximately proportional to S_{AB}.

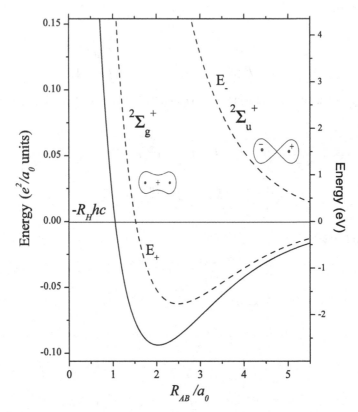

Fig. 8.2 Energy curve for the ground and first excited state of Hydrogen molecule ion as a function of the inter-proton distance R_{AB}, according to MO-LCAO orbital (*dotted lines*), with the classification of the electronic states in the separated-atoms scheme (see Sect. 7.3.3) and sketchy forms of the correspondent molecular orbitals. The bonding character of the $\sigma_g 1s$ state grants a minimum of the energy (in qualitative agreement with the exact calculation, *solid line*) while the $\sigma_u 1s$ orbital, for which $E_- > -R_H hc \equiv E(R_{AB} \to \infty)$, is *anti-bonding*. The exact result for E_- (not reported in figure) is well above the approximate energy E_- (*dotted line*)

From Eq. (8.6) and the expressions for H_{AA}, S_{AB} and H_{AB}, the energy curves $E_{\pm}(R_{AB})$ are obtained. In Fig. 8.2 $E_+(R_{AB})$ is compared to the exact eigenvalue for the ground-state that could be obtained from the solution of Eq. (8.1) through elliptic coordinates.

The minimum in E_+ indicates that when the electron occupies the lowest energy state (the $\sigma_g 1s$ according to Sect. 7.5) bonding does occur.

Starting from atomic orbitals pertaining to excited states, e.g. the $2p_x$ Hydrogen states, one can obtain the molecular orbitals for the excited states, as sketched below (see also Fig. 7.5):

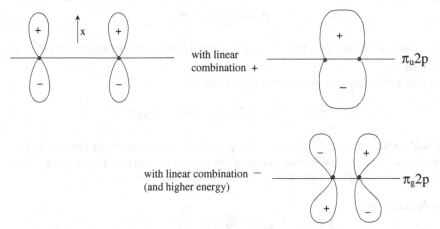

A better evaluation of eigenvalues and eigenfunctions (although still approximate) could be obtained by using more refined atomic orbitals. For instance, in order to take into account the polarization of the atomic orbitals due to the proton charge nearby, one could assume a wavefunction $\phi^{(A)}$ of the form

$$\phi^{(A)} = \phi_{1s}^{(A)} + a\,z\,e^{-Z_e\,r_A/a_0}, \tag{8.11}$$

with Z_e an effective charge. Along these lines of procedure one could derive values of the bonding energy and of the equilibrium interatomic distance R_{AB}^{eq} close to the experimental ones, which are

$$E(R_{AB}^{eq}) = -2.79\,\text{eV}, \quad R_{AB}^{eq} = 1.06\,\text{Å}. \tag{8.12}$$

Rather than pursuing a quantitative numerical agreement with the experimental data, now we shall move to the discussion of the physical aspects of the bonding mechanism.

Problems

Problem 8.1 Consider a muon-molecule formed by two protons and a muon. In the assumption that the muon behaves as the electron in the H$_2^+$ molecule, by means of scaling arguments evaluate the order of magnitude of the internuclear equilibrium distance, of the bonding energy and of the zero-point energy. (The zero-point energy $h\nu/2$ in H$_2^+$ is 0.14 eV and it is reminded that $\nu = 1/2\pi\sqrt{k/M}$, with k the elastic constant and M the reduced mass.)

Solution: R_{AB}^{eq} is controlled by the analogous of the Bohr radius a_0, which is reduced with the mass by factor m_μ/m_e. Then

$$R_{AB}^{eq} \approx \frac{1}{200} R_{AB}^{eq}(H_2^+) \simeq 5 \times 10^{-3} \text{Å}.$$

and

$$E \approx 200 \, E^{(H_2^+)} \simeq 500 \, \text{eV}.$$

The force constants can approximately be written $k \approx \frac{e^2}{R^3}$ and then $k(\mu) = k(H_2^+) \times 8 \times 10^6$.

The vibrational energies scale with $\sqrt{k}$, so that

$$E_{v=0}^{vib} \approx 2.8 \cdot 10^3 \, E_{v=0}^{vib}(H_2^+) \simeq 396 \, \text{eV}.$$

Problem 8.2 Write the behavior of the probability density ρ for the electron in H_2^+ at the middle of the molecular axis as a function of the inter-proton distance R_{AB}, for the ground MO-LCAO state, and in the assumption that $S_{AB} \ll 1$.

Solution: From

$$\rho = |\phi_+|^2 \propto 2e^{-\frac{r_A+r_B}{a_0}} + e^{-\frac{2r_A}{a_0}} + e^{-\frac{2r_B}{a_0}},$$

for $r_A = r_B = R_{AB}/2$, $\rho \propto 4\exp[-R_{AB}/a_0]$.

Problem 8.3 In the harmonic approximation the vibrational frequency of a diatomic molecule is given by

$$\frac{1}{2\pi}\sqrt{\frac{(d^2E/dR^2)_{R_e}}{\mu}},$$

where μ is the reduced mass and R the interatomic distance (for detail see Sect. 10.3). Derive the vibrational frequency for H_2^+ in the ground-state.

Solution: From $E(R) = (H_{AA} + H_{AB})/(1 + S_{AB})$ (see Eq. (8.6))

$$\frac{\partial^2 E}{\partial R^2} = \frac{[\frac{\partial^2(H_{AA}+H_{AB})}{\partial R^2}(1 + S_{AB}) - \frac{\partial^2 S_{AB}}{\partial R^2}(H_{AA} + H_{AB})](1 + S_{AB})}{(1 + S_{AB})^3} -$$

$$- \frac{2[\frac{\partial(H_{AA}+H_{AB})}{\partial R}(1 + S_{AB})\frac{\partial S_{AB}}{\partial R} - (\frac{\partial S_{AB}}{\partial R})^2(H_{AA} + H_{AB})]}{(1 + S_{AB})^3}$$

From Eqs. (8.8) to (8.10), for $x = R/a_0$ and $k_1 = e^2/a_0$, one writes

$$S_{AB}(x) = e^{-x}\left(1 + x + \frac{x^2}{3}\right),$$

$$H_{AA}(x) = k_1 e^{-2x}\left(1 + \frac{1}{x}\right) - \frac{k_1}{2},$$

$$H_{AB}(x) = k_1 e^{-x} \left(\frac{1}{x} - \frac{1}{2} - \frac{7x}{6} - \frac{x^2}{6} \right).$$

Since $\partial E/\partial R = (\partial E/\partial x)(1/a_0)$ and $\partial^2 E/\partial R^2 = (\partial^2 E/\partial x^2)(1/a_0^2)$, one can conveniently express the second derivative of $E(R)$ in terms of x.

The curves for $E(R)$ and for $(d^2 E/dR^2)$ are reported below (dashed line $(d^2 E/dR^2)$, in $e^2/a^3{}_0$ unit, dotted line $E(R)$ referred to $-R_H hc$, see also Fig. 8.2).

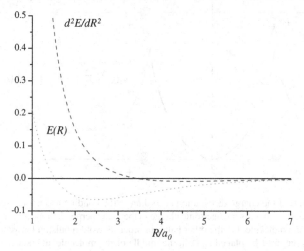

At $R_{eq} = 2.49 a_0$ one finds $\partial^2 E/\partial R^2 = 0.054 e^2/a_0^3 = 0.839 \times 10^5$ dyne/cm, yielding a vibrational frequency $\nu = 5.04 \times 10^{13}$ Hz (return to Problem 7.1). The experimental value is $\nu = 6.89 \times 10^{13}$ Hz for $R_{eq} = 1.06$ Å.

8.1.2 Bonding Mechanism and the Exchange of the Electron

How the bonded state of the Hydrogen molecule ion is generated? Why the bonding orbital is the $\sigma_g 1s$ while $\sigma_u 1s$ is antibonding? Which is the substantial role of the resonance integral H_{AB}?

A first way to answer to these questions is to look at the electronic charge distribution, controlled by $\rho_\pm = |\phi_\pm|^2$, where ϕ can be taken as in Eq. (8.7). The intersection of ρ with a plane containing the molecular axis is sketched in Fig. 8.3.

In order to minimize the Coulomb energy one has to place the electron in the middle of the molecule. Thus one understands why only the $\sigma_g 1s$ state has a minimum in the energy E versus. R_{AB}.

It may be remarked that this consideration of forces between nuclei according to "classical" Coulomb-like estimate of the energies is not in contrast with the quantum character of the system. In fact, as stated by the *Hellmann-Feynman theorem* the forces can actually be evaluated "classically" provided that the charge is distributed according to the quantum description.

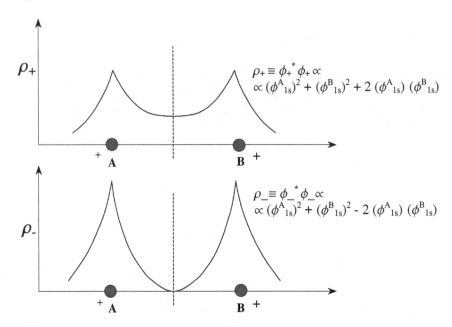

$$\rho_+ \equiv \phi_+^* \phi_+ \propto$$
$$\propto (\phi^A_{1s})^2 + (\phi^B_{1s})^2 + 2\,(\phi^A_{1s})\,(\phi^B_{1s})$$

$$\rho_- \equiv \phi_-^* \phi_- \propto$$
$$\propto (\phi^A_{1s})^2 + (\phi^B_{1s})^2 - 2\,(\phi^A_{1s})\,(\phi^B_{1s})$$

Fig. 8.3 Sketches of the charge distribution according to the bonding and antibonding molecular orbitals in H_2^+. For ϕ_- there is no electronic charge in the plane perpendicular to the molecular axis at the center of the molecule. On the other hand, in order to avoid repulsion between the protons, the negative charge must be placed right in the middle of the molecule, as indicated by classical considerations (see Problem 8.4)

Now we are going to discuss the role of the resonance integral (Eq. (8.10)) that through H_{AB} is the source of the minimum in the energy at a given inter-proton distance (see Fig. 8.2). A suggestive interpretation of the role of H_{AB} can be given in terms of the exchange of the electron between the two equivalent $1s$ states centered at the proton A and at the proton B.

According to the model developed in Appendix 1.2, by considering the basis states $|1\rangle$ and $|2\rangle$, as sketched below

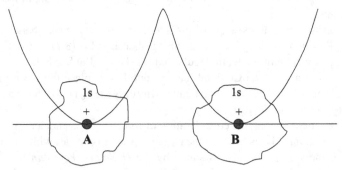

and by writing the generic state in term of linear combination $|\psi\rangle = c_1|1\rangle + c_2|2\rangle$, the coefficients obey the equations

$$i\hbar\dot{c}_1 = H_{11}c_1 + H_{12}c_2 \tag{8.13}$$
$$i\hbar\dot{c}_2 = H_{21}c_1 + H_{22}c_2$$

with $H_{11} = H_{22} = E_o$. H_{12} is the probability amplitude that the electron moves from state $|1\rangle$ to state $|2\rangle$.

By labeling A the value (negative) of H_{12}, from Eq. (8.13) by taking sum and difference, one has

$$c_1(t) = \frac{a}{2}e^{-i(\frac{E_o-A}{\hbar})t} + \frac{b}{2}e^{-i(\frac{E_o+A}{\hbar})t} \tag{8.14}$$
$$c_2(t) = \frac{a}{2}e^{-i(\frac{E_o-A}{\hbar})t} - \frac{b}{2}e^{-i(\frac{E_o+A}{\hbar})t}.$$

It is noted that for the choice of the integration constant $a = 0$ or $b = 0$, stationary states $|\pm\rangle$ are obtained, correspondent to $\sigma_g 1s$ and to $\sigma_u 1s$, i.e.

$$|+\rangle = \frac{1}{\sqrt{2}}\left[|1\rangle + |2\rangle\right], \quad |-\rangle = \frac{1}{\sqrt{2}}\left[|1\rangle - |2\rangle\right]$$

with energies $E = E_o \pm A$.

The constants a and b in Eq. (8.14) can be written in terms of the initial conditions for $c_1(t)$ and $c_2(t)$. By setting $c_1(0) = 1$ and $c_2(0) = 0$, one has

$$c_1(t) = e^{-i\frac{E_o}{\hbar}t}\cos(A t/\hbar)$$
$$c_2(t) = ie^{-i\frac{E_o}{\hbar}t}\sin(A t/\hbar)$$

with the behavior of the correspondent probabilities of presence $P_{1,2} = |c_{1,2}|^2$ shown in Fig. 8.4.

Thus the formation of the molecule can be idealized as due to the exchange of the electron from left to right and back, with the related decrease of the energy.

This description has some correspondence in classical systems, such as two weakly-coupled mechanical oscillators or LC circuits, with their two normal modes and the correspondent exchange of energy. Scattering experiments of protons on Hydrogen atoms confirm that the exchange process of the electron is real. When a proton is in the neighborhood (distance of the order of a_0) of an Hydrogen atom for a time of the order of $\hbar/2A$, with $A = (E_+ - E_-)$ (or multiple), an Hydrogen atom comes out after the scattering process.

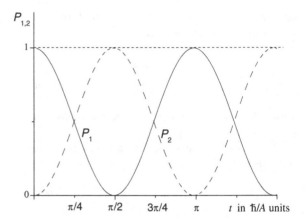

Fig. 8.4 Time dependence
of the probability of presence
of the electron on the sites A
and B according to the
description of two-levels
states for H_2^+

8.2 Homonuclear Molecules in the MO Scenario

From the MO description of the states in H_2^+ it is now possible to analyze multi-
electron *homonuclear* diatomic molecules. In a way analogous to the *aufbau* method
in atoms, to build up the molecule in a first approximation one has to accommodate
the electrons on the one-electron states derived for the prototype. This procedure
is particularly simple if *a priori* one does not take into account the inter-electron
interactions (e^2/r_{ij}), thus ideally assuming independent electrons. Then the energy
is evaluated on the basis of the complete Hamiltonian, for $\Phi_{total} = \prod_i \phi_{MO}(r_i)$,
by considering the dynamical equivalenceof the electrons when different states are
hypothesized. At Sect. 8.4 we shall discuss the hydrogen molecule to some extent,
by taking into account the spin states and the antisymmetry requirement. For the
moment, let us proceed to a qualitative description of some homonuclear diatomic
molecules by referring most to the ground states.

For H_2 the ground state has the electronic configuration $(\sigma_g 1s)^2$, it is labelled
$^1\Sigma_g^+$ (see Sect. 7.2.3) and the MO wavefunction is

$$\phi_{(\sigma_g 1s)^2}(r_1, r_2) = \sigma_g 1s(r_1)\, \sigma_g 1s(r_2), \tag{8.15}$$

that in the LCAO approximation is written (see Eq. (8.7))

$$\phi_{(\sigma_g 1s)^2}(r_1, r_2) = \frac{1}{2(1 + S_{AB})} \left[\phi_{1s}^{(A)}(r_1) + \phi_{1s}^{(B)}(r_1) \right] \cdot \left[\phi_{1s}^{(A)}(r_2) + \phi_{1s}^{(B)}(r_2) \right]. \tag{8.16}$$

The energy $E(R_{AB})$, evaluated by including in the Hamiltonian the term (e^2/r_{12}) by means of calculations strictly similar to the ones detailed for H_2^+ at Sect. 8.1, is sketched below:

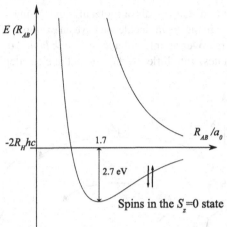

In He_2^+ the ground state has the electronic configuration $(\sigma_g 1s)^2(\sigma_u 1s)$ and the notation is $^2\Sigma_u$. The third electron has to be of u character, because of the Pauli principle.

The He_2 molecule cannot exist instate a stable state.[2] In fact, the electronic configuration should be $(\sigma_g 1s)^2(\sigma_u 1s)^2$, with the pictorial representation sketched below:

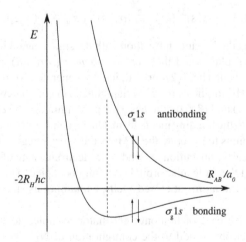

---

[2] Van der Waals interactions (described at Sect. 13.2.2), leading to very weak bonds at large distances, are not considered here.

Since for $R_{AB} \simeq R_{AB}^{eq.}$ one has $E_- > |E_+|$ (see Eq. (8.6)) the two antibonding electrons force the nuclei apart in spite of the bonding role of the electrons placed in the ground energy state.

Now we are going to discuss a pair of molecules exhibiting some aspects not yet encountered until now. In the N_2 molecule we have an example of "*strong bond*" due to σ MO orbital at large overlap integral and of "*weak bond*" due to π MO orbitals involving p atomic states, with little overlap. In fact one can depict the formation of the molecule as below

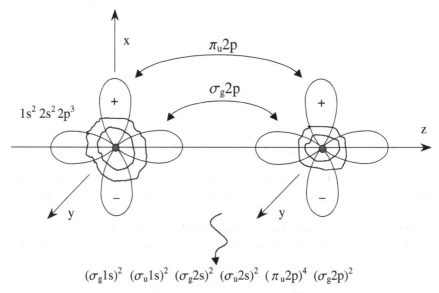

$$(\sigma_g 1s)^2 \ (\sigma_u 1s)^2 \ (\sigma_g 2s)^2 \ (\sigma_u 2s)^2 \ (\pi_u 2p)^4 \ (\sigma_g 2p)^2$$

where it is noted that the linear combination with the sign $+$ again implies electronic charge in the central plane (and therefore is a *bonding orbital*) although now the inversion symmetry is u. The $\sigma_g 2p$ orbital, ideally generated from the combination of $2p_z$ atomic orbitals, implies strong overlap. Since H_{AB} is somewhat proportional to S_{AB} (see Eqs. (8.9) and (8.10)) one has a deep minimum in the energy and then a strong contribution to the bonding mechanism. On the contrary, from the combination of $2p_{x,y}$ atomic orbitals to generate the π MO's the overlap region is small and then one can expect a weak contribution to bond. The electronic state of the N_2 molecule is labelled $^1\Sigma_g^+$ and the molecular orbitals are fully occupied. Thus the molecule is somewhat equivalent to atoms at closed shells, explaining its stability and scarcely reactive character.

Another instructive case of homonuclear diatomic molecule is O_2. Here there are two further electrons to add to the configuration of N_2. These electrons must be set on the $\pi_g 2p$ orbital, in view of the Pauli principle. The $\pi_g 2p$ orbital is not fully occupied and one has to deal with **LS** coupling procedure, similar to the one discussed for atoms for non-closed shells. In principle there are the possibilities sketched below:

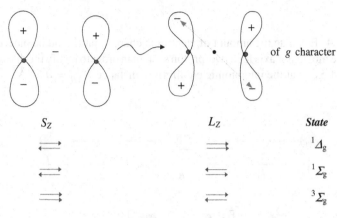

S_Z	L_Z	State
$\rightleftarrows$	$\Longrightarrow$	$^1\Delta_g$
$\rightleftarrows$	$\Longleftarrow$	$^1\Sigma_g$
$\Longrightarrow$	$\rightleftarrows$	$^3\Sigma_g$

According to Hund rules, that hold also in molecules, the ground state is $^3\Sigma_g^-$ corresponding to the maximization of the total spin. The g and $-$ characters can be understood by inspection: in Fig. 8.5 it is shown how the property under the reflection in a plane containing the molecular axis results from the symmetry of the π orbitals.

The molecule is *paramagnetic* and because of the partially empty external orbital has a certain reactivity, at variance with N_2. In fact the O_3 molecule (ozone) is known to exist.

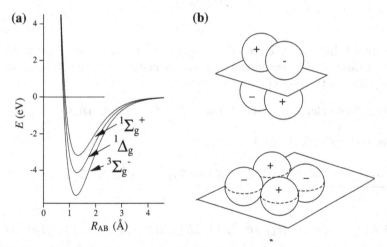

Fig. 8.5 Energy curves for the low energy states in the O_2 molecule (**a**) and sketchy illustration of the $(+-)$ symmetries for π^+ and π^- orbitals (**b**). The Σ state requires a label to characterize the behavior under reflection with respect to a plane containing the molecular axis. Since the two electrons occupy different π orbitals, one of them is $+$ and the other $-$, implying the overall $-$ character of the configuration

Problems

Problem 8.4 Evaluate the amount of electronic charge that should be placed at the center of the molecular axis for a two-proton system in order to justify the dissociation energy ($\simeq 4.5$ eV) at the interatomic equilibrium distance $R_{AB}^{eq} = 0.74$ Å.

Solution:

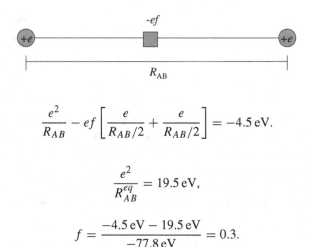

$$\frac{e^2}{R_{AB}} - ef \left[\frac{e}{R_{AB}/2} + \frac{e}{R_{AB}/2} \right] = -4.5 \, \text{eV}.$$

From

$$\frac{e^2}{R_{AB}^{eq}} = 19.5 \, \text{eV},$$

$$f = \frac{-4.5 \, \text{eV} - 19.5 \, \text{eV}}{-77.8 \, \text{eV}} = 0.3.$$

Problem 8.5 Indicate the electronic configuration and the spectroscopic terms according to Sect. 7.2.3 for the ground states and the first excited states of the molecules H_2, Li_2, B_2, N_2, C_2 and Br_2.

Solution: Molecule;   ground-state;   first excited states

H_2; $(\sigma_g 1s)^2$, $^1\Sigma_g^+$; $(\sigma_g 1s)(\sigma_u 1s)$, $^3\Sigma_u^+$, $^1\Sigma_u^+$.

Li_2; $\underbrace{(\sigma_g 1s)^2 (\sigma_u 1s)^2}_{KK} (\sigma_g 2s)^2$, $^1\Sigma_g^+$; $KK(\sigma_g 2s)(\sigma_u 2s)$, $^3\Sigma_u^+$, $^1\Sigma_u^+$.

B_2; $KK(\sigma_g 2s)^2 (\sigma_u 2s)^2 (\pi_u 2p)^2$, $^3\Sigma_g^-$; $KK(\sigma_g 2s)^2(\sigma_u 2s)^2(\pi_u 2p)(\sigma_g 2p)$, $^3\Pi_u$.

N_2; ......$(\pi_u 2p)^4 (\sigma_g 2p)^2$, $^1\Sigma_g^+$; ......$(\pi_u 2p)^4 (\sigma_g 2p)(\pi_g 2p)$, $^3\Pi_g$, $^1\Pi_g$.
  For C_2 the proper sequence of the energy levels has to be taken into account (one electron could be promoted from π_u to the σ_g state, see Fig. 7.4). However the electronic configuration $(\sigma_g 2s)^2 (\sigma_u 2s)^2 (\pi_u 2p)^4$ seems to be favored and the ground state term is $^1\Sigma_g^+$.

$$\text{Br}_2 \text{ (atoms in } ^2P \text{ states),} \qquad ^1\Sigma_g^+$$

$$\text{excited states} \qquad ^1\Sigma_u^-, \, ^1\Pi_g, \, ^1\Pi_u, \, ^1\Delta_g.$$

It is noted that in some cases the exact sequence of the levels is poorly known because of the possible modifications of the energy upon excitation and for the correlation effects. Thus the sequence of the eigenvalues reported in Fig. 7.4 for a given R_{AB} distance might be altered. Only elaborate computational descriptions can lead to quantitative deductions.

8.3 H$_2$ as Prototype of the VB Approach

In the framework of the *valence bond (VB)* method, where the molecule results from the interaction of atoms dressed by their electrons, the prototype is the Hydrogen molecule.

The Hamiltonian is written (see Fig. 8.6)

$$\mathcal{H} = \left[-\frac{\hbar^2}{2m}\nabla_1^2 - \frac{e^2}{r_{A1}} \right] + \left[-\frac{\hbar^2}{2m}\nabla_2^2 - \frac{e^2}{r_{B2}} \right] + \left[-\frac{e^2}{r_{A2}} - \frac{e^2}{r_{B1}} \right] + \left[\frac{e^2}{r_{AB}} + \frac{e^2}{r_{12}} \right] \equiv$$

$$\equiv [a] + [b] + [c] + [d] \tag{8.17}$$

A tentative wavefunction could be $\phi(\mathbf{r}_1, \mathbf{r}_2) = \phi_{1s}^A(\mathbf{r}_1)\,\phi_{1s}^B(\mathbf{r}_2)$, corresponding to the situation in which the two electrons keep their atomic character and only Coulomb-like interactions with classical analogies are supposed to occur. However, this wave-function does not lead to the formation of the real bonded state. In that case, in fact, for the [a] and [b] terms in the Hamiltonian one obtains $-2R_H hc$ and for the interaction terms [c] and [d] one has

$$J = \frac{e^2}{R_{AB}} + \int |\phi_{1s}^A(\mathbf{r}_1)|^2 \frac{e^2}{r_{12}} |\phi_{1s}^B(\mathbf{r}_2)|^2 \, d\tau_1 \, d\tau_2 - 2\int \frac{e^2}{r_{A2}} |\phi_{1s}^B(\mathbf{r}_2)|^2 \, d\tau_2. \tag{8.18}$$

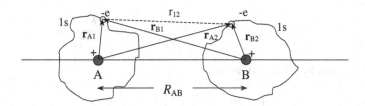

Fig. 8.6 Definition of the coordinates involved in the Hamiltonian for the H$_2$ molecule

The latter term in Eq. (8.18) is twice the attractive interaction between the electron in B and the proton A, as sketched below

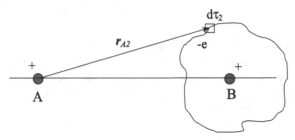

All the terms in Eq. (8.18) correspond to classical electrostatic interactions and therefore J is usually called *Coulomb integral*. From the evaluation of J through elliptic coordinates, as described for ε_{AA} at Sect. 8.1.1, one could figure out that the energy curve $E(R_{AB})$ displays only a slight minimum, around 0.25 eV, in large disagreement with the experimental findings (see Fig. 8.7). On the other hand, by recalling the description of the two electrons in Helium atom (Sect. 2.2) the inadequacy of the wavefunction $\phi_{1s}^A(r_1)\phi_{1s}^B(r_2)$ can be expected, since the indistinguishability of the electrons, once that the atoms are close enough to form a molecule, is not taken into account.

Then one rather writes

$$\phi_{VB}(r_1, r_2) = c_1\,\phi_{1s}^A(r_1)\,\phi_{1s}^B(r_2) + c_2\,\phi_{1s}^A(r_2)\,\phi_{1s}^B(r_1) \equiv c_1\,|1\rangle + c_2\,|2\rangle \quad (8.19)$$

By deriving the energy function with the usual variational procedure (see Eqs. (8.3)–(8.6)) one obtains $c_1 = \pm c_2$ and

$$E_\pm = \frac{H_{11} \pm H_{12}}{1 \pm S_{12}}, \qquad (8.20)$$

where $H_{11} \equiv \langle 1|H|1\rangle = \langle 2|H|2\rangle$, $H_{12} \equiv \langle 2|H|1\rangle$, $S_{12} = S_{AB}^2$ and

$$\phi_\pm = \frac{1}{\sqrt{2\,(1 \pm S_{12})}}\Big[|1\rangle \pm |2\rangle\Big]. \qquad (8.21)$$

The eigenvalues turn out

$$E_\pm(R_{AB}) = -2R_H hc + \frac{J}{1 \pm S_{AB}^2} \pm \frac{K}{1 \pm S_{AB}^2} \qquad (8.22)$$

where J is given by Eq. (8.18), while

$$K = \int \phi_{1s}^{A*}(r_1)\,\phi_{1s}^{B*}(r_2)\left[-\frac{e^2}{r_{A2}} - \frac{e^2}{r_{B1}} + \frac{e^2}{R_{AB}} + \frac{e^2}{r_{12}}\right]\phi_{1s}^A(r_2)\,\phi_{1s}^B(r_1)\,d\tau_1\,d\tau_2 \qquad (8.23)$$

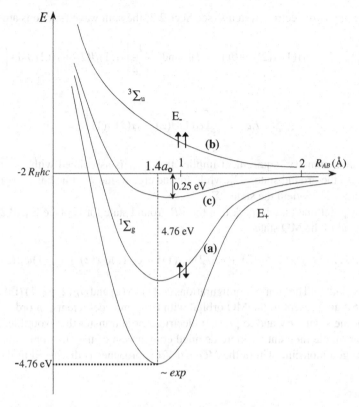

Fig. 8.7 Sketch of the energy curves of the Hydrogen molecule in the VB scheme as a function of the interatomic distance R_{AB}. The real curve (reconstructed by a variety of experiments) is indicated as *exp*, while curve *c* illustrates the behavior expected from the Coulomb integral only (Eq. (8.18) in the text). Curves *a* and *b* illustrate the approximate eigenvalues $E_\pm$ in Eq. (8.22)

is the *extended exchange integral*, with no classical analogy and related to the quantum character of the wavefunction. K can be rewritten

$$K = \frac{e^2}{R_{AB}} S_{AB}^2 - 2 S_{AB}\, \epsilon_{AB} + \int \phi_{1s}^{A*}(r_1)\, \phi_{1s}^{B*}(r_2)\, \frac{e^2}{r_{12}}\, \phi_{1s}^{A}(r_2)\, \phi_{1s}^{B}(r_1)\, d\tau_1\, d\tau_2 \tag{8.24}$$

where again one finds the resonance integral ϵ_{AB} (Eq. (8.10)) and a *reduced exchange integral*

$$K_{red} = \int \phi_{1s}^{A*}(r_1)\, \phi_{1s}^{B*}(r_2)\, \frac{e^2}{r_{12}}\, \phi_{1s}^{A}(r_2)\, \phi_{1s}^{B}(r_1)\, d\tau_1\, d\tau_2 \tag{8.25}$$

analogous to the one in Helium atom and *positive*.

From the evaluation of J and K the energy curves can be obtained, as depicted in Fig. 8.7. It should be remarked that most of the bond strength is due to the exchange integral K.

As for any two electron systems (see Sect. 2.2) the spin wave functions are

$$\chi_{symm}^{S=1} \quad i.e. \quad \alpha(1)\alpha(2), \quad \beta(1)\beta(2) \quad \text{and} \quad \frac{1}{\sqrt{2}}\Big[\alpha(1)\beta(2) + \alpha(2)\beta(1)\Big]$$

$$\chi_{ant}^{S=0} \quad i.e. \quad \frac{1}{\sqrt{2}}\Big[\alpha(1)\beta(2) - \alpha(2)\beta(1)\Big]$$

and the antisymmetry requirement implies that χ_{ant} is associated with ϕ_+, corresponding to the ground state $^1\Sigma_g$, while for the eigenvalue E_- one has to associate χ_{symm} with ϕ_-, to yield the state $^3\Sigma_u$.

At this point one may remark that the *VB* ground state for H_2 (see Eq. (8.21)) is proportional to the MO state

$$\phi_{MO}(1,2) \propto [\phi_A(1)\phi_B(2) + \phi_A(2)\phi_B(1)] + \phi_A(1)\phi_A(2) + \phi_B(1)\phi_B(2)$$

(1,2 for r_1 and r_2) The "ionic" configurations $\phi_A(1)\phi_A(2)$ and $\phi_B(2)\phi_B(1)$ (Eqs. (8.7) and (8.16)) are present in the MO orbital with the same coefficients, in order to account for the symmetry and to prevent electric charge transfer that would lead to a molecular dipole moment. A more detailed comparison of the electronic states for the Hydrogen molecule within the *MO* and *VB* approaches is discussed in the next section.

Problems

Problem 8.6 Reformulate the description of the H_2 molecule in the VB approach in the assumption that the two Hydrogen atoms in their ground state are at a distance R so that exchange effects can be neglected. Prove that for large distance the interaction energy takes the dipole-dipole form and that by using the second order perturbation theory an attractive term going as R^{-6} is generated (see Sect. 13.2.2 for an equivalent formulation). Then remark that for degenerate $n = 2$ states the interaction energy would be of the form R^{-3}.

Solution: From Fig. 8.6 and Eq. (8.17) the interaction is written
$V = e^2/R - e^2/r_{A2} - e^2/r_{B1} + e^2/r_{12}$.

Expansions in spherical harmonics (see Problem 2.5) yield

$$\frac{1}{r_{B1}} = \frac{1}{|R\rho - r_{1A}|} = \sum_{\lambda=0} \frac{r_{A1}^\lambda}{R^{\lambda+1}} P_\lambda(\cos\theta) = \frac{1}{R} + \frac{(r_{A1}\cdot\rho)}{R^2} + \frac{3(r_{A1}\cdot\rho)^2 - r_{A1}^2}{2R^3} + \cdots,$$

$$\frac{1}{r_{12}} = \frac{1}{|R\rho + r_{B2} - r_{A1}|} = \frac{1}{R} + \frac{(r_{A1} - r_{B2})\cdot\rho}{R^2} +$$

$$+ \frac{3[(r_{A1} - r_{B2})\cdot\rho]^2 - (r_{A1} - r_{B2})^2}{2R^3} + \cdots,$$

$$\frac{1}{r_{A2}} = \frac{1}{R} + \frac{(\mathbf{r}_{B2} \cdot \boldsymbol{\rho})}{R^2} + \frac{3(\mathbf{r}_{B2} \cdot \boldsymbol{\rho})^2 - r_{B2}^2}{2R^3} + \cdots$$

($\boldsymbol{\rho}$ unit vector along the interatomic axis).

Thus the dipole-dipole term (see Sect. 13.2.2) is obtained

$$V = -\frac{2z_1 z_2 - x_1 x_2 - y_1 y_2}{R^3} e^2,$$

the z-axis being taken along $\boldsymbol{\rho}$. By resorting to the second-order perturbation theory and taking into account the selection rules (Appendix 1.3 and Sect. 3.5), the interaction energy turns out

$$E^{(2)} = \frac{e^4}{R^6} \sum_{mn} \frac{4z_{0m}^2 z_{0n}^2 + x_{0m}^2 x_{0n}^2 + y_{0m}^2 y_{0n}^2}{2E_0 - E_m - E_n}$$

where x_{om}, x_{on} etc. are the matrix elements connecting the ground state (energy E_0) to the excited states (energies E_m, E_n). $E^{(2)}$ being negative, the two atoms attract each other (*London* interaction, see Sect. 13.2.2 for details). For the states at $n = 2$ the perturbation theory for degenerate states has to be used. From the secular equation a first order energy correction is found (see the similar case for Stark effect at Sect. 4.2). Thus the interaction energy must go as R^{-3}.

8.4 Comparison of MO and VB Scenarios in H₂: Equivalence from Configuration Interaction

Going back to the MO description for the H₂ molecule, by considering the possible occurrence of the first excited σ_u one-electron state and by taking into account the indistinguishability, four possible wavefunctions are:

$$\Phi_I(g, g) \equiv \phi_g(1)\phi_g(2) \qquad \mathbf{gg} \qquad a)$$
$$\Phi_{II}(u, u) \equiv \phi_u(1)\phi_u(2) \qquad \mathbf{uu} \qquad b)$$
$$\Phi_{III}(g, u) \equiv \frac{1}{\sqrt{2}}\left[\phi_g(1)\phi_u(2) + \phi_g(2)\phi_u(1)\right] \quad \mathbf{ug^{symm}} \quad c)$$
$$\Phi_{IV}(g, u) \equiv \frac{1}{\sqrt{2}}\left[\phi_g(1)\phi_u(2) - \phi_g(2)\phi_u(1)\right] \quad \mathbf{ug^{ant}} \quad d) \qquad (8.26)$$

In view of the four spin wavefunctions χ^{ant} and χ^{symm}, in principle 16 spin-molecular orbitals could be constructed. Due to the Pauli principle, in the H₂ molecule one finds only 6 states, the ones of antisymmetric character.

The ground MO state $(\sigma_g 1s)(\sigma_g 1s)\chi^{ant}$ can be detailed by referring to the LCAO specialization, so that the complete spin-MO is

$$\phi_{TOT}^{MO}(1, 2) = \chi_{S=0}^{ant}\left[\phi_{VB} + \phi_A(1)\phi_A(2) + \phi_B(1)\phi_B(2)\right] \tag{8.27}$$

namely the VB form with the "ionic" states, as already mentioned.

To find the excited MO state corresponding to the $(VB)^-$ wavefunction, in Eq. (8.26) one can look for the one that without the ionic states does correspond to 8.21 ϕ_- without the ionic states. From Eq. (8.26d) with the LCAO specialization it is found that

$$\Phi_{IV} = \frac{1}{\sqrt{2}}\left[\phi_A(1)\phi_B(2) - \phi_A(2)\phi_B(1)\right] \tag{8.28}$$

is the same as ϕ_{VB}^-.

From another point of view, now one understands why the $^3\Sigma_u$ state is unstable: it corresponds to have one electron in the g bonding MO orbital and one in the u antibonding MO (see Sect. 8.2), this latter being strongly repulsive. In Fig. 8.8 the lowest energy levels in H_2 corresponding to 8.26 are sketched.

For a more quantitative comparison of the MO and the VB descriptions in H_2, let us look at the values for the dissociation energies and the equilibrium distances (see also Fig. 8.9) in the ground state:

$$\phi_{VB} \qquad\qquad E_{diss} \simeq 3.14\,\text{eV} \qquad\qquad R_{AB}^{eq.} = 1.7\,a_0$$

$$\phi_{MO} \qquad\qquad E_{diss} \simeq 2.7\,\text{eV} \qquad\qquad R_{AB}^{eq.} = 1.7\,a_0$$

$$Experimental \qquad E_{diss} \simeq 4.75\,\text{eV} \qquad\qquad R_{AB}^{eq.} = 1.4\,a_0$$

One should remark that the VB orbital does not include the ionic states while the MO-LCAO overestimates their weight. In fact, the energy to remove the electron from the Hydrogen atom (13.56 eV) is much higher than the energy gain Δ in setting it on the configuration H^-. The energy gain Δ (sometimes called *electron affinity*) in principle could be estimated from the Coulomb integral in Helium atom (Sect. 2.2), with $Z = 1$ for the nuclear charge (however, see Problem 2.8). From accurate estimates one actually would find $\Delta = 0.75\,\text{eV}$. Therefore the ionic states cannot be weighted as much as they are in the MO-LCAO orbital. This observation suggests a tentative wavefunction of the form

$$\phi_{VB} + \lambda\phi_{ionic}, \tag{8.29}$$

namely a mixture of the covalent VB and of ionic states with a coefficient λ, for instance to be estimated variationally. From the derivative of the energy function

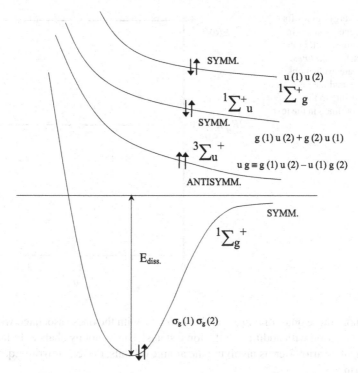

Fig. 8.8 Schematic energy curves for H_2 corresponding to the wavefunctions in Eq. (8.26). More accurate forms of the energies for the $^1\Sigma_g^+$ and $^3\Sigma_u^+$ states are reported in Fig. 8.9, in comparison with the VB eigenvalues

$E(\lambda)$ one could find that the minimum corresponds to $\lambda = 0.25$. Therefore, from the normalization of the wavefunction the weight of the ionic states is given by $\lambda^2/(\lambda^2 + 1)$, about 6 percent.

How could the MO description of the ground state in H_2 be improved? Since the wavefunctions 8.26 involve the ionic states with different coefficients, it is conceivable that a better approximation is obtained if a proper combination of the wavefunctions correspondent to different configurations is attempted. This procedure is an example of the approach called *configuration interaction (CI)*. In the combination one has to take into account that the mixture must involve states with the same symmetry properties and same spin. Thus one should combine the **gg** state with the **uu** one, both coupled to χ^{ant}:

$$\phi_{CI}(1, 2) = \phi_I + k\phi_{II} \tag{8.30}$$

From this wavefunction, as usual, one generates two energy levels, one of them having energy $E < E_+$, E_+ being the energy for ϕ_I. In this way one could find a dissociation energy and equilibrium distance close to the experimental values.

Fig. 8.9 Energy curves for
the lowest energy states in
H$_2$: *dotted lines*, within the
VB approach; *solid lines*,
more accurate evaluations
for the $^1\Sigma_g^+$ and the $^3\Sigma_u^+$
states according to the
procedure outlined in the text

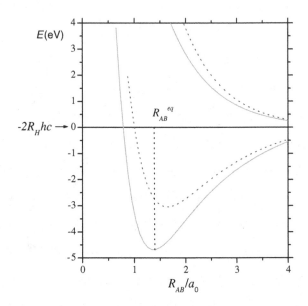

Furthermore those quantities are found to *coincide* with the ones associated with the
VB wavefunction with addition of the ionic states! This is not by chance. In fact, by
collecting the various terms involving the atomic orbitals, one can rewrite Eq. (8.30)
in the form

$$\phi_{CI}(1,2) = (1-k)\phi_{VB} + (1+k)\phi_{ionic}$$

and by defining $\lambda = (1+k)/(1-k)$ one sees that it coincides with Eq. (8.29).

This is an example of a more general issue: *the MO-LCAO method with interaction
of the configurations is equivalent to the VB approach with addition of the ionic states
to the covalent wavefunctions.*

8.5 Heteronuclear Molecules and the Electric
Dipole Moment

In the following we shall recall some novel aspects present in diatomic molecules
when the two atoms are different.

First of all one remarks that the inversion symmetry, with the Hamiltonian $\mathcal{H}(\mathbf{r})$
equal to $\mathcal{H}(-\mathbf{r})$, no longer holds. Therefore, within the separated atoms scheme one
cannot longer classify the states as *g* or *u* and the one-electron states become (see
Fig. 7.6) $\sigma 1s_A$, $\sigma 1s_B$, $\sigma 2s_A$, $\sigma 2s_B$,...

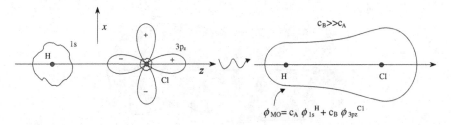

Fig. 8.10 Sketchy illustration of the polarized MO-LCAO orbital in HCl. The p_x and p_y atomic orbitals are scarcely involved in the formation of the molecule since they imply small overlap integral S_{AB} and resonance integral H_{AB} (see text at Sect. 8.2)

Within the MO-LCAO scheme the one-electron orbital is written

$$\phi = c_A \phi_A + c_B \phi_B$$

with $c_A \neq c_B$. Equivalently, in the normalized form

$$\phi_{MO}^{LCAO} = \frac{1}{(1 + \lambda^{*2} + 2\lambda^* S_{AB})^{\frac{1}{2}}} \left[\phi_A + \lambda^* \phi_B \right]. \qquad (8.31)$$

Here λ^* can vary from $-\infty$ to $+\infty$ and it characterizes the *polarization* of the orbital, namely measuring the *electronic charge transfer* from one atom to the other. As illustrative example in Fig. 8.10 the molecular orbital for the HCl molecule is sketched.

In the VB description the only way to account for the charge transfer is to add the ionic states in the molecular orbital, no longer with the same weight as for the homonuclear molecules (see Eq. (8.29)). In practice only the ionic configuration favoured by the polarity of the molecule can be included. Then

$$\phi_{VB}^{het} = \phi_{VB} + \lambda \phi_{ionic}$$

In HCl, for instance, the large contribution to the polar character described by λ is due to the ionic function representing $H^+ Cl^-$.

The parameters λ in the above definition and λ^* in Eq. (8.31) are difficult to evaluate from first principles. They have been empirically related to the electronegativity of the atoms or to the difference between the ionization energy with respect to the one pertaining to the purely covalent configuration.

An illustrative relationship of λ and λ^* to molecular properties is the one involving the electric dipole moment μ_e. By referring to the sketched schematization for a given molecular orbital with two electrons (Sect. 8.3),

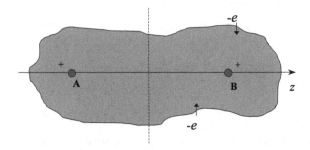

the dipole moment is written $\mu_e = 2e <z>$, with $<z>$ the expectation value of the coordinate, corresponding to the first moment of the electronic charge distribution.

For an MO-LCAO orbital as in Eq. (8.31), one has

$$<z> = \frac{1}{(1 + \lambda^{*2} + 2\lambda^* S_{AB})} \int z \left[|\phi_A|^2 + \lambda^{*2} |\phi_B|^2 + 2\lambda^* \phi_A \phi_B \right] d\tau \quad (8.32)$$

The mixed term $<A|z|B>$ is usually negligible. By assuming for simplicity $S_{AB} \ll 1$ one obtains

$$<z> = \frac{1}{(1 + \lambda^{*2})} \left[-\frac{1}{2} R_{AB} + \frac{1}{2} R_{AB} \lambda^{*2} \right] \quad (8.33)$$

By defining $g = \mu_e / e R_{AB}$ as *degree of ionicity* (g being the unit for total charge transfer and dipole moment $\mu_e^{max} = e R_{AB}$), λ^* can be expressed in terms of a relevant property of the molecule:

$$g = \frac{\lambda^{*2} - 1}{\lambda^{*2} + 1} \quad (8.34)$$

In analogous way in the VB framework, where g is evidently given by the weight of the ionic structure, one has

$$g = \frac{\lambda^2}{\lambda^2 + 1}. \quad (8.35)$$

As for the homonuclear molecules, the energy curve $E(R_{AB})$ in principle could be evaluated in terms of the overlap and resonance integrals.

Direct understanding of the mechanism leading to the bonded state can easily be achieved by referring to a model of *totally ionic molecule*, i.e. $\phi_{MO} = \phi_B$ (or configuration $A^+ B^-$) and in the assumption of Coulombic interaction between point charge ions. This is an oversimplified way to derive the eigenvalue as a function of the interatomic distance, still allowing one to grasp the main source of the bonding.

For numerical clarity let us refer to the NaCl molecule (Fig. 8.11). One observes that for distances R_{AB} above about 10 Å the energy of the neutral atoms is below the one for ions. When the distance is smaller than the R_{AB}^* for which $e^2 / R_{AB}^* \simeq (E_I - E_A)$, the ionic configuration is favoured and the system reduces the energy by decreasing the interatomic distance.

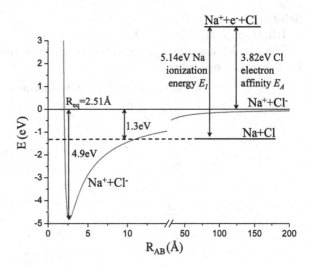

Fig. 8.11 Energies of the neutral atoms and of the ionic configuration in the NaCl molecule. E_I is the ionization energy of Na, about 5.14 eV while $E_A = 3.82$ eV is the electron affinity in Cl and it corresponds to the energy to remove an electron from Cl^-

At short distance a repulsive term is acting. Its phenomenological form can be written

$$E_{rep} \sim B \, exp \, [-R_{AB}/\rho],\qquad(8.36)$$

an expression known as *Born-Mayer* repulsion. Thus the energy curve depicted as solid line in the Fig. 8.11 is generated.

The dissociation energy $E(R_{AB}^{min})$ can be evaluated by estimating the distance where the energy minimum occurs. A detailed calculation of this type will be used for the cohesive energy in ionic crystals (Sect. 13.2.1).

Finally, for some polar diatomic molecules the electric dipole moment μ_e, the degree of ionicity and the value of λ^* according to Eq. (8.32) are reported below (having used for S_{AB} a value around 0.3).

	μ_e (Debye)	μ_e/eR_{AB}	λ^*
HF	1.82	0.43	1.88
HCl	1.08	0.17	1.28
HBr	0.78	0.11	1.19
KF	8.5	0.67	2.93
KCl	10.27	0.77	3.36

(1 Debye $= 10^{-18}$ u.e.s cm). In H_2O, $\mu_e = 1.85$ Debye.

Problems

Problem 8.7 By using the united atoms classification scheme (Fig. 7.6) write the electronic configuration and the spectral terms for HB, LiH, CH and NO. Why NO is paramagnetic ? Figure out in detail how the HB molecule correlates to the C atom in the united atoms scheme.

Solution:

HB; $(1s\sigma)^2(2s\sigma)^2(2p\sigma)^2$, $^1\Sigma^+$.

The detailed correlation diagram for the HB molecule with Carbon atom (see Fig. 7.6) is reported below

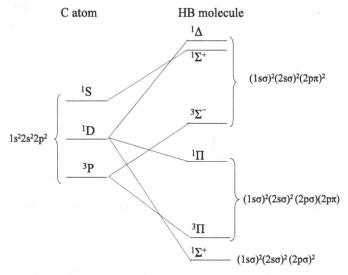

LiH; $(1s\sigma)^2(2s\sigma)^2$, $^1\Sigma^+$;

first excited configuration $(1s\sigma)^2(2s\sigma)(2p\sigma)$, $^3\Sigma^+$ or $^1\Sigma^+$.

CH; $(1s\sigma)^2(2s\sigma)^2(2p\sigma)^2(2p\pi)$, $^2\Pi$;

first excited configuration $(1s\sigma)^2(2s\sigma)^2(2p\sigma)(2p\pi)^2$, $^4\Sigma^-, {}^2\Delta, {}^2\Sigma^+, {}^2\Sigma^-$.

NO, from the atomic configurations $1s^22s^22p^3$ for N and $1s^22s^22p^4$ for O, the most plausible ground-state configuration could be

NO$[KK(2p\sigma)^2(3s\sigma)^2(3p\sigma)^2(2p\pi)^4(3p\pi)]$, $^2\Pi$. Since there is one unpaired $3p\pi$ electron NO is paramagnetic ($S = 1/2$). As already mentioned the actual sequence of the eigenvalues is rather uncertain (see Problem 8.5).

Problem 8.8 In the ionic bond approximation assume for the eigenvalue in the NaCl molecule the expression

$$E(R_{AB}) = -\frac{e^2}{R_{AB}} + \frac{A}{R_{AB}^n}.$$

From the equilibrium interatomic distance $R_{AB}^{eq} = 2.51\,\text{Å}$ and knowing that the vibrational frequency is $1.14 \cdot 10^{13}$ Hz, obtain A and n and estimate the dissociation energy.

Solution: At the minimum

$$\left(\frac{dE}{dR_{AB}}\right)_{R_{AB} = R_{AB}^{eq.} \equiv R_e} = \frac{e^2}{R_e^2} - nA\frac{1}{R_e^{n+1}} = 0$$

thus

$$\frac{e^2}{n\,A} = \frac{1}{R_e^{n-1}} .$$

The elastic constant is (see Problem 8.3)

$$k = \left(\frac{d^2 E}{d R_{AB}^2}\right)_{R_{AB}=R_e} = -\frac{2e^2}{R_e^3} + A\,n(n+1)\frac{1}{R_e^{n+2}}$$

Then

$$k = \frac{e^2}{R_e^3}(n-1) .$$

For the reduced mass

$$\mu = 2.3 \cdot 10^{-23}\,g$$

the elastic constant takes the value

$$k = 4\pi^2 \mu \nu_0^2 = 1.18 \cdot 10^5 \text{ dyne/cm}$$

(see Sect. 10.3).
Then

$$n - 1 = \frac{k\,R_e^3}{e^2} \simeq 8.$$

From

$$A = \frac{e^2 R_e^{n-1}}{n}$$

the energy at R_e is

$$E_{min} = -\frac{e^2}{R_e} + \frac{A}{R_e^n} = -\frac{e^2}{R_e}\left(1 - \frac{1}{n}\right) = -5.1\,\text{eV}.$$

and then the dissociation energy turns out

$$E_{diss} = -\left[E_{min} + \frac{1}{2} h\nu_0\right] \simeq 5\,\text{eV}.$$

Problem 8.9 The first ionization energy in the K atom is 4.34 eV while the electron affinity for Cl is 3.82 eV. The interatomic equilibrium distance in the KCl molecule is 2.79 Å. Assume for the characteristic constant in the Born-Mayer repulsive term $\rho = 0.28$ Å. In the approximation of point-charge ionic bond, derive the energy required to dissociate the molecule in neutral atoms.

Solution: From

$$V(R) = -\frac{e^2}{R} + B e^{-\frac{R}{\rho}}$$

and the equilibrium condition

$$\left(\frac{dV}{dR}\right)_{R=R_e} = 0 = \frac{e^2}{R_e^2} - \frac{B}{\rho} e^{-\frac{R_e}{\rho}}$$

one obtains

$$V(R_e) = -\frac{e^2}{R_e}\left(1 - \frac{\rho}{R_e}\right) \simeq -4.66\,\text{eV}.$$

The energy for the ionic configuration $K^+ + Cl^-$ is $(4.34 - 3.82)\,\text{eV} = 0.52\,\text{eV}$ above the one for neutral atoms. Then the energy required to dissociate the molecule is

$$E_{diss} = (+4.66 - 0.52)\,\text{eV} \simeq 4.12\,\text{eV}.$$

Problem 8.10   In the molecule KF the interatomic equilibrium distance is 2.67 Å and the bonding energy is 0.5 eV smaller than the attractive energy of purely Coulomb character. By knowing that the electron affinity of Fluorine is 4.07 eV and that the first ionization potential for potassium is 4.34 V, derive the energy required to dissociate the molecule in neutral atoms.

Solution: Since

$$E_{Coulomb} = \frac{e^2}{R_e} = 8.6 \cdot 10^{-12}\,\text{erg} = 5.39\,\text{eV}$$

the energy required for the dissociation in ions is $E_i = (5.39 - 0.5) = 4.89\,\text{eV}$. For the dissociation in neutral atoms $E_a = E_i + A_f - P_{ion} = 4.89 + 4.07 - 4.34\,\text{eV} = 4.62\,\text{eV}$.

Problem 8.11   Derive the structure of the hyperfine magnetic states for the ground-state of the Hydrogen molecular ion. Then numerically evaluate their energy separation in the assumption of $\sigma_g 1s$ molecular orbital in the form of linear combination of $1s$ atomic orbitals (the interatomic equilibrium distance can be assumed $2a_0$).

Solution: From the extension of Eq. (5.3) $H_{mag}^{hyp} = A_{H_2^+}(\mathbf{I}_A + \mathbf{I}_B) \cdot \mathbf{s}$, with $A_{H_2^+}$ the hyperfine coupling constant. From $\mathbf{I} = \mathbf{I}_A + \mathbf{I}_B$, $I = 0$ or $I = 1$, namely states with $F = 1/2, 3/2$ and $F = 1/2$ are obtained.
  Since $\mathbf{I} \cdot \mathbf{s} = (1/2)[F(F + 1) - I(I + 1) - s(s + 1)]$
the $F = 3/2$ and $F = 1/2$ levels are separated by $\Delta E = (3/2)A_{H_2^+}$.
$A_{H_2^+}$ can be obtained from

$$\phi_{\sigma_g 1s} = \frac{1}{\sqrt{2(1 + S_{AB})}} \frac{1}{\sqrt{\pi a_0^{3/2}}} \left[e^{-r_A/a_0} + e^{-r_B/a_0} \right]$$

considering $r_A = 0$ and $r_B = R_{AB}$. Then

$$|\phi_{\sigma_g 1s}(0)|^2 = \frac{1}{\pi a_0^3} \frac{\left[1 + e^{-R_{AB}/a_0} \right]^2}{2(1 + S_{AB})} \simeq \frac{0.41}{\pi a_0^3}$$

for $S_{AB} \approx 0.58$ (see Eq. (8.9)).

In atomic Hydrogen where $|\phi_{1s}(0)|^2 = 1/\pi a_0^3$ the separation between the $F = 1$ and $F = 0$ hyperfine levels is $A_H/h = 1421.8$ MHz. Then in H_2^+ one deduces $\Delta E/h = (3/2)0.41 A_H \simeq 810$ MHz. (For the difference between the ortho-states at $I = 1$ and the para-state at $I = 0$ read Sect. 10.9.)

Problem 8.12 In the assumption that an electric field $\mathcal{E}$ applied along the molecular axis of H_2^+ can be considered as a perturbation, evaluate the electronic contribution to the electric polarizability (for rigid molecule and for molecular orbital LCAO).

Solution:
$$\mathcal{H}_P = ez\mathcal{E}$$

At first order $< g|\mathcal{H}_P|g > = 0$ (where $|g > = (1/\sqrt{2(1 + S_{AB})})(\phi_A + \phi_B)$), since it corresponds to the first moment of the electronic distribution, evidently zero for a homonuclear molecule (see Sect. 8.5).

At the second order, involving only the first excited state $|u > = (1/\sqrt{2(1 - S_{AB})})(\phi_A - \phi_B)$, one has

$$E^{(2)} = e^2 \mathcal{E}^2 \frac{| < u|z|g > |^2}{E_+ - E_-}$$

From

$$
\begin{aligned}
< u|z|g > &= \frac{1}{2} \frac{1}{\sqrt{1 + S_{AB}}\sqrt{1 - S_{AB}}} \int (\phi_A + \phi_B) z (\phi_A - \phi_B) d\tau \\
&= \frac{1}{2} \frac{1}{\sqrt{1 - S_{AB}^2}} \left[-\frac{R_{AB}}{2} - \frac{R_{AB}}{2} \right],
\end{aligned}
$$

$$E^{(2)} = -\frac{e^2 \mathcal{E}^2 R_{AB}^2}{4(1 - S_{AB}^2)(E_- - E_+)}$$

and then

$$\alpha_{H_2^+} = -\frac{1}{\mathcal{E}} \frac{\partial E^{(2)}}{\partial \mathcal{E}} = \frac{e^2 R_{AB}^2}{2(1 - S_{AB}^2)(E_- - E_+)}$$

From $R_{AB}^{eq} \simeq 2a_0$, $S_{AB}^2 \ll 1$ and $(E_- - E_+) \simeq 0.1e^2/a_0$ one has $\alpha_{H_2^+} \simeq (R_{AB}^{eq})^3/0.4$, of the expected order of magnitude (see Problem 10.8).

Problem 8.13 For two atoms A and B with $J = S = 1/2$, in the initial spin state $\alpha_A\beta_B$, *spin-exchange collision* is the process by which at large distances (no molecule is formed) they interact and end up in the final spin state $\beta_A\alpha_B$ (This process is often used in atomic optical spectroscopy to induce polarization and *optical pumping*). From the extension of the VB description of H_2 (Sect. 8.3) one can assume a spin-dependent interaction $\mathcal{H} = -2K(R)S_A \cdot S_B$, where $K(R)$ is the *negative*, R-dependent exchange integral favouring the $S = 0$ ground-state.

Discuss the condition for the spin-exchange process by making reasonable assumptions for the collision time R_c/v, R_c being an average interaction distance and v the relative velocity of the two atoms.

Solution: In the singlet ground-state the interaction is

$$E(R) = -2K(R)\left[\frac{S^2 - S_A^2 - S_B^2}{2}\right] = \frac{6K(R)}{4}$$

An approximate estimate of the time required to shift from $\alpha_A\beta_B$ to $\beta_A\alpha_B$ can be obtained by referring to the Rabi equation (Appendix 1.2), in a way somewhat analogous to the exchange of the electron discussed at Sect. 8.1.2 for the H_2^+ molecule. Here the Rabi frequency has to be written $\Gamma \approx E(R)/\hbar$, for R around R_c. Then the time for spin exchange is of the order of $(\pi/3)(\hbar/|K(R_c)|)$ while the time for interaction is $\tau_c = R_c/v$ (v can be considered the thermal velocity at room temperature in atomic vapors, i.e. $\approx 3 \cdot 10^5$ cm/s, for light atoms). Thus one derives $-3K(R_c)R_c \simeq \pi\hbar v$.

For an order of magnitude estimate one can assume that at large distance $K(R_c)$ is in the range $10^{-3} - 10^{-4}$eV thus yielding R_c in the range 6–60 Å.

These crude estimates for the spin-exchange process and the limits of validity are better discussed at Chap. 5 of the book by *Budker, Kimball* and *De Mille*, where a more rigorous analysis of this problem can be found.

Specific References and Further Reading

1. D. Budker, D.F. Kimball and D.P. De Mille, *Atomic Physics - An Exploration Through Problems and Solutions*, (Oxford University Press, 2004).
2. A. Balzarotti, M. Cini, M. Fanfoni, *Atomi, Molecole e Solidi. Esercizi risolti*, (Springer Verlag, 2004).
3. B.H. Bransden and C.J. Joachain, *Physics of atoms and molecules*, (Prentice Hall, 2002).
4. C.A. Coulson, *Valence*, (Oxford Clarendon Press, 1953).
5. W. Demtröder, *Molecular Physics*, (Wiley-VCH, 2005).
6. H. Eyring, J. Walter and G.E. Kimball, *Quantum Chemistry*, (J. Wiley, New York, 1950).
7. H. Haken and H.C. Wolf, *Molecular Physics and Elements of Quantum Chemistry*, (Springer Verlag, Berlin, 2004).

8. G. Herzberg, *Molecular Spectra and Molecular Structure*, Vol. I, II and III, (D. Van Nostrand, New York, 1964–1966, reprint 1988–1991).

9. C.S. Johnson and L.G. Pedersen, *Quantum Chemistry and Physics*, (Addison-Wesley, 1977).

10. N. Manini, *Introduction to the Physics of Matter*, (Springer, 2014).

11. M.A. Morrison, T.L. Estle and N.F. Lane, *Quantum States of Atoms, Molecules and Solids*, (Prentice-Hall Inc., New Jersey, 1976).

12. J.C. Slater, *Quantum Theory of Matter*, (McGraw-Hill, New York, 1968).

Chapter 9
Electronic States in Selected Polyatomic Molecules

Topics

Polyatomic Molecules from Bonds Between Pairs of Atoms
Hybrid Atomic Orbitals
Geometry of Some Molecules
Bonds for Carbon Atom
Electron Delocalization and the Benzene Molecule

In this chapter some general aspects of the electronic states in polyatomic molecules shall be discussed. Some more detail will be given for typical molecules where novel phenomena not encountered in diatomic molecules occur.

The electronic structure in polyatomic molecules is based on principles analogous to the ones described for diatomic molecules. As already mentioned (see note at Sect. 8.1.1), a general theory somewhat equivalent to the Slater theory for many-electron atoms (Sect. 3.4) can be developed. The steps of that approach are the following. Molecular orbitals of the form $\phi(\mathbf{r}_i, \mathbf{s}_i) = \sum_p c_p^{(i)} \phi_p(\mathbf{r}_i) \chi_{spin}^{(i)}$ are assumed as a basis, in terms of linear combination of atomic spin-orbitals centered at the various sites with unknown coefficients $c_p^{(i)}$. The determinantal wavefunction for all the electrons is then built up and the energy function is constructed from the full Hamiltonian $(T + V_{ne} + V_{ee})$ (see Sect. 7.1). The Hartree-Fock variational procedure is then carried out in order to derive the coefficients $c_p^{(i)}$. This approach is known as *MO.LCAO.SCF* (self-consistent field). Advanced computational methods are required and the one developed by *Roothaan* is one of the most popular. More recently the *density functional theory* is often applied in *ab-initio* procedures, based on the idea that the energy can be written in terms of electron probability density, thus becoming a *functional* of the charge distribution, while the *local density approximation* is used to account for the exchange-correlation corrections. Configuration

© Springer International Publishing Switzerland 2015
A. Rigamonti and P. Carretta, *Structure of Matter*,
UNITEXT for Physics, DOI 10.1007/978-3-319-17897-4_9

interaction (see Sect. 8.4) is usually taken into account. We will not deal with these topics, essentially belonging to the realm of computational quantum chemistry.

We shall see how in poliatomic molecules qualitative aspects can be understood simply in terms of the idealization of *independent bonds*, by considering the molecule as resulting from pairs of atoms, each pair corresponding to a given bond. In this way the main aspects worked out in diatomic molecules (Chap. 8) can be extended to polyatomic molecules. Typical illustrative example is the NH_3 molecule.

At Sect. 9.2 we will discuss the molecular bonds involving *hybrid atomic orbitals* and giving rise to particular geometries of the molecules, typically the ones related to the variety of bonds involvingthe carbon atom. In Sect. 9.3 the *delocalization* of the electrons will be addressed, with reference to the typical case of the benzene molecule.

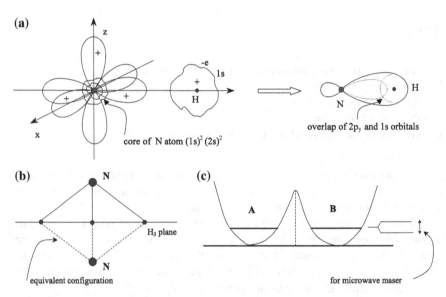

Fig. 9.1 Pictorial view of the formation of the NH_3 molecule in terms of combination of localized $1s$ Hydrogen and of $2p_{x,y,z}$ N atomic orbitals, with the criterium of the *maximum overlap* to grant the largest contribution to the bonding energy from each bond (**a**). The electronic configuration of the molecule can be written $N(1s)^2(2s)^2[N(2p) + H(1s), \sigma]^6$. The equivalent configuration is shown in part (**b**), where the molecule can be thought to result from the approach of the H atoms along the opposite directions of the coordinate axes. The evolution of the two level states is sketched in part **c** with the *inversion doublet* resulting from the removal of the degeneracy. The separation energy of the doublet is related to the exchange integral. These two states were used to obtain the first *maser* operation (see Appendix 9.1)

9.1 Qualitative Aspects of NH₃ and H₂O Molecules

In the spirit of the simplified picture of orbitals localized between pair of atoms and independent bonds, one can sketch the formation of the NH_3 molecule as resulting from three mutually perpendicular σ MO orbitals involving LCAO combination of $2p$ N atomic orbitals and $1s$ Hydrogen orbital (see Fig. 9.1).

Similar qualitative picture can be given for the H_2O molecule (Fig. 9.2).

From those examples one can understand how the geometry of the molecules, with certain angles between bonds, is a consequence of the maxima for the probability of presence of the electrons controlled by atomic orbitals coupled with the criterium of *strong overlap*, in order to maximize the resonance integral.

However it should be remarked that this qualitative picture is incomplete. In fact the angles between bonds are far from being 90°, in general. For instance in H_2O the angle between the two OH bonds is about 104.31° (see Fig. 9.2). As we shall see in the next section, the geometry of the molecules is consistent with the assumption of *hybrid atomic orbitals* involved in the formation of the MO's.

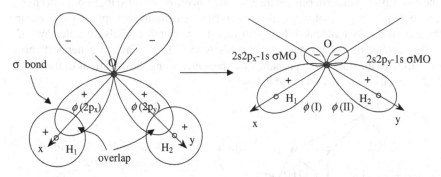

Fig. 9.2 Schematic view of the H_2O molecule as resulting from two σ MO's involving $2p$ O and $1s$ H atomic orbitals, with strong overlap of the wavefunctions when the Hydrogen atoms approach the Oxygen along the directions of the x and y axes. The electronic configuration can be written $O(1s)^2(2s^2)(2p_z)^2[O(2p_{x,y}) + H(1s), \sigma]^4$. The increase of the angle between the bonds with respect to the idealized situation of $\pi/2$ can be accounted for by the formation of non-equivalent hybrid orbitals leading to a shift of the center of the electronic cloud (or by non-localized one-electron orbitals)

9.2 Bonds Due to Hybrid Atomic Orbitals

By naively referring to the electrons available to form bonds by occupying the molecular orbitals, the Be, B and C atoms would be characterized by valence numbers $n_V = 0, 1$ and 2, correspondent to the electrons outside the closed shells. The common experimental findings ($n_V = 2, 3$ and 4, respectively) could qualitatively be understood by assuming that when molecules are formed one electron in those atoms is promoted to an excited state. The increase in the number of bonds, with the related decrease of the total energy upon bonding, would account for the energy required to promote the electron to the excited state. This argument by itself cannot justify the experimental evidence. At the sake of clarity, let us refer to the CH_4 molecule: its structure, with four equivalent C-H bonds, with angles 109°28′ in between, can hardly be justified by assuming for the carbon atom one electron in each $2s, 2p_x, 2p_y$ and $2p_z$ atomic orbitals. A related consideration, claiming for an explanation of the molecular geometry, is the one aforementioned for the angles between bonds in the H_2O molecule.

Again referring to CH_4 and in the light of the equivalence of the four C-H bonds, one can still keep the criterium of the *maximum overlap* provided that atomic orbitals, not corresponding to states of definite angular momentum, are supposed to occur in the atom when the molecule is being formed. These atomic orbitals are called *hybrid*.

To account for the geometry of the bonds in CH_4 we have to generate hybrid orbitals with maxima in the probability of presence along the directions of the *tetrahedral* environment, as sketched below

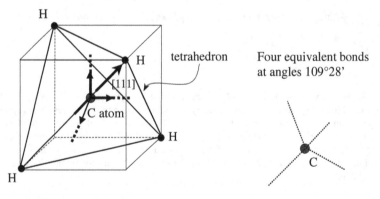

From the linear combination

$$\phi_C = a\phi_{2s} + b\phi_{2p_x} + c\phi_{2p_y} + d\phi_{2p_z}$$

by resorting to the orthonormality condition, to the requirement of electronic charge displaced along the tetrahedral directions and by considering that for symmetry reasons the s electron has to be equally distributed on the four hybrid orbitals, one can figure out that the coefficients must be

$$a^2 = 1/4 \qquad b^2 + c^2 + d^2 = 3/4 \qquad b^2 = c^2 = d^2,$$

yielding maxima for the probability of presence along the directions

$$(1, 1, 1), \quad (1, -1, -1), \quad (-1, 1, -1), \quad (-1, -1, 1).$$

Thus the hybrid orbitals of the C atom are

$$\phi_I = \frac{1}{2}\left[2s + 2p_x + 2p_y + 2p_z\right]$$

$$\phi_{II} = \frac{1}{2}\left[2s + 2p_x - 2p_y - 2p_z\right]$$

$$\phi_{III} = \frac{1}{2}\left[2s - 2p_x + 2p_y - 2p_z\right]$$

$$\phi_{IV} = \frac{1}{2}\left[2s - 2p_x - 2p_y + 2p_z\right]$$

The individual bonds with the H atoms can then be thought to result from σ MO, given by linear combinations of the C hybrids and of the $1s$ H atomic orbitals, as sketched below

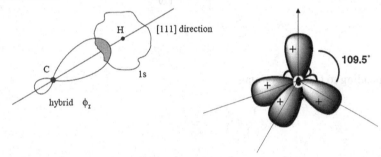

That type of hybridization is called (sp^3) or *tetragonal*. Besides the methane molecule, is the one that can be thought to occur in the molecular-like bonding in some crystals, primarily in diamond (C) and in semiconductors such as Ge, Si and others (see Chap. 11).

Another type of hybridization involving the carbon atom is (sp^2) or *trigonal* one, giving rise to planar geometry of the molecule, with three equivalent bonds forming angles of $120°$ between them, such as in the ethylene molecule, C_2H_4. The hybrid orbital can be derived in a way similar to the tetragonal hybridization, from a linear combination of $2s, 2p_x$ and $2p_y$. By taking into account that the coefficients b and c are proportional to the cosine of the related angles and that $a^2 = 1/3, b^2 + c^2 = 2/3$, one has the following picture

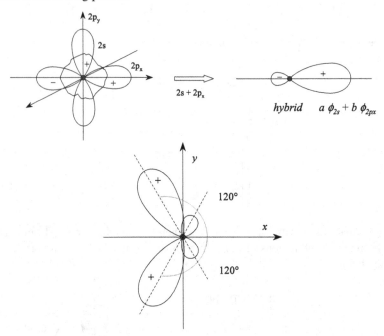

corresponding to C hybrid orbitals

$$\phi_I = \frac{1}{\sqrt{3}} \phi_{2s} + \frac{\sqrt{2}}{\sqrt{3}} \phi_{2px}$$

$$\phi_{II} = \frac{1}{\sqrt{3}} \phi_{2s} - \frac{1}{\sqrt{6}} \phi_{2px} + \frac{1}{\sqrt{2}} \phi_{2py}$$

$$\phi_{III} = \frac{1}{\sqrt{3}} \phi_{2s} - \frac{1}{\sqrt{6}} \phi_{2px} - \frac{1}{\sqrt{2}} \phi_{2py}$$

Therefore the σ MO bonds are generated from the linear combination of the $1s$ H orbitals or of the equivalent hybrid orbital of the othercarbon atom, as sketched below

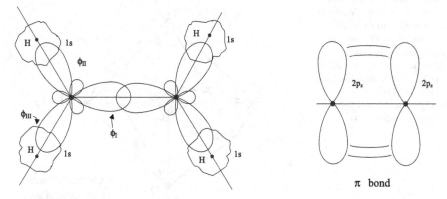

π bond

The $2p$ electron described by the atomic orbital $2p_z$, perpendicular to the plane of the molecule, is not involved in the hybrid and therefore it can form a π C-C MO of the type already seen in diatomic molecules, leading to an additional *weak bond* (see the N_2 molecule at Sect. 8.2).

Another interesting hybrid orbital for the carbon atom and leading to linear molecule, such as acetylene (C_2H_2) is the *digonal* (sp) hybrid. It mixes the s electron and one p electron only. The electronic configuration sketched below is derived

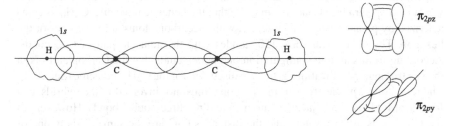

The C-C bond here is a triple one (one strong bond and two weak bonds).

More complex hybrid orbitals are generated in other multi-atoms molecules, with particular geometries. For its importance and at the sake of illustration we mention the (d^2sp^3) atomic orbitals, occurring in atoms with incomplete d shells. The hybridization implies *six bonds*, along the positive and negative directions of the Cartesian axes. By combining with $2p$ oxygen orbitals the octahedral structure depicted in Fig. 9.3 originates, for example for $BaTiO_3$.

We shall come back to this relevant atomic configuration, characteristic of the perovskite-type ferroelectric titanates such as $BaTiO_3$, at Sect. 13.3 when dealing with the *CuO_6 octahedron*,which is the structural core of high-temperature superconductors.

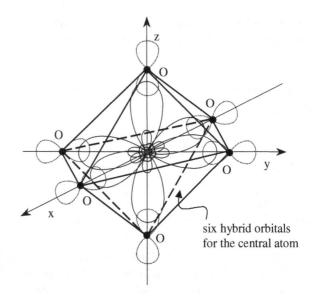

six hybrid orbitals
for the central atom

9.3 Delocalization and the Benzene Molecule

Experimental evidences, such as X-ray diffraction (in the solid state) and roto-
vibrational spectra (see Chap. 10) indicate that the benzene molecule, C_6H_6, is characterized by planar hexagonal structure, with thecarbon atoms at the vertices of the
hexagon. The C-H bonds form 120° angles with the adjacent pair of C-C bonds.
According to this atomic configuration one understands that the Carbon atom is in
the sp^2 trigonal hybridization, as the one discussed for the C_2H_4 molecule (Sect. 9.2).
The remaining $2p_z$ electrons of the Carbon atoms, not involved in the hybrids, can
form a πMO between adjacent C atoms, yielding three double bonds. However, all
the C-C bonds are equivalent and the distances C-C are the same. This is one of
the evidences that the simplified picture of localized electrons, with "independent"
bonds between pairs of atoms, in some circumstances has to be abandoned. We shall
see that the structure of the benzene molecule, as well as of other molecules with
π-bonded atoms like the polyenes, can be justified only by *delocalizing* the $2p_z$
electrons all along the carbon ring. Thus the one-electron orbitals are not necessarily localized between pairs of carbon atoms. The delocalization process is a further
mechanism of bonding, since the total energy is decreased, as it will be shown. At
Chap. 12 we shall see that the electronic states in crystals can be described as related
to the delocalization of the electrons. Thus, for certain aspects the benzene molecule
can also be regarded as a prototype for the electronic states in crystals.

By extending to the six Carbon atoms in the benzene ring the MO.LCAO description, the one-electron orbital is written

$$\phi_{MO}(i) = \sum_r c_r \phi_{2p_z}^{(r)}$$ (9.1)

where $\phi_{2p_z}^{(r)}$ are C $2p$ orbitals centered at the rth site of the hexagon (r runs from 1 to 6). Then the energy function, by referring only to the Hamiltonian $\mathcal{H}$ for the $2p_z$ C electrons, is

$$E(c_r) = \frac{\int \phi_{MO}^* \mathcal{H} \phi_{MO} \, d\tau}{\int \phi_{MO}^* \phi_{MO} \, d\tau}$$ (9.2)

By resorting to the concepts already used in diatomic molecules (Sect. 8.1) we label $\beta_{rs} = \int \phi_{2p_z}^{*(s)} \mathcal{H} \phi_{2p_z}^{(r)} \, d\tau$ as resonance integral, while $S_{rs} = \int \phi_{2p_z}^{*(s)} \phi_{2p_z}^{(r)} \, d\tau$ is the overlap integral. One has $\int \phi_{2p_z}^{*(r)} \mathcal{H} \phi_{2p_z}^{(r)} \, d\tau = E_0$, energy of $2p$ electron in the C atom and $\int \phi_{2p_z}^{*(r)} \phi_{2p_z}^{(r)} \, d\tau = 1$.

It is conceivable to assume $S_{rs} = 0$ for $r \neq s$ and to take into account the resonance integral only between adjacent C atoms: $\beta_{rs} = \beta$ for $r = s \pm 1$ and zero otherwise (this criterion was first proposed by *Hückel*).

Then the secular Equation for the energy function $E(c_r)$ reads

$$\begin{pmatrix} (E_0 - E) & \beta & 0 & 0 & 0 & \beta \\ \beta & (E_0 - E) & \beta & 0 & 0 & 0 \\ 0 & \beta & (E_0 - E) & \beta & 0 & 0 \\ 0 & 0 & \beta & (E_0 - E) & \beta & 0 \\ 0 & 0 & 0 & \beta & (E_0 - E) & \beta \\ \beta & 0 & 0 & 0 & \beta & (E_0 - E) \end{pmatrix} = 0.$$

The roots are $E_0 + 2\beta$, $E_0 \pm \beta$ (twice), $E_0 - 2\beta$ (note that $\beta < 0$).

The lowest energy delocalized π orbital, correspondent to the eigenvalue $E_0 + 2\beta$, can accommodate two electrons, while on the state at energy $E_0 + \beta$ one can place four electrons, as sketched below

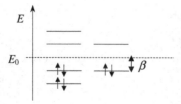

The bonding energy turns out $2(2\beta) + 4\beta = 8\beta$, lower than the energy 6β that one would obtain for localized electrons. The energy 2β can be considered the contribution to the ground state energy due to the delocalization.

In correspondence to the root $(E_0 + 2\beta)$ the coefficients c_r in Eq. (9.1) are equal. The normalization yields $c_r = 1/\sqrt{6}$ and therefore the molecular orbital is

$$\phi_{MO}(\mathbf{r}_i) = \frac{1}{\sqrt{6}}\left[\phi_{2p_z}(\mathbf{r}_i - \mathbf{l}_1) + ... + \phi_{2p_z}(\mathbf{r}_i - \mathbf{l}_6)\right] \qquad (9.3)$$

where $\mathbf{l}_r$ indicate $n_r\mathbf{a}$ and specifies the position of the Carbon atom along the ring of step $\mathbf{a}$. The wavefunction 9.3 is sketched in Fig. 9.4.

In correspondence to the root $(E_0 + \beta)$ different choices for the coefficients c_r are possible (see Problem 9.2 for a similar situation). One choice is

$$\phi_{MO}(\mathbf{r}_i) = \frac{1}{\sqrt{12}}\Bigg[2\phi_{2p_z}(\mathbf{r}_i - \mathbf{l}_1) + \phi_{2p_z}(\mathbf{r}_i - \mathbf{l}_2) - \phi_{2p_z}(\mathbf{r}_i - \mathbf{l}_3) -$$

$$2\phi_{2p_z}(\mathbf{r}_i - \mathbf{l}_4) - -\phi_{2p_z}(\mathbf{r}_i - \mathbf{l}_5) + \phi_{2p_z}(\mathbf{r}_i - \mathbf{l}_6)\Bigg]. \qquad (9.4)$$

The eigenvalues can be written in the form

$$E_p = E_o + 2\beta cos[(2\pi/6a)pa], \qquad (9.5)$$

while for the coefficients

$$c_p^{(l)} = (e^{2\pi ilp/6})/(6)^{1/2}, \qquad (9.6)$$

where $p = 0, \pm1, \pm2, 3$.

The benzene ring can be considered the cyclic repetition of a "crystal" of six Carbon atoms. The eigenvalues and the coefficients in the forms 9.5 and 9.6 are somewhat equivalent to the band states in a one-dimensional crystal (see Chap. 12).

The quantitative evaluation (by means of numerical methods or by resorting to approximate radial parts of the wavefunctions) of the electronic eigenvalue as a function of the interatomic distance a yields a minimum for $a \simeq 1.4$ Å, in between the values $a' = 1.34$ Å and $a'' = 1.54$ Å pertaining to double and to simple C-C bond,

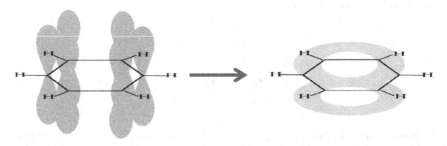

Fig. 9.4 Pictorial view of the πMO delocalized orbital correspondent to Eq. (9.3) (eigenvalue $E_0 + 2\beta$)

respectively. From spectroscopic and thermodynamic considerations the values of E_0 and β turn out $E_0 \simeq -1.52$ eV and $\beta = -0.87$ eV.

The structural anisotropy of the molecule is reflected, for instance, in the strong dependence of the diamagnetic susceptibility χ_{dia} on the orientation. In fact, by extending the arguments discussed for atoms (Sect. 4.5) one can expect $\chi_{dia} \propto \sum_i < r_i^2 sin^2\theta_i >$, with θ angle between the magnetic field and the positional vector of a given electron. Then, in the benzene molecule, $\chi_{dia}^{\parallel} \ll \chi_{dia}^{\perp}$ (with $\parallel$ and $\perp$ to the plane of the hexagon) (see Problem 9.3).

Appendix 9.1 Ammonia Molecule in Electric Field and the Ammonia Maser

According to Fig. 9.1 the Ammonia molecule can be found in two equivalent config-urations, depending on the position of the N atom above (state $|1 >$) or below (state $|2 >$) the xy plane of the H atoms. By considering the molecule in its ground elec-tronic state and neglecting all other degrees of freedom, let us discuss the problem of the position of the N atom along the z direction perpendicular to the xy plane, therefore involving the vibrational motion in which N oscillates against the three coplanar H atoms (for details on the vibrational motions see Sects. 10.3 and 10.6).

The potential energy $V(z)$, that in the framework of the Born-Oppenheimer separation (Sect. 7.1) controls the nuclear motions and that is the counterpart of the energy $E(R_{AB})$ in diatomic molecules, has the shape sketched below

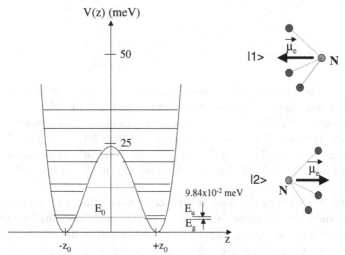

The distance of the N atom from the xy plane corresponding to the minima in $V(z)$ is $z_o = 0.38$ Å, while the height of the potential energy for $z = 0$ is $V_o \simeq 25$ meV. In the state $|1 >$ the molecule has an electric dipole moment μ_e along the negative z direction, while in the state $|2 >$ the dipole moment is parallel to the reference z-axis. Within each state the N atoms vibrate around $+z_o$ or $-z_o$. As for any molecular oscillator the ground state has a zero-point energy different from zero, that we label E_o (correspondent to the two levels A and B sketched in Fig. 9.1c). The vibrational eigenfunction in the ground state is a Gaussian one, centered at $\pm z_o$ (see Sect. 10.3). The effective mass of the molecular oscillator is $\mu = 3M_H M_N/(3M_H + M_N)$.

Thus the system is formally similar to the H_2^+ molecule discussed at Sect. 8.1, the $|1>, |2>$ states corresponding to the electron hydrogenic states $1s_A$ and $1s_B$, while the vibration zero-point energy corresponds to $-R_H hc$. Therefore, the generic state of the system is written

$$|\psi> = c_1|1> + c_2|2> \qquad\qquad (A.9.1.1)$$

with coefficients c_i obeying to Eqs. (8.13). Here $H_{12} = -A$ is the probability amplitude that because of the quantum tunneling the N atom jumps from $|1>$ to $|2>$ and *vice versa*, in spite of the fact that $E_o \ll V_o$. Two stationary states are generated, say $|g>$ and $|u>$, with eigenvalues $E_o - A$ and $E_o + A$, respectively. The correspondent eigenfunctions are linear combinations of the Gaussian functions describing the oscillator in its ground state (see Sect. 10.3):

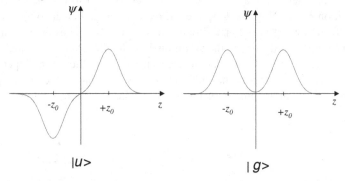

$$|u> \qquad\qquad\qquad\qquad |g>$$

The degeneracy of the original states is thus removed and the vibrational levels are in form of doublets (*inversion doublets*).For the ground-state the splitting $E_g - E_u = 2A$ corresponds to $0.793\,\mathrm{cm}^{-1}$, while it increases in the excited vibrational states, owing to the increase of H_{12}. For the first excited state $2A' = 36.5\,\mathrm{cm}^{-1}$ and for the second excited state $2A'' = 312.5\,\mathrm{cm}^{-1}$. It can be remarked that the vibrational frequency (see Sect. 10.3) of N around the minimum in one of the wells is about $950\,\mathrm{cm}^{-1}$.

The inversion splitting are drastically reduced in the deuterated Ammonia molecule ND_3 where for the ground-state $2A = 0.053\,\mathrm{cm}^{-1}$. Thus the tunneling frequency, besides being strongly dependent on the height of the effective potential barrier V_o, is very sensitive to the reduced mass μ. For instance, in the AsH_3 molecule,the time required for a complete tunneling cycle of the As atom is estimated to be about two years. These marked dependences on V_o and μ explains why in most molecules the inversion doublet is too small to be observed.

In NH_3 the so-called *inversion spectrum* was first observed (*Cleeton* and *Williams*, 1934) as a direct absorption peak at a wavelength around 1.25 cm, by means of microwave techniques. This experiment opened the field presently known as *microwave spectroscopy*.

The typical experimental setup is schematically shown below

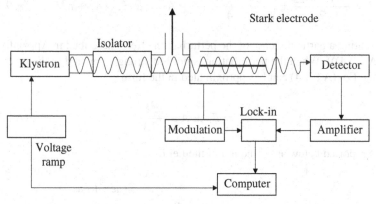

Finally it should be remarked that the rotational motions of the molecule (Sect. 10.2), as well as the magnetic and quadrupolar interactions (Chap. 5), in general cause fine and hyperfine structures in the inversion spectra.

As already mentioned the $|g>$ and $|u>$ states of the inversion doublet in NH_3 have been used in the first experiment (*Townes* and collaborators) of *microwave amplification* by *stimulated emission of radiation* (see Problem 1.24). The *maser* action requires that the statistical population N_u is maintained larger than N_g while a certain number of transitions from $|u>$ to $|g>$ take place.

Now we are going to discuss how the Ammonia molecule behaves in a static electric field. Then we show how by applying an electric field gradient (*quadrupolar electric lens*) one can select the Ammonia molecule in the upper energy state.

In the presence of a field $\mathcal{E}$ along z the eigenvalue for the states $|1>$ and $|2>$ become

$$H_{11} = E_o + \mu_e\mathcal{E} \quad \text{and} \quad H_{22} = E_o - \mu_e\mathcal{E}$$

The rate of exchange can be assumed approximately the same as in absence of the field, namely $H_{12} = -A$. The analogous of Eqs. (8.13) for the coefficients c_i in Eq. (A.9.1.1) are then modified in

$$i\hbar\frac{dc_1}{dt} = (E_o + \mu_e\mathcal{E})c_1 - Ac_2 \qquad \text{(A.9.1.2)}$$

$$i\hbar\frac{dc_2}{dt} = (E_o - \mu_e\mathcal{E})c_2 - Ac_1 \qquad \text{(A.9.1.3)}$$

The solutions of these equations must be of the form $c_i = a_i exp(-iEt/\hbar)$, with E the unknown eigenvalue. The resulting equations for a_i are

$$(E - E_o - \mu_e \mathcal{E})a_1 + Aa_2 = 0$$
$$Aa_1 + (E - E_o + \mu_e \mathcal{E})a_2 = 0$$

and the solubility condition yields

$$E_\pm = \frac{H_{11} + H_{22}}{2} \pm \sqrt{\frac{(H_{11} - H_{22})^2}{4} + A^2} = E_o \pm \sqrt{A^2 + \mu_e^2 \mathcal{E}^2} \qquad (A.9.1.4)$$

(representing a particular case of the perturbation effects described in Appendix 1.2 (Eq. (A.1.2.4))). When the perturbation is not too strong compared to the inversion splitting, Eq. (A.9.1.4) can be approximated in the form

$$E_\pm = E_o \pm A \pm \frac{\mu_e^2 \mathcal{E}^2}{2A}. \qquad (A.9.1.5)$$

$E_\pm$ are reported below as a function of the field.

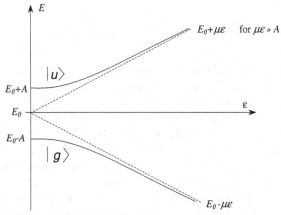

Equation (A.9.1.5) can be read in terms of induced dipole moments $\mu_{ind}^\pm = -dE_\pm/d\mathcal{E} = \mp \mu_e^2 \mathcal{E}/A$. Therefore, if a collimated beam of molecules passes in a region with an electric field gradient across the beam itself, molecules in the $|u>$ and $|g>$ states will be deflected along *opposite directions* (this effect is analogous to the one observed in the Rabi experiment at Sect. 6.2). In particular, the molecules in the $|g>$ state will be deflected towards the region of stronger $\mathcal{E}^2$, owing to the force $-\mathbf{V}[-(\mu_e \mathcal{E})^2/2A]$.

In practice, to obtain a beam with molecules in the upper energy state one uses quadrupole electric lenses, providing a radial gradient of $\mathcal{E}^2$. The square of the electric field varies across the beam. Passing through the lens the beam is enriched in molecules in the excited state and once they enter the microwave cavity the maser action becomes possible. The experimental setup of the Ammonia maser is sketched in the following figure.

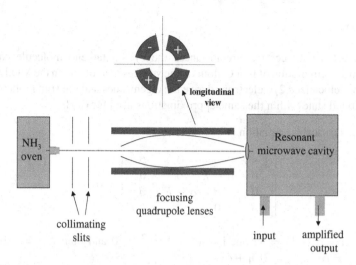

The basic principles outlined above for the Ammonia maser are also at work in other type of atomic or solid-state masers. In the Hydrogen or Cesium atomic maser the stimulated transition involves the hyperfine atomic levels (see Chap. 5). For the line at 1420 MHz, for instance, the selection of the atoms in the upper hyperfine state with $F = 1$ is obtained by a magnetic multipolar lens. Then the atomic beam enters a microwave cavity tuned at the resonance frequency. The resolution (ratio between the linewidth and 1420 MHz) can be improved up to 10^{-10}, since the atoms can be kept in the cavity up to a time of the order of a second. The experimental value of the frequency of the $F = 1 \rightarrow 0$ transition in Hydrogen is presently known to be $(1420405751.781 \pm 0.016\,\text{Hz})$, while for ^{133}Cesium the $F = 4 \rightarrow 3$ transition is estimated $9192631770\,\text{Hz}$, which is the frequency used to calibrate the unit of time (see Sect. 5.2).

Solid state masers are usually based on crystals with a certain number of paramagnetic transition ions, kept in a magnetic field and at low temperature, in order to increase the spin-lattice relaxation time T_1 and to reduce the linewidth associated with the life-time broadening (see Chap. 6) (as well as to reduce the spontaneous emission acting against the population inversion). A typical solid state maser involves ruby, a single crystal of Al_2O_3 with diluted Cr^{3+} ions (electronic configuration $3d^3$). The crystal field removes the degeneracy of the $3d$ levels (details will be given at Sect. 13.3) and the magnetic field causes the splitting of the $M_J = \pm 3/2, \pm 1/2$ levels. The population inversion between these levels is obtained by microwave irradiation of proper polarization.

Here we have presented only a few aspects of the operational principles of masers, which nowadays have a wide range of applications, due to their resolution (which can be increased up to 10^{-12}) and sensitivity (it can be recalled that maser signals reflected on the surface of Venus have been detected).

Problems

Problem 9.1 Under certain circumstances the cyclobutadiene molecule can be formed in a configuration of four C atoms at the vertices of a square. In the MO.LCAO picture of delocalized $2p_z$ electrons derive the eigenvalues and the spin molteplicity of the ground state within the same approximations used for C_6H_6.

Solution: The secular equation is

$$\begin{vmatrix} \alpha - E & \beta & 0 & \beta \\ \beta & \alpha - E & \beta & 0 \\ 0 & \beta & \alpha - E & \beta \\ \beta & 0 & \beta & \alpha - E \end{vmatrix} = 0.$$

By setting $\alpha - E = x$, one has $x^4 - 4\beta^2 x^2 = 0$ and then $E_1 = -2|\beta| + \alpha$, $E_{2,3} = \alpha$, $E_4 = 2|\beta| + \alpha$.

Ground state: $4\alpha - 4|\beta|$. Since the Hund rules hold also in molecules (see Sect. 8.2), the ground-state is a triplet.

Problem 9.2 Refer to the C_3H_3 molecule, with carbon atoms at the vertices of an equilateral triangle. Repeat the treatment given for C_6H_6, deriving eigenfunctions and the energy of the ground-state. Then release the assumption of zero overlap integral among orbitals centered at different sites and repeat the derivation. Estimate, for the ground-state configuration, the average electronic charge per C atom.

Solution: For $S_{ij} = 0$ for $i \neq j$, the secular equation is

$$\begin{vmatrix} E_0 - E & \beta & \beta \\ \beta & E_0 - E & \beta \\ \beta & \beta & E_0 - E \end{vmatrix} = 0$$

so that

$$E_I = E_0 + 2\beta \qquad\qquad E_{II,III} = E_0 - \beta$$

and the ground-state energy is

$$E_g = 3E_0 + 4\beta - \beta = 3E_0 + 3\beta$$

The eigenfunctions turn out

$$\phi_I = \frac{1}{\sqrt{3}}[\phi_1 + \phi_2 + \phi_3] \equiv \frac{A}{\sqrt{3}}$$

$$\phi_{II} = \frac{1}{\sqrt{2}}[\phi_1 - \phi_3] \equiv \frac{B}{\sqrt{2}}$$

$$\phi_{III} = \frac{1}{\sqrt{6}}[-\phi_1 + 2\phi_2 - \phi_3] \equiv \frac{C}{\sqrt{6}}$$

The total amount of electronic charge on a given atom (e.g. atom 1) is given by the sum of the squares of the coefficient pertaining to ϕ_1 in $\phi_{I,II,III}$:

$$q = 2\left(\frac{1}{\sqrt{3}}\right)^2 + \frac{1}{2}\left[\left(\frac{1}{\sqrt{2}}\right)^2 + \left(\frac{1}{\sqrt{6}}\right)^2\right] = 1$$

(having taken the average of the two degenerate states).

For $S_{ij} \equiv S \neq 0$, the secular equation becomes

$$\begin{vmatrix} E_0 - E & \beta - SE & \beta - SE \\ \beta - SE & E_0 - E & \beta - SE \\ \beta - SE & \beta - SE & E_0 - E \end{vmatrix} = 0$$

and the eigenvalues are

$$E_I = \frac{E_0 + 2\beta}{1 + 2S} \qquad\qquad E_{II,III} = \frac{E_0 - \beta}{1 - S}$$

The ground-state energy is

$$E_g = 2E_I + E_{II} = 3\frac{E_0 + \beta(1 - 2S)}{(1 + 2S)(1 - S)}$$

with normalized eigenfunctions

$$\phi_I = \frac{A}{\sqrt{3(1 + 2S)}}$$

$$\phi_{II} = \frac{B}{\sqrt{2(1 - S)}}$$

$$\phi_{III} = \frac{C}{\sqrt{6(1 - S)}}$$

Again, by estimating the squares of the coefficients the charge at a given atom turns out

$$q' = \frac{1}{(1 - S)(1 + 2S)}.$$

The charge in the region "in between" two atoms (e.g. atoms 1 and 2) is obtained by evaluating the sum of the coefficients $c_1 c_2$ (for ϕ_1 and ϕ_2) in $\phi_{I,II,III}$, multiplied by the overlap integral. Thus

$$q'' = \frac{S(1 - 2S)}{(1 - S)(1 + 2S)}.$$

Problem 9.3 Estimate the order of magnitude of the diamagnetic contributions to the susceptibilityin benzene, for magnetic field perpendicular to the molecular plane.

Solution: The diamagnetic susceptibility (per molecule) can approximately be written

$$\chi_\psi = -\frac{n_\psi e^2}{6mc^2} < r^2 >_\psi,$$

where n_ψ is the number of electrons in a molecular state ψ and $< r^2 >_\psi$ is the mean square distance.

In benzene there are 12 $1s$ electrons of C, with $< r^2 >_{1s} \simeq a_0^2/Z^2$ ($Z = 6$). Then there are 24 electrons in σ bonds for which, approximately,

$$< r^2 >_\sigma \simeq \int_{-\frac{L}{2}}^{\frac{L}{2}} \frac{dx}{L} x^2 = \frac{L^2}{12},$$

the length of the σ bond being $L = 1.4$ Å.

Finally there are 6 electrons in the delocalized bond π_z, where one can assume $< r^2 >_{\pi_z} \simeq L^2$. The diamagnetic correction at the center of the molecule is in large part due to the delocalized electrons and from that value of $< r^2 >_{\pi_z}$ one can crudely estimate $\chi_\pi \approx -0.49 \cdot 10^{-28}$ cm^3 . The experimental values for the single-molecule susceptibility are $\chi_{dia}^\perp = -1.52 \cdot 10^{-28}$ cm^3 and $\chi_{dia}^\parallel = -0.62 \cdot 10^{-28}$ cm^3 , for magnetic field perpendicular and parallel to the plane of the molecule.

Further Reading

1. A. Balzarotti, M. Cini, M. Fanfoni, *Atomi, Molecole e Solidi. Esercizi risolti*, (Springer Verlag, 2004).
2. B.H. Bransden and C.J. Joachain, *Physics of atoms and molecules*, (Prentice Hall, 2002).
3. C.A. Coulson, *Valence*, (Oxford Clarendon Press, 1953).
4. W. Demtröder, *Molecular Physics*, (Wiley-VCH, 2005).
5. H. Eyring, J. Walter and G.E. Kimball, *Quantum Chemistry*, (J. Wiley, New York, 1950).
6. C.S. Johnson and L.G. Pedersen, *Quantum Chemistry and Physics*, (Addison-Wesley, 1977).
7. J.C. Slater, *Quantum Theory of Matter*, (McGraw-Hill, New York, 1968).
8. S. Svanberg, *Atomic and Molecular Spectroscopy*, (Springer Verlag, Berlin, 2003).

Chapter 10
Nuclear Motions in Molecules and Related Properties

Topics

Rotations and Vibrations in Diatomic Molecules
How the Rotational and Vibrational States are Studied
The Normal Modes in Polyatomic Molecules
Basic Principles of Raman Spectroscopy
Nuclear Spin Statistics and Symmetry-Related Effects

10.1 Generalities and Introductory Aspects for Diatomic Molecules

In the framework of the Born-Oppenheimer separation (Sect. 7.1), once that the electronic state has been described and the eigenvalue $E_e(\mathbf{R})$ and wavefunction $\phi_e(\mathbf{r}, \mathbf{R})$ have been found, then the motions of the nuclei are described by a function $\phi_\nu^{(g)}(\mathbf{R})$, where g represents the quantum label for the electrons and ν are the quantum numbers (to be found) for the nuclei. This wavefunction is solution of the equation

$$\left\{ -\sum_\alpha \frac{\hbar^2}{2M_\alpha} \nabla_\alpha^2 + E_e(\mathbf{R}) \right\} \phi_\nu^{(g)}(\mathbf{R}) = E_{g,\nu} \, \phi_\nu^{(g)}(\mathbf{R}) \qquad (10.1)$$

(note that V_{nn} in Eqs. (7.3) and (7.5) has been included in $E_e(\mathbf{R})$, see for example Eq. (8.1)).

Let us refer to a diatomic molecule in the ground electronic state, for which we assume $\Lambda = 0$ and $S = 0$ ($^1\Sigma$ state) and let us indicate the effective potential energy, resulting from the electronic eigenvalue and the nucleus-nucleus repulsion, with $V(R)$, R being the interatomic distance (previously often indicated by R_{AB}). It is reminded that $V(R)$ has the form sketched below

© Springer International Publishing Switzerland 2015
A. Rigamonti and P. Carretta, *Structure of Matter*,
UNITEXT for Physics, DOI 10.1007/978-3-319-17897-4_10

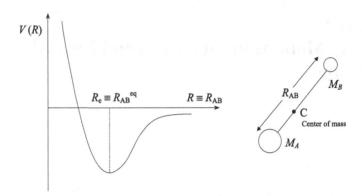

By introducing the reduced mass $\mu = M_A M_B/(M_A+M_B)$ the molecule becomes equivalent to a single particle. By recalling the approach used for the Hydrogen atom, Eq. (10.1) is rewritten

$$\left\{ -\frac{\hbar^2}{2\mu}\left[\frac{1}{R^2}\frac{\partial}{\partial R}\left(R^2\frac{\partial}{\partial R}\right)\right] + T_\theta + T_\varphi + V(R) \right\} \phi(R,\theta,\varphi) = E\,\phi(R,\theta,\varphi)$$
(10.2)

where the polar coordinates R, θ and φ have been introduced and $(T_\theta + T_\varphi)$ involves the angular momentum operator $\mathbf{L}^2$. The difference with respect to the radial part of the Schrödinger equation for Hydrogen is in the potential energy $V(R)$, obviously different from the Coulombic form although still of central character. Thus the factorization of the wavefunction follows:

$$\phi(R,\theta,\varphi) = \mathcal{R}(R)Y(\theta,\varphi),$$
(10.3)

$Y_{KM}(\theta,\varphi)$ being the spherical harmonics characterized by quantum numbers K and M (the analogous of l and m in the H atom), related to the eigenvalues for $\mathbf{L}^2$ and L_z.
The radial part of the wavefunction, $\mathcal{R}(R)$, obeys the equation

$$T_R\mathcal{R} + \left[V(R) + \frac{K(K+1)\hbar^2}{2\mu R^2}\right]\mathcal{R} = E\mathcal{R}$$
(10.4)

and corresponds to the one-dimensional probability of presence along a given direction under a potential energy including the centrifugal term, as sketched below

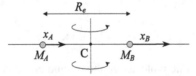

By indicating with Q the internuclear distance R with respect to the equilibrium distance R_e, in terms of the local displacements x_A and x_B one has

$$Q = R - R_e = x_B + R_e - x_A - R_e \equiv x_B - x_A. \qquad (10.5)$$

Thus Q is a *non-local* coordinate (we shall return to this point when discussing the vibrational motions in polyatomic molecules, Sect. 10.6). Then the centrifugal term in Eq. (10.4) can be written

$$\frac{K(K+1)\hbar^2}{2\mu R_e^2} \left[\frac{1}{1+\frac{Q}{R_e}} \right]^2 \simeq \frac{K(K+1)\hbar^2}{2\mu R_e^2} \left(1 - \frac{2Q}{R_e} \right) \qquad (10.6)$$

having taken into account that $(Q/R_e) \ll 1$.

In Eq. (10.6) the term $2Q/R_e$ couples the vibrational and the rotational motions. In a first approximation this term can be considered as perturbation and one can deal with the rotational part of the Schrodinger equation only. After the analysis of the vibrational part and the derivation of the correspondent wavefunction $\mathcal{R}(R) \equiv \phi_{vib}(R)$, it will be possible to take into account the roto-vibrational coupling by referring to unperturbed states described by

$$\phi(R, \theta, \varphi) = \phi_{vib}(R)\,\phi_{rot}(\theta, \varphi) \qquad (10.7)$$

with $\phi_{rot}(\theta, \varphi) \equiv Y_{KM}(\theta, \varphi)$, with the perturbation term given by $-(2Q/R_e)$.

10.2 Rotational Motions

10.2.1 Eigenfunctions and Eigenvalues

From Sect. 10.1 it follows that the contribution to the energy of the molecule from the rotational motion is

$$E_{rot} = K\,(K+1)\,\hbar^2/2\,\mu\,R_e^2 \qquad (10.8)$$

This result can be thought to derive directly from the quantization of the angular momentum $\mathbf{P}$ in the classical expression of the rotational energy $P^2/2I$, I being the moment of inertia $I = R_e^2\mu$.

The eigenfunctions $\phi_{rot}(\theta, \varphi)$, so that $\phi_{rot}^*\phi_{rot}\,d\Omega$ yields the probability that the molecular axis is found inside the elemental solid angle $d\Omega = sin\theta\,d\theta\,d\varphi$, coincide with the spherical harmonics. In the light of the classical relation $|\mathbf{P}| = I\omega$, to a given quantum state with eigenvalue $|\mathbf{P}| = [K(K+1)]^{1/2}\,\hbar$, one can associate a frequency of rotation $\nu_{rot} = (h/4\pi^2 I)\,[K(K+1)]^{1/2}$.

A fundamental *rotational frequency*

$$\nu_{rot} = \hbar/2\pi\mu R_e^2 = \hbar/2\pi I \qquad (10.9)$$

Fig. 10.1 Levels diagram
for the lowest-energy
rotational states

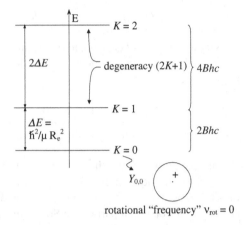

rotational "frequency" $\nu_{rot} = 0$

or equivalently a *rotational constant* (in cm^{-1}) $B = (\hbar/4\pi I c)$, are usually defined $(B \equiv (\hbar^2/2\mu R_e^2)/hc)$. The energy diagram for the first rotational states is reported in Fig. 10.1.

The probability distribution of the molecular axis involves $Y^*_{KM} Y_{KM}$. Since the φ-dependence of the spherical harmonics goes as $exp(\pm i M \varphi)$, the distribution of the molecular axes is characterized by rotational symmetry with respect to a given z direction. For $M = K$ and large value of the quantum numbers the distribution tends to the classical one, as expected from the *correspondence principle*.

10.2.2 Principles of Rotational Spectroscopy

The rotational states are experimentally studied by means of spectroscopic techniques involving the microwave range (typically $10^{-1} \div 10$ cm^{-1}) or the far infrared range $(10 \div 500$ cm$^{-1})$ of the electromagnetic spectrum (see Appendix 1.1). Usually the sample is a gas at reduced pressure, since frequent collisions would prevent the definition of a precise quantum state (which is hard to define in the liquid state, for instance).

The generators of the radiation are often metals at high temperature or arc lamps while the detectors are semiconductor devices (for wavenumbers typically larger than 10 cm^{-1}). When low frequencies are required (say below 150 GHz), klystrons, magnetrons or Gunn diodes (usually fabricated with GaAs) are the microwave sources. Wave guides, resonant cells and again semiconductor detectors are commonly used.

Without going into details of technical character, we shall devote attention to the selection rules for electric dipole transitions between the states (K', M') and (K'', M''). The electric dipole matrix element reads

$$\mathbf{R}_{1 \to 2} = \int Y^*_{K''M''}(\theta, \varphi)\, \boldsymbol{\mu}_e\, Y_{K'M'}(\theta, \varphi)\, sin\theta\, d\theta\, d\varphi \qquad (10.10)$$

where μ_e is the dipole moment of the molecule. Therefore only *heteronuclear polar molecules*, where $|\mu_e| \neq 0$ can be driven into transitions between different rotational states. Homonuclear molecules cannot interact with the radiation. From the matrix element in Eq. (10.10), in a way similar to the deduction of the selection rules for atomic transitions (see Sect. 3.5 and Appendix 1.3), the selection rules for electric dipole transitions between rotational states in polar molecules are

$$\Delta K = \pm 1 \qquad \Delta M = 0, \pm 1 \, , \tag{10.11}$$

the latter being relevant when a static electric field is applied (see Sect. 10.2.4).

The energy difference between the states K and $(K + 1)$ is

$$\Delta E_{K+1, K} = \frac{\hbar^2}{2I}\left[(K+2)(K+1) - K(K+1)\right] = \frac{\hbar^2}{I}(K+1). \tag{10.12}$$

Then in principle one expects rotational transitions at frequencies $\nu = n\,\nu_{rot}$ (Eq. (10.9)), with n integer.

The intensities of the lines, to a good approximation, are controlled by the statistical populations of the rotational levels. According to the Maxwell-Boltzmann statistics (the molecules are distinguishable), the number of molecules on a given K state, at the thermal equilibrium, is

$$N_K = A\, g_K\, e^{-E_K/k_B T} \tag{10.13}$$

with $g_K = (2K + 1)$. The normalization constant can be expressed in terms of the population on the $K = 0$ level and thus one writes

$$N_K(T) = N_{K=0}(T)\,(2K + 1)\, e^{-\frac{K(K+1)\hbar^2}{2Ik_B T}}, \tag{10.14}$$

a function of the "variable" K of the form sketched below

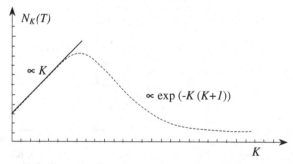

and implying typical absorption spectrum of the form shown in Figs. 10.2 and 10.3.

The fundamental rotational constants $B = \hbar/4\pi I c$ for some diatomic molecules are reported below

Molecule	$B(\text{cm}^{-1})$
H_2	60.8
N_2	2.01
O_2	1.45
HCl	10.6
NaCl	0.19

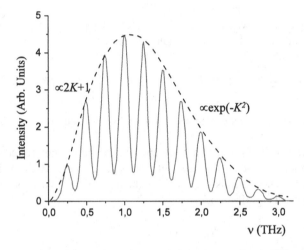

Fig. 10.2 Sketch of the expected absorption rotational spectrum for the DBr molecule on the basis of Eqs. (10.14) and (10.9). The intensities of the lines are normalized to the one of the $K = 0 \rightarrow K = 1$ line. The rotational frequency $\bar{\nu} = \nu_{rot}/c$ is around 8 cm^{-1}, corresponding to *rotational temperature* $h\nu_{rot}/k_B \approx 12$ K. The separation between adjacent lines is $2B$ and θ_{rot} is often defined as $\theta_{rot} = \hbar^2/2Ik_B = Bhc/k_B$ (see Sect. 10.2.1)

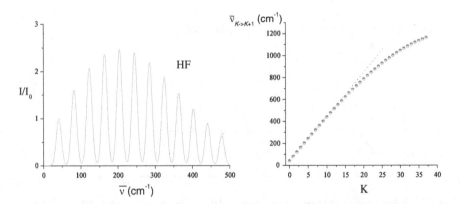

Fig. 10.3 (*Left*) Absorption rotational spectrum of the HF molecule. The intensities of the lines are normalized to the $K = 0 \rightarrow K = 1$ line. In the right panel the wavenumbers associated with the $K \rightarrow K + 1$ transitions are reported as a function of K. A departure from the interval rule is observed at large K, owing to the increased strength of the coupling between rotational and vibrational motions (see Sect. 10.5)

10.2.3 Thermodynamical Energy from Rotational Motions

Once the structure of the quantum levels is known, from the statistical distribution function it is possible to derive the thermodynamical energy U_{rot} and the specific heat. One has

$$U_{rot} = \sum_K N_K E_K \qquad (10.15)$$

with N_K given by Eq. (10.14), where $N_{K=0}(T)$ can be written

$$N_{K=0} = N/Z_{rot} \qquad (N \text{ total number of molecules})$$

The rotational partition function Z_{rot} is

$$Z_{rot} = \sum_K (2K+1) e^{-E_K/k_B T} \qquad (10.16)$$

It is noted that the single molecule energy in terms of Z_{rot} (see Problem 6.4) is

$$U_{rot} = k_B T^2 \frac{d}{dT} \ln Z_{rot}.$$

In the high temperature limit $T \gg \theta_{rot} \equiv \hbar^2/2Ik_B$ ($\theta_{rot} \sim 5 \div 10\,K$ for most molecules, with the exception of H_2 where $\theta_{rot} \simeq 87\,K$), the sum over K can be transformed to an integral:

$$Z_{rot} \approx \int_0^\infty 2K\, e^{-K^2 \theta_{rot}/T}\, dK = \frac{T}{\theta_{rot}}. \qquad (10.17)$$

For one mole ($N = N_A$), $U_{rot} \approx N_A k_B T$ and the specific heat turns out $C_V \approx R$, as expected from classical statistics.

The temperature behavior of the molar specific heat (see Problem 6.7) is schematically reported below

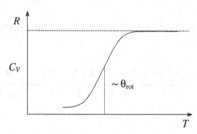

For relevant aspects occurring in homonuclear molecules see Problems 10.19 and 10.20.

10.2.4 Orientational Electric Polarizability

Let us outline how one can describe the effect of a static electric field on an assembly of dipolar diatomic molecules and how the polarizability is evaluated.

The perturbation to the rotational states is given by the Hamiltonian

$$\mathcal{H}_p = -\boldsymbol{\mu}_e \cdot \mathcal{E} \equiv -|\mu_e| \mathcal{E} \cos\theta,$$

with $\mathcal{E}$ electric field along the z direction.

From parity argument one notes that the first order contribution to the energy is zero:

$$\langle K', M' \,|\, \mathcal{H}_p \,|\, K', M' \rangle = 0$$

Thus a correction term to the eigenvalues of the form $\Delta E \propto \mathcal{E}^2$ and $(\mu_{eff})_z \propto \mathcal{E}$ is expected, implying an *effective induced dipole moment* and therefore *positive* polarizability, somewhat similar to the paramagnetic susceptibility derived at Sect. 4.4.

For the ground-state at $K = M = 0$ the second-order perturbation correction

$$\Delta E_0^{(2)} = \sum_{(K,M) \neq 0} \frac{<0|\mathcal{H}_p|K, M><K, M|\mathcal{H}_p|0>}{E_0 - E_{K,M}}$$

reduces to

$$\Delta E_0^{(2)} = -\frac{I}{\hbar^2} \mu_e^2 \mathcal{E}^2 | \int \sin\theta d\theta d\phi \frac{1}{\sqrt{4\pi}} \cos\theta \frac{\sqrt{3}}{\sqrt{4\pi}} \cos\theta|^2 = -\frac{1}{3} \mu_e^2 \mathcal{E}^2 \frac{I}{\hbar^2}, \quad (10.18)$$

having taken into account that the only matrix element different from zero is the one connecting the state $K = 0$ to the state $K = 1, M = 0$, with eigenfunctions $1/\sqrt{4\pi}$ and $\sqrt{3/4\pi}\cos\theta$, respectively. Then

$$\alpha(0, 0) = -\frac{1}{\mathcal{E}} \frac{\partial \Delta E_0^{(2)}}{\partial \mathcal{E}} = \frac{2\mu_e^2 I}{3\hbar^2} \quad (10.19)$$

For the states at $K \neq 0$ we report the result of the estimate similar to the one given above:

$$E(K, M, \mathcal{E}) = E_K^0 + \frac{\mu_e^2 \mathcal{E}^2 I}{\hbar^2} \left[\frac{K(K+1) - 3M^2}{K(K+1)(2K-1)(2K+3)} \right], \quad (10.20)$$

for $K \neq 0$.

From the sum over M in a given state $|K, M>$ (in first approximation the energy can be considered to depend only on K), Eq. (10.20) yields

$$\alpha(K) \simeq \sum_M \alpha(K, M) = 0 \quad \text{for} \quad K \neq 0$$

(note that $\sum_{K=-M}^{M} M^2 = K(K+1)(2K+1)/3$).

Then only the ground state ($K = M = 0$) contributes to the orientational *effective* electric moment along the field. The polarizability is temperature dependent, since the population of this state is affected by the temperature.

The thermal average reads

$$\langle \alpha \rangle_T = \frac{\alpha(0, 0)}{Z_{rot}} \tag{10.21}$$

When for Z_{rot} the sum over K can be transformed into an integral (see Eq. (10.17)), by taking into account Eq. (10.18) with $\theta_{rot} = \hbar^2/2I k_B$, the single-molecule polarizability becomes

$$\langle \alpha \rangle_T = \frac{\mu_e^2}{3 k_B T}, \tag{10.22}$$

similar to the classical form

$$\langle \mu_e \rangle_z = \mu_e \langle \cos\theta \rangle$$

for

$$\langle \cos\theta \rangle = \frac{\int e^{\mu \mathcal{E} \cos\theta / k_B T} \cos\theta \, d\Omega}{\int e^{\mu \mathcal{E} \cos\theta / k_B T} \, d\Omega} = ctnh\, x - \frac{1}{x} \equiv L(x),$$

with $L(x)$ Langevin function, that for $x = \mu\mathcal{E}/k_B T \ll 1$ becomes $L(x \ll 1) \simeq x/3$.

10.2.5 Extension to Polyatomic Molecules and Effect of the Electronic Motion in Diatomic Molecules

In the following we sketch how the rotational eigenvalues can be obtained in polyatomic molecules when particular symmetries allows one to extend the quantum rules for the angular momentum already recalled in diatomic molecules.

The classical rotational Hamiltonian reads

$$\mathcal{H}_{rot} = \frac{P_A^2}{2 I_A} + \frac{P_B^2}{2 I_B} + \frac{P_C^2}{2 I_C} \tag{10.23}$$

where $I_{A,B,C}$ are the moments of inertia with respect to the principal axes of the tensor of inertia (conventionally $I_A < I_B < I_C$).

When the molecule is a *prolate* rotator, namely $I_A < I_B = I_C$, as for instance in CH_3F sketched below

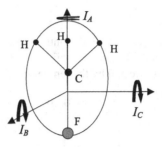

then the Hamiltonian can be rewritten

$$\mathcal{H}_{rot} = \frac{P_A^2}{2\,I_A} + \frac{P_A^2}{2\,I_B} - \frac{P_A^2}{2\,I_B} + \frac{P_B^2 + P_C^2}{2\,I_B} = \frac{P^2}{2\,I_B} + P_A^2\left(\frac{1}{2\,I_A} - \frac{1}{2\,I_B}\right) \quad (10.24)$$

Therefore from the quantization rules

$$P^2 = K(K+1)\,\hbar^2 \qquad P_A = M\hbar$$

the eigenvalues of the rotational energy turn out

$$E(K,M) = \frac{K(K+1)\,\hbar^2}{2\,I_B} + M^2\,\hbar^2\left(\frac{1}{2\,I_A} - \frac{1}{2\,I_B}\right), \qquad (10.25)$$

where now M refers to the component *along the molecular axis A*.

Equivalently, for an *oblate* rotator, where $I_A = I_B < I_C$ (as for instance in C_6H_6 (see Sect. 9.3)), one has a similar result.

In the general case, when $I_A \neq I_B \neq I_C$, no simple expressions can be derived for the eigenvalues and therefore reference to limit situations is usually made.

Up to now, in discussing the rotational motions, the electronic motions have been disregarded. In fact, for diatomic molecules it has been assumed the most common case of $^1\Sigma$ ground-state, where the components of the orbital and spin moments along the molecular axis are zero.

The derivation of the rotational eigenvalues carried out for the prolate polyatomic rotator and leading to Eq. (10.25), can be used to include the effect of the electron motion for diatomic molecules in electronic state $\Lambda \neq 0$. In a simplified picture, in fact, the electronic clouds can be regarded as a rigid charge distribution rotating around the molecular z-axis. Thus the diatomic molecule can be considered somewhat equivalent to a prolate rotator, with moments of inertia $I_A \equiv I_{elec}$ and $I_B = I_C = I_{nucl.}$, with $I_A \ll I_{B,C}$. Then, from extension of Eq. (10.25), at variance with Eq. (10.8) the rotational eigenvalues for diatomic molecules in an electronic state different from Σ turn out

$$E_{rot}(K, \Lambda) = \frac{\hbar^2}{2\mu R_e^2}[K(K+1) - \Lambda^2] + \frac{\hbar^2}{2I_{elec}}\Lambda^2, \qquad (10.26)$$

Λ being the quantum number for L_z (or for J_z in the case of strong spin-orbit coupling). The last term in Eq. (10.26), much larger than the first one and independent from the rotational levels, is the one involved in the electron kinetic energy. When the molecule is in a state at $\Lambda \neq 0$ the roto-vibrational structure (see Sect. 10.5) in the spectra involving electronic states display an *extra line* correspondent to a transition at $\Delta K = 0$, called *Q-branch*. This line is frequently observed in the electronic lines of band spectra (Sect. 10.8) or in Raman spectroscopy (Sect. 10.7), when transitions between electronic states are involved (see for example Figs. 10.8 and 10.9). This line is obviously absent when transition involves two Σ states.

Problems

Problem 10.1 As sketched in the following scheme the emission spectrum in the far infrared region from HBr molecules displays a series of lines regularly shifted by about $15\,\text{cm}^{-1}$.

Derive the statistical populations of the rotational levels for $T = 12, 36$ and $120\,\text{K}$. Estimate the interatomic equilibrium distance and obtain the relationship between temperature and rotational number K_{max} corresponding to the line of maximum intensity.

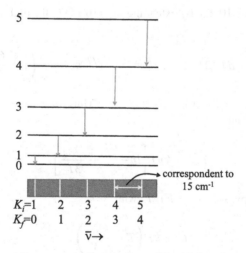

Solution: The separation among adjacent lines (Fig. 10.1 and Eq. (10.12)) is $2Bhc$, then $\hbar^2/2I = 1.06\,\text{meV}$, yielding $R_e = 1.41\,\text{Å}$.

The maximum intensity implies $(\partial N(K)/\partial K)_{K_{max}} = 0$. Then, from Eq. (10.14), $T = \hbar^2(2K_{max} + 1)^2/4k_B I$.

The statistical populations as a function of K are reported below:

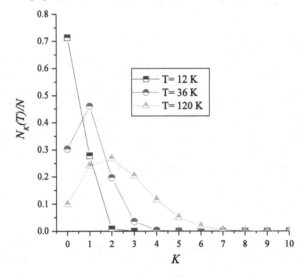

Problem 10.2 For an ensemble of diatomic molecules at the thermal equilibrium write the contribution from rotational motions to the free energy F and to the entropy S in the limits $T \gg \theta_{rot}$ and $T \to 0$.

Solution: From Sect. 10.2.3, by extending Eq. (10.17) in the high temperature limit one has

$$Z \simeq \int_0^\infty dK (2K + 1) e^{-K(K+1)\Theta_{rot}/T} \simeq \frac{T}{\Theta_{rot}} \left(1 + \frac{\Theta_{rot}}{3T} \right)$$

while for $T \to 0$ $Z \simeq 1 + 3\, exp(-2\Theta_{rot}/T)$. Thus for $T \gg \Theta_{rot}$ the free energy per molecule turns out

$$F \simeq -k_B T \left[ln \frac{T}{\Theta_{rot}} + \frac{\Theta_{rot}}{3T} \right],$$

while for the entropy $S = (U - F)/T$, with

$$U \simeq k_B \left(T - \frac{\Theta_{rot}}{3} \right),$$

so that

$$S \simeq k_B + k_B ln \frac{T}{\Theta_{rot}}.$$

For $T \to 0$ one finds

$$U \simeq 6 k_B \Theta_{rot} e^{-2\Theta_{rot}/T},$$

$$F \simeq -k_B T ln(1+3e^{-2\Theta_{rot}/T}) \ , \quad S \simeq \frac{6k_B \Theta_{rot}}{T} e^{-2\Theta_{rot}/T} + k_B ln(1+3e^{-2\Theta_{rot}/T}).$$

Problem 10.3 By referring to the three rotational levels depicted in Fig. 10.1, plot the splittings induced in the $H^{35}Cl$ molecule by a static electric field $\mathcal{E}$ (*Stark effect*), indicating the transition that can be observed in rotational spectroscopy. Then assume for the field the value $\mathcal{E} = 10^4$ V/cm and estimate the splitting induced in the $K = 1$ states, giving an order of magnitude of the resolution of the spectrometer required to evidence the doublet associated with $K = 0 \to K = 1$ transition.

Solution: From Eq. (10.20) the $K = 1$ state is opened in a doublet, with a splitting among $M = 0$ and $M = \pm 1$ levels of $(3/20)\mu_e^2 \mathcal{E}^2/Bhc$. From Eq. (10.18) the shift of the ground-state is $\mu_e^2 \mathcal{E}^2/6Bhc$. The transitions follow the selection rule $\Delta K = \pm 1$, $\Delta M = 0, \pm 1$. Since the doublet associated with $K = 0 \to K = 1$ transition is split by the amount $(3/20)\mu_e^2 \mathcal{E}^2/Bhc$, for $\mu_e \simeq 10^{-18}$ u.e.s. cm and $B \simeq 10.56$ cm^{-1} (correspondent to $I = 2.68 \times 10^{-40}$ g.cm^2), one finds that the resolution required is

$$\frac{\Delta \bar{\nu}}{\bar{\nu}} = \frac{3}{20} \frac{\mu_e^2 \mathcal{E}^2}{Bhc} \frac{1}{2Bhc} \simeq 1.9 \times 10^{-5}$$

Problem 10.4 In the rotational spectrum of $H^{35}Cl$ two lines are detected with the same strength at 106 cm^{-1} and at 233.2 cm^{-1}. Derive the temperature of the gas (remind that $B \simeq 10.56$ cm^{-1}).

Solution: From

$$\bar{\nu} = 2B(K + 1),$$

with $B = 10.6$ cm^{-1}, one has for $\bar{\nu}_1 = 106.0$ cm^{-1} $K_1 = 4$ and for $\bar{\nu}_2 = 233.2$ cm^{-1} $K_2 = 10$. For intensity proportional to the population of the rotational levels,

$$(2K_1 + 1)e^{-\frac{hcBK_1(K_1+1)}{k_B T}} = (2K_2 + 1)e^{-\frac{hcBK_2(K_2+1)}{k_B T}}$$

and

$$T = \frac{90\, hcB}{k_B\, ln(2.33)} \simeq 1620\, K.$$

10.3 Vibrational Motions

10.3.1 Eigenfunctions and Eigenvalues

Going back to the radial part of the Schrodinger equation (Eq. (10.4)), again disregarding the term $-2Q/R_e$ and without including the rotational terms $K(K+1)\hbar^2/2I$ which does not depend on $(R - R_e)$, the function $U(R) = R\mathcal{R}(R)$ is introduced, so that the equation becomes

$$\frac{d^2U}{dR^2} + \frac{2\mu}{\hbar^2}\left[E - V(R)\right] = 0 \qquad (10.27)$$

While the full expression of the potential energy $V(R)$ is unknown, the vibration involves small displacements around the equilibrium position R_e and then one can write

$$V(R) = V(R_e) + \left(\frac{dV}{dR}\right)_{R_e}(R - R_e) + \frac{1}{2}\left(\frac{d^2V}{dR^2}\right)_{R_e}(R - R_e)^2 + \cdots \quad (10.28)$$

Since $(\frac{dV}{dR})_{R_e}$ is zero, by omitting the constant $V(R_e)$ (namely the energy E has to be added to the electronic energy at the equilibrium position, see Fig. 7.2), in terms of the non-local coordinate $Q = x_A - x_B$ (Eq. (10.5)), one has

$$V(R) = \frac{1}{2}kQ^2 + \cdots, \qquad (10.29)$$

where k is the curvature of $V(R)$ around the equilibrium position. In the *harmonic approximation* higher order terms in the expansion (10.28) are neglected. The equation for the vibration of the nuclei around the equilibrium position is thus written

$$-\frac{\hbar^2}{2\mu}\frac{d^2U}{dQ^2} + \frac{1}{2}kQ^2U = EU, \qquad (10.30)$$

the well known form for the harmonic oscillator (with $-R_e \leq Q < \infty$). Then the eigenfunctions are related to the *Hermite polynomials* and the eigenvalues are

$$E_v = (v + 1/2)h\nu_o \qquad (10.31)$$

with quantum number $v = 0, 1, 2, 3\ldots$, while $\nu_o = (1/2\pi)\sqrt{k/\mu}$ corresponds to the frequency of the classical oscillator with same mass and elastic constant.

The eigenvalues and eigenfunctions for low energy states are depicted in Fig. 10.4. The ground state ($v = 0$) is described by the wavefunction

$$U(Q) = \left(\frac{b}{\pi}\right)^{1/4} e^{-Q^2 b/2}, \quad \text{with } b = \frac{2\pi\nu_o\mu}{\hbar} \qquad (10.32)$$

implying a behavior significantly different from the one expected for classical oscillator for which the maxima of probability of presence are at the boundaries of the motion. Other relevant differences with respect to classical oscillator are the extension of the "motion" outside the extreme elongations and the occurrence of zero-point energy $E_{v=0} = (1/2)h\nu_o$.

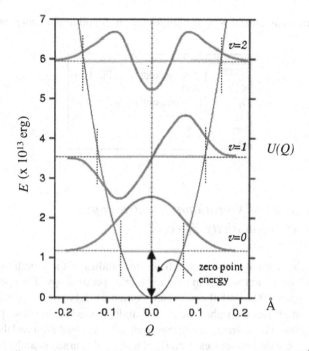

Fig. 10.4 First vibrational states in a diatomic molecule, having assumed $k = 5 \times 10^5$ dyne/cm and effective mass $\mu = 10^{-23}$ g, corresponding to vibrational frequency $\nu_0 = 3.6 \cdot 10^{13}$ Hz. The *dotted lines* correspond to the maxima elongations according to the classical oscillator in the parabolic potential energy indicated by the solid line. Typical values for the force constants are (i) in H_2 $k \simeq 5 \times 10^5$ dyne/cm; (ii) in O_2 (where a double bond is present) $k \simeq 11 \times 10^5$ dyne/cm; (iii) in N_2 (triple bond) $k \simeq 23 \times 10^5$ dyne/cm; (iv) in NaCl(ionic bond) $k \simeq 1.2 \times 10^5$ dyne/cm

The mean square displacement from the equilibrium position reads

$$< Q^2 >_v = \int U_v^* \, Q^2 \, U_v \, dQ \qquad (10.33)$$

and from the expressions of the Hermite polynomials one finds

$$< Q^2 >_v = \frac{\hbar}{\sqrt{\mu k}}(v + 1/2) = E_v/k \qquad (10.34)$$

implying $E_v = k < Q^2 >_v$, the same relation holding for the classical oscillator.

To give a few representative examples, in the HCl molecule the vibrational frequency is $\nu_0 = 8.658 \times 10^{13}$ Hz, corresponding to a force constant $k = 4.76 \times 10^5$ dyne/cm, while in CO $\nu_0 = 6.51 \times 10^{13}$ Hz, corresponding to a force constant $k = 18.65 \times 10^5$ dyne/cm.

Some vibrational constants $\bar{\nu}_0$ for homonuclear diatomic molecules are reported below:

Molecule	(cm^{-1})
H_2	4159 (see caption in Fig. 10.4)
N_2	2330
O_2	1556 (see caption in Fig. 10.4)
Li_2	246
Na_2	158
Cs_2	42

10.3.2 Principles of Vibrational Spectroscopy and Anharmonicity Effects

In regards of the main aspects the spectroscopic studies of the vibrational states in molecules are similar to the ones in optical atomic spectroscopy. The spectral range typically is within $100 - 4000$ cm^{-1} (see Appendix 1.1) and the devices are no longer based on glasses but rather use alkali halides, to reduce the absorption of the *infrared radiation*. The diffraction gratings grant a better resolution and the detectors are usually semiconductor devices. Details of technical character can be found in the exhaustive book by *Svanberg*.

As for rotational spectroscopy, being more interested into the fundamental aspects, we turn our attention to the transition probability due to the electric dipole mechanism. For two vibrational states at quantum numbers v' and v'', the component along the molecular axis of the electric dipole matrix element reads

$$(\mathbf{R}_{v' \to v''})_z = \int U_{v''}^* \mu_e U_{v'} dQ \qquad (10.35)$$

where $\mu_e(Q)$ is a complicate function of the interatomic distance R.

The sketch of a plausible dependence of μ_e with R is given below

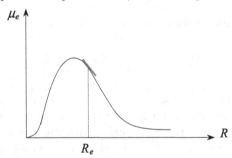

One can expand μ_e around R_e in terms of Q:

$$\mu_e = \mu_e(0) + \left(\frac{d\mu_e}{dQ}\right)_0 Q + \cdots , \qquad (10.36)$$

where the first term is involved in the rotational selection rules, while the expansion has been limited to the linear term in Q (often called *linear electric approximation*).

From Eqs. (10.35) and (10.36) one concludes that only *heteronuclear molecules*, with $\mu_e \neq 0$, can be driven to transitions among vibrational states. Furthermore, from the term linear in Q one deduces that only states of different parity imply a matrix element different from zero. Indeed, as it can be seen from inspection to the Hermite polynomials, only transitions between adjacent states are allowed : $\Delta v = \pm 1$. One can also remark that the frequency emitted or absorbed is the one expected for a classical Lorentz-like oscillator.

Thus in the harmonic approximation (Eq. (10.29)) and in the linear dipole approximation (Eq. (10.36)) one expects a *single absorption line* at the frequency ν_o. The line yields the curvature of the energy of the molecule at the equilibrium interatomic distance. The intensity of each component, to a large extent, is controlled by the statistical population on the vibrational levels:

$$N_v(T) = Ae^{-(v+1/2)h\nu_o/k_BT} \equiv N_0(T)e^{-vh\nu_o/k_BT} \qquad (10.37)$$

with

$$N_0(T) = (N/Z_{vib})e^{-h\nu_o/2k_BT}, \qquad (10.38)$$

N total number of molecules and Z_{vib} the vibrational partition function.

For $k_BT \ll h\nu_o$, as it is often the case, the ground state is by large the most populated and therefore the absorption line is practically related to the transition $v = 0 \rightarrow v = 1$.

Now a brief discussion of the *anharmonicity* effects is in order. The electrical anharmonicity originates from the term in Q^2, neglected in the expansion (10.36). According to the correspondent matrix element in Eq. (10.35), because of parity characters of the operator and of the Hermite polynomials, that term implies transitions between states at the same parity. Therefore the selection rule $\Delta v = \pm 2$ results, for states pertaining to the mechanical harmonic approximation (Eq. (10.29)).

The qualitative effect of the terms proportional to Q^3 and Q^4 in the expansion (10.28) (*mechanical anharmonicity*) is to cause a progressive reduction in the separation between the states at high quantum number v, as sketched below.

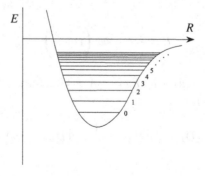

Then from the matrix element of the form correspondent to Eq. (10.35), transitions at frequencies different from ν_o have to be expected.

The anharmonic terms can be analyzed as perturbation of the vibrational states described by the wavefunctions $U_v(Q)$. The term in Q^3 must be considered up to the second order, its expectation value being zero for unperturbed states. Thus the eigenvalues turn out of the form

$$E_v = (v + 1/2)h\nu_o - a(v + 1/2)^2 h\nu_o \qquad (10.39)$$

where the constant a, much smaller than unit, is related to the ratio $[(d^3V/dR^3)_{R_e}]^2/k^{5/2}$. To give an idea, for the Hydrogen molecule H_2, $a = 0.027$. In $H^{35}Cl$, $\nu_0/c = 2885.6$ cm^{-1} and $a = 0.0176$.

For an heuristic potential $V(R)$ in the Morse-like form (see Sect. 10.4) one derives $a = h\nu_0/4D_e$, with $D_e \equiv -V(R_e)$, as we shall discuss in a subsequent section (see Eqs. (10.41) and (10.47)).

Problems

Problem 10.5 Consider the H_2 molecule in the vibrational ground state and in the first excited rotational state $(K = 1)$. Evaluate the number of oscillations occurring during one rotation.

Solution: From

$$\nu_{rot} = \frac{[K(K+1)]^{1/2}\hbar}{2\pi I} = \frac{\sqrt{2}\hbar}{2\pi\mu R_e^2}$$

and $\nu_{vib} = (1/2\pi)\sqrt{k/\mu}$ the number of oscillations is

$$n_{osc} = \frac{1}{2\pi}\sqrt{\frac{k}{\mu}} \cdot \frac{2\pi\mu R_e^2}{\sqrt{2}\hbar} = \frac{\sqrt{k\mu}R_e^2}{\sqrt{2}\hbar} \simeq 25 \,.$$

Problem 10.6 The dissociation energy in the D_2 molecule is increased by 0.08 eV with respect to the one in H_2. Estimate the zero point energy for both molecules.

Solution: From

$$E = -A + \hbar\omega\left(v + \frac{1}{2}\right)$$

the dissociation energy is given by $E_d = +A - \frac{1}{2}\hbar\omega$, for $v = 0$.
Since $A(H) = A(D)$,

$$E_d(D) - E_d(H) = -\frac{1}{2}\hbar[\omega(D) - \omega(H)].$$

Hence

$$\frac{\hbar\omega(H)[1 - \omega(D)/\omega(H)]}{2} = \frac{\hbar\omega(H)[1 - 1/\sqrt{2}]}{2} = 0.08 \text{ eV},$$

and the zero-point energy of H_2 turns out $\hbar\omega(H)/2 = 0.27$ eV while for D_2 one has $\hbar\omega(D)/2 = (0.27/\sqrt{2})$ eV $= 0.19$ eV.

Problem 10.7 The infrared spectrum of a gas of diatomic molecules displays lines equally spaced by about 10^{11} Hz. A static electric field of 3 kV/cm is applied. The lowest frequency line, with intensity 2.7 times smaller than the adjacent one, splits in a doublet, with 1 MHz of separation between the lines. Derive the molar polarizability.

Solution: From the ratio of the intensities $I(1)/I(0) = 2.7 = 3\,exp(-h\nu_{rot}/k_B T)$, with $\nu_{rot} = 10^{11}$ Hz, the temperature is deduced: $T \simeq 45.6$ K.

Since $k_B T \gg h\nu_{rot}$ the molar polarizability reads (see Eq. (10.22)) $\alpha = N_A \mu_e^2/3 k_B T$.

The electric field partially removes the degeneracy of the $K = 1$ level. The separation between levels at $M_K = 0$ and $M_K = \pm 1$ turns out (see Eq. (10.20))

$$\Delta\nu = 10^6 \text{ Hz} = \frac{\mu_e^2 \mathcal{E}^2}{h^2 \nu_{rot}} \frac{3}{10}$$

Then $\mu_e^2 = 14.6 \times 10^{-38}$ u.e.s.^{2}cm^2 and $\alpha(T = 45.6K) = 4.7$ emu/mole.

Problem 10.8 Evaluate the order of magnitude of the electronic, rotational and vibrational polarizabilities for the HCl molecule at $T = 1000$ K (the elastic constant is $k = 4.76 \times 10^5$ dyne/cm and the internuclear distance $R_e = 1.27$ Å). From the Clausius-Mossotti relation (see Sect. 16.2) estimate the dielectric constant of the gas, at ambient pressure.

Solution: For order of magnitude estimates one writes (see Problem 8.13 and Eq. (10.22))

$$\alpha_{el} \simeq 8a_0^3 \sim 10^{-24} \text{ cm}^3,$$

$$\alpha_{rot} = \frac{e^2 R_e^2}{3k_B T} \simeq 10^{-22} \text{ cm}^3.$$

For the vibrational contribution (see Problem 10.16)

$$\alpha_v \simeq \frac{e^2}{k} \simeq 5 \times 10^{-25} \text{ cm}^3.$$

From the equation of state $PV = RT$, by taking into account that at ambient pressure and temperature the molar volume is $V = 2.24 * 10^4$ cm^3, at $T = 1000°$ K one finds $V \sim 8.2 \times 10^4$ cm^3, corresponding to the density $N \simeq 10^{19}$ molecules cm^{-3}.

From $\Delta\epsilon = (4\pi N\alpha_{rot})/[1 - 4\pi N\alpha_{rot}/3] \simeq 4\pi N\alpha_{rot}$, one derives $\epsilon \approx 1.01$.

10.4 Morse Potential

An heuristic energy curve for diatomic molecules that can be assumed as potential energy for the vibrational motion of the nuclei (Eq. (10.27)) is the one suggested by *Morse*:

$$V_M = D_e \left[1 - exp[-\beta(R - R_e)] \right]^2 , \tag{10.40}$$

of the form sketched below.

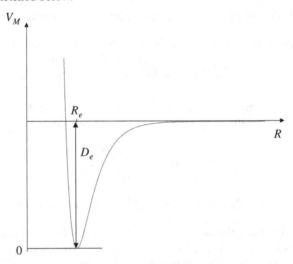

This expression retains a satisfactory validity for R around the interatomic equilibrium distance R_e. D_e corresponds to the energy of the molecule for $R = R_e$ (the real dissociation energy being D_e minus the zero-point vibrational energy $(1/2)h\nu_0$), while β is a characteristic constant.

It is noted that for R close to R_e Eq. (10.40) yields $V_M \simeq D_e Q^2 \beta^2$, namely the harmonic potential with elastic constant $k = 2D_e \beta^2$ and $\nu_0 = (\beta/\sqrt{2}\pi)\sqrt{D_e/\mu}$.

The Morse potential, often useful for approximate expression of the electronic eigenvalue $E(R_{AB})$ in diatomic molecules, has the advantage that the Schrödinger equation for the vibrational motion (Eq. (10.27)) can be solved analytically, although with cumbersome calculations. The eigenvalues turn out

$$E_M = h\nu_0[(v + 1/2) - a(v + 1/2)^2] \tag{10.41}$$

with $a = (h\nu_0/4D_e)$ (see Eq. (10.39)). The eigenfunctions are no longer even or odd functions for v even or odd, respectively, at variance with the ones derived in the harmonic approximation. Therefore one has transitions at $\Delta v \neq \pm 1$ without having to invoke electrical anharmonicity.

Problems

Problem 10.9 From the approximate expression for the energy of diatomic molecules

$$V(R) = A \left(1 - exp[-B(R - C)]\right)^2$$

with $A = 6.4$ eV, $B = 0.7 \times 10^8$ cm^{-1} and $C = 10^{-8}$ cm, derive the properties of the rotational and vibrational motions.

Sketch the qualitative temperature dependence of the specific heat. Assume for the reduced mass the one pertaining to HF molecule.

Solution: The elastic constant is $k = 2AB^2 = 10^5$ dyne/cm. For the reduced mass $\mu = 0.95M$ (with M the proton mass), the fundamental vibrational frequency is $\nu_0 = 4 \times 10^{13}$ Hz, corresponding to the vibrational temperature $\Theta_v = h\nu_0/k_B \simeq 1846$ K.

For an equilibrium distance $R_e = C$, the moment of inertia is $I = 1.577 \times 10^{-40}$ g cm^2. The separation between adjacent lines in the rotational spectrum is $\Delta\bar{\nu} = 35.6$ cm^{-1} and therefore $\Theta_r = Bhc/k_B = 25.6$ K.

The temperature dependence of the molar specific heat is sketched below.

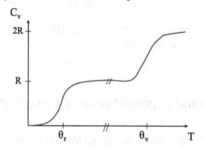

Problem 10.10 In the RbH molecule ($R_e = 2.36$ Å) the fundamental vibrational frequency is $\nu_0 = 936.8$ cm^{-1} and the dissociation energy in wavenumber is $D_e = 15505$ cm^{-1}. Derive the Morse potential and the correction due to the rotational term for $K = 40$ and $K = 100$. Discuss the influence of the rotation on the dissociation energy.

Solution: The parameter β for Eq. (10.40) is $\beta = 2\pi\nu_0\sqrt{\mu/2D_e}$ and from the reduced mass $\mu = 1.65 \cdot 10^{-24}$ g one has $\beta = 9.14 \cdot 10^7$ cm^{-1}.

At the equilibrium distance R_e the rotational constant turns out

$$B = \frac{h}{8\pi^2 c\mu R_e^2} \simeq 3 \text{ cm}^{-1}.$$

To account for the rotational contribution the energy has to be written in terms of the R-dependent rotational constant (see Eq. (10.4)). Therefore

$$E_{\text{rot}}(R) = hcBK(K + 1) \cdot \frac{R_e^2}{R^2},$$

so that the effective potential becomes

$$V_{\text{eff}}(R) = V_M(R) + E_{\text{rot}}(R),$$

plotted below.

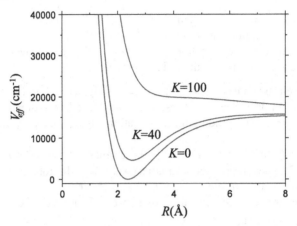

On increasing the rotational number the equilibrium distance is increased and the strength of the energy bond is reduced, as expected (see Eq. (10.47)).

10.5 Roto-Vibrational Eigenvalues and Coupling Effects

In high resolution an absorption line involving transitions between vibrational states evidences the fine structure related to simultaneous transitions between rotational states (See Fig. 10.5).

Still assuming weak roto-vibrational coupling, the wavefunction is $\phi_{rot}^{K,M}\mathcal{R}_v(R)$ and the eigenvalues are

$$E_{K,v} = \left(v + \frac{1}{2}\right)h\nu_o + \frac{\hbar^2 K(K+1)}{2\mu R_e^2} \qquad (10.42)$$

The electric dipole matrix element connecting two states (K', v') and (K'', v'') reads

$$\mathbf{R}_{K',v'\to K'',v''} \propto \int \phi_{rot}^{K'',M''*}\mathcal{R}_{v''}(R)^* \cdot (\mu_e + c\mathbf{j}Q)\phi_{rot}^{K',M'}\mathcal{R}_{v'}(R)\sin\theta d\theta d\phi dQ \qquad (10.43)$$

where $\mathbf{j}$ is a unitary vector along the molecular axis. The term involving μ_e drives the purely rotational transitions while

Fig. 10.5 Rotational structure in the vibrational spectral line of the HF molecule and illustration of the transitions generating the P and R branches

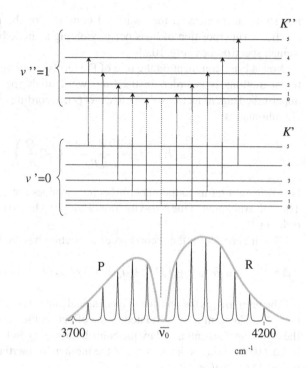

$$c \int \phi_{rot}^{K'',M''} \mathbf{j} \phi_{rot}^{K',M'} sin\theta d\theta d\phi \int \mathcal{R}_{v''}^{*}(R) Q \mathcal{R}_{v'}(R) dQ \qquad (10.44)$$

implies transitions with the selection rules $\Delta K = \pm 1$ and $\Delta v = \pm 1$ (see Eq. (10.11) and Sect. 10.3.2). When in the $v' \rightarrow v''$ transition the quantum number K increases then the correspondent line is found at a frequency $\nu > \nu_o$ (branch R in Fig. 10.5) while when K decreases one has $\nu < \nu_0$ (branch P).

It is noted that the line at $\nu = \nu_0$ is no longer present. When electronic states at $\Lambda \neq 0$ are involved in a transition a component at ν_0 can be observed (called Q branch), usually in form of a broad line (see Sect. 10.2.5, Eq. (10.26) and examples at Figs. 10.8 and 10.9).

From Eqs. (10.42) and (10.44) the wavenumbers associated with the roto-vibrational transitions are

$$\bar{\nu}_R = \bar{\nu}_o + B_{v''}(K+1)(K+2) - B_{v'} K(K+1) \qquad (10.45)$$
$$\bar{\nu}_P = \bar{\nu}_o + B_{v''} K(K-1) - B_{v'} K(K+1) \qquad (10.46)$$

Since $B_{v'} \simeq B_{v''} \simeq B_v$ the separation between the adjacent lines turns out about $2B_v$, as shown in Fig. 10.5. The wavenumbers of the Q branch are

$$\bar{\nu}_Q = \bar{\nu}_o + B_{v'} K(K+1) - B_{v''} K(K+1).$$

For slight differences in the rotational constants of the two vibrational levels in practice a superposition of lines occur, yielding a single broad line, as observed in Raman spectroscopy (Fig. 10.8).

Now a brief comment on the role of the terms coupling the rotational and vibrational motions is in order. In the framework of the perturbative approach, with unperturbed eigenfunctions $Y_{K,M}(\theta, \phi)\mathcal{R}(Q)$, according to Eq. (10.6) the perturbing Hamiltonian is

$$\mathcal{H}_P = \frac{\hbar^2 K(K+1)}{2\mu_e R_e}\left(-2\frac{Q}{R_e}\right).$$

No correction terms to the unperturbed eigenvalues are expected at the first order. Thus the evaluation of the roto-vibrational coupling has to be carried out at the second order in $\mathcal{H}_P$.

The final result for the second order correction has the form

$$\Delta E^{(2)} = -a_1 h\nu_0(v+1/2)^2 + a_2(v+1/2)K(K+1) - a_3 K^2(K+1)^2 \quad (10.47)$$

The term in a_1 is the one due to mechanical anharmonicity, already discussed (Eq. (10.39)). The term in a_2 is related to the effect on the elastic constant produced by the centrifugal potential and by the contribution in Q and Q^3. Finally the term in a_3 in Eq. (10.47) reflects the increase of the moment of inertia due to the rotation of the molecule (*centrifugal distortion*).

The detailed expressions for the coefficients a_i in Eq. (10.47) are $a_1 = (\hbar/384\,\pi\,\mu k\,\nu_0^3)\,(5\,\alpha^2 - 3\,k\,\beta)$, $a_2 = (\hbar^3/k^2\,R_e^2)\,(R_e\,\alpha + 3\,k)$ and $a_3 = \hbar^4/2\,k\,\mu^2\,R_e^6$ (as it can be obtained also classically by writing $kQ = \mu\omega^2\,R$ and then evaluating the rotational energy). α and β are the coefficients of the terms in Q^3 and in Q^4 in the perturbative Hamiltonian resulting in the expansion of Eq. (10.6).

Problems

Problem 10.11 The rotovibrational absorption spectrum for the HCl molecule is shown below.

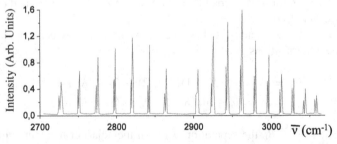

Derive the equilibrium distance and the elastic constant for the molecule. Which is the origin of the doublets observed at each peak? How does the spacing between adjacent lines change for the deuterated molecule?

Solution: From the spacing among adjacent lines $\bar{\Delta}\nu \simeq 21\,\text{cm}^{-1} = 2Bhc$, the moment of inertia being $I = 2.67 \times 10^{-40}\,\text{g\,cm}^2$, the equilibrium distance turns out $R_e = 1.27\,\text{Å}$ for the reduced mass $\mu = 0.972\,M_H = 1.6 \times 10^{-24}\,\text{g}$. The vibrational frequency is about $\nu_0 = 2885\,\text{cm}^{-1} \times c \simeq 8.65 \times 10^{13}\,\text{Hz}$, implying an elastic constant $k = 4\pi^2\nu_0^2\mu = 4.56 \times 10^5\,\text{dyne/cm}$.

The doublet arises from the spectra of the $H^{35}Cl$ and $H^{37}Cl$ molecules.

Deuteration implies an increase of the reduced mass μ by about a factor 2, leading to a spacing of the lines about $\bar{\Delta}\nu \simeq 10.5\,\text{cm}^{-1}$. The vibrational frequency is reduced by a factor close to $\sqrt{2}$.

Problem 10.12 The fundamental vibrational frequency of the NaCl molecule is $\nu_0 = 1.14 \cdot 10^{13}\,\text{Hz}$. Report in a plot the temperature dependence of mean-square displacement from the equilibrium interatomic distance.

Solution: From Eqs. (10.34) and (10.37) (see Problem 10.22 for the average energy) the plot results as below

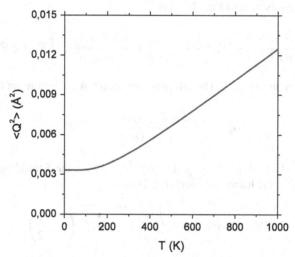

Problem 10.13 Derive the equilibrium distance and the vibrational frequency of a diatomic molecule in the assumption of interatomic effective potential $V(R) = 4U((a/R)^{12} - (a/R)^6)$, with $a = 3.98\,\text{Å}$ and $U = 0.02\,\text{eV}$, for reduced mass $\mu = 10^{-22}\,\text{g}$.

Solution: From $\partial V/\partial R\big|_{R=R_{eq}} = 0$ one has $R_{eq} = a \cdot (2)^{\frac{1}{6}} = 4.47\,\text{Å}$, while

$$V(R_{eq}) = 4U\left[\frac{1}{4} - \frac{1}{2}\right] = -U = -0.02\,\text{eV}.$$

By deriving $V(R)$ twice, one finds

$$k = \frac{4U}{a^2}\left(12 \cdot 13 \cdot 2^{-7/3} - 6 \cdot 7 \cdot 2^{-4/3}\right) = 57.144 \frac{U}{a^2} = 11.558 \cdot 10^2 \text{ dyne/cm},$$

and

$$\nu = \frac{1}{2\pi}\sqrt{\frac{k}{\mu}} = 5.18 \cdot 10^{11} \text{ Hz}.$$

Problem 10.14 In a diatomic molecule the eigenvalue $E(R)$ for the ground state is approximated in the form

$$E(R) = -2V_0\left[\frac{1}{\rho} - \frac{1}{2\rho^2}\right]$$

(with $\rho = R/a$ and a a characteristic length). Derive the rotational, vibrational and roto-vibrational energy levels in the harmonic approximation.

Solution: The equivalent of Eq. (10.4) is

$$-\frac{\hbar^2}{2\mu}\frac{d^2\mathcal{R}}{dR^2} + \left[-2V_0\left(\frac{1}{\rho} - \frac{1}{2\rho^2}\right) + \frac{K(K+1)\hbar^2}{2\mu a^2\rho^2}\right]\mathcal{R} = E\mathcal{R},$$

where μ is the reduced mass. The effective potential has the minimum for

$$\rho_0 \equiv 1 + \frac{K(K+1)\hbar^2}{2\mu a^2 V_0} \equiv 1 + B$$

For $V(\rho) = -V_0(1+B)^{-1} + V_0(1+B)^{-3}(\rho-\rho_0)^2$ the Schrödinger equation takes the form for the harmonic oscillator. Then

$$E + V_0(1+B)^{-1} = \hbar\sqrt{\frac{2V_0}{\mu a^2}(1+B)^{-3}}\left(v+\frac{1}{2}\right).$$

For $B \ll 1$ (small quantum number K),

$$E = -V_0 + \frac{K(K+1)\hbar^2}{2\mu a^2} + h\nu_0\left(v+\frac{1}{2}\right) - \frac{3}{2}\frac{\hbar^3 K(K+1)\left(v+\frac{1}{2}\right)}{\mu^2 a^4 2\pi\nu_0}$$

where $\nu_0 = (1/2\pi)\sqrt{2V_0/\mu a^2}$ (see Eq. (10.47)).

Problem 10.15 From the data reported in the figure at Problem 10.11 for the HCl molecule, estimate the vibrational contribution to the molar specific heat, at room temperature.

Solution: From the thermodynamical energy $<E> = \sum_v E_v N_v(T)$ (see Eq. (10.37)) the molar specific heat turns out

$$C_V = R\left(T_D/T\right)^2 \frac{e^{T_D/T}}{\left(e^{T_D/T} - 1\right)^2} \; ,$$

where T_D is the *vibrational temperature* $T_D = h\nu_0/k_B$, given by $T_D = 4.15 \cdot 10^3\,K$ for $\nu_0 = 0.87 \cdot 10^{14}$ Hz.

At room temperature $T \ll T_D$ and $C_V \simeq R(T_D/T)^2 exp(-T_D/T) = 1.5 \cdot 10^4$ erg/K mole.

Problem 10.16 A static and homogeneous electric field $\mathcal{E}$ is applied along the molecular axis of an heteronuclear diatomic molecule. In the harmonic approximation for the vibrational motion, by assuming an effective mass μ and an effective charge $-ef$ (with $0 < f \le 1$, see Sect. 8.5) derive the contribution to the electrical polarizability, in the perturbative approach.

Prove that the result derived in this way is the *exact one*.

Solution: For

$$\mathcal{H}_p = fez\mathcal{E}$$

the first order correction is $<v|z|v> = 0$ because of the definite parity of the vth eigenfunction of the oscillator. The matrix elements $\mathcal{H}_{vv'}$ for z are

$$\mathcal{H}_{vv'} = <v|z|v'> = \left(\frac{v+1}{2\alpha}\right)^{1/2} \quad \text{for} \quad v' = v+1$$

$$\left(\frac{v}{2\alpha}\right)^{1/2} \quad \text{for} \quad v' = v-1 ,$$

where $\alpha = \frac{\mu 2\pi\nu_0}{\hbar}$ and $\nu_0 = \frac{1}{2\pi}\sqrt{k/\mu}$, with k force constant.

The second order correction to the energy E_v^0 turns out

$$E_v^{(2)} = (fe\mathcal{E})^2 \left\{ \frac{|\mathcal{H}_{v,v+1}|^2}{-h\nu_0} + \frac{|\mathcal{H}_{v,v-1}|^2}{h\nu_0} \right\} =$$

$$= (fe\mathcal{E})^2 \left\{ \frac{\frac{v+1}{2\alpha}}{-h\nu_0} + \frac{\frac{v}{2\alpha}}{h\nu_0} \right\} = -(fe\mathcal{E})^2 \frac{1}{8\pi^2\mu\nu_0^2} .$$

Then the electric polarizability is

$$\chi = N\alpha = N\frac{1}{\mathcal{E}}\left(-\frac{\partial E_v^{(2)}}{\partial \mathcal{E}}\right) = N\,(fe)^2\frac{1}{k}$$

independent of the state of the oscillator (N number of molecule for unit volume).

The result for the single molecule polarizability α is the exact one. In fact, going back to the Hamiltonian of the oscillator in the presence of the field

$$\mathcal{H} = -\frac{\hbar^2}{2\mu}\frac{d^2}{dz^2} + \frac{1}{2}kz^2 + fe\mathcal{E}z$$

by the substitution $z = z' - (fe\mathcal{E}/k)$ it becomes

$$\mathcal{H}' = -\frac{\hbar^2}{2\mu}\frac{d^2}{dz'^2} + \frac{1}{2}k(z')^2 - \frac{1}{2k}(fe\mathcal{E})^2$$

implying the eigenvalues

$$E'_v = E^0_v - \frac{(fe\mathcal{E})^2}{2k}$$

and therefore $\alpha = (fe)^2/k$.

10.6 Polyatomic Molecules: Normal Modes

In a polyatomic molecule with S atoms, $(3S - 6)$ degrees of freedom involve the oscillations of the nuclei around the equilibrium positions. If q_i indicate generalized local coordinates expressing the displacement of a given atom, as sketched below,

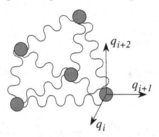

for small displacements the potential energy, in the harmonic approximation, can be written

$$V = V_o + \sum_i \left(\frac{\partial V}{\partial q_i}\right)_o q_i + \frac{1}{2}\sum_{i,j}\left(\frac{\partial^2 V}{\partial q_i \partial q_j}\right)_o q_i q_j \simeq \sum_{i,j} b_{ij} q_i q_j \qquad (10.48)$$

Similarly, the kinetic energy is $T = \sum_{i,j} a_{ij}\dot{q}_i\dot{q}_j$. The classical equations of motion become

$$\frac{d}{dt}\frac{\partial \mathcal{L}}{\partial \dot{q}_i} - \frac{\partial \mathcal{L}}{\partial q_i} = \sum_j a_{ij}\ddot{q}_j + \sum_j b_{ij} q_j = 0 \ , \qquad (10.49)$$

namely $(3S - 6)$ coupled equations, corresponding to complex motions that can hardly be formally described. Before moving to the quantum mechanical formulation it is necessary to introduce a new group of coordinates $Q = \sum_j h_j q_j$ (and a group of constants c_1, c_2, etc.) so that, by multiplying the first Eq. (10.48) by c_1, the second by c_2, etc. and adding up, one obtains equations of the form $d^2 Q/dt^2 + \lambda Q = 0$. This is the classical approach to describe small displacements around the equilibrium positions. The conditions to achieve such a new system of equations are

$$\sum_i c_i a_{ij} = h_j \tag{10.50}$$

$$\sum_i c_i b_{ij} = \lambda h_j \ . \tag{10.51}$$

Therefore, in terms of the constants c_i

$$\sum_i c_i (\lambda a_{ij} - b_{ij}) = 0 \ , \tag{10.52}$$

implying

$$|\lambda a_{ij} - b_{ij}| = 0 \tag{10.53}$$

This secular equation yields the roots $\lambda^{(1)}, \lambda^{(2)}, \ldots$ corresponding to the conditions allowing one to find h_j so that the equations of motions become

$$\frac{d^2}{dt^2} Q_i + \lambda^{(i)} Q_i = 0 \tag{10.54}$$

These equations in terms of the *non-local, collective* coordinates

$$Q_i = \sum_j h_{ij} q_j \tag{10.55}$$

correspond to an Hamiltonian in the normal form

$$\frac{1}{2} \sum_i \dot{Q}_i^2 + \frac{1}{2} \sum_i \lambda^{(i)} Q_i^2, \tag{10.56}$$

where

$$\lambda^{(i)} \equiv \left(\frac{\partial^2 V}{\partial Q_i^2} \right)_o \tag{10.57}$$

The Q's are called *normal coordinates*. The normal form of the Hamiltonian will allow one to achieve a direct quantum mechanical description of the vibrational motions in polyatomic molecules and in crystals (see Chap. 14).

A few illustrative comments about the role of the normal coordinates are in order. From the inverse transformation the local coordinates are written

$$q_i = \sum_j g_{ij} Q_j \tag{10.58}$$

and therefore, from Eq. (10.54),

$$q_i = \sum_j g_{ij} A_j \cos\left[\sqrt{\lambda^{(j)}}\, t + \varphi_j\right] \tag{10.59}$$

Thus the *local motion is the superposition of normal modes* of vibration. Each normal mode corresponds to an harmonic motion of the full system, with all the S atoms moving with the same frequency $\sqrt{\lambda^{(j)}}$ and the same phase. The amplitudes of the local oscillations change from atom to atom, in general.

Taking a look back to a diatomic molecule and considering the vibration along the molecular axis (see Sect. 10.3) it is now realized that the normal coordinate is $Q = (x_A - x_B)$. The root of the secular equation analogous to Eq. (10.53) yields the frequency $\omega = \sqrt{k/\mu}$ and the (single) normal mode implies the harmonic oscillation, in phase opposition, of each atom, with relationship in the amplitudes given by $x_A = -x_B (M_B/M_A)$.

The formal derivation of the normal modes in polyatomic molecules in most cases is far from being trivial and the symmetry operations are often used to find the detailed form of the normal coordinates. For a linear molecule with three atoms, as CO_2, the description of the longitudinal vibrational motions in the harmonic approximation is straightforward (see for an illustrative example Problem 10.26). Figure 10.6 provides the illustration of the four normal modes.

The coupled character of the motions is hidden in the collective frequency $\lambda^{(i)} \equiv (\frac{\partial^2 V}{\partial Q_i^2})_o$, namely in the curvature of the potential energy under the variation of the i-th normal coordinate.

It is noted that the *stability* of the system is related to the *sign* of λ. Structural and ferroelectric *phase transitions* in crystals, for instance, are associated with the temperature dependence of the frequency of a normal mode, so that at a given temperature the structure becomes unstable ($\lambda^{(i)}$ is approaching zero) and a transition to a new phase, restoring large and positive $\lambda^{(i)}$, is driven.

Once that the vibrational motions are described by normal coordinates Q_i, the quantum formulation is straightforward. In fact, in view of the form of the classical Hamiltonian, the eigenfunction $\Phi(Q_1, Q_2, ...)$ is the solution of the equation

$$\sum_i \left(-\frac{\hbar^2}{2}\frac{\partial^2}{\partial Q_i^2} + \frac{1}{2}\lambda^{(i)} Q_i^2\right)\Phi(Q_1, Q_2, ...) = E\Phi(Q_1, Q_2, ...) \tag{10.60}$$

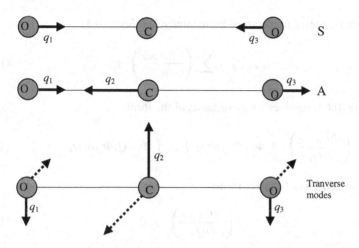

Fig. 10.6 Normal modes (longitudinal and transverse) in the CO_2 molecule. The symmetric mode S is not *active* in the infrared absorption spectroscopy, the selection rule requiring that the normal mode causes the variation of the *electric dipole moment* (see Eq. (10.66)), while the antisymmetric mode A is active. The inverse proposition holds for Raman spectroscopy (see Sect. 10.7) where the variation of the *polarizability* rather than of the dipole moment is required to allow one to detect the normal mode of vibration

(the nuclear masses are included in mass-weighted coordinates Q's). Therefore the wavefunctions and the eigenvalues are

$$\Phi(Q_1, Q_2, ...) = \prod_i \Phi(Q_i) \qquad (10.61)$$

$$E = \sum_i (n_i + 1/2)\hbar\sqrt{\lambda^{(i)}}, \qquad n_i = 0, 1, ... \qquad (10.62)$$

where $\Phi(Q_i)$ and $E_i = (n_i + 1/2)\hbar\sqrt{\lambda^{(i)}}$ are the single normal oscillator eigen-functions and eigenvalues. Thus, by recalling the results for the diatomic molecule (Sect. 10.3), the vibrational state is described by a *set of numbers* $n_1, n_2, ...$, labelling the state of each normal mode.

Now we are going to show that within the harmonic approximation any normal oscillator interacts individually with the electromagnetic radiation, in other words the normal modes are *spectroscopically independent*.

The electric dipole matrix element for a transition from a given initial state to a final one, reads

$$\mathbf{R}_{in\to f} \propto \int \Phi^*_{n_1^f}(Q_1)\Phi^*_{n_2^f}(Q_2)...\mu_e(Q_1, Q_2, ...)\Phi_{n_1^{in}}(Q_1)\Phi_{n_2^{in}}(Q_2)... \ . \qquad (10.63)$$

In the approximation of electrical harmonicity (see Sect. 10.3)

$$\mu_{ex,y,z} \propto \sum_i \left(\frac{\partial \mu_{ex,y,z}}{\partial Q_i} \right)_o Q_i \tag{10.64}$$

Equation (10.63) involves a sum of terms of the form

$$\left(\frac{\partial \mu_{ex,y,z}}{\partial Q_i} \right)_o \int \Phi^*_{n_1^f} \Phi_{n_1^{in}} dQ_1 \int \cdots \int \Phi^*_{n_i^f} Q_i \Phi_{n_i^{in}} dQ_i \cdots \tag{10.65}$$

This term is different from zero when

$$\left(\frac{\partial \mu_{ex,y,z}}{\partial Q_i} \right)_o \neq 0, \tag{10.66}$$

meaning that the i-th normal mode must imply a variation of the electric dipole moment of the molecule. At the same time it is necessary that

$$n_j^f = n_j^{in}, \qquad \text{for } j \neq i$$

$$n_i^f = n_i^{in} \pm 1. \tag{10.67}$$

Therefore each normal oscillator interacts with the electromagnetic radiation independently from the others, with absorption spectrum displaying lines in correspondence to the eigenfrequencies of the various modes.

When the selection rule in Eq. (10.66) is verified the mode is said to be *infrared active*. As a consequence of this condition, one can infer that in the CO_2 molecule only the antisymmetric longitudinal mode can interact with the radiation while the symmetric one is silent (see Fig. 10.6).

Finally we just mention that in the harmonic approximation the contribution to the thermodynamic energy in polyatomic molecules is obtained by adding the contributions expected from each mode, of the form derived in diatomic molecules (see Problem 10.15).

10.7 Principles of Raman Spectroscopy

As it has been remarked, by means of infrared or microwave absorption spectroscopies some rotational or vibrational motions cannot be directly studied. This is the case of rotations and vibrations in homonuclear molecules or for normal modes which do not comply with the selection rule given by Eq. (10.66). Motions of those types can often be investigated by means of a spectroscopic technique based on the analysis of *diffuse radiation*: the *Raman* spectroscopy.

Phenomenologically the Raman effect can be described by referring to the experimental set-up schematically reported below

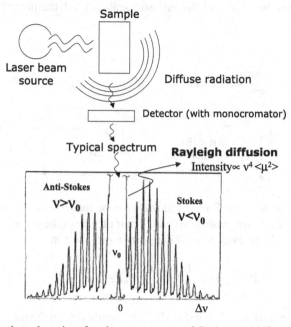

The classical explanation for the occurrence of Stokes and of anti-Stokes lines in the diffuse radiation, although not appropriate in some respects, still it enlightens the physical basis of the phenomenon. A normal mode of vibration can be thought to cause a time dependence of the molecular polarizability:

$$\alpha = \alpha_o + \alpha^* cos(\omega_i t) \tag{10.68}$$

Therefore the electric component of the radiation $\mathbf{E}(t) = \mathbf{E}_o cos(\omega_o t)$ (the wavelength is much larger than the molecular size) induces an electric dipole moment

$$\mu_{ind} = \mathbf{E}(t)\alpha(t) = \alpha_o \mathbf{E}_o cos(\omega_o t) + \frac{1}{2}\alpha^* \mathbf{E}_o cos(\omega_o - \omega_i)t + \frac{1}{2}\alpha^* \mathbf{E}_o cos(\omega_o + \omega_i)t \tag{10.69}$$

From the phenomenological picture of *oscillating dipoles* as source of radiation one can realize that components of the diffuse light at frequencies $\omega_o \pm \omega_i$ have to be expected.

The inadequacy of the classical description can be emphasized by observing that the experimental findings indicate that the anti-Stokes lines, in general, are less intense than the Stokes lines. The interpretation based on the oscillating dipole as in Eq. (10.69), would predict intensities proportional to the fourth power of the frequency and then the anti-Stokes lines should be more intense than the Stokes ones.

The quantum description of the Raman effect is based on the process of *scattering of photons* and provides a satisfactory description of all the aspects of the phenomenon. The intensity of the lines, in fact, are controlled by the statistical populations on the ground state and on excited vibrational states, as it can be grasped from the sketch of the inelastic scattering of the photon ($h\nu_i$) given below:

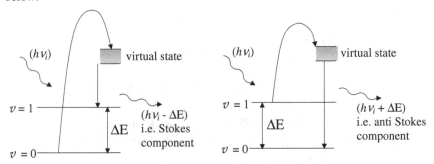

The basic aspects of the Raman radiation can be realized by extending the idea of electric dipole moment associated with a pair of states (already used in a variety of cases) to include the *field induced dipole moment*. Then in

$$R_{n\to m}(t) = \left[\int \Phi_m^* \mu_e \Phi_n d\tau\right] e^{\frac{i(E_m - E_n)t}{\hbar}}$$

the dipole moment $\alpha E_0 \cos\omega_o t$, induced by the electric field of the incident radiation, is included:

$$R_{n\to m}(t) = \left[E_o \int \Phi_m^* \alpha \Phi_n d\tau\right] e^{-i\left(\omega_o - \frac{E_m - E_n}{\hbar}\right)t}. \qquad (10.70)$$

Again interpreting this expression as a kind of microscopic source of radiation somewhat equivalent to irradiating dipoles, one sees that lines at the frequencies $\omega_o \pm \omega_{mn}$ have to be expected.

The amplitude of the matrix element of the *polarizability* in Eq. (10.70) controls the strength of the Raman components and therefore the selection rules. By referring for simplicity to scalar polarizability, in a first order approximation the analogous of Eq. (10.64) can be written

$$\alpha = \alpha_o + \sum_i \left(\frac{\partial \alpha}{\partial Q_i}\right)_o Q_i \qquad (10.71)$$

Thus to have Raman radiation the conditions

$$\left(\frac{\partial \alpha}{\partial Q_i}\right)_o \neq 0 \qquad (10.72)$$

$$v_i^m = v_i^n \pm 1 \qquad (10.73)$$

must be fulfilled (see Eqs. (10.66) and (10.67)).

Going back to Fig. 10.6 now one realizes that the S mode, which is not active in direct infrared absorption, can give Raman diffusion and *vice versa*.[1]

Raman spectroscopy can also be used to study the rotational motions. In this case the fundamental aspect to pay attention to is the tensorial character of the molecular polarizability. The rotation of the molecule implies the rotation of the frame of reference Σ^P of the principal axes of the polarizability tensor $\bar{\bar{\alpha}}$, thus modulating the component along the direction of the electric field in the laboratory frame Σ^L, as sketched below

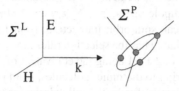

Therefore the incident radiation interacts with a time-dependent molecular polarizability, "modulated" at a frequency $2\nu_{rot}$ (the tensor being symmetric). For a molecule to be active in rotational Raman spectroscopy is not required to have a dipole moment. Any molecule not spherically symmetric and thus having *anisotropic polarizability*, is Raman active, in principle. The selection rule in terms of the quantum number K is $\Delta K = 0, \pm 2$, according to parity arguments, at variance with the selection rules 10.11 for the direct electric dipole transition between rotational states.

Problems

Problem 10.17 For a gas of diatomic molecules the roto-vibrational energy diagram is sketched below

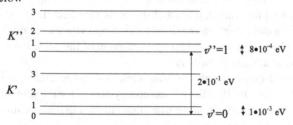

Figure out which lines can be detected in infrared and in Raman spectroscopies when the two nuclei are non-identical (heteronuclear molecules).

What do you expect if the molecule has two identical nuclei, with spin $I = 0$ or $I = 1/2$?

Solution: The solution follows directly from Fig. 10.5 and from the selection rules $\Delta K = \pm 1$, $\Delta v = \pm 1$ for infrared absorption and $\Delta K = \pm 2$ for Raman lines. For identical nuclei read Sect. 10.9.

---

[1] This statement regarding the alternative role of symmetric and antisymmetric modes in Raman and infrared activity is a general one, holding in any molecule with *inversion symmetry*. It is related to the fact that the polarizability upon inversion transforms as a second order tensor while the dipole moment is a vector.

10.8 Electronic Spectra and Franck—Condon Principle

The *band spectra* or the *electronic spectra* (usually in the visible or in the UV ranges of the electromagnetic spectrum (see Appendix 1.1)) associated with simultaneous transitions between electronic and roto-vibrational states in molecules involve rather complex selection rules.

If no coupling to the nuclear motions is taken into account, the selection rules for transitions among electronic states can be directly obtained by referring to the matrix elements for the electric dipole operator $-\sum_l e\mathbf{r}_l$ and to the symmetry properties. In the **LS** scheme and weak spin-orbit interaction, from factorization of the total eigenfunction one immediately has the selection rules $\Delta S = 0$ and $\Delta M_S = 0$, while for the spatial part $\Delta \Lambda = 0, \pm 1$ (Λ being the equivalent of M_L for atoms along the molecular axis). By referring to diatomic molecules, from parity arguments for the components of the total electric dipole, the following rules are derived:

$$\Sigma^+ \to \Sigma^+ \quad \text{and} \quad \Sigma^- \to \Sigma^- \quad \text{allowed}$$

$$\Sigma^+ \to \Sigma^- \quad \text{and} \quad \Sigma^- \to \Sigma^+ \quad \text{forbidden.}$$

In fact, $< \phi_{fin}|z|\phi_{in} > \neq 0$ for the first case, while it must be zero in the second case in order to not change sign upon reflection with respect to a plane containing the molecular axis. Also $< \phi_{fin}|x|\phi_{in} >$ and $< \phi_{fin}|y|\phi_{in} >$ must be zero, otherwise they would change sign upon reflection with respect to the xz and yz planes. $\sum_l \mathbf{r}_l$ being an ungerade operator, it is evident that the $g \to u$ transitions are allowed, while the $u \to u$ and $g \to g$ transitions are forbidden for homonuclear molecules having equal nuclei for charge, mass and spin state (see Sect. 10.9). In the presence of coupling of the electronic states with rotational and vibrational motions (with the related so-called *vibronic* transitions) the derivation of the selection rules becomes really complex, as already mentioned.

In diatomic molecules we only illustrate a relevant and general aspect: *the Franck-Condon principle*. In Fig. 10.7 the typical energy curves for the ground and the first excited states are sketched and some transitions involving the vibrational states are indicated.

The classical description of the principle (given by Franck) was based on the following arguments. The nuclei-electron coupling is weak, the electronic transitions occur in very short times (typically $10^{-15} \div 10^{-16}$ s in comparison to the typical periods, around 10^{-13} s, of the vibrational motions). Therefore the interatomic distance can hardly change while the electrons are carried from one electronic state to the other. Since for the classical oscillator the probability to find the atoms at a given distance is large in correspondence to the maxima elongations, it is conceivable to expect a certain prevalence of the end-to-end transitions, as the one indicated in Fig. 10.7 by the arrow on the right side.

The basic aspect of the quantum description is outlined hereafter. The transition probability is controlled by the matrix element

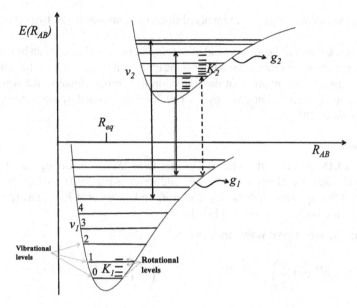

Fig. 10.7 Energy curves for the electronic ground state g_1 and the first excited state g_2 and sketches of the transitions involving the vibrational v' and v'' levels. The *solid lines* indicate transitions with large Franck-Condon factors, at variance with the classical prediction. The *dashed line* refers to a transition with small Franck-Condon factor

$$\mathbf{R}_{g_1 \to g_2, v_1 \to v_2, K_1 \to K_2} = \int \phi_e^{g_2*} \phi_{vib}^{v_2*} \phi_{rot}^{K_2*} \left[\boldsymbol{\mu}_{ele} + \boldsymbol{\mu}_N \right] \phi_e^{g_1} \phi_{vib}^{v_1} \phi_{rot}^{K_1} d\tau_e d\tau_N$$

$$(10.74)$$

with $\boldsymbol{\mu}_{ele} = -e \sum_i \mathbf{r}_i$ and $\boldsymbol{\mu}_N = e \sum_\alpha Z_\alpha \mathbf{R}_\alpha$. The rotational part of the wavefunction involves only the angles θ and ϕ and therefore it can be considered separately. Thus one is left with

$$\mathbf{R}_{g_1 \to g_2, v_1 \to v_2} = \int \phi_e^{g_2*} \phi_{vib}^{v_2*} \left[\boldsymbol{\mu}_{ele} + \boldsymbol{\mu}_N \right] \phi_e^{g_1} \phi_{vib}^{v_1} d\tau_e d\tau_N \qquad (10.75)$$

This term can be separated in two, the one involving $\boldsymbol{\mu}_N$ being zero since the electronic wavefunctions for g_1 and g_2 are orthogonal. Then only the term involving $\boldsymbol{\mu}_e$ has to be considered and by assuming that the electronic wavefunctions are only slightly modified when the interatomic distance is varied, the matrix element is written

$$\mathbf{R}_{g_1 \to g_2, v_1 \to v_2} = \int \phi_{vib}^{v_2*} \phi_{vib}^{v_1} d\tau_N \int \phi_e^{g_2*} \left[\boldsymbol{\mu}_{ele} \right] \phi_e^{g_1} d\tau_e = S_{FC} \int \phi_e^{g_2*} \left[\boldsymbol{\mu}_{ele} \right] \phi_e^{g_1} d\tau_e.$$

$$(10.76)$$

Thus the matrix element appears as the usual electronic term multiplied by the *Franck-Condon factor* S_{FC}. For $v_1 \neq v_2$ S_{FC} can be different from zero since two different electronic states are involved in correspondence to the two vibrational levels.

The Franck-Condon factor is a kind of *overlap integral* and now it can be realized why for large quantum vibrational numbers the empirical formulation of the principle is again attained. The intensity of the transition line, proportional to the square of the transition dipole moment given by Eq. (10.76), is controlled by the factor $|S_{FC}|^2$ (see Problem 10.18).

Problems

Problem 10.18 Evaluate the Franck-Condon term $|S_{FC}|^2$ involving two $v = 0$ vibrational states for electronic states g_1 and g_2 (see Fig. 10.7) having the same curvature at the equilibrium distances, one at R_e and the other at $R_e + \Delta R_e$ (problem inspired by the book of Atkins and Friedman).

Solution: The vibrational wavefunctions are

$$\phi_0^{(1)} = \left(\frac{b}{\pi}\right)^{1/4} e^{-bQ^2/2}, \quad \phi_0^{(2)} = \left(\frac{b}{\pi}\right)^{1/4} e^{-b(Q-\Delta R_e)^2/2}$$

where $b = \mu\omega_0/\hbar$ (see Eq. (10.32)). The overlap integral is

$$S_{FC}(0,0) = \left(\frac{b}{\pi}\right)^{1/2} \int_{-\infty}^{+\infty} e^{-b(Q^2/2)-[b(Q-\Delta R_e)^2/2]} dQ =$$

$$= \left(\frac{b}{\pi}\right)^{1/2} e^{-b(\Delta R_e/2)^2} \int_{-\infty}^{+\infty} e^{-b(Q-\Delta R_e/2)^2} dQ = e^{-b(\Delta R_e/2)^2}$$

and

$$|S_{FC}|^2 = e^{-b(\Delta R_e)^2/2},$$

as plotted below:

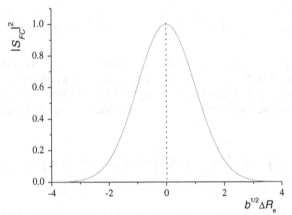

10.9 Effects of Nuclear Spin Statistics in Homonuclear Diatomic Molecules

Now we turn to an impressive demonstration of quantum principles, without any classical counterpart: the influence of the nuclear spins on the statistics, the related selection of molecular states and the occurrence of *zero-temperature rotations*.

Let us consider the total wavefunction of a homonuclear diatomic molecule

$$\phi_T = \phi_e \, \phi_{vib} \, \phi_{rot} \, \chi_{spin} \tag{10.77}$$

upon exchange of the nuclei, each having nuclear spin I. The total number of spin wavefunctions is $(2I+1)^2$. $(2I+1)$ of them are symmetric, since the magnetic quantum numbers m_I are the same for both nuclei. Half of the remaining wavefunctions are symmetric and half antisymmetric. Thus $[(2I + 1)^2 - (2I + 1)]/2 + (2I + 1) = (I + 1)(2I + 1)$ are *symmetric* and the remaining $I(2I + 1)$ *antisymmetric*. Therefore the ratio of the *ortho* (symmetric) to *para* (antisymmetric) molecules is $(N_{para}/N_{ortho}) = I/(I + 1)$. For example, for Hydrogen 75 % of the molecules belong to orthohydrogen type and 25 % to parahydrogen.

Let P indicate the operator exchanging spatial and spin coordinates of the nucleus A with the ones of the nucleus B. One has $P\phi_T = +\phi_T$ for nuclei with integer I (*bosons*) while $P\phi_T = -\phi_T$ for nuclei with half integer spin (*fermions*).

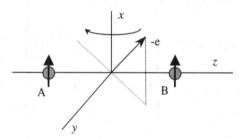

For the electronic wavefunction ϕ_e the exchange of the nuclei is equivalent to:

(i) rotation by 180 degrees around the x axis;
(ii) inversion of the electronic coordinates with respect to the origin;
(iii) reflection with respect to the yz plane.

For the most frequent case of electronic ground state Σ_g^+ one concludes

$$P\phi_e = +\phi_e \tag{10.78}$$

The vibrational wavefunction is evidently symmetrical, i.e. $P\phi_{vib} = +\phi_{vib}$, since it depends only on $(R - R_e)$. For the rotational wavefunctions one has

$$P\phi_{rot} = (-1)^K \phi_{rot} \tag{10.79}$$

namely they are symmetrical (positive parity) upon rotation when the number K is even while are antisymmetric (negative parity) the ones having odd rotational numbers K. By taking into account Eqs. (10.77)–(10.79) one deduces the requirement that for half integer nuclear spin (total wavefunction antisymmetric upon exchange of the nuclei) *ortho molecules* (having symmetric spin functions) can be found only in rotational states with odd K. On the contrary *para molecules* (having antisymmetric spin functions) can be found only in rotational states with K even. For integer nuclear spins the propositions are inverted. It is noted that ortho to para transitions are hardly possible, for the same argument used to discuss the (almost) lack of transitions from singlet to triplet states in the Helium atom (see Sect. 2.2).

A relevant spectroscopic consequence of the symmetry properties in diatomic molecules, for instance, is the fact that in Raman spectra in H_2 the lines associated to transitions starting from rotational states at K odd (see the illustrative plot in the following figure) are approximately three time stronger than the ones involving the states at K even, once that thermal equilibrium is established between the two species ortho and para (see Problem 10.19). For D_2 an opposite alternation in the intensities occurs (by a factor of two), the nuclear spin of deuterium being $I = 1$.

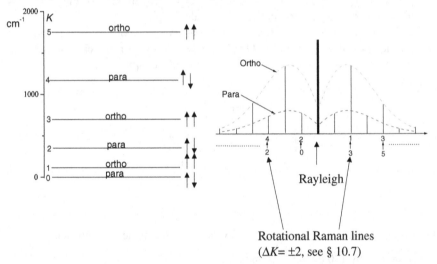

Rotational Raman lines
($\Delta K = \pm 2$, see § 10.7)

For O_2, the electronic ground state being $^3\Sigma_g^-$, the nuclear spin for ^{16}O is zero and χ_{spin} is necessarily symmetric. Then only odd K states are allowed. Thus only the rotational lines corresponding to $\Delta K = \pm 2$ and involving odd K states are observed in Raman spectroscopy (see Fig. 10.8). If of the nuclei is substituted by its isotope ^{17}O, all the rotational lines are detected.

For N_2, the nuclear spin being $I = 1$, the roto-vibrational structure shows the same alternation in the intensities expected for D_2 (see Fig. 10.9).

Analogous spectroscopic effects are observed in polyatomic molecules having inversion symmetry, such as CO_2 or C_2H_2. In particular CO_2 is a nice counterpart of the O_2 molecule. While in this latter the ground state is $^3\Sigma_g^-$, in CO_2 the ground

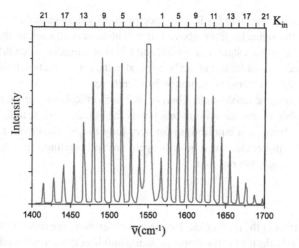

Fig. 10.8 Raman spectrum of $^{16}O_2$ displaying the rotational structure. At $\bar{\nu} \simeq 1556$ cm^{-1} the Q branch (Sect. 10.5), for $\Delta K = 0$, is observed (broad line). The lines with even K are missing (see the book by Haken and Wolf (2004))

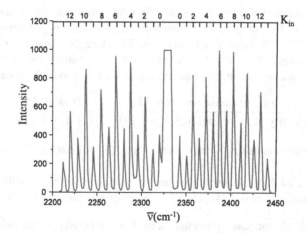

Fig. 10.9 Roto-vibrational Raman spectrum in N_2 (for ^{14}N -^{14}N). The alternation in the line intensities is in the ratio 1:2 (see the book by Haken and Wolf). The symmetry of the wavefunction does not change in the Raman transitions $\Delta K = 0, \pm 2$, as well as for the spin function. It can be mentioned that the bosonic character of ^{14}N nucleus has been claimed for the first time by *Heitler* and *Herzberg* in 1929 just from the alternation in the line intensities, before the discovery of the neutron which three years later explained why $I = 1$

electronic state is $^1\Sigma_g^+$. Thus in view of the change of the symmetry with respect to the xz plane, in CO_2 a variety of spectroscopic studies has evidenced that only the rotational states at even K do occur.

It should be stressed that for Raman spectroscopy, where virtual electronic states are involved, the remarks given above imply that these states retain the symmetry properties of the ground state. For optical and UV transitions between different electronic states these considerations can be applied to the roto-vibrational fine structure. For detail see the extensive presentation by *Herzberg*.

As a final remark one should observe that for ortho-Hydrogen molecules the lowest accessible rotational state in practice is the one at $K = 1$, unless one waits for the thermodynamical equilibrium for very long times. Thus even at the lowest temperature the molecules (in solid hydrogen) are still rotating. This is an example of the so called *quantum rotators*.

Problems

Problem 10.19 For the molecules $^3\text{He}_2$ and $^4\text{He}_2$ (existing in excited states, assumed of $\sum_g^+$ character), derive the rotational quantum numbers that are allowed. By assuming thermal equilibrium, obtain the ratio of the intensities of the absorption lines in the roto-vibrational spectra, at high temperature.

Solution: The nuclear spin for ^3He is $I = 1/2$ while for ^4He is $I = 0$. Therefore in $^3\text{He}_2$ only states at K even are possible for total spin 0 and only states at K odd for total spin 1. The intensity of the lines for $E_K \ll k_B T$ is proportional to the degeneracy $e^{-E_K/k_B T}$ being practically unit. Therefore, by remembering that the rotational degeneracy is $(2K + 1)$, while $(2I + 1)(I + 1)$ spin states are symmetric and $(2I + 1)I$ are antisymmetric, one can expect for the ratio of the intensities

$$\frac{\text{Intensity of transitions from } 2K}{\text{Intensity of transitions from } (2K - 1)} = \frac{I(4K + 1)}{(I + 1)(4K - 1)}.$$

For large rotational numbers the ratio reduces to $I/(I+1)$, namely 1:3, as discussed at Sect. 10.9.

For $^4\text{He}_2$, I being 0, no antisymmetric nuclear spin functions are possible, only the rotational states at $K = 0, 2, 4...$ are allowed and every other line is absent.

Problem 10.20 In the assumption that in the low temperature range only the rotational states at $K \le 2$ contribute to the rotational energy of the H_2 molecule, derive the contribution to the molar specific heat.

Solution: The ortho-molecules (on the $K = 1$ state) in practice cannot contribute to the increase or the internal energy U_{rot} upon a temperature stimulus. Then, only the partition function Z_{rot}^{para} of the para-molecules (on the $K = 0$ and $K = 2$ rotational states) has to be considered in

$$U_{rot} = N k_B T^2 \frac{d}{dT} \ln Z_{rot}.$$

From $\sum_K (2K + 1) \exp[-K(K + 1)\theta_{rot}/T]$ (with $\theta_{rot} \equiv \hbar^2/2Ik_B \simeq 87$ K in H_2), $Z_{rot}^{para} \simeq 1 + 5 \exp[-6\theta_{rot}/T]$. Then $U_{rot} \simeq Nk_B 30\theta_{rot} \cdot \exp[-6\theta_{rot}/T]$.

Since the number of para-molecules in a mole can be considered $N_A/4$,

$$C_V \simeq N_A/4 \cdot 180(\theta_{rot}/T)^2 \exp[-6\theta_{rot}/T].$$

Problem 10.21 Estimate the fraction of para-Hydrogen molecules in a gas of H_2 at temperature around the rotational temperature $\theta_{rot} = 87$ K and at $T \simeq 300$ K, in the assumption that thermal equilibrium has been attained. (Note that after a thermal jump it may take very long times to attain the equilibrium, see text). Then evaluate the fraction of para molecules in D_2 at the same temperature.

Solution: For $T \simeq 300$ K the fraction of para molecules is controlled by the spin statistical weights (see Problem 10.19). Thus, for H_2, $f_{para} \simeq 1/(1 + 3) \simeq 0.25$.

Around the rotational temperature one writes

$$f_{para} = \frac{\sum_{K even}(2K + 1) \exp[-K(K + 1)\theta_{rot}/T]}{Z_{para} + Z_{ortho}} \simeq$$

$$\simeq \frac{1 + 5e^{-6} + \cdots}{1 + 5e^{-6} + \cdots + 3[3e^{-2} + 7e^{-12} + \cdots]} \simeq 0.46$$

For D_2 the rotational temperature is lowered by a factor 2 and the spin statistical weights are $I + 1 = 2$ for K even and $I = 1$ for states at K odd. At $T \simeq 300$ K $f_{para} = 0.33$. At $T \simeq 87$ K one writes

$$f_{para} = \frac{\sum_{K odd}(2K + 1) \exp[-K(K + 1)\theta_{rot}/2T]}{2\sum_{K even}\cdots + 1\sum_{K odd}\cdots} \simeq \frac{3 \exp[-\theta_{rot}/T]}{2 + 1[3e^{-1}]} \simeq 0.35$$

Problem 10.22 Derive the temperature dependence of the mean square amplitude $< (R - R_e)^2 > \equiv < Q^2 >$ of the vibrational motion in a diatomic molecule of reduced mass μ and effective elastic constant k. Then evaluate the mean square amplitude of the vibrational motion for the $^1H^{35}Cl$ molecule at room temperature, knowing that the fundamental absorption frequency is $\nu_0 = 2990$ cm^{-1}.

Solution: From the virial theorem $< E > = 2 < V >$ and then $< E > = 2 \cdot (k/2) < Q^2 > \equiv \mu\omega^2 < Q^2 >$ (see also Eq. (10.34)).

For the thermal average

$$< \overline{E} > = \hbar\omega \left(\frac{1}{2} + < n >\right), \quad \text{with} \quad < n > = 1/(e^{\frac{\hbar\omega}{k_B T}} - 1)$$

one writes

$$< \overline{Q^2} > = \frac{\hbar}{\mu\omega} \left(\frac{1}{2} + \frac{1}{e^{\hbar\omega/k_B T} - 1}\right).$$

For $k_B T \gg \hbar\omega$,

$$< \overline{Q^2} > \simeq const + \frac{k_B T}{\mu\omega^2}.$$

(see plot in Problem 10.12).

For $h\nu_0 \gg k_B T$ $< \overline{Q^2} > \simeq (1/2)h\nu_0/k$ and $\sqrt{< \overline{Q^2} >} \simeq \sqrt{h/8\pi^2\nu_0\mu} \simeq$ 0.076 Å.

Problem 10.23 At room temperature and at thermal equilibrium condition the most populated rotational level for the CO_2 molecule is found to correspond to the rotational quantum number $K_M = 21$. Estimate the rotational constant B.

Solution: From Eq. (10.14), by deriving with respect to K one finds

$$2K_M + 1 = \left(\frac{2k_B T}{Bhc}\right)^{1/2}$$

Then $Bhc = 2k_B T/(2K_M + 1)^2 = 4.48 \cdot 10^{-17}$ erg.

Problem 10.24 When a homogeneous and static electric field $\mathcal{E} = 1070$ V cm^{-1} is applied to a gas of the linear molecule OCS the rotational line at 24,325 MHz splits in a doublet, with frequency separation $\Delta\nu = 3.33$ MHz. Evaluate the rotational eigenvalues for $K = 1$ and $K = 2$ in the presence of the field, single out the transitions originating the doublet and derive the electric dipole moment of the molecule.

Solution: From Eq. (10.20)

$$E(1, M, \mathcal{E}) = 2Bhc + \mu^2\mathcal{E}^2(2 - 3M^2)/20Bhc$$
$$E(2, M, \mathcal{E}) = 6Bhc + \mu^2\mathcal{E}^2(2 - M^2)/84Bhc$$

The transitions at $\Delta K = \pm 1$ and $\Delta M = 0$ yield the frequencies $\nu = \nu_0 + \delta\nu(M)$, where $\nu_0 = 4Bc$ while the correction due to the field is

$$\delta\nu(M) = \frac{\mu^2\mathcal{E}^2}{Bh^2c}\frac{(29M^2 - 16)}{210}$$

Then $\delta\nu(0) = -(8/105)(\mu^2\mathcal{E}^2)/(Bh^2c)$, $\delta\nu(1) = (13/210)(\mu^2\mathcal{E}^2)/(Bh^2c)$ and the separation between the lines is

$$\Delta\nu = \delta\nu(1) - \delta\nu(0) = \frac{29}{210}\frac{\mu^2\varepsilon^2}{Bh^2c} = \frac{58}{105}\frac{\mu^2\mathcal{E}^2}{h^2\nu_0}$$

The dipole moment of the molecule turns out

$$\mu_e = \frac{h}{\mathcal{E}}\sqrt{\frac{105}{58}\nu_0\Delta\nu} = 2.37 \cdot 10^{-21} \text{ erg cm V}^{-1} = 0.71 \text{ Debye}$$

Problem 10.25 Derive an approximate expression, valid in the low temperature range, for the rotational contribution to the specific heat of a gas of HCl molecules.

Solution: At low temperature only the first two rotational levels E_0 and E_1 can be considered and the rotational partition function is written

$$Z \simeq 1 + 3e^{-\frac{E_1}{k_B T}}.$$

The energy is $U_{rot} = -\partial \ln Z / \partial \beta$, $(\beta = 1/k_B T)$ and then

$$U_{rot}(T \to 0) = 3E_1 e^{-E_1/k_B T}$$

and the specific heat (per molecule) becomes

$$(C_V)_{T \to 0} = \frac{3E_1^2}{k_B T^2} e^{-E_1/k_B T}.$$

This expression can actually be used only for

$$k_B T \ll \frac{\hbar^2}{\mu R_e^2} \equiv 2Bhc, \quad \text{where} \quad B = 10.6 \, \text{cm}^{-1}$$

(see Sect. 10.2.2), i.e. for $T_{val} \ll 30 \, \text{K}$.

Problem 10.26 Derive the longitudinal normal modes for the system sketched below, assuming that the force constant of the spring in between the two masses is twice the ones for the lateral springs, that are stuck at fixed points (the springs have negligible mass). (Problem suggested in the book by Eyring, Walter and Kimball).

Solution: In terms of local coordinates (see Sect. 10.6)

$$T = \frac{M}{2}(\dot{q}_1^2 + \dot{q}_2^2), \qquad V = \frac{k}{2}q_1^2 + \frac{2k}{2}(q_2 - q_1)^2 + \frac{k}{2}q_2^2$$

The equations of motion in Lagrangian form are

$$M\ddot{q}_1 + 3kq_1 + 2kq_2 = 0 \quad \text{and} \quad M\ddot{q}_2 + 3kq_2 - 2kq_1 = 0$$

By multiplying by $c_{1,2}$ and summing

$$M(c_1\ddot{q}_1 + c_2\ddot{q}_2) + q_1(3kc_1 - 2kc_2) + q_2(-2kc_1 + 3kc_2) = 0$$

The normal form in terms of the coordinates Q_i is obtained for

$$Mc_1 = \frac{1}{\lambda}(3kc_1 - 2kc_2) = h_1 \text{ and } Mc_2 = \frac{1}{\lambda}(-2kc_1 + 3kc_2) = h_2$$

yielding the secular equation

$$\begin{vmatrix} \lambda M - 3k & 2k \\ 2k & \lambda M - 3k \end{vmatrix} = 0,$$

with roots $\lambda_1 = 5k/M$ (implying $c_1 = -c_2$) and $\lambda_2 = k/M$ ($c_1 = c_2$), so that

$$Q_1 = (Mq_1 - Mq_2) \text{ and } Q_2 = (Mq_1 + Mq_2)$$

The equations of motion in the normal form are

$$\ddot{Q}_1 + \frac{5k}{M}Q_1 = 0, \quad \ddot{Q}_2 + \frac{k}{M}Q_2 = 0$$

One normal mode corresponds to $q_1 = q_2$, and frequency $\omega_2 = \sqrt{k/M}$, while the second one corresponds to $q_1 = -q_2$ and frequency $\omega_1 = \sqrt{5k/M}$.

Problem 10.27 For the NH_3 the rotational temperatures are $\Theta_A = \Theta_B = 14.3$ K and $\Theta_C = 9.08$ K (oblate rotator) (see Sect. 10.2.5). Evaluate the molar rotational energy and the specific heat at room temperature (neglect the effects of the nuclear spin statistics).

Solution: From

$$E(K, M) = k_B \Theta_B K(K + 1) + k_B M^2 (\Theta_C - \Theta_A)$$

the rotational partition function can be written

$$Z_{rot}(T) = \sum_{K=0}^{\infty} \sum_{M=-K}^{K} (2K + 1)exp[-\Theta_B K(K + 1)/T - M^2(\Theta_C - \Theta_A)/T]$$

and by changing the sums to integrals and recalling that

$$U(T) = RT^2 \frac{d}{dT} \ln Z_{rot}$$

one has $U(T = 300) \simeq (3/2)RT$ and $C_V \simeq (3/2)R$.

It can be remarked that when the effects of the spin statistics are taken into account the high temperature partition function should be scaled according to the amount of the occupied rotational levels.

Specific References and Further Reading

1. S. Svanberg, *Atomic and Molecular Spectroscopy*, (Springer Verlag, Berlin, 2003).
2. H. Eyring, J. Walter and G.E. Kimball, *Quantum Chemistry*, (J. Wiley, New York, 1950).
3. P.W. Atkins and R.S. Friedman, *Molecular Quantum Mechanics*, (Oxford University Press, 252 Oxford, 1997).
4. G. Herzberg, *Molecular Spectra and Molecular Structure, Vol. I, II and III*, (D. Van Nostrand, New York, 1964–1966, reprint 1988–1991).
5. H. Haken and H.C. Wolf, *Molecular Physics and Elements of Quantum Chemistry*, (Springer Verlag, Berlin, 2004).
6. B.H. Bransden and C.J. Joachain, *Physics of atoms and molecules*, (Prentice Hall, 2002).
7. J.A. Cronin, D.F. Greenberg, V.L. Telegdi, *University of Chicago Graduate Problems in Physics*, (Addison-Wesley, 1967).
8. W. Demtröder, *Molecular Physics*, (Wiley-VCH, 2005).
9. R.N. Dixon, *Spectroscopy and Structure*, (Methuen and Co LTD, London, 1965).

Chapter 11
Crystal Structures

Topics

Elementary Crystallography
Translational Invariance
Reciprocal Lattice
The Bragg Law
Brillouin Zone
Typical Crystal Structures

In this chapter and in the following three chapters we shall be concerned with the general aspects of the solid state of the matter, namely the atomic arrangements where the interatomic interactions are strong enough to keep the atoms bound at well defined positions. We will address the bonding mechanisms leading to the formation of the crystals, the electronic structure and the vibrational dynamics of the atoms. The liquid and solid states are similar in many respects, for instance in regards of the density, short range structure and interactions. The difference between these two states of the matter relies on the fact that in the former the thermal energy is larger than the cohesive energy and the atoms cannot keep definite equilibrium positions.

Before the advent of quantum mechanics the solid-state physics was practically limited to phenomenological descriptions of macroscopic character, thus involving quantities like the compressibility, electrical resistivity or other mechanical, dielectric, magnetic and thermal constants. After the application of quantum mechanics to a model system of spatially ordered ions (the *crystal lattice*, indicated by *Laue X-ray diffraction* experiments) quantitative studies of the microscopic properties of solids began.

© Springer International Publishing Switzerland 2015
A. Rigamonti and P. Carretta, *Structure of Matter*,
UNITEXT for Physics, DOI 10.1007/978-3-319-17897-4_11

During the last forty years the study of the condensed matter has allowed one to develop the *transistors*, the *solid state lasers*, novel devices for *opto-electronics*, the *SQUID, superconducting magnets* based on new materials, *etc.* As regards the development of the theory, solid state physics has triggered monumental achievements for *many-body systems*, such as the theories for *superconductivity* or of *quantum magnetism* for strongly correlated electrons, as well as the explanation of the *fractional quantum Hall effect.*

Besides the spatially ordered crystalline structures there are other types of solids, as polymers, amorphous and glassy materials, Fibonacci-type quasi-crystals, which are not characterized by regular arrangement of the atoms. Our attention shall be devoted to the simplest model, the *ideally perfect crystal*, with no defects and/or surfaces, where the atoms occupy spatially regular positions granting *translational invariance*. In the Chap. 1 we shall present some aspects of elementary crystallographic character in order to describe the crystal structures and to provide the support for the quantum description of the fundamental properties. Many solid-state physics books (and in particular the texts by *Burns*, by *Kittel* and by *Aschcroft* and *Mermin*) report in the introductory chapters more complete treatments of *crystallography*, the "geometrical" science of crystals.

11.1 Translational Invariance, Bravais Lattices and Wigner-Seitz Cell

In an ideal crystal the physical properties found at the position **r**

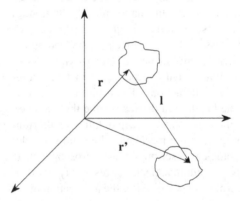

are also found at the position $\mathbf{r}' = \mathbf{r} + \mathbf{l}$, where

$$\mathbf{l} \equiv m\mathbf{a} + n\mathbf{b} + p\mathbf{c} \tag{11.1}$$

with m, n, p integers and **a**, **b**, **c** *fundamental translational vectors* which characterize the crystal structure.

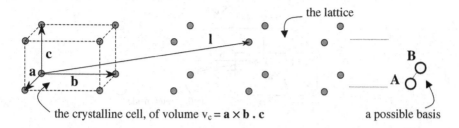

the crystalline cell, of volume $v_c = \mathbf{a} \times \mathbf{b} \cdot \mathbf{c}$ a possible basis

This property is called *translational symmetry* or *translational invariance*. As we shall see in Chap. 12, it is a symmetry property analogous to the ones utilized for the electronic states in atoms and molecules.

The extremes of the vectors $\mathbf{l}$, when the numbers m, n, p in Eq. (11.1) are running, identify the points of a geometrical network in the space, called *lattice*. By placing at each lattice point an atom or an identical group of atoms, called the *basis*, the real crystal is obtained. Thus one can ideally write *crystal = lattice + basis*.

The lattice and the fundamental translational vectors $\mathbf{a}, \mathbf{b}, \mathbf{c}$ are called *primitive* when Eq. (11.1) holds for any arbitrary pair of lattice points. Accordingly, in this case one has the maximum density of lattice points and the basis contains the minimum number of atoms, as it can be realized from the sketchy example reported below for a two-dimensional lattice:

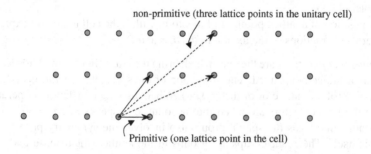

The geometrical figure resulting from vectors $\mathbf{a}, \mathbf{b}, \mathbf{c}$ is called the *crystalline cell*. The lattice originates from the repetition in space of this fundamental *unitary cell* when the numbers m, n and p run. The unitary cell is called *primitive* when it is generated by the primitive translational vectors. The primitive cell has the smallest volume among all possible unitary cells and it contains just *one lattice point*. Therefore it can host one basis only.

Instead of referring to the cell resulting from the vectors $\mathbf{a}, \mathbf{b}, \mathbf{c}$ one can equivalently describe the structural properties of the crystal by referring to the *Wigner-Seitz (WS) cell*. The WS cell is given by the region included within the planes bisecting the vectors connecting a lattice point to its neighbors, as in the example sketched below.

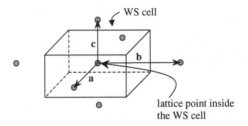

lattice point inside
the WS cell

The lattice points are then at the center of the WS cells.

The translation of the WS cell by all the vectors **l** belonging to the group T of the *translational operations* (see Eq. (11.1)) generates the whole lattice.

A few statements of geometrical character are the following:

1. The orientation of a plane of lattice points is defined by the *Miller indexes* (hkl), namely by the set of integers without common factors, inversely proportional to the intercepts of the plane with the crystal axes. The reason of such a definition will be clear after the discussion of the properties of the reciprocal lattice (Sect. 11.2).
2. A direction in the crystal is defined by the smallest integers $[hkl]$ having the same ratio of its components along the crystal axes. For example, in a crystal with a cubic unitary cell the diagonal is identified by $[111]$. One should observe that the direction $[hkl]$ is perpendicular to the plane having Miller indexes (hkl) (see Problem 11.1).
3. The position of a lattice point, or of an atom, within the cell is usually expressed in terms of fractions of the axial lengths a, b and c.

The *symmetry operations* are the ones which bring the lattice into itself, while leaving a particular lattice point fixed. The collection of the symmetry operations is called *point group* (of the lattice or of the crystal). When also the translational operations through the lattice vectors are taken into account, one speaks of *space group*. For non-monoatomic basis the spatial group also involves the symmetry properties of the basis itself. The point groups are groups in the mathematical sense and are at the basis of an elegant theory (the *group theory*) which can predict most symmetry-related properties of crystal just from the geometrical arrangement of the atoms.

Crystal System	Bravais Lattice	Unit Cell Dimensions
Triclinic	Primitive (P)	$a \neq b \neq c$ $\alpha \neq \beta \neq \gamma \neq 90°$
Monoclinic	Primitive (P) Base-Centered (C)	$a \neq b \neq c$ $\alpha = \gamma = 90° \neq \beta$
Orthorhombic	Primitive (P) Base-Centered (C) Body-Centered (I) Face-Centered (F)	$a \neq b \neq c$ $\alpha = \beta = \gamma = 90°$
Trigonal	R-Centered (R)	$a = b \neq c$ $\alpha = \beta = \gamma \neq 90° < 120°$
Hexagonal	Primitive (P)	$a = b \neq c$ $\alpha = \beta = 90° \ \gamma = 120°$
Tetragonal	Primitive (P) Body-Centered (I)	$a = b \neq c$ $\alpha = \beta = \gamma = 90°$
Cubic	Primitive (P) Body-Centered (I) Face-Centered (F)	$a = b = c$ $\alpha = \beta = \gamma = 90°$

The crucial point is that the requirement of *translational invariance limits the number of symmetry operations* that can be envisaged to define the crystal structures. To illustrate this restriction it is customary to recall that in a plane the unitary cell cannot be a pentagon (which is characterized by a rotational invariance after a rotation by an angle $2\pi/5$) since in that case one cannot achieve translational invariance.

In three dimensions (3D) there are *32 point groups* and *230 space groups* collecting all the symmetry operations compatible with translational invariance and with the symmetry of the basis. These groups define *14 fundamental lattices*, called the *Bravais lattices*. These lattices are shown in Fig. 11.1, where the *unitary conventional cell* generally used is indicated. It is noted that some cells might appear non-primitive, since there is more than one lattice point within them (see for instance the *bcc* lattice). However, one can easily identify the fundamental lattice vectors defining the

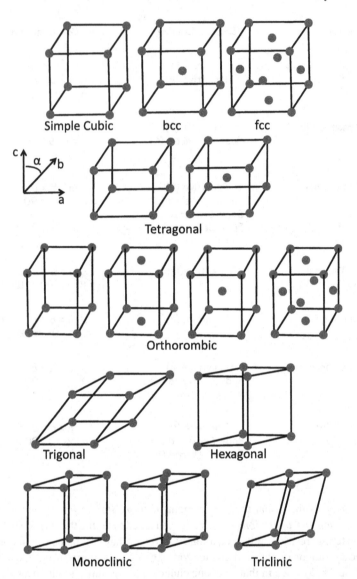

Fig. 11.1 Bravais crystal lattices with the conventional unitary cells, with the relations among the lattice lengths and among the characteristic angles (see table in the previous page)

primitive cell of the body-centered-cubic (*bcc*) Bravais lattice, in terms of the more frequently used non-primitive cubic lattice vectors **a**, **b**, **c** shown in the figure. For the analogous case of the *fcc* (face-centered cubic) lattice, see Fig. 11.4 and Problem 11.4.

11.2 Reciprocal Lattice and Brillouin Cell

As a consequence of the translational invariance in the ideal crystal, any local function $f(\mathbf{r})$ of physical interest (for instance, the energy or the probability of presence of electrons) must be spatially periodic, in other words invariant under the translation $\mathbf{T_l}$ by a vector belonging to the translational group:

$$\mathbf{T_l} f(\mathbf{r}) = f(\mathbf{r} + \mathbf{l}) = 1 \cdot f(\mathbf{r}). \tag{11.2}$$

Then one can abide by the Fourier expansion of $f(\mathbf{r})$ and by referring for simplicity to a crystal with orthogonal axes $\mathbf{a}$, $\mathbf{b}$ and $\mathbf{c}$ and choosing x, y and z along these axes, one writes

$$f(\mathbf{r}) = \sum_{-\infty \, n_x}^{+\infty} A_{n_x}(y, z) e^{[i n_x x (2\pi/a)]} = \sum_{-\infty \, n_x}^{+\infty} A_{g_x} e^{[i g_x x]},$$

where n_x is an integer and $g_x = n_x(2\pi/a)$ are reciprocal lattice lengths. The coefficients A_{n_x} can be Fourier-expanded along y and z and so one can put the function $f(\mathbf{r})$ in the form

$$f(\mathbf{r}) = \sum_{\mathbf{g}} A_{\mathbf{g}} e^{i \mathbf{g} \cdot \mathbf{r}} \tag{11.3}$$

where

$$A_{\mathbf{g}} = \frac{1}{v_c} \int_{-\infty}^{+\infty} f(\mathbf{r}) e^{-i \mathbf{g} \cdot \mathbf{r}} d\mathbf{r}, \tag{11.4}$$

v_c being the volume of the unitary cell. $\mathbf{g}$ is a *reciprocal lattice vector* built up by linear combination, with integer numbers $n_{x,y,z}$, of the *fundamental reciprocal vectors*, i.e.

$$\mathbf{g} = n_x (2\pi/a)\hat{x} + n_y (2\pi/b)\hat{y} + n_z (2\pi/c)\hat{z}. \tag{11.5}$$

It follows that for any reciprocal lattice vector $\mathbf{g}$ and for any translational vector $\mathbf{l}$, given by Eq. (11.1), one has

$$e^{i \mathbf{g} \cdot \mathbf{l}} = 1, \tag{11.6}$$

corresponding to the necessary and sufficient condition to allow the Fourier expansion of local functions.

The above arguments can be generalized for non-orthogonal crystal axes by defining the fundamental reciprocal vectors $\mathbf{a}^*$, $\mathbf{b}^*$ and $\mathbf{c}^*$ in the form

$$\mathbf{a}^* = \frac{2\pi}{(\mathbf{a} \times \mathbf{b} \cdot \mathbf{c})} (\mathbf{b} \times \mathbf{c}) = \frac{2\pi}{v_c} (\mathbf{b} \times \mathbf{c}),$$

$$\mathbf{b}^* = \frac{2\pi}{v_c} (\mathbf{c} \times \mathbf{a}),$$

$$\mathbf{c}^* = \frac{2\pi}{v_c} (\mathbf{a} \times \mathbf{b}). \tag{11.7}$$

The set of points, in the reciprocal space, reached by the vectors

$$\mathbf{g} = h\mathbf{a}^* + k\mathbf{b}^* + l\mathbf{c}^* \tag{11.8}$$

with h, k and l integers, defines the *reciprocal lattice*:

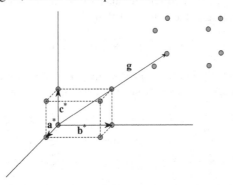

Instead of referring to the reciprocal lattice cell defined by $\mathbf{a}^*$, $\mathbf{b}^*$ and $\mathbf{c}^*$, it is often convenient to use its Wigner-Seitz equivalent, having a reciprocal lattice point at the center. This cell is called the *Brillouin cell* and it is shown schematically below for orthogonal axes:

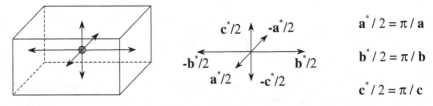

$$\mathbf{a}^* / 2 = \pi / \mathbf{a}$$

$$\mathbf{b}^* / 2 = \pi / \mathbf{b}$$

$$\mathbf{c}^* / 2 = \pi / \mathbf{c}$$

For instance, the Brillouin cell for the fcc lattice is obtained by taking eight reciprocal lattice vectors (bcc lattice, see Problem 11.4) bisected by planes perpendicular to such vectors and when the six next-shortest reciprocal lattice vectors are also bisected. This Brillouin cell is depicted in Fig. 11.2.

Fig. 11.2 Brillouin cell for fcc lattice

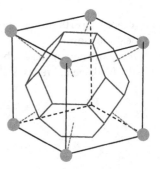

From the definitions of reciprocal lattice and of fundamental reciprocal vectors, one can derive the following properties (see Problem 11.1):

(i) $\mathbf{g}(h, k, l)$ is perpendicular to the planes with Miller indexes (hkl);
(ii) $|\mathbf{g}|$ is inversely proportional to the distance among the lattice planes (hkl).

The reciprocal lattice plays a relevant role in solid state physics. Its importance was first evidenced in diffraction experiments when it was noticed that each point of the reciprocal lattice corresponds to a diffraction spot. When the momentum of the electromagnetic wave (or of the De Broglie neutron wave) as a consequence of the scattering process changes by any reciprocal lattice vector, then the wave does not propagate through the crystal but undergoes Bragg reflection, as sketched below:

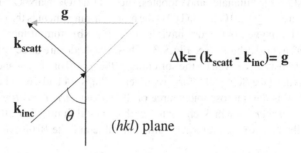

This condition corresponds to the Bragg law in the form

$$n\lambda = 2d\sin\theta \tag{11.9}$$

for the constructive interference of the radiation diffused by adjacent planes (d separation between the planes, $n = 1, 2, 3 \ldots$, X-ray beam incident at the angle θ the planes). In fact $\Delta\mathbf{k} = \mathbf{g}$ is equivalent to $2\pi/|\Delta\mathbf{k}| = d(hkl)$, while $|\mathbf{k}_{inc}| = |\mathbf{k}_{scatt}| = 2\pi/\lambda$ (for elastic scattering) and $\Delta\mathbf{k} = (4\pi/\lambda)\sin\theta$.

Furthermore, as we shall see at Chap. 12, the generators of the Brillouin cell, cut in a way related to the number of the cells in a reference volume, define the generators of a three-dimensional network in the reciprocal space. These vectors correspond to the wave-vectors of the excitations that can propagate through the crystal. Meantime they set the quantum numbers of the electron states.

11.3   Typical Crystal Structures

CsCl is the prototype of a family of cubic primitive (P) crystals with the basis formed by two atoms, one at position (0,0,0) and the other at (1/2,1/2,1/2). As sketched below the coordination number, i.e. the number of nearest neighbors around the Cs (or Cl) atoms is 8.

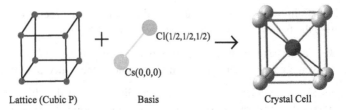

Other diatomic crystals with the same structure are TlBr, TlI, AgMg, AlNi and BeCu. Elements having the simple cubic (the basis being formed by one atom) Bravais lattice are P and Mn.

A group of interesting crystals having a P cubic lattice with a more complex basis are the perovskite-type titanates and niobates, such as $BaTiO_3$, $NaNbO_3$, $KNbO_3$. At high temperature ($T \geq 120$ C for $BaTiO_3$) the atomic arrangement is the one reported in Fig. 11.3. The oxygen octahedra having the Ti (or Nb) atom at the center result from the d^2sp^3 hybrid orbitals (see Fig. 9.3). These octahedra are directly involved in the structural transitions driven by the softening of the $q = 0$ or of the zone-boundary vibrational modes (see Sect. 10.6 for a comment, Chaps. 14 and 16). The distortion of the cubic cell is the microscopic source of the *ferroelectric transition* and of the *electro-optical properties* which characterize that crystal family. For all the crystal lattices described above the reciprocal lattice is cubic and the Brillouin cell is also cubic.

NaCl crystal is a typical example of face-centered cubic (fcc) lattice. The non-primitive, conventional, unitary cell and the primitive cell are shown in Fig. 11.4. The basis is formed by two atoms at the positions (0,0,0) and (1/2,0,0). The coordination number is 6. The fcc lattice characterizes also the structure of KBr, AgBr and LiH and of several metal elements such as Al, Ca, Cu, Au, Pb, Ni, Ag and Sr.

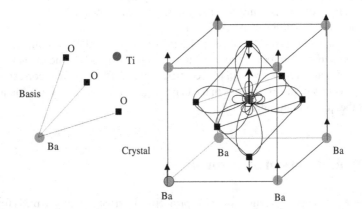

Fig. 11.3 Sketch of the crystal cell in BaTiO_3 (in the cubic phase). At $T_c \simeq 120$ C a displacive phase transition occurs, to a structure of tetragonal symmetry. The arrows indicate the directions of the displacements of the ions, having taken the oxygen ions at $c/2$ fixed (also a slight shrinkage in the ab plane occurs). The displacement of the positive and negative ions in opposite directions are responsible for the *spontaneous polarization* arising as a consequence of the transition from the cubic to the tetragonal phase (*ferroelectric state* see Chap. 16)

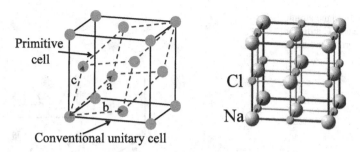

Fig. 11.4 Conventional and primitive cells for NaCl. The basis is formed by a Na atom and by a Cl atom

The fcc lattice also characterizes the *diamond* (C) and the semiconductors Si, Ge, GaAs and InSb. In these cases the basis is given by two atoms (both C for diamond, Si and Ge) at the positions (0,0,0) and (1/4,1/4,1/4). Each atom has a tetrahedral coordination that may be thought to result from the formation of sp^3 hybrid atomic orbitals (Sect. 9.2), as sketched below:

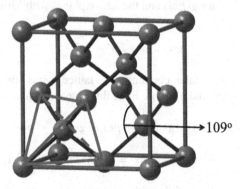

Carbon is known to crystallize also in the form of *graphite*, where the sp^2 hybridization of the C atomic orbitals yields a planar (2D) atomic arrangement. The 2D lattice is formed by two interpenetrating triangular lattices (see Fig. 11.5).

It should be mentioned that carbon can also crystallize in other forms, as for example in the fcc *fullerene*, where at each fcc lattice site there is a C_{60} molecule, with the shape of truncated icosahedron (a cage of hexagons and pentagons).

Another relevant crystalline form is the one having the *hexagonal close-packed* lattice, with the densest packing of hard spheres placed at the lattice points. The arrangement is obtained by placing the atoms at the vertexes of planar hexagons and then creating a second layer with "spheres" superimposed in contact with the three spheres of the underlying layer. The crystal lattice is the P hexagonal and the basis is given by two atoms placed at (0,0,0) and at (2/3,1/3,1/2).

In the hard sphere model 74 % of the volume is occupied and the ratio c/a is 1.633. In real crystals with this structure one has values of c/a slightly different, as 1.85 for Zn and 1.62 for Mg.

Fig. 11.5 In-plane atomic
arrangement of C atoms in
graphite, corresponding to
graphene

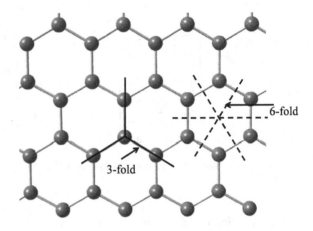

Problems

Problem 11.1 From geometrical considerations derive the relationships between
the reciprocal lattice vector $\mathbf{g}(hkl)$ and the lattice planes with Miller indexes (hkl).

Solution: For

$$\mathbf{g} = h\mathbf{a}^* + k\mathbf{b}^* + l\mathbf{c}^*.$$

let us take a plane perpendicular, containing the lattice points $m\mathbf{a}$, $n\mathbf{b}$ and $p\mathbf{c}$. Then,
since $m\mathbf{a} - n\mathbf{b}$, $m\mathbf{a} - p\mathbf{c}$ and $n\mathbf{b} - p\mathbf{c}$ lie in this plane, one has

$$\mathbf{g} \cdot (m\mathbf{a} - n\mathbf{b}) = \mathbf{g} \cdot (m\mathbf{a} - p\mathbf{c}) = \mathbf{g} \cdot (n\mathbf{b} - p\mathbf{c}) = 0.$$

Then $hm - kn = 0$, $mh = pl$ and $nk = pl$, yielding $m = 1/h$, $n = 1/k$ and
$p = 1/l$.

From the definition of the Miller indexes one finds that the plane perpendicular
to $\mathbf{g}$, passing through the lattice points $m\mathbf{a}$, $n\mathbf{b}$ and $p\mathbf{c}$ is the one characterized by
(hkl).

Now it is proved that the distance $d(hkl)$ between adjacent (hkl) planes is
$2\pi/|\mathbf{g}(hkl)|$. Let us consider a generic vector $\mathbf{r}$ connecting the lattice points of
two adjacent (hkl) planes. Since $\mathbf{g}(hkl)$ is perpendicular to these planes one has
$\mathbf{r} \cdot \hat{g}(hkl) = d(hkl)$. One can arbitrarily choose $\mathbf{r} = \mathbf{a}/h$. Then $\mathbf{a} \cdot \mathbf{g}(hkl) = 2\pi h$
and since $\hat{g} = \mathbf{g}/|\mathbf{g}|$ one has $\mathbf{r} \cdot \hat{g} = 2\pi/|g|$. Therefore

$$d(hkl) = \frac{2\pi}{|\mathbf{g}(hkl)|}$$

Problem 11.2 Derive the density of the following compounds from their crystal
structure and lattice constants:

Iron (bcc, $a = 2.86$ Å), Lithium (bcc, $a = 3.50$ Å), Palladium (fcc, $a = 3.88$ Å),
Copper (fcc, $a = 3.61$ Å), Tungsten (bcc, $a = 3.16$ Å).

Solution:

$$\text{Fe}: \qquad \rho = \frac{\text{atomic mass} \cdot 2}{v_c} = 7.93 \text{ g cm}^{-3}.$$

$$\text{Li}: \qquad \rho = \frac{2 \cdot 1.660 \cdot 10^{-24} \cdot 6.939}{(3.5 \cdot 10^{-8})^3} = 0.537 \text{ g cm}^{-3}.$$

$$\text{Pd}: \qquad \rho = 12.095 \text{ g cm}^{-3}.$$

$$\text{Cu}: \qquad \rho = 8.968 \text{ g cm}^{-3}.$$

$$\text{W}: \qquad \rho = 19.344 \text{ g cm}^{-3}.$$

Problem 11.3 Estimate the order of magnitude of the kinetic energy of the neutrons used in diffraction experiments to obtain the crystal structures. By assuming that the neutron beam arises from a gas, estimate the order of magnitude of the temperature required to have diffraction.

Solution: The neutron wavelength has to be of the order of the lattice spacing, i.e. of the order of 1 Å. Then $E_{kin} = h^2/2M_n\lambda^2 \simeq 80\,\text{meV}$. The corresponding velocity is around 4×10^5 cm/s. Since $E_{kin} = 3k_BT/2$, one has $T \simeq 630$ K.

Problem 11.4 Show that the reciprocal lattice for the fcc lattice is a bcc lattice and *vice-versa*.

Solution: In terms of the side a of the conventional cubic cell the *primitive* lattice vectors of the fcc structure are (Fig. 11.4):

$$\mathbf{a}_1 = \frac{a}{2}(\mathbf{i} + \mathbf{j})$$
$$\mathbf{a}_2 = \frac{a}{2}(\mathbf{i} + \mathbf{k})$$
$$\mathbf{a}_3 = \frac{a}{2}(\mathbf{j} + \mathbf{k})$$

($\mathbf{i}, \mathbf{j}, \mathbf{k}$ orthogonal unit vectors parallel to the cube edges). Note that $|\mathbf{a}_i| = a/\sqrt{2}$ and therefore the volume of the primitive cell is $(\mathbf{a}_1 \times \mathbf{a}_2) \cdot \mathbf{a}_3 = a^3/4$. Then the primitive vectors of the reciprocal lattice are

$$\mathbf{a}_1^* = \frac{2\pi \mathbf{a}_2 \times \mathbf{a}_3}{a^3/4}$$

and similar expressions for $\mathbf{a}_2^*$ and $\mathbf{a}_3^*$ (Eq. (11.7)) (in the unit cube of volume a^3 there are four lattice points). Thus

$$\mathbf{a}_1^* = \frac{2\pi}{a}(-\mathbf{i} - \mathbf{j} + \mathbf{k})$$

$$\mathbf{a}_2^* = \frac{2\pi}{a}(-\mathbf{i} + \mathbf{j} - \mathbf{k})$$

$$\mathbf{a}_3^* = \frac{2\pi}{a}(\mathbf{i} - \mathbf{j} - \mathbf{k})$$

The shortest (non-zero) reciprocal lattice vectors are given by the eight vectors $(2\pi/a)(\pm\mathbf{i} \pm \mathbf{j} \pm \mathbf{k})$ which generate the bcc (reciprocal) lattice.

A similar procedure applied to the primitive translational vectors of the bcc lattice

$$\mathbf{a}_1 = \frac{a}{2}(\mathbf{i} + \mathbf{j} + \mathbf{k})$$
$$\mathbf{a}_2 = \frac{a}{2}(-\mathbf{i} + \mathbf{j} + \mathbf{k})$$
$$\mathbf{a}_3 = \frac{a}{2}(-\mathbf{i} - \mathbf{j} + \mathbf{k})$$

(yielding for the volume of the primitive cell $(\mathbf{a}_1 \times \mathbf{a}_2) \cdot \mathbf{a}_3 = a^3/2$) implies

$$\mathbf{a}_1^* = \frac{2\pi}{a}(\mathbf{i} + \mathbf{k})$$
$$\mathbf{a}_2^* = \frac{2\pi}{a}(-\mathbf{i} + \mathbf{j})$$
$$\mathbf{a}_3^* = \frac{2\pi}{a}(-\mathbf{j} + \mathbf{k})$$

as primitive vectors of fcc lattice.

The Brillouin cell of the bcc lattice is shown below (compared to the one in Fig. 11.2).

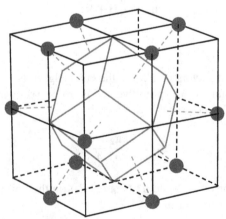

Specific References and Further Readings

1. G. Burns, *Solid State Physics*, (Academic Press Inc., 1985).
2. C. Kittel, *Introduction to Solid State Physics*, 8th Edition, (J. Wiley and Sons, 2005).
3. N.W. Ashcroft and N.D. Mermin, *Solid State Physics*, (Holt, Rinehart and Winston, 1976).
4. J.M. Ziman, *Principles of the Theory of Solids*, (Cambridge University Press, 1964).
5. A. Balzarotti, M. Cini and M. Fanfoni, *Atomi, Molecole e Solidi. Esercizi risolti*, (Springer, Verlag, 2004).
6. J.S. Blakemore, *Solid State Physics*, (W.B. Saunders Co., 1974).
7. G. Grosso and G. Pastori Parravicini, *Solid State Physics*, 2nd Edition, (Academic Press, 2013).

Chapter 12
Electron States in Crystals

Topics

Bands of Energy Levels
Bloch Orbital and Crystal Momentum
Effective Mass of the Electron
Density of States
Free-Electron Model
Magnetic and Thermal Properties of Metals
Perturbative Effects on Free-Electron States and Energy Gaps
Tight-Binding Model
Bands Overlap and Intrinsic Semiconductors

12.1 Introductory Aspects and the Band Concept

A fundamental issue in solid state physics is the structure of the electronic states. Transport, magnetic and optical properties, as well as the very nature (*metal, insulator* or *semiconductor*) of the crystals, are indeed controlled by the arrangement of the energy levels.

The complete form of the Schrödinger equation for electrons and nuclei can hardly be solved, even by means of computational approaches. Therefore to describe the electron states in a crystal it is necessary to rely on approximate methods applied to model systems.

Usually the crystal is ideally separated into ions (the atoms with the core electrons practically keeping their atomic properties) and the *valence* electrons, which are affected by the crystalline arrangement. The Born-Oppenheimer separation

© Springer International Publishing Switzerland 2015
A. Rigamonti and P. Carretta, *Structure of Matter*,
UNITEXT for Physics, DOI 10.1007/978-3-319-17897-4_12

(Sect. 7.2) is usually the starting point, often in the adiabatic approximation.[1] From the many-body problem for the electrons, by means of Hartree-Fock description one can devise the one-electron effective potential that takes into account the interaction with the positive ions, the Coulomb-like repulsion among the electrons as well as the generalized exchange integrals. We shall not derive the potential energy in detail on the basis of that type of approach. Rather, similarly to atoms and molecules, we shall address the main aspects of the electronic structure in crystals on the basis of the fundamental symmetry property, namely the *translational invariance* for the potential energy:

$$V(\mathbf{r} + \mathbf{l}) = V(\mathbf{r}) \tag{12.1}$$

with $\mathbf{l}$ lattice vector (Eq. (11.1) and Sect. 11.2).

First we shall derive the general properties and the classification of the electronic states in terms of a *pseudo-momentum* vector in the reciprocal space. Then a deeper description will be made on the basis of particular models, at the sake of illustration of the generalities, meantime illustrating the properties of typical groups of solids.

Henceforth, by extending the molecular orbital approach (Sect. 8.1) in the LCAO form, one can express the one-electron wave function as *Bloch orbital*. This is somewhat equivalent to the delocalized MO introduced for the benzene molecule (Sect. 9.3).

Referring to an ideal crystal formed by a chain of N one-electron atoms

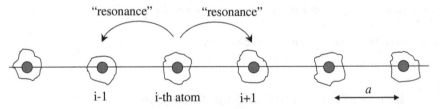

and generalizing the concepts used for H_2 molecule (Sect. 8.2), the formation of the *band* of electron levels can be understood as resulting from the removal of the degeneracy of the atomic levels. In fact, by taking into account the resonance of one electron among neighboring atoms (see sketch above), the wave function of the electron centered at ith site is written

$$i\hbar\frac{d\psi_i}{dt} = E_o\psi_i + A\psi_{i-1} + A\psi_{i+1}, \tag{12.2}$$

where $A < 0$ is the *resonance integral between adjacent sites* (equivalent to H_{AB} in Sect. 8.1). From what has been learned for the H_2^+ molecule, we look for a solution of Eq. (12.2) in the form

---
[1] As already mentioned (Sect. 7.1) several relevant phenomena belonging to the realm of solid state physics, for instance electrical resistivity and superconductivity, require to go beyond the adiabatic approximation.

$$\psi_i = \phi_i e^{-iEt/\hbar} \tag{12.3}$$

where E is the unknown eigenvalue, while ϕ_i is the electron eigenfunction for the atom centered at the site i. Then

$$E\phi_i = E_o\phi_i + A(\phi_{i-1} + \phi_{i+1}), \tag{12.4}$$

with $\phi_i = \phi(x_i)$ and $\phi_{i\pm 1} = \phi(x_i \pm a)$. By looking for a solution of the form $exp(ikx_i)$, typical of the difference equations and already used for the benzene molecule (Sect. 9.3), Eq. (12.4) is rewritten

$$Ee^{ikx_i} = E_o e^{ikx_i} + A\left[e^{ik(x_i+a)} + e^{ik(x_i-a)} \right], \tag{12.5}$$

yielding

$$E = E_o + 2A cos(ka) \tag{12.6}$$

The formation of a *band* of electronic levels, each level labelled by k, as a consequence of the removal of the degeneracy existing for non-interacting atoms, is illustrated below

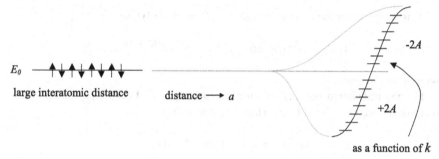

The band of N electron levels is the generalization of the g and u levels in the H_2 molecule or of the four levels in the C_6H_6 molecule. The energy interval between two adjacent bands, related to different atomic eigenvalues E_o, will be called *energy gap*.

We shall come back to the problem of labelling the electron states and to the mechanisms leading to the appearance of the gap, after the discussion of suitable crystal models (Sect. 12.7).

12.2 Translational Invariance and the Bloch Orbital

In ideal crystals, with no defects and without surfaces, the translation operator T_l (see Eq. (11.2)) commutes with the Hamiltonian:

$$T_l \mathcal{H}(\mathbf{r})\phi(\mathbf{r}) = \mathcal{H}(\mathbf{r}+\mathbf{l})\phi(\mathbf{r}+\mathbf{l}) = \mathcal{H}(\mathbf{r})T_l\phi(\mathbf{r})$$

Then the one-electron eigenfunction $\phi(\mathbf{r})$ must be eigenfunction of T_l also, with eigenvalues c_l satisfying the condition $|c_l|^2 = 1$, since

$$|\phi(\mathbf{r}+\mathbf{l})|^2 \equiv |T_l\phi(\mathbf{r})|^2 = |\phi(\mathbf{r})|^2.$$

On the other hand, two translations $T_{l_1} T_{l_2} \equiv T_{l_1+l_2}$, must yield the same result of the translation by $l_1 + l_2$:

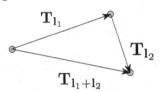

This suggests for the eigenvalue the form $c_l = exp(i\lambda_l)$, so that

$$T_{l_1+l_2}\phi = T_{l_1}e^{i\lambda_2}\phi = e^{i\lambda_2}e^{i\lambda_1}\phi = e^{i(\lambda_1+\lambda_2)}\phi,$$

with λ_l *real number*.

For any translation vector $\mathbf{l}$ a vector $\mathbf{k}$ so that $\lambda_l = \mathbf{k}\cdot\mathbf{l}$ can be picked up in the reciprocal space (see Sect. 11.2). Therefore one writes

$$T_l\phi(\mathbf{r}) = \phi(\mathbf{r}+\mathbf{l}) = e^{i\mathbf{k}\cdot\mathbf{l}}\phi(\mathbf{r}),$$

and by multiplying by $e^{-i\mathbf{k}\cdot\mathbf{r}}$

$$e^{-i\mathbf{k}\cdot\mathbf{r}}\phi(\mathbf{r}) = e^{-i\mathbf{k}\cdot(\mathbf{r}+\mathbf{l})}\phi(\mathbf{r}+\mathbf{l}).$$

This condition shows that the function $u_\mathbf{k}(\mathbf{r}) = exp(-i\mathbf{k}\cdot\mathbf{r})\phi(\mathbf{r})$ has the *periodicity of the lattice*, as described by Eq. (11.2).

Then the one-electron wave function can be written as *Bloch orbital*, i.e.

$$\phi_\mathbf{k}(\mathbf{r}) = u_\mathbf{k}(\mathbf{r})e^{i\mathbf{k}\cdot\mathbf{r}}$$
$$u_\mathbf{k}(\mathbf{r}+\mathbf{l}) = u_\mathbf{k}(\mathbf{r}), \tag{12.7}$$

which couples the free-electron wave function $exp(i\mathbf{k}\cdot\mathbf{r})$ (characteristic of the *empty lattice*, namely in the limit $V(\mathbf{r}) \rightarrow 0$) with an unknown wave function $u_\mathbf{k}(\mathbf{r})$ having the *lattice periodicity*.

It can be remarked that up to now $\mathbf{k}$ in the Bloch orbital is just a vector in the reciprocal space used to label the one-electron states in a periodic potential. In the next section the role and the physical properties of $\mathbf{k}$ shall be discussed.

In order to illustrate the Bloch orbital we will take into consideration a particular form for the function $u_{\mathbf{k}}(\mathbf{r})$. $u_{\mathbf{k}}(\mathbf{r})$ can be found from the one-electron Schrödinger equation $\mathcal{H}\phi(\mathbf{r}) = E\phi(\mathbf{r})$ by writing for $\phi(\mathbf{r})$ the Bloch orbital according to Eq. (12.7):

$$\left[\frac{-\hbar^2}{2m}(\mathbf{\nabla} + i\mathbf{k})^2 + V(\mathbf{r})\right]u_{\mathbf{k}}(\mathbf{r}) = E_{\mathbf{k}}u_{\mathbf{k}}(\mathbf{r}) \qquad (12.8)$$

Under the assumption that $u_{\mathbf{k}}(\mathbf{r})$ is weakly $\mathbf{k}$-dependent, i.e. $u_{\mathbf{k}}(\mathbf{r}) \simeq u_{\mathbf{k}=0}(\mathbf{r})$, one can write $\phi_{\mathbf{k}}(\mathbf{r}) = u_{\mathbf{k}=0}(\mathbf{r})e^{i\mathbf{k}\cdot\mathbf{r}}$. From Eq. (12.8) one sees that for $\mathbf{k} = 0$ $u_{\mathbf{k}=0}(\mathbf{r})$ is the solution of the atomic-type Schrödinger equation. The only difference is in the boundary conditions which impose the continuity at the border of the Wigner-Seitz cell (see Sect. 11.1). In Fig. 12.1 the function $u_{\mathbf{k}=0}(\mathbf{r})$ for the 3s electron, in the Na crystal derived under these constraints (this procedure is the core of the so-called *cellular method*), is sketched. The corresponding Bloch orbitals are schematically depicted in Fig. 12.2.

Fig. 12.1 Sketch of $u_{\mathbf{k}=0}(\mathbf{r})$ for 3s electron in the Na crystal, derived by *Wigner* and *Seitz* by means of the cellular method (by approximating the WS cell to a sphere; see also the book by *Slater*)

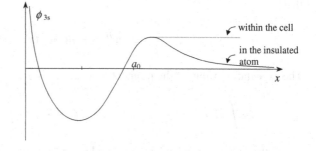

Fig. 12.2 Sketch of the real part of Bloch orbitals in a one-dimensional crystal for different values of k, with a interatomic distance. The dashed lines are the real parts of the corresponding plane waves

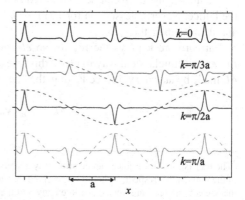

12.3 Role and Properties of k

The reciprocal space vector **k**, labelling the eigenvalues of the translational operator which commutes with the Hamiltonian, is a *constant of motion*: its components k_x, k_y and k_z have to be considered as good quantum numbers for the one-electron states. Hence, as far as the translational invariance condition holds, the electron remains in a given state **k**.[2]

A first illustration of the role of **k** can be provided by considering the limiting case of vanishing potential energy $V(\mathbf{r})$, often called the *empty lattice* condition, as already mentioned. Then the eigenfunctions are

$$\phi_{\mathbf{k}}(\mathbf{r}) \propto e^{i\mathbf{k}\cdot\mathbf{r}} \tag{12.9}$$

with eigenvalues

$$E_{\mathbf{k}} = \frac{\hbar^2 k^2}{2m}. \tag{12.10}$$

Therefore for the empty lattice, **k** represents the *momentum of the electron*, in $\hbar$ units.

When $V(\mathbf{r}) \neq 0$ $\hbar\mathbf{k}$ is no longer the momentum of the electron (it is not the eigenvalue of $-i\hbar\nabla$). In fact, by referring for simplicity to the x direction, one sees that

$$-i\hbar\frac{\partial}{\partial x}u_{k_x}(x)e^{ik_x x} \neq \hbar k_x u_{k_x}(x)e^{ik_x x}.$$

The expectation value of the momentum is given by

$$-i\hbar\int u_{k_x}^* e^{-ik_x x}\frac{\partial}{\partial x}u_{k_x}e^{ik_x x}dx = \hbar k_x + (-i\hbar)\int u_{k_x}^*\frac{\partial}{\partial x}u_{k_x}, \tag{12.11}$$

where the second term can be considered as an "average momentum" transferred to the lattice. Nevertheless, even for $V(\mathbf{r}) \neq 0$, **k** continues to be a constant of motion and then it labels the state.

Furthermore **k** plays the role of an *electron momentum in regards of external forces*. A semiclassical way to prove this role of **k** is to consider the elemental work δL made by an external force $\mathbf{F}_e$ (e.g. the one due to an external electric field). Since

$$\delta L = \mathbf{F}_e \cdot \mathbf{v}_g \delta t$$

---

[2]The translational invariance can be broken by defects, free surfaces or by the vibrational motions of the ions. In this respect, it should be observed that, at variance with the states in molecules, here the **k**-electron states are very close in energy and the vibrational motions of the ions may cause variation of the electron state. These processes contribute to the electrical *resistivity* (see Sect. 13.4 for remarks on these aspects).

with the group velocity $\mathbf{v}_g = (1/\hbar)\partial E_\mathbf{k}/\partial\mathbf{k}$, one has

$$\delta L = \mathbf{F}_e \cdot \frac{1}{\hbar}\frac{\partial E_\mathbf{k}}{\partial\mathbf{k}}\delta t.$$

By equating the elemental work δL to $\delta E = (\partial E_\mathbf{k}/\partial\mathbf{k}) \cdot \delta\mathbf{k}$, one derives

$$\frac{\delta\mathbf{k}}{\delta t}\hbar = \hbar\dot{\mathbf{k}} = \mathbf{F}_e, \qquad (12.12)$$

illustrating how $\hbar\mathbf{k}$ behaves as a momentum. Thus it can be defined as *pseudo-momentum* or *crystal momentum*.

Up to now $\mathbf{k}$ is a continuous vector in the reciprocal space. As already seen in atoms and in molecules, the boundary conditions determine discrete eigenvalues and then discrete values for $\mathbf{k}$. In this respect one possibility would be to fix the nodes of the wavefunctions at the surface of the crystal. Quantum conditions similar to the ones for a particle in a box can be expected. However, this procedure would imply the transformation of the wavefunctions from running waves to stationary waves and surface effects would arise. It is often more convenient to impose *periodic boundary conditions* (*Born-Von Karman* procedure), as we shall see in the next section.

Problems

Problem 12.1 For k-dependence of the electron eigenvalues given by

$$E(k) = Ak^2 - Bk^4$$

derive the eigenvalue $E(k^*)$ for which phase and group velocities of the electrons are the same. Give the proper orders of magnitude and units for the coefficients A and B.

Solution: From $v_{ph} = \omega/k = (Ak - Bk^3)/\hbar$ and $v_g = \partial\omega/\partial k = (2Ak - 4Bk^3)/\hbar$, one has $2A - 4Bk^{*2} = A - Bk^{*2}$, yielding

$$k^* = \left(\frac{A}{3B}\right)^{\frac{1}{2}}$$

and $E^* = k^{*2}(A - Bk^{*2}) = 2A^2/9B$. The orders of magnitude of A and B are $A \sim$ eV Å^2 and $B \sim$ eV Å^4.

Problem 12.2 Discuss the trajectory of an electron under the Lorentz force due to an external magnetic field along the z-direction, for energy eigenvalues of the form $E_k = \alpha k_x^2 + \beta k_y^2$.

Solution: According to the extension of Eq. (12.12) to the Lorentz force, from

$$\hbar\frac{d\mathbf{k}}{dt} = \frac{-e\mathbf{v}_g}{c} \times \mathbf{H},$$

with $\mathbf{v}_g$ the group velocity, one writes

$$\frac{d\mathbf{k}}{dt} = -\frac{e}{\hbar^2 c}(\nabla_k E_k \times \mathbf{H}).$$

For magnetic field along the z-direction one has

$$\frac{d\mathbf{k}}{dt} = \frac{2eH}{\hbar^2 c}(\alpha k_x \mathbf{j} - \beta k_y \mathbf{i})$$

or

$$\dot{k}_x = -\frac{2eH}{\hbar^2 c}\beta k_y \qquad \dot{k}_y = +\frac{2eH}{\hbar^2 c}\alpha k_x$$

yielding

$$k_x = k_{x0}\cos(\omega t + \phi), \quad k_y = k_{y0}\sin(\omega t + \phi),$$

where

$$\omega = \frac{2eH}{\hbar^2 c}(\alpha\beta)^{1/2}$$

The trajectory in the k plane is an ellipse. From the integration of the group velocity

$$\mathbf{v}_g = (2\alpha/\hbar)k_{x0}\cos(\omega t + \phi)\mathbf{i} + (2\beta/\hbar)k_{y0}\sin(\omega t + \phi)\mathbf{j},$$

it is found that also in the real space the trajectory of the motion is an ellipse, for a given value of the energy, so that $\alpha k_{x0}^2 = \beta k_{y0}^2 = E_{k_0}$. The motion induced by the magnetic field is called *cyclotron motion* (see Appendix 13.1 for details).

Problem 12.3 In a cubic crystal the $\mathbf{k}$-dependence of the electron eigenvalues is

$$E(\mathbf{k}) = C - 2V_1[\cos k_x a + \cos k_y a + \cos k_z a]$$

(a form that can be obtained in the framework of the tight-binding model, see Sect. 12.7.3). Derive the acceleration of an electron due to an external electric field.

Solution: From the time derivative of the group velocity $\mathbf{v}_g = (1/\hbar)\nabla_{\mathbf{k}}E_{\mathbf{k}}$, by considering that $\dot{\mathbf{k}} = \mathbf{F}_e/\hbar$, the tensor describing the relationship between the electric field $\mathcal{E}$ and the acceleration $\dot{\mathbf{v}}_g$ turns out

$$\begin{pmatrix} A\cos k_x a & 0 & 0 \\ 0 & A\cos k_y a & 0 \\ 0 & 0 & A\cos k_z a \end{pmatrix},$$

with $A = \frac{2V_1a^2}{\hbar^2}$. Then, for $\boldsymbol{\mathcal{E}} = \mathcal{E}_x\mathbf{i} + \mathcal{E}_y\mathbf{j} + \mathcal{E}_z\mathbf{k}$ the acceleration is

$$\dot{\mathbf{v}}_g = -A[(e\mathcal{E}_x \cos k_x a)\,\mathbf{i} + \left(e\mathcal{E}_y \cos k_y a\right)\mathbf{j} + (e\mathcal{E}_z \cos k_z a)\,\mathbf{k}].$$

Since no off-diagonal elements of the tensor are present, the acceleration is along the same direction of the field. The ratio between the external force and the acceleration leads to the concept of *effective mass* (see Sect. 12.6).

12.4 Periodic Boundary Conditions and Reduction to the First Brillouin zone

Let us refer to a region of macroscopic size in an ideal crystal containing N cells, N_1 along the $\mathbf{a}$ direction, N_2 along $\mathbf{b}$ and N_3 along $\mathbf{c}$. The reference volume is Nv_c, with $v_c = (\mathbf{a} \times \mathbf{b}).\mathbf{c}$. The electron wavefunctions $\phi_\mathbf{k}$ have to be identical in equivalent points of that region and of a replica region. By assuming for simplicity that the crystal axes are perpendicular and considering the vector $\mathbf{L} = N_1\mathbf{a} + N_2\mathbf{b} + N_3\mathbf{c}$, then according to Eq. (12.7) one has to write

$$e^{i\mathbf{k}\cdot\mathbf{r}} = e^{i\mathbf{k}\cdot(\mathbf{r}+\mathbf{L})}, \tag{12.13}$$

the equality of $u_\mathbf{k}(\mathbf{r})$ in the replica region being obviously granted. Then the conditions

$$k_x = n_1\frac{2\pi}{aN_1}, \quad k_y = n_2\frac{2\pi}{bN_2}, \quad k_z = n_3\frac{2\pi}{cN_3} \tag{12.14}$$

with n_i integers are obtained. By referring to the reciprocal lattice vectors (Sect. 11.3) $\mathbf{a}^*, \mathbf{b}^*, \mathbf{c}^*$, thus extending the above arguments to non-perpendicular crystal axes, the periodic boundary conditions yield

$$\mathbf{k} = n_1\frac{\mathbf{a}^*}{N_1} + n_2\frac{\mathbf{b}^*}{N_2} + n_3\frac{\mathbf{c}^*}{N_3}. \tag{12.15}$$

It should be noticed that $\mathbf{k}$ can be outside of the Brillouin cell. However, as we shall see in the following, an electron state $\mathbf{k}$ outside the Brillouin cell (or first Brillouin cell) is equivalent to a given state within the cell. Therefore, one can classify the states by means of the set of discrete N vectors $\mathbf{k}$ given by Eq. (12.15), with n_i such that $\mathbf{k}$ lies within the *Brillouin zone* (BZ). This statement can be understood with the aid of the planar reciprocal lattice, as sketched below:

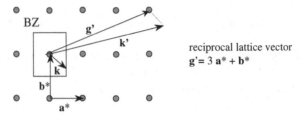

reciprocal lattice vector
$\mathbf{g}' = 3\,\mathbf{a}^* + \mathbf{b}^*$

$\mathbf{g}'$ is a reciprocal lattice vector that from $\mathbf{k}'$ outside the BZ brings to a point inside it. Thus $\mathbf{k} = \mathbf{k}' - \mathbf{g}'$ and the wavefunction $\phi_{\mathbf{k}'}$ can be written

$$\phi_{\mathbf{k}'} = u_{\mathbf{k}'}(\mathbf{r})e^{i\mathbf{k}'\cdot\mathbf{r}} = e^{i\mathbf{k}\cdot\mathbf{r}}e^{i\mathbf{g}'\cdot\mathbf{r}}u_{\mathbf{k}'}(\mathbf{r}). \qquad (12.16)$$

Now one can observe that $e^{i\mathbf{g}'\cdot\mathbf{r}}u_{\mathbf{k}'}(\mathbf{r})$ has the lattice periodicity since, according to Eq. (11.6), $e^{i\mathbf{g}'\cdot\mathbf{l}} = 1$. Hence $e^{i\mathbf{g}'\cdot\mathbf{r}}u_{\mathbf{k}'}(\mathbf{r}) = u_{\mathbf{k}}(\mathbf{r})$ is the function which makes $\phi_{\mathbf{k}}$ in the form of a Bloch orbital. Then

$$\phi_{\mathbf{k}'} = \phi_{\mathbf{k}} \quad \text{and} \quad E_{\mathbf{k}'} = E_{\mathbf{k}} \qquad (12.17)$$

and the electron states *can be classified by means of N vectors* $\mathbf{k}$ *inside the BZ.* The states $\mathbf{k}'$ outside this zone merely correspond to equivalent states, in a representation called *extended zone representation.* This representation has to be compared to the *reduced zone representation* where all states are reported inside the BZ. The details of the electron states in the framework of specific crystal models (see Sect. 12.7) will better clarify this aspect.

For a one-dimensional (1D) crystal one has the illustrative plots reported below, for a band of the form as in Eq. (12.6).

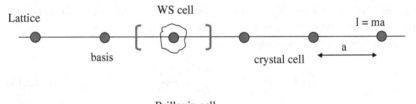

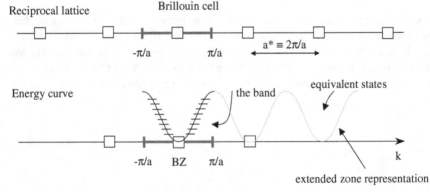

12.5 Density of States, Dispersion Relations and Critical Points

As discussed in the previous section, the electron states can be described by referring to the reciprocal space and in particular to the first Brillouin zone. The state of the whole crystal can be thought to result from the assignment of two electrons, with opposite spins, to each state **k**, in a way similar to the *aufbau* principle used in atoms and in molecules. For the moment we shall refer for simplicity to the condition of zero temperature, so that one can disregard the thermal excitations to higher energy states.

One can sketch the situation as below,

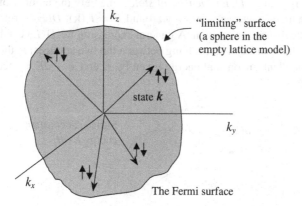

with a limit surface in the reciprocal space including all the occupied electron states. This surface, corresponding to a sphere in the empty lattice model (Eq. (12.10)), is called *Fermi surface*.

The following points should be remarked:

(i) if one increases the reference volume Nv_c, by increasing the number of crystal cells, the total number of **k** states increases;

(ii) if the crystal cell is expanded (v_c increases) the BZ volume decreases;

(iii) for monoatomic crystals, with the basis formed by a single atom with one valence electron, the BZ is half filled by occupied states;

(iv) again for monoatomic crystal, when each atom contributes with two valence electrons, the BZ is fully occupied (the surface of the Brillouin cell not necessarily coincides with the Fermi surface).

The density of **k** states $D(\mathbf{k})$ can be derived once it is noticed that within the BZ there are N states, equally spaced in the reciprocal volume. Then, the BZ volume being $v_c^* = 8\pi^3/v_c$, one has

$$D(\mathbf{k}) = \frac{N}{v_c^*} = \frac{Nv_c}{8\pi^3}. \tag{12.18}$$

The reference volume is often assumed $1\,cm^3$. Since for such a volume the number of states within the BZ is around 10^{22}, although $\mathbf{k}$ in principle is a discrete variable, in practice it is often convenient to treat it as a continuous variable, so that

$$\sum_{\mathbf{k}'} \rightarrow \int D(\mathbf{k})d\mathbf{k} \equiv \frac{Nv_c}{8\pi^3} \int d\mathbf{k} \tag{12.19}$$

The sequence of energy levels $E(\mathbf{k})$ is the *band*, that we have already introduced qualitatively in Sect. 12.1. In analogy to wave optics, the $\mathbf{k}$-dependence of the eigenvalues is called *dispersion relation*.

An important quantity characterizing the structure of the energy levels is the density of energy states $D(E)$ (*density of states*), namely the number of electronic states within a unitary interval of energy around $E = E(\mathbf{k})$. $D(E)$ is related both to $D(\mathbf{k})$ and to the dispersion relation. A general expression for $D(E)$ can be obtained by estimating the number of states lying between the two surfaces, in the reciprocal space, correspondent to constant energy given by E and $E + dE$, respectively (see sketch below).

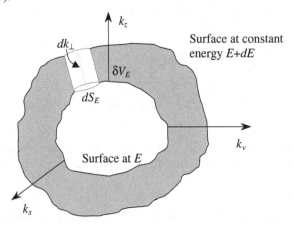

The number of states in the volume δV_E is

$$D(E)dE = \frac{Nv_c}{8\pi^3}.2.\delta V_E,$$

the factor 2 accounting for the spin degeneracy. For the volume δV_E in between the two surfaces one has

$$\delta V_E = \int_S dS_E dk_\perp = \int_S dS_E \frac{dE}{|\nabla_{\mathbf{k}} E(\mathbf{k})|}$$

since $dk_\perp = dE/|\partial E/\partial \mathbf{k}|$. Therefore

$$D(E) = \frac{Nv_c}{4\pi^3} \int_S dS_E \frac{1}{|\nabla_\mathbf{k} E(\mathbf{k})|}. \tag{12.20}$$

From the above expression it is evident that $D(E)$ has singularities (*Van Hove singularities*) whenever the gradient of $E(\mathbf{k})$ in the reciprocal space vanishes. The points, in the reciprocal space, where this condition is fulfilled are called *critical points*. These critical points are particularly relevant for the optical and transport properties since they imply a marked denseness of states. As it will be shown in the next section, electrons around a critical point behave as if they had particular *effective masses*.

12.6 The Effective Electron Mass

As shown in Sect. 12.3 the $\mathbf{k}$-dependence of the energy controls the behavior of the electron under external forces. In fact $\hbar\dot{\mathbf{k}} = \mathbf{F}_e$, while the group velocity is $\mathbf{v}_g = (1/\hbar)(\partial E(\mathbf{k})/\partial \mathbf{k})$. By differentiating $\mathbf{v}_g$ one has

$$\mathbf{a} = \frac{d\mathbf{v}_g}{dt} = \frac{1}{\hbar} \frac{\partial^2 E(\mathbf{k})}{\partial \mathbf{k}^2} \frac{\partial \mathbf{k}}{\partial t} = \frac{1}{\hbar^2} \frac{\partial^2 E(\mathbf{k})}{\partial \mathbf{k}^2} \mathbf{F}_e \tag{12.21}$$

On the basis of the classical analogy, the relationship between the force and the acceleration points out that the electron reacts to the external force as if it had a mass

$$\widetilde{m^*} = \hbar^2 \left(\frac{\partial^2 E(\mathbf{k})}{\partial \mathbf{k}^2} \right)^{-1}. \tag{12.22}$$

In the empty lattice limit, or free electron model (see Sect. 12.7.1), the *effective mass* coincides with the real electron mass: $m^* = \hbar^2/[\partial^2(\hbar^2 k^2/2m)/\partial k^2] \equiv m$.

In order to illustrate the concept of effective mass let us refer to the dispersion curve derived in Sect. 12.1 by applying to a linear chain of atoms (1D) the idea of resonance among adjacent atoms: $E(k) = 2A\cos(ka)$, with k along x axis and $A < 0$. Then, from Eq. (12.22), the effective mass turns out (see Fig. 12.3)

$$m^* = -\frac{\hbar^2}{2Aa^2} \frac{1}{\cos(ka)}. \tag{12.23}$$

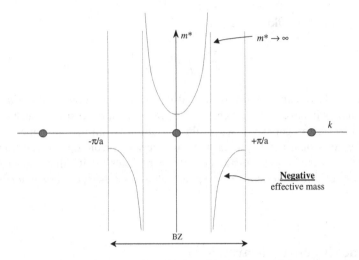

Fig. 12.3 Effective mass m^* as a function of k for a 1D model crystal, in correspondence to the dispersion relation $E(k) = 2A\cos(ka)$, with $A < 0$

Finally, as it appears from Eq. (12.22), the effective mass $\widetilde{m^*}$ has a tensorial character, with components (see Problem 12.3)

$$m^*_{\alpha\beta} = \hbar^2 \left(\frac{\partial^2 E(\mathbf{k})}{\partial k_\alpha \partial k_\beta} \right)^{-1}. \qquad (12.24)$$

Problems

Problem 12.4 For a one-dimensional crystal the dispersion relation

$$E(k) = E_1 + (E_2 - E_1) \sin^2 \left(\frac{ka}{2} \right),$$

is assumed, with lattice step $a = 1$ Å. By referring to a single electron in the band and by neglecting any scattering process (with defects, boundaries or impurities) derive the effective mass, the velocity and the motion of the electron in the real space, under the action of a constant electric field $\mathcal{E}$. For $\mathcal{E} = 100$ V/m and $(E_2 - E_1) = 1$ eV, obtain the period and the amplitude of the oscillatory motion.

Solution: The group velocity is $v_g = \left[a(E_2 - E_1)/2\hbar \right] \sin(ak_x)$. The effective mass is $m^* = \hbar^2 [d^2 E/dk_x^2]^{-1} = m_0 \sec(ak_x)$, where $m_0 = [2\hbar^2/a^2(E_2 - E_1)]$ is the mass at the bottom of the band. m^* becomes infinite for $k_x = \pm\pi/2a$ (see plots).

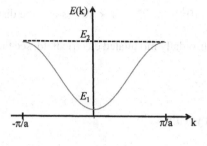

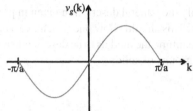

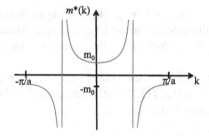

For a single non-scattered electron in a time-independent electric field $\mathcal{E}_x$ the force implies $dk_x/dt = (-e\mathcal{E}_x/\hbar)$. Then k_x scans repetitively through the Brillouin zone, with period $t^* = (2\pi\hbar/ae\mathcal{E}_x)$.

In the assumption that at $t = 0$ $E = E_1, m^* = m_o, k_x = 0$, the electron has finite positive mass for some time, becoming infinite at $t = t^*/4$.

At $t = t^*/2$ the electron arrives at $k_x = -(\pi/a)$. The equivalence of this state with the one at $k_x = +\pi/a$ corresponds to the return into the BZ (this corresponds to the Bragg reflection of the De Broglie wave, see also Sect. 12.7.2). Then k_x decreases again and the mass divergence is reached at $t = (3/4)t^*$.

From the velocity

$$v_g(t) = \left[a(E_2 - E_1)/2\hbar\right]\sin(-2\pi t/t^*) = \left[a(E_2 - E_1)/2\hbar\right]\sin(-ae\mathcal{E}_x t/\hbar)$$

it is found that in the real space an oscillatory motion occurs:

$$x(t) = \int v_g dt = \left[(E_2 - E_1)/2e\mathcal{E}_x\right]\cos(-ae\mathcal{E}_x t/\hbar)$$

For $a = 1\text{Å}$ and $\mathcal{E}_x = 10^2\,\text{V/m}$, $t^* \simeq 4 \times 10^{-7}\,\text{s}$ and the distance covered would be about 1 cm.

For the case of a sinusoidally modulated electric field, see the Problem 3.30 in the book by *Blakemore*.

12.7 Models of Crystals

Now we are going to apply the general description given in previous sections to particular models of crystals. This should allow one to achieve a better understanding of the physical concepts. Meantime the models to be described, to a good approximation correspond to particular groups of solids.

12.7.1 Electrons in Empty Lattice

The condition of potential energy $V(\mathbf{r})$ going to zero has already been occasionally addressed. Now we shall explore in more detail this ideal situation and derive some finite-temperature properties which reflect the thermal excitations and the statistical effects.

When $V(\mathbf{r}) \to 0$ the electrons delocalize in the reference volume $N v_c$ and are described by Bloch orbitals (Eq. (12.7)) with constant $u_\mathbf{k}(\mathbf{r})$. According to the one-electron Schrödinger equation one has

$$\phi_\mathbf{k} = \frac{1}{\sqrt{N v_c}} e^{i\mathbf{k}\cdot\mathbf{r}} \tag{12.25}$$

and

$$E(\mathbf{k}) = \frac{\hbar^2 k^2}{2m}. \tag{12.26}$$

The valence electrons can be thought to move freely in the reference volume and they become responsible for the electric conduction. This model is suited to describe the *metals*.

The theory of metals in the framework of the *free electron model* was actually developed before the advent of quantum mechanics. Significant successes were achieved, as the derivation of *Ohm law* and of the relationship between thermal and electrical conductivity (*Wiedemann-Franz law*). At variance, the behaviour of other quantities, such as the heat capacity and the magnetic susceptibility, requiring in the derivation the use of Fermi-Dirac distribution, could hardly be explained in the early theories. On the other hand, in spite of the successful quantum mechanical description, the limits of the free electron model become obvious when one recalls the huge

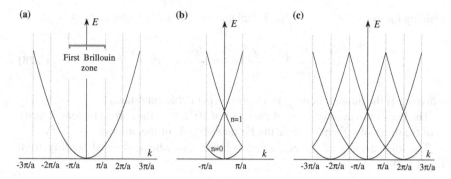

Fig. 12.4 Dispersion curves for the empty lattice model, in a crystal of lattice step a, within: **a** the extended zone scheme, **b** the reduced zone scheme, **c** the repeated zone scheme. The indexes (in **b**) indicate the number of reciprocal lattice vectors **a*** required for the reduction to the first BZ

change in the electrical conductivity from metals to insulators or the existence of semiconductors. In these compounds the role played by a non-zero lattice potential is crucial (see next section).

The dispersion curve for electrons in empty lattice (Eq. (12.26)) is reported in Fig. 12.4 in the *extended*, *reduced* and *repeated zone* representations, along a reciprocal space axis.

The constant energy surfaces in **k** space are spherical,

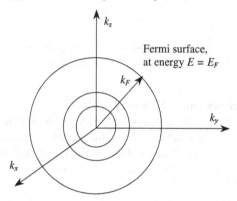

At $T = 0$ the electrons fill all the states up to a given wavevector of modulus k_F, called the *Fermi wavevector*, which corresponds to the radius of the *Fermi surface* (see Sect. 12.6). In a crystal with N cells and Z electrons per cell, k_F can be directly derived by considering the volume of the Fermi sphere, the density of states $D(\mathbf{k})$ (Eq. (12.18)) and the spin variable for each **k** state:

$$ZN = \frac{N v_c}{8\pi^3} . 2 . \frac{4\pi}{3} k_F^3,$$

(12.27)

yielding $k_F = (3\pi^2 Z/v_c)^{1/3}$. The Fermi energy $E_F = \hbar^2 k_F^2/2m$ turns out

$$E_F = \frac{\hbar^2}{2m}\left(3\pi^2 \frac{Z}{v_c}\right)^{2/3} = \frac{\hbar^2}{2m}(3\pi^2)^{2/3} n_d^{2/3}, \qquad (12.28)$$

where n_d is the number density of electrons (per cubic centimeter).

The Fermi wavevector k_F is of the order of 10^8 cm^{-1}, the correspondent velocity is of the order of 10^8 cm/s, while the Fermi energy is of the order of $1 - 10$ eV.

The total density of states for the volume Nv_c can be derived starting from Eq. (12.20):

$$D(E) = \frac{Nv_c}{4\pi^3}\int_S dS_E \frac{1}{|\nabla_\mathbf{k} E(\mathbf{k})|} = \frac{Nv_c}{2\pi^2}\left(\frac{2m}{\hbar^2}\right)^{3/2} E^{1/2} = \frac{3NZ}{2}\frac{E^{1/2}}{E_F^{3/2}}.$$

The density of states per unit cell is reported below:

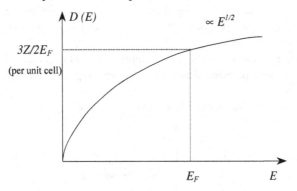

$D(E)$ is often defined per unit volume or per atom.

Now we briefly discuss the situation occurring at finite temperature, when the statistical excitation of the electrons above the Fermi level has to be taken into account. The probability of occupation of the level at energy E is given by the Fermi function

$$f(E) = \frac{1}{e^{\frac{E-\mu}{k_B T}} + 1}, \qquad (12.29)$$

where for temperatures much lower than the Fermi temperature $T_F = E_F/k_B$ the chemical potential μ can be considered to coincide with the Fermi energy E_F (of the order of 10^4 K) (see Problem 12.7).

Then the distribution function and the density of occupied states take the forms plotted below:

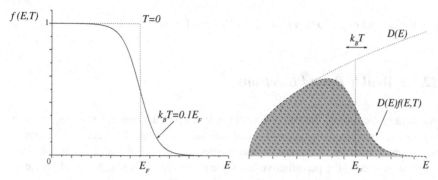

The average energy is

$$< E > = \int ED(E) f(E) dE,$$

and for $T \to 0$

$$< E > = \int_0^{E_F} ED(E) dE = \frac{3}{5} N Z E_F, \qquad (12.30)$$

while at finite temperatures (Problem 12.14) it turns out

$$< E > = \simeq \frac{3}{5} N Z E_F + \frac{\pi^2}{4} N Z k_B T \frac{T}{T_F}. \qquad (12.31)$$

It is noted that the contribution to the energy at $T \neq 0$ takes a form similar to the classical energy $3k_B T/2$ (per electron) times the "fraction" $\sim T/T_F$ of electrons in the neighborhood of the Fermi level.

The specific heat C_V and the magnetic susceptibility χ_P can be derived as illustrated in the Problems 12.14 and 12.10.

A simple way to estimate the order of magnitude of C_V and χ_P is to consider that only a fraction T/T_F of all the electrons can be thermally or magnetically excited. In fact, the states at $E \ll E_F$ are all occupied and the Pauli principle prevents double occupancies. Then, from the classical expressions for Boltzmann statistics one can approximately write

$$C_V \simeq \frac{\partial}{\partial T} \left(\frac{3}{2} n_d k_B T \right) \frac{T}{T_F} = \gamma T, \qquad (12.32)$$

with $\gamma = 3 n_d k_B / T_F$ (the correct expression is $\gamma = \pi^2 D(E_F) k_B^2 /3$, with $D(E_F)$ the density of states at the Fermi energy per unit volume, see Problem 12.14), while

$$\chi_P \simeq \frac{n_d \mu_B^2}{3k_B T} \frac{T}{T_F} = \frac{n_d \mu_B^2}{3k_B T_F}, \tag{12.33}$$

(for the correct expression $\chi_P = \mu_B^2 D(E_F)$ see Problem 12.10).

12.7.2 Weakly Bound Electrons

As already mentioned the free electron model cannot account for the properties of crystals different from metals, as for instance the semiconductors, not even at a qualitative level. In order to explain the basic aspects of those solids one has to take into account, at least in the perturbative limit, the effects of the lattice potential, in the so called *nearly free electron approximation*. Even a weak perturbation causes relevant modifications with respect to the empty lattice situation and yields the appearance of the *gap*, namely the energy interval where no electron states can exist. In particular, a marked gap arises for the electrons at De Broglie wavelength (of the order of the inverse of $|\mathbf{k}|$) close to the lattice step, in analogy with the diffraction phenomenon in optics.

The simplest way to account for the effect of the lattice potential $V(\mathbf{r})$ in modifying the electron dispersion curve $E(\mathbf{k})$ is to consider the perturbative correction to empty-lattice states $E^o(\mathbf{k})$:

$$E(\mathbf{k}) = E^o(\mathbf{k}) + <\mathbf{k}|V(\mathbf{r})|\mathbf{k}> + \sum_{\mathbf{k}' \neq \mathbf{k}} \frac{|<\mathbf{k}|V(\mathbf{r})|\mathbf{k}'>|^2}{E^o(\mathbf{k}) - E^o(\mathbf{k}')}, \tag{12.34}$$

where

$$|<\mathbf{k}|V(\mathbf{r})|\mathbf{k}'> \equiv \int e^{-i(\mathbf{k}-\mathbf{k}')\cdot\mathbf{r}} V(\mathbf{r}) d\mathbf{r}. \tag{12.35}$$

For $\mathbf{k} - \mathbf{k}' \neq \mathbf{g}$, with $\mathbf{g}$ reciprocal lattice vectors, the integral vanishes due to the fast oscillations with $\mathbf{r}$ of the function $e^{-i(\mathbf{k}-\mathbf{k}')\cdot\mathbf{r}}$. Whereas for $\mathbf{k} - \mathbf{k}' = \mathbf{g}$ the matrix element reads

$$|<\mathbf{k}|V(\mathbf{r})|\mathbf{k} - \mathbf{g}> \equiv \int e^{-i\mathbf{g}\cdot\mathbf{r}} V(\mathbf{r}) d\mathbf{r} = V_{\mathbf{g}}, \tag{12.36}$$

which is non zero since it corresponds to the coefficient $V_{\mathbf{g}}$ of the Fourier expansionof the periodic lattice potential (see Eq. (11.3)):

$$V(\mathbf{r}) = \sum_{\mathbf{g}} V_{\mathbf{g}} e^{i\mathbf{g}\cdot\mathbf{r}}. \tag{12.37}$$

It can be remarked that for degenerate states, where $E^o(\mathbf{k}) = E^o(\mathbf{k}')$ at the denominator in Eq. (12.34), one should rely on the perturbation theory for degenerate states and still $< \mathbf{k}|V(\mathbf{r})|\mathbf{k}' >= 0$, for $\mathbf{k}' \neq \mathbf{k} + \mathbf{g}$.

Therefore Eq. (12.34) is rewritten

$$E(\mathbf{k}) = E^o(\mathbf{k}) + V^o + \sum_{\mathbf{g} \neq 0} \frac{|V_{\mathbf{g}}|^2}{E^o(\mathbf{k}) - E^o(\mathbf{k} - \mathbf{g})}, \qquad V^o = \int V(\mathbf{r}) d\mathbf{r} \quad (12.38)$$

modifying the dispersion curve for free electrons, at the second order. The validity of Eq. (12.38) requires the rapid convergence of the series of Fourier components $|V(\mathbf{g})|^2$, which should be granted by a plausible lattice potential (often a *pseudo-potential*). In addition it requires that

$$E^o(\mathbf{k}) \neq E^o(\mathbf{k} - \mathbf{g}),$$

which corresponds to avoid the wavevectors $\mathbf{k}$ at the BZ boundary. In fact, recalling that $E^o(\mathbf{k}) = \hbar^2 k^2/2m$, the condition $E^o(\mathbf{k}) = E^o(\mathbf{k} - \mathbf{g})$ implies $(\mathbf{k})^2 = (\mathbf{k} - \mathbf{g})^2$ and then

$$\mathbf{k} \cdot \mathbf{g} = \frac{g^2}{2}, \qquad (12.39)$$

corresponding to $\mathbf{k}$ at the BZ boundary, as depicted below for a 2D lattice.

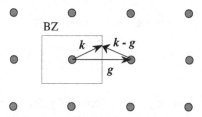

Thus at the BZ boundaries, where the states $\phi_{\mathbf{k}}$ and $\phi_{\mathbf{k}-\mathbf{g}}$ have the same energy, one has the breakdown of the perturbative approach leading to Eq. (12.38).

The situation arising at the zone boundaries can be deduced by mean of arguments based on the perturbation theory for degenerate states. An illustrative example is easily carried out for a *one-dimensional* lattice, with perturbative periodic potential of the form:

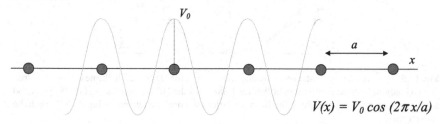

$$V(x) = V_0 \cos (2\pi x/a)$$

The zero-order wave function is

$$\phi_{\mathbf{k}}^{(1)} = c_1\phi_{\mathbf{k}} + c_2\phi_{\mathbf{k}-\mathbf{g}}$$

and the secular equation becomes

$$\begin{pmatrix} < \mathbf{k}|V(\mathbf{r})|\mathbf{k} > -\varepsilon & < \mathbf{k}|V(\mathbf{r})|\mathbf{k}-\mathbf{g} > \\ < \mathbf{k}-\mathbf{g}|V(\mathbf{r})|\mathbf{k} > & < \mathbf{k}-\mathbf{g}|V(\mathbf{r})|\mathbf{k}-\mathbf{g} > -\varepsilon \end{pmatrix} = 0$$

The choice of the potential implies $< \mathbf{k}|V(\mathbf{r})|\mathbf{k} > = < \mathbf{k}-\mathbf{g}|V(\mathbf{r})|\mathbf{k}-\mathbf{g} > = 0$ and

$$< \mathbf{k}|V(\mathbf{r})|\mathbf{k}-\mathbf{g} > = \frac{1}{2a}\int_0^a e^{-igx} V_o (e^{\frac{i2\pi x}{a}} + e^{\frac{-i2\pi x}{a}})dx = \frac{1}{2}V_o. \qquad (12.40)$$

Thus the correction to the unperturbed eigenvalues turns out $\varepsilon_{\pm} = \pm V_o/2$, implying a *gap* for the states around the BZ boundaries, as schematically shown in Fig. 12.5 (to be compared to Fig. 12.4).

The gap can be thought to arise from the *Bragg reflection* occurring when the De Broglie wavelength is $\lambda = 2a$. In fact, in this case (see Eq. (11.9)) the Bragg reflected wave, travelling in opposite direction, induces standing waves, as sketched below:

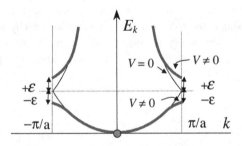

The cosine and sine standing waves formed by the $\pm$ linear combination of $\exp[(\pm ikx)]$, with $k = \pi/a$, yield different distributions of probability density.

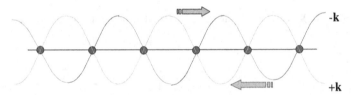

Fig. 12.5 Schematic representation of the dispersion curve for 1D crystal, in the nearly free electron approximation, by taking into account that for **k** far from the BZ boundaries Eq. (12.38) is a good approximation, while approaching the BZ boundaries the correction given by Eq. (12.40) has to be considered

Thus the electron charge densities $\rho(x)$ around the lattice sites in the two cases imply different energies:

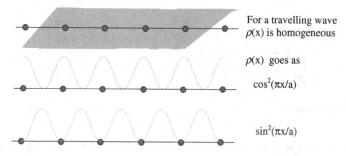

For a travelling wave
$\rho(x)$ is homogeneous

$\rho(x)$ goes as

$\cos^2(\pi x/a)$

$\sin^2(\pi x/a)$

The characteristic feature of the gap generation for perturbed electrons can be derived by constructing the complete **k**-dependence of the eigenvalues in a periodic square-well potential, in one dimension. The potential energy in the Schrödinger equation is assumed $V(x) = 0$ for $0 < x \leq a$ and $V(x) = V_0$ for $a < x \leq a + b$, the lattice parameter being $(a + b)$. *Kronig* and *Penney* solved this artificial model and derived the k-dependent eigenvalues. In the limit where $V(x)$ is characterized by Dirac δ functions separated by distance a (the product $V_0 b$ remaining finite) the dispersion curve in the extended zone scheme has the form sketched below:

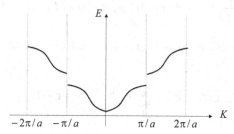

12.7.3 Tightly Bound Electrons

In this model the electrons are assumed to keep, to a large extent, the properties they have in the neighborhood of the atoms. Only in the region in between the atoms sizeable effects occur and the atomic levels are thus spread in a band. The model allows one to understand how the Bloch orbitals are related to the atomic states, in a way similar to the case discussed for the benzene molecule (Sect. 9.3).

Let us refer to the lattice potential reported in Fig. 12.6, along a given direction in the crystal.

By extending the idea of the molecular orbital used for the delocalization of the $2p$ electrons along the C_6H_6 ring, we shall assume a one-electron wavefunction of the form

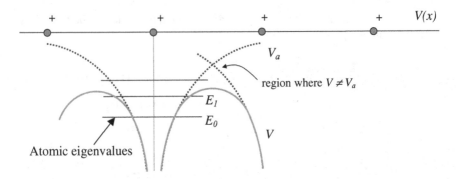

Fig. 12.6 Schematic form of the potential energy for tightly bound electrons, along a given direction in the crystal

$$\phi_{\mathbf{k}} = \sum_{\mathbf{l}} e^{i\mathbf{k}\cdot\mathbf{l}}\phi_a(\mathbf{r} - \mathbf{l}). \tag{12.41}$$

$\phi_a(\mathbf{r} - \mathbf{l})$ is an atomic wavefunction centered at the l-th site and an eigenfunction of the equation

$$\left\{-\frac{\hbar^2}{2m}\nabla^2 + V_a(\mathbf{r} - \mathbf{l})\right\}\phi_a(\mathbf{r} - \mathbf{l}) = E_a\phi_a(\mathbf{r} - \mathbf{l}). \tag{12.42}$$

To show that $\phi_{\mathbf{k}}$ in the form as in Eq. (12.41) is a Bloch orbital, one multiplies by $exp(i\mathbf{k}\cdot\mathbf{r}).exp(-i\mathbf{k}\cdot\mathbf{r})$:

$$\phi_{\mathbf{k}}(\mathbf{r}) = e^{i\mathbf{k}\cdot\mathbf{r}}\sum_{\mathbf{l}} e^{-i\mathbf{k}\cdot(\mathbf{r}-\mathbf{l})}\phi_a(\mathbf{r} - \mathbf{l}).$$

Then it can be observed that the term multiplying the plane wave function has the lattice periodicity and plays the role of $u_{\mathbf{k}}(\mathbf{r})$ in Eq. (12.7), as requested. One also notices that $\phi_{\mathbf{k}}$ in the form 12.41 is a combination of localized atomic orbitals and in the neighborhood of an atom the orbital behaves in a way similar to the one for insulated atoms. The phase factor $exp(i\mathbf{k}\cdot\mathbf{l})$ modifies the orbital from site to site, while $|\phi_{\mathbf{k}}|^2$ is unaffected.

To obtain the eigenvalues $E_{\mathbf{k}}$, the eigenfunction in Eq. (12.41) is inserted in the one-electron Schrödinger equation $(-\hbar^2\nabla^2/2m + V)\phi_{\mathbf{k}} = E_{\mathbf{k}}\phi_{\mathbf{k}}$.

By recalling Eq. (12.42) one obtains

$$(E_{\mathbf{k}} - E_a)\sum_{\mathbf{l}} e^{i\mathbf{k}\cdot\mathbf{l}}\phi_a(\mathbf{r} - \mathbf{l}) = \sum_{\mathbf{l}}(V - V_a)e^{i\mathbf{k}\cdot\mathbf{l}}\phi_a(\mathbf{r} - \mathbf{l}). \tag{12.43}$$

By multiplying both sides of this equation by $\phi_a^*(\mathbf{r} - \mathbf{l})$ and integrating, one has

$$(E_\mathbf{k} - E_a) \sum_\mathbf{l} e^{i\mathbf{k} \cdot \mathbf{l}} \int \phi_a^*(\mathbf{r} - \mathbf{l}')\phi_a(\mathbf{r} - \mathbf{l})d\mathbf{r} =$$

$$= \sum_\mathbf{l} \int \phi_a^*(\mathbf{r} - \mathbf{l}')(V - V_a)e^{i\mathbf{k} \cdot \mathbf{l}}\phi_a(\mathbf{r} - \mathbf{l})d\mathbf{r}. \qquad (12.44)$$

When the orthogonality condition for $\mathbf{l} \neq \mathbf{l}'$ is assumed

$$\int \phi_a^*(\mathbf{r} - \mathbf{l}')\phi_a(\mathbf{r} - \mathbf{l})d\mathbf{r} = 0, \qquad (12.45)$$

by taking into account that the sum in Eq. (12.44) only depends on the difference $\mathbf{h} = \mathbf{l} - \mathbf{l}'$, one finds

$$E_\mathbf{k} = E_a + \sum_\mathbf{h} e^{i\mathbf{k} \cdot \mathbf{h}} \int \phi_a^*(\mathbf{r} + \mathbf{h})V_1\phi_a(\mathbf{r})d\mathbf{r}. \qquad (12.46)$$

In the matrix element in this equation, somewhat analogous to the resonance integral (Sect. 8.1.2), V_1 is the difference between the local $V(\mathbf{r})$ and the atomic potential energy V_a (see Fig. 12.6). The matrix element is *negative*.

For cubic crystal, with atoms of the same species,

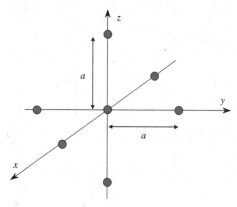

assuming that the matrix element for V_1 is different from zero only when nearest neighbors are involved, Eq. (12.46) takes the form

$$E_\mathbf{k} = E_a + V_o + 2 < V_1 > \left[cos(k_{xa}) + cos(k_{ya}) + cos(k_{za}) \right], \qquad (12.47)$$

depicted in Fig. 12.7, along the k_x direction.

Fig. 12.7 Dispersion
relation $E(\mathbf{k})$ for $\mathbf{k}$ along one
of the reciprocal lattice axis
in a cubic monoatomic
crystal, according to
Eq. (12.47). V_0 is often
negligible

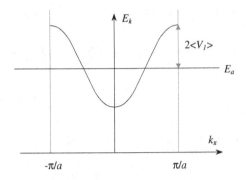

The band $E(\mathbf{k})$ results from the spread of the atomic energy level when the inter-atomic distance in the crystal is reduced. The gap is the direct consequence of the discrete character of the atomic eigenvalues E_a's. One also realizes that the number of states in a single band is $N(2l+1)$, for N atoms in the reference volume of the crystal (l quantum number for the atomic orbital momentum). The band width is proportional to $< V_1 >$ and, therefore, to the *overlap integral*, in a way somewhat equivalent to the molecules (see Sect. 8.1). This explains why the internal bands are narrow and why the cores states are little affected by the formation of the crystal, as sketched below:

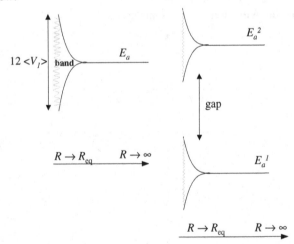

In the framework of the tight binding model the *effective mass* (see Sect. 12.6) of the electron can be derived from Eq. (12.47). For small $\mathbf{k}$, by expanding E_k, one obtains

$$E_\mathbf{k} = E_o + V_o + 6 < V_1 > - < V_1 > a^2 k^2,$$

yielding

$$m^* = \frac{-\hbar^2}{2a^2 < V_1 >} > 0,$$

For $k_x, k_y, k_z \to (\pi/a)$ one has

$$m^* = \frac{\hbar^2}{2a^2 < V_1 >} < 0,$$

and the electron responds to external forces as a positive charge.

When the spread of the atomic levels leads to the superposition of adjacent bands related to different states, one has a *degenerate band* that can be thought to result from hybrid atomic orbitals:

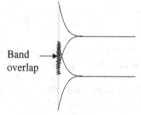

Band overlap

This happens, for instance, in the case of diamond, Si and Ge, as shown in Fig. 12.8.

The energy bands are usually labelled by referring to the atomic orbitals which lead to their formation. Furthermore, since the **k**-dependence of the energy in the reciprocal space reflects all the symmetry properties of the *point group* (see Sect. 11.1), one could classify the electron states in a crystal on the basis of the symmetry properties.

Problems

Problem 12.5 For a one-dimensional crystal the Fourier components of a perturbative potential energy are V_G (Eq. (12.38)), with **G** reciprocal lattice vectors. Evaluate the effective mass m^* for $k = 0$ in terms of the lattice step a. Reformulate the evaluation for $V(x) = 2V_1 \cos(2\pi x/a)$.

Solution: From

$$E_k = E_k^0 + V_0 + \sum_{G \neq 0} \frac{|V_G|^2}{(E_k^0 - E_{k-G}^0)},$$

with $E_k^0 = \hbar^2 k^2/2m$, one has

$$E_k = \frac{\hbar^2 k^2}{2m} + V_0 - \frac{2m}{\hbar^2} \sum_{G \neq 0} \frac{|V_G|^2}{G(G - 2k)}.$$

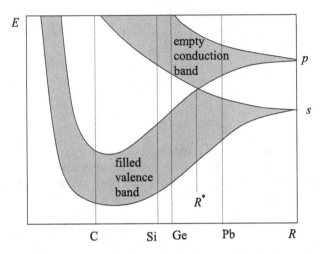

Fig. 12.8 Sketchy picture of the energy bands for the $(ns)^2(np)^2$ electronic configurations showing how, below a certain interatomic distance R, the p and s bands overlap and change the electronic structure of the crystal. While for $R > R^*$ one has a partially filled p band and the possibility of charge transport (see Chap. 13) (this is the case of Pb, in regards of the $6p$ and $6s$ electrons), for $R < R^*$ one has an entirely filled valence band and therefore an insulator. Note that for those elements there are two atoms for each unitary cell. Thus for $R < R^*$ when the zones overlap, the lower zone system is exactly filled by eight electrons per unit cell. When the gap to the upper band (which is empty at $T = 0$) is comparable to the thermal energy $k_B T$, then the electrons can be promoted to the upper conducting band. In this case one can have an intrinsic *semiconductor*, as it happens for Si and Ge, in terms of the $n = 3$ and $n = 4$ electrons (see Sect. 13.1) (figure inspired by Alonso and Finn book)

From Eq. (12.22)

$$\frac{1}{m^*} = \frac{1}{\hbar^2} \frac{\partial^2 E}{\partial k^2} = \frac{1}{m} - \frac{16m}{\hbar^4} \sum_{G \neq 0} \frac{|V_G|^2}{G(G - 2k)^3}.$$

and $G = n2\pi/a$. For $k = 0$ one finds

$$\frac{1}{m^*} = \frac{1}{m} - \frac{ma^4}{\hbar^4 \pi^4} \sum_{n=1}^{\infty} \frac{|V_{G_n}|^2}{n^4}.$$

For $V(x) = 2V_1 \cos(2\pi x/a)$ only V_G for $n = \pm 1$ is non-zero and

$$m^* = m \left(1 - \frac{2m^2 a^4 V_1^2}{\hbar^4 \pi^4} \right)^{-1}.$$

Problem 12.6 For a one-dimensional crystal of lattice step a generalize the result obtained at Sect. 12.7.2 exactly at the zone boundary, in order to obtain the energy $E(k)$ as a function of k for k close to π/a.

Solution: Near $k = \pi/a$ the wavefunction can be written as the linear combination of the two degenerate unperturbed eigenfunctions (see Eq. (12.40)):

$$\psi = c_1 e^{ikx} + c_2 e^{i(k-2\pi/a)x}.$$

By substituting this tentative wavefunction into the Schrödinger equation

$$-\frac{\hbar^2}{2m}\frac{d^2\psi}{dx^2} + V\psi = \varepsilon\psi,$$

first multiply by e^{-ikx} and integrate over all space. Then multiply by $e^{-i(k-2\pi/a)x}$ and again integrate. From the secular equation for c_1 and c_2 (see the equivalent at Sect. 12.7.2), the eigenvalues turns out

$$E = \frac{\hbar^2 k^2}{2m} + \frac{\hbar^2 \pi}{ma}\left\{\left(\frac{\pi}{a}-k\right) \pm \left[\left(\frac{\pi}{a}-k\right)^2 + \left(\frac{amV_0}{2\pi\hbar^2}\right)^2\right]^{1/2}\right\}.$$

Problem 12.7 Consider a metal with one electron per unit cell in geometric dimension $n = 1, 2$ and 3 and derive the density of states $D(E)$ as a function of n. Then give a general expression for $D(E)$ in terms of the Fermi energy. Finally derive the chemical potential μ (Hint: write the total number of electrons in terms of the Fermi distribution and use the identity $\int_{-\infty}^{+\infty} f(t)\frac{e^t dt}{(1+e^t)^2} = f(0) + \frac{\pi^2}{6}f''(0)$).

Solution: From $E_F = \hbar^2 k_F^2/2m$, since the total number of states including the spin degeneracy is

$$N = 2D(\mathbf{k})\frac{4\pi k_F^3}{3} \quad \text{for } n = 3, \quad N = 2D(\mathbf{k})\pi k_F^2 \quad \text{for } n = 2,$$

$$N = 2D(\mathbf{k})2k_F \quad \text{for } n = 1,$$

one finds $E_F = (\hbar^2/2m)(3\pi^2 N/V)^{2/3}$ for $n = 3$, $E_F = \pi\hbar^2 N/mA$ for $n = 2$ and $E_F = (\hbar^2/2m)(N/2L)^2$ for $n = 1$.

One can write

$$D(\mathbf{k})d^n k = 2\left(\frac{\sqrt{2m}}{2\pi\hbar}\right)^n d^n x$$

with $x = \sqrt{E}$. From $D(E)dE = \int D(\mathbf{k})d^n k$

$$D(E) = \frac{3N}{2}\frac{E^{1/2}}{E_F^{3/2}} \quad \text{for } n = 3, \quad D(E) = \frac{mA}{\pi\hbar^2} \quad \text{for } n = 2,$$

$$D(E) = \frac{N}{2}\frac{E^{-1/2}}{E_F^{1/2}} \quad \text{for } n = 1.$$

In general $D(E)dE = Nd(E/E_F)^{n/2}$.

The total number of occupied states at finite temperature is

$$N = \int_0^\infty \frac{D(E)}{e^{\beta(E-\mu)}+1}dE = N\int_0^\infty \frac{1}{e^{\beta(E-\mu)}+1}d\left(\frac{E}{E_F}\right)^{n/2} =$$

$$= N\int_{-\beta\mu}^\infty \frac{1}{e^t+1}d\left(\frac{\mu+(t/\beta)}{E_F}\right)^{n/2} = N\int_{-\beta\mu}^\infty \frac{e^t}{(e^t+1)^2}\left(\frac{\mu+(t/\beta)}{E_F}\right)^{n/2}dt$$

In the low temperature limit ($T \ll T_F$, $-\beta\mu \to -\infty$) one has

$$1 = \left(\frac{\mu}{E_F}\right)^{n/2} + \frac{\pi^2}{6}\frac{n(n-2)}{4}\left(\frac{T}{T_F}\right)^2\left(\frac{\mu}{E_F}\right)^{(n/2)-2} + \dots$$

yielding

$$\mu = E_F\left(1 - \frac{\pi^2}{12}(n-2)\left(\frac{T}{T_F}\right)^2 + \dots\right)$$

Problem 12.8 The specific mass (density) of alluminum is $d = 2.7\,\text{g/cm}^3$. Evaluate the Fermi energy, the Fermi velocity, the average velocity of the conduction electrons and the quantum pressure (for $T \to 0$).

Solution: The number of atoms per cubic cm turns out $N = 0.54 \cdot 10^{23}$. For three free-electrons per atom from Eq. (12.28)

$$E_F = \frac{\hbar^2}{2m}(3\pi^2 N_e/V)^{2/3} = 11.7\,\text{eV},$$

(with $N_e = 3N$) and $v_F = \sqrt{2E_F/m} = 2.03 \cdot 10^8\,\text{cm/s}$. The distribution function for the velocities is

$$p(v)dv = D(E)dE = N_e d\left(\frac{v}{v_F}\right)^3 = 3N_e\frac{v^2}{v_F^3}dv$$

and then

$$< v > = \int_0^{v_F} v p(v) dv = \frac{3}{4} v_F = 1.5 \cdot 10^8 \text{ cm/s}.$$

From Eq. (12.30) and $P = -\partial < E > /\partial V$ one has $P = (2/3)(3/5) (N_e/V)E_F \simeq 1.2 \cdot 10^{12}$ dyne/cm^2.

Problem 12.9 In the assumption that the electrons in a metal can be described as a classical free-electron gas, show that no magneticsusceptibility would arise from the orbital motion.

Solution: The magnetization is $M = N k_B T d(\ln Z/dH)_T$. The classical partition function is

$$Z \propto \int_{-\infty}^{\infty} d_x d_y d_z \int_{-\infty}^{\infty} dp_x dp_y dp_z exp\left[-\frac{E}{k_B T}\right]$$

with

$$E = \frac{(m\mathbf{v})^2}{2m} = \frac{(\mathbf{p} + \frac{e}{c}\mathbf{A})^2}{2m}$$

with $\mathbf{A}$ the vector potential and $\mathbf{p} = m\mathbf{v} - (e/c)\mathbf{A}$ the canonical moment. By transforming the volume element in the phase space from canonical to kinetic moments one has

$$Z \propto \int_0^{\infty} d(m\mathbf{v})(m\mathbf{v})^2 exp\left[-\frac{(m\mathbf{v})^2}{2mk_B T}\right],$$

field independent, thus implying $M = 0$. This is the physical content of the *Bohr-van Leeuwen theorem*.

Problem 12.10 Derive the paramagnetic susceptibility due to the free electrons in a metal (*Pauli susceptibility*).

Solution: In the absence of a magnetic field the number of electrons with spin up N_+ is equal to the number of electrons with spin down N_- and the total magnetization $M = \mu_B(N_+ - N_-)$ is zero.

From the field H the energy of spins up is lowered by an amount $\mu_B H$, while the one of the spins down is increased by the same amount and the unbalance in the populations yields the magnetization. The number of spins up is

$$N_+ = \int_{-\mu_B H}^{\infty} f(E, T) \frac{D(E + \mu_B H)}{2} dE,$$

(the factor $1/2$ in the density of states $D(E)$ takes into account that only the electrons with spin up are considered). Introducing $E' = E + \mu_B H$ one can write

$$N_+ = \int_0^\infty f(E' - \mu_B H, T) \frac{D(E')}{2} dE'.$$

In the weak-field limit $f(E' - \mu_B H, T) \simeq f(E', T) - \mu_B H (\partial f / \partial E)_{E'}$ and

$$N_+ = \frac{1}{2} \int_0^\infty f(E', T) D(E') dE' - \frac{1}{2} \int_0^\infty \mu_B H \left(\frac{\partial f}{\partial E} \right)_{E'} D(E') dE'.$$

In the same way

$$N_- = \frac{1}{2} \int_0^\infty f(E', T) D(E') dE' + \frac{1}{2} \int_0^\infty \mu_B H \left(\frac{\partial f}{\partial E} \right)_{E'} D(E') dE'$$

For $H \to 0$ $\chi_P = M/H$ and therefore, from $M = \mu_B (N_+ - N_-)$,

$$\chi_P = \mu_B^2 \int_0^\infty \left(\frac{-\partial f}{\partial E} \right)_{E'} D(E') dE'$$

For $E_F \gg k_B T$ and provided that the density of states varies smoothly around E_F, one writes $(-\partial f / \partial E)_{E'} \simeq \delta(E' - E_F)$, so that

$$\chi_P = \mu_B^2 D(E_F)$$

(see Eq. (12.33)). $D(E)$ is the density of states per unit volume and thus χ_P is dimensionless.

According to the above equation one finds for the electron contribution to the spin susceptibility in alkali metals:

$$\text{Li}:\ 1 \times 10^{-5}, \quad \text{Na}:\ 0.83 \times 10^{-5}, \quad \text{K}:\ 0.67 \times 10^{-5},$$

$$\text{Rb}:\ 0.63 \times 10^{-5}, \quad \text{Cs}:\ 0.58 \times 10^{-5}.$$

The experimental data are 2.5, 1.4, 1.1, 1 and 1 (in 10^{-5} units), respectively.

Problem 12.11 For a cubic metal at lattice step $a = 5$ Å and electron density 2×10^{-2} electrons per cell, evaluate the temperature at which the electron gas can be considered degenerate and write the approximate form for the specific heat well above that temperature.

Solution: The electron density is

$$n = \frac{2 \times 10^{-2}}{a^3} = 1.6 \times 10^{20} \text{ cm}^{-3}$$

and the average spacing among the electrons is $d \simeq (3/4\pi n)^{1/3} \simeq 13$ Å.

The electron gas can be considered degenerate when $d \leq \lambda_{DB}$, the *De Broglie* wavelength. Since $\lambda_{DB} \simeq h/\sqrt{3mk_BT}$, the gas can be considered degenerate for $T < h^2/(3d^2mk_B) \simeq 8600$ K. Above that temperature the gas is practically a classical one and the specific heat is $C_V \simeq (3/2)k_Bn$.

Problem 12.12 The bulk modulus $B = -V(\partial P/\partial V)_T$ of potassium crystal at low temperature is $B = 0.28 \times 10^{11}$ dyne/cm². Discuss this result in the assumption that B is entirely due to the electron gas.

Solution: The pressure of the electron Fermi gas is $P = (2/5)nE_F$, with n electron density (see Problem 12.8). Then

$$B = -V\frac{\partial P}{\partial V} = \frac{2}{3}nE_F$$

For a specific mass of 0.86 g/cm³, the electron density is $n = 1.4 \times 10^{22}$ cm⁻³ and the Fermi energy is $E_F = 2.1$ eV. Then $B \simeq 0.32 \times 10^{11}$ dyne/cm², in rather good agreement with the experimental finding.

Problem 12.13 Prove that in a semiconductor at thermal equilibrium the concentration of electrons and of vacant states, called *holes*, in the valence band are given by

$$n \simeq N_c e^{-(E_c-E_F)/k_BT}, \quad p \simeq N_v e^{-(E_F-E_v)/K_BT}$$

where

$$N_c = 2\left(\frac{2\pi m_e k_B T}{h^2}\right)^{3/2} \quad N_v = 2\left(\frac{2\pi m_h k_B T}{h^2}\right)^{3/2},$$

E_F is the Fermi level (in the middle of the gap), E_c the bottom of the conduction band and E_v the top of the valence band (m_e and m_h are the effective masses of electrons and of holes). Assume parabolic bands, going as $(k - k_c)^2$ and $(k - k_v)^2$.

Then evaluate

(a) the value of N_c for $m_c = m$ (m the electron mass) and $T = 300\,°$K;
(b) the carriers concentration in Si, at $T = 300\,°$K, assuming $m_c = m_h = m$, and a gap of 1.14 eV.

Solution: At thermal equilibrium the concentration of electrons is given by

$$n = \int_{E_{\min}}^{E_{\max}} D_c(E)f(E)dE$$

the density of states $D_c(E)$ per unit volume being

$$D_c(E) = (4\pi/h^3)(2m_c)^{3/2}(E - E_c)^{1/2}.$$

The Fermi-Dirac distribution for $(E - E_F) \gg k_B T$ can be approximated as

$$f(E) = \frac{1}{1 + e^{(E-E_F)/k_B T}} \simeq e^{-(E-E_F)/k_B T}$$

Then

$$n = \frac{4\pi}{h^3}(2m_c)^{3/2} e^{-(E_c-E_F)/k_B T} \int_0^{E_{\max}-E_c} x^{1/2} e^{-x/k_B T} dx$$

$$\simeq \frac{4\pi}{h^3}(2m_c)^{3/2} e^{-(E_c-E_F)/k_B T} \int_0^{\infty} x^{1/2} e^{-x/k_B T} dx$$

$$= \frac{4\pi}{h^3}(2m_c)^{3/2} [e^{-(E_c-E_F)/k_B T}] \frac{1}{2}\pi^{1/2}(k_B T)^{3/2}$$

yielding

$$n = 2\left(\frac{2\pi m_e k_B T}{h^2}\right)^{3/2} e^{-(E_c-E_F)/k_B T}$$

In analogous way the hole concentration can be derived.

Since at 300 K $N_c = 2.5 \times 10^{19}$ cm^{-3} the electron and hole concentrations in Si turn out $n = p = 3.14 \times 10^9$ cm^{-3}. It can be remarked that this value refers to pure Si (intrinsic semiconductor), while in practice impurities induce larger carrier concentrations.

Problem 12.14 Derive the contribution to the specific heat associated with the conduction electrons in a metal, for temperature small compared to E_F/k_B.

Solution: In a way analogous to the derivation of the Pauli susceptibility (see Problem 12.10) the increase of the electron energy when the temperature is brought from 0 to T is written in the form

$$U(T) = \int_{E_F}^{\infty} (E - E_F)f(E)D(E)dE - \int_0^{E_F} (E - E_F)(1 - f(E))D(E)dE.$$

In the second integral $(1 - f(E))$ gives the probability that an electron is removed from a state at energy below E_F. Then

$$C_V = \int_0^\infty (E - E_F)\frac{\partial f}{\partial T}D(E)dE \simeq D(E_F)\int_0^\infty (E - E_F)\frac{\partial f}{\partial T}dE.$$

Since

$$\frac{\partial f}{\partial T} = \frac{(E - E_F)}{k_B T^2}\frac{e^{(E-E_F)/k_B T}}{[e^{(E-E_F)/k_B T} + 1]^2}$$

by utilizing $\int_{-\infty}^\infty x^2 e^x dx/(e^x + 1)^2 = (\pi^2/3)$, one obtains

$$C_V = \frac{\pi^2}{3}D(E_F)k_B^2 T$$

This result can be read as the derivative of the product $k_B T$ times the fraction T/T_F of the electrons in the energy range $k_B T$ around E_F (see Eqs. (12.31) and (12.32)).

Problem 12.15 Derive the equation of state (relation between P, V and T) for the Fermi gas, in the limit $T \to 0$.

Solution: From the energy (see Eq. (12.31))

$$U = (3/5)NE_F\left(1 + \frac{5\pi^2}{12}\left(\frac{k_B T}{E_F}\right)^2 + \cdots\right)$$

with

$$E_F = (\hbar^2/2m)\left(3\pi^2\frac{N}{V}\right)^{2/3}$$

(Eq. (12.28)), one writes

$$P = -\frac{\partial U}{\partial V} = \frac{2}{5}\frac{NE_F}{V}\left(1 - \frac{5\pi^2}{18}\left(\frac{k_B T}{E_F}\right)^2 + \cdots\right)$$

i.e.

$$PV = (2/5)NE_F\left(1 - \frac{5\pi^2}{18}\left(\frac{T}{T_F}\right)^2 + \cdots\right).$$

Problem 12.16 The temperature dependence of the specific heat in Gallium is reported in the figure

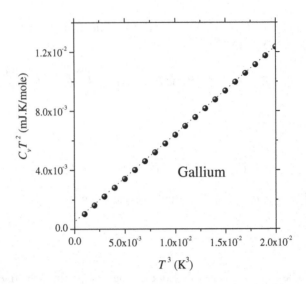

Noticing that at low temperature the contribution to the specific heat due to lattice vibrations (see Sect. 14.5) can be neglected, from the data derive the Fermi energy and the electric field gradient at the nucleus (assume for simplicity that ^{69}Ga with $I = 3/2$ and $Q = 0.168$ barn is the only isotope).

Solution: From the experimental data one deduces the straight line $C_v T^2 = (a + bT^3)$ with $a \simeq 4 \cdot 10^{-4}$mJ · K/mole and $b \simeq 0.6$ mJ/moleK^2. The second contribution is associated with conduction electrons. From $C_v = (\pi^2/3) \cdot k_B{}^2 T D(E_F)$, one derives $E_F \simeq 5.6$ eV (see Problem 12.14 and Sect. 12.7.1).

The first term for C_V, going as $1/T^2$, is the high-temperature tail of the Schottky-like specific heat C_v^Q associated with the hyperfine split by quadrupolar inter-action, with energy separation E. Since for $k_B T \gg E$, for a mole one has $C_v^Q = (1/4)N_A k_B (E/k_B T)^2$ (see Problem 5.12), one finds $E \simeq 6.3 \times 10^{-20}$ erg. For $I = 3/2$ the splitting between the $M_I = \pm 1/2$ and $M_I = \pm 3/2$ levels due to quadrupole interaction is $E = eQV_{zz}/2$, with V_{zz} the principal component of the electric field gradient (see Sect. 5.3). Then one obtains $V_{zz} \simeq 16 \times 10^{14}$ u.e.s./cm^3.

Problem 12.17   Consider two cubic clusters of Lithium (lattice step $a = 3.5$ Å and bcc structure) formed by 1.6×10^7 and 16,000 atoms, respectively. Evaluate the Fermi energy for each cluster and estimate the separation among the electronic levels in proximity of the center of the Brillouin zone.

Solution: The electron density is $n = 2/a^3 = 4.6 \times 10^{22}$ cm^{-3} and the Fermi energy $E_F = 4.7$ eV, size independent. The size affects the spacing among k states. The first cluster is a cube of size $L_1 = 200a$, while the second one of size $L_2 = 20a$. Then the separation among the lowest energy levels is

$$\Delta E_{1,2} = \frac{\hbar^2}{2m}\left(\frac{\pi}{L_{1,2}}\right)^2,$$

namely $\Delta E_1 = 0.11 \times 10^{-15}$ erg and $\Delta E_2 = 1.1 \times 10^{-14}$ erg $\simeq 6.9$ meV, corresponding to $T \simeq 80$ K. Quantum size effects can be expected at low temperature.

Problem 12.18 The density of Lithium is 0.53 g/cm^3. Evaluate the contribution to the bulk modulus due to electrons, in the low temperature range. Compare the estimated value with the experimental result $B \simeq 0.12 \times 10^{12}$ dyne/cm^2.

Solution: From Problem 12.12, the electron density being $n = 4.7 \times 10^{22}$ cm^{-3} and the Fermi energy $E_F = 4.74$ eV, then $B = 2.4 \times 10^{11}$ dyne/cm^2, not far from the experimental result.

Problem 12.19 In semiconductors the concentration of itinerant electrons is low and one can expect that the Pauli susceptibility turns to a Curie-like susceptibility characteristic of localized electrons. Discuss the derivation of the Pauli susceptibility for semiconductors (neglect the electron-electron Coulomb interaction).

Solution: The Pauli susceptibility is

$$\chi_P = \mu_B^2 \int_0^\infty \left(\frac{-\partial f}{\partial E} \right)_{E'} D(E') dE'$$

(see Problem 12.10).

For diluted Fermi gas, at room temperature the statistical distribution function can be written $f(E) \simeq e^{-(E-E_F)/k_B T}$ (see Problem 12.13).

Thus $-\partial f / \partial E = f/k_B T$ and the susceptibility turns out

$$\chi_P = \frac{\mu_B^2}{k_B T} \int_0^\infty f(E') D(E') dE' = n \frac{\mu_B^2}{k_B T}$$

with n concentration of conduction electrons.

Specific References and Further Reading

1. J.C. Slater, *Quantum Theory of Matter*, (McGraw-Hill, New York, 1968).
2. J.S. Blakemore, *Solid State Physics*, (W.B. Saunders Co., 1974).
3. J.M. Ziman, *Principles of the Theory of Solids*, (Cambridge University Press, 1964).
4. G. Grosso and G. Pastori Parravicini, *Solid State Physics*, 2nd Edition, (Academic Press, 2013).
5. N.W. Ashcroft and N.D. Mermin, *Solid State Physics*, (Holt, Rinehart and Winston, 1976).
6. G. Burns, *Solid State Physics*, (Academic Press Inc., 1985).
7. H. Ibach and H. Lüth, *Solid State Physics: an Introduction to Theory and Experiments*, Springer Verlag (1990).
8. M. Alonso and E.J. Finn, *Fundamental University Physics Vol. III- Quantum and Statistical Physics*, (Addison Wesley, 1973).
9. C. Kittel, *Introduction to Solid State Physics*, 8th Edition, (J. Wiley and Sons, 2005).
10. H.J. Goldsmid (Editor), *Problems in Solid State Physics*, (Pion Limited, London, 1972).
11. H. Kuzmany, *Solid-State Spectroscopy*, (Springer-Verlag, Berlin, 1998).
12. L. Mihály and M.C. Martin, *Solid State Physics - Problems and Solutions*, (J. Wiley, 1996).

Chapter 13
Miscellaneous Aspects Related to the Electronic Structure

Topics

Covalent, Metallic, Ionic and Molecular Crystals
Cohesive Energies and Bonding Mechanisms
Lennard-Jones Potential
Crystal-field Effects in Magnetic Ions
Electric Current Flow
Magnetic Properties of Itinerant Electrons

13.1 Typology of Crystals

In the light of the main aspects involving the electronic properties, a classification of crystalline solids can be devised. This can be done either in a valence-bond scenario by looking at the bonding mechanisms or by referring to the electric conduction and the band structure.

In the first case the crystals can be divided in *covalent, metallic, ionic* and *molecular*. In *covalent crystals* the bonding mechanism and the strength of bonds are similar to the ones in covalent molecules. In other words, the crystal can be conceived as a "macroscopic"molecule with marked directional bonds between pairs of atoms where spin-paired electrons can be placed. Therefore, covalent crystals are stiff, scarcely plastic and fragile. Illustrative examples can be found in carbon-based crystals, such as diamond and graphite. Diamond, as well as the isostructural Ge, Si, Sn and Pb crystals, result from an ideally infinite network of sp^3 hybrid orbitals (Sect. 9.2). On the other hand, in graphite the sp^2 hybridization yields a planar atomic arrangement, with weak interaction among adjacent planes (see Sect. 11.3).

Metallic crystals are somewhat equivalent to large molecules with electrons delocalized through all the volume, an extension of what discussed in benzene (Sect. 9.3). The description of these systems in a VB-like framework would require

© Springer International Publishing Switzerland 2015
A. Rigamonti and P. Carretta, *Structure of Matter*,
UNITEXT for Physics, DOI 10.1007/978-3-319-17897-4_13

the superposition of a large number of equivalent configurations. It is evident that a Bloch-like approach is more convenient for the metallic crystals. A suitable way to describe these solids is to refer to a model of positive ions at the lattice sites embedded in a *sea* of electrons, with a nearly uniform charge distribution. In general the bonds are not saturated. For instance, in Li metal (bcc structure) each ion has 8 nearest neighbors and in a molecular-like picture one can think that there is $1/4$ of electron on each orbital.

In *ionic crystals* the electrons are characterized by molecular-like orbitals centered at the atoms having larger electronegativity, as in the case of strongly heteronuclear molecules (see Sect. 8.5). The attractive interaction may often be approximated to the one for point charge ions and in a crude approximation the ions can be assumed to have the closed shells configurations. For example, in LiF crystal, the $(1s)^2$ shell for Li^+ and the $(2p)^6$ shell for F^-. From the X-ray diffraction peaks one can estimate the actual number of electrons at a given site. For instance, in NaCl it turns out that there are 17.85 electrons at Cl site. Thus the order of magnitude of the bond energy *per pair* is $-(0.85e)^2/R$, with R interatomic distance.

The hydrogen bond O-H-O typical of hydrides, of the ferroelectric KDP (potassium dihydrogen phosphate) and of other organic compounds, can be considered as a type of ionic bond. The hydrogen atom can be thought in a local double-well potential. Several electric and elastic properties of these crystals are rather well explained within this simple model.

In molecular-like scenarios one can hardly devise any bonding mechanism for neutral molecules at high ionization energy or for closed shell atoms, such as inert gases. In these cases the aggregation into a solid state can occur because of an interaction that we have not directly considered in molecules: the *Van der Waals* forces, associated with fluctuating electric dipoles. This mechanism yields a weak attractive potential decreasing as R^{-6} and leads to the formation of *molecular crystals* (a mention has been given in Problem 8.6 and that interaction shall be described in some detail at Sect. 13.2.2).

The classification scheme based on the bonding mechanisms is not very suited to describe the properties involving the electrical transport. This aim is better achieved by referring to the band scheme, in the framework of Bloch orbitals. Let us remind that a band arising from s atomic states can be occupied by $2N$ electrons (N the number of atoms in a reference volume) while in the p band this number is $3 \times 2N$. As already mentioned, the electrical conductivity originates from the "acceleration"(namely from the change of state) induced by the electric field, for a single electron described by the equation $\hbar d\mathbf{k}/dt = -e\mathbf{E}$ (see Sect. 12.3). A few observations can be made in regards of the current flow (for some more detail see Sect. 13.4). In a fully filled band each $\mathbf{k}$ state is occupied by two electrons and a neat flow of current is not possible (unless the electric field is so strong to alter the unperturbed bands) and one has an *insulator*. For a partially filled band electrical transport can occur and one has a *conductor*. When the gap between full *valence band* and an empty *conduction band* is of the order of 0.1–1 eV, then one has an intrinsic *semiconductor*. These crystals are insulators for $T \to 0$, while progressive increase in the conductivity with increasing temperature occurs, as a consequence of the partial filling up of the conduction band. At variance, in a metal the conductivity decreases with increasing temperature due to the increase in the scattering rate between the electrons and the ionic vibrational modes.

In such a scenario one can predict that alkali crystals as Li, Na, Rb, etc. and transition metals as Cu, Ag and Au are metallic conductors, since they have an odd number of electrons per unit cell. This rule however, is not quite valid and often one has to pay attention to other details. For instance, although As, Sb and Bi atoms convey five electrons in the conduction band, they generate a crystal which is essentially insulator. Without going into the real aspects of the electronic band structure, we only mention that the reason for the quasi-insulating character is related to the generation of five bands that are completely filled by the valence electrons of the two atoms present in the unit cell.

In some crystals there is also the possibility of a tiny superposition of bands, giving a small metallic character and causing electrical conduction with a particular temperature dependence: these are the *semimetals*.

Strong *band overlap* (see Sect. 12.7.3 and Fig. 12.8) can drastically change the simplified picture given above. For instance, according to the previous statements Be crystal (atomic configuration $(1s)^2(2s)^2$) should be insulator. This is not the case: the overlap between s and p orbitals generates a partially occupied *hybrid band*, a situation similar to the one in diamond (Fig. 12.8). The overlap of these bands yields a fully occupied valence band and an empty conduction band in this latter crystal. At the equilibrium distance characteristic of Si and Ge the gap between the two band diminishes and a semiconducting behavior can be observed. On the other hand, Sn can undergo a transition from metallic to semiconductor, in view of the proximity to the overlap condition. Finally Pb is a metal, since the $6p$ band is only partially filled (Fig. 12.8).

Semiconducting behavior can be expected for a class of materials with tetrahedral structure generated by sp^3 hybridization. The so-called III-V semiconductors are the crystals in which the basis, instead of being formed by the same atoms at $(0,0,0)$ and $(1/4,1/4,1/4)$ in the fcc structure (as in C, Si, Ge and Sn) involves one element of the third group (Ga for instance) and one of the fifth group (As, for example). The covalent "transfer" of one electron from As to Ga gives rise to the $s^2 p^2$ configuration in both atoms, as in Ge or Si, thus triggering the sp^3 hybrid bands and the semiconducting behavior. The band gaps in Ge (0.75 eV) and in Si (1.14 eV) (indirect gaps, the maximum of the valence band and the minimum of the conduction band occurring at different points of the Brillouin zone) are of the same order of magnitude in GaAs (1.52 eV) and in GaSb (0.81 eV).

In ionic crystals the gap between the fully occupied valence band and the empty conduction band can be larger than about 5 eV, thus explaining their insulating behavior.

13.2 Bonding Mechanisms and Cohesive Energies

The *cohesive energy* is defined as the difference between the energies of the atoms for interatomic distance $R \to \infty$ and the one for $R = R_e$, the interatomic equilibrium distance. From thermodynamical and spectroscopic measurements the order of magnitude of the cohesive energies turn out:

(i) around 5 eV/atom in covalent crystals (e.g. 7.36 eV for diamond);

(ii) around 1 eV/atom in metallic alkali crystals;

(iii) around 5-10 eV/(pair of atoms) in ionic crystals;

(iv) from 10^{-2} to 10^{-1} eV in molecular crystals, with a sizeable increase in the binding energy with increasing atomic number for inert atoms crystals.

Quantitative estimates of the cohesive energy are evidently difficult, since in principle they correspond to the derivation of the eigenvalues in the Schrödinger equation for the electronic states. It is possible to achieve satisfactory descriptions of the relevant aspects of the binding mechanisms and to obtain rather good estimates of the cohesive energies by referring to limit ideal situations. For instance, one usually refers to molecular-like scenarios or to ionic atomic configurations. In covalent crystals, where the bonds are similar to the ones in molecules, the cohesive energy per molecule is expected around the one described at Chap. 8. The bonding mechanism in metals can be considered as due to the attractive term related to the electron delocalization (favored by the band overlap) and the repulsive term arising from the increase of the Fermi energy when the electron density increases (see Eq. (12.28)).

More quantitative descriptions of the binding energies for ionic and molecular crystals shall be given in the subsequent subsections.

13.2.1 Ionic Crystals

Let us refer to a crystal with N positive and N negative ions, per cubic cm. In the point-charge approximation the interaction between two ions is written

$$V_{ij} = \pm \frac{e^2}{R_{ij}} + Be^{-R_{ij}/\rho}, \tag{13.1}$$

where the sign of the first term depends on the signs of the charges at the ith and jth ions. The second term has the *Born-Mayer* form used to take into account the short-range repulsion in heteronuclear molecules (Eq. (8.36)).

For a given ith ion the energy is $V_i = \sum_j' V_{ij}$ and by writing the distance $R_{ij} = p_{ij}R$ (R being the nearest neighbour distance) one has

$$V_i = \sum_j' V_{ij} = \frac{e^2}{R} \sum_j' \frac{(\pm 1)}{p_{ij}} + zBe^{-R/\rho}, \tag{13.2}$$

where in view of its short-range character the repulsive term has been limited to the z nearest neighbours. Then the total energy becomes

$$V_T = N V_i = -\frac{Ne^2}{R}\alpha + NzBe^{-R/\rho}, \tag{13.3}$$

with

$$\alpha = \sum_j \frac{(\pm 1)}{p_{ij}}, \tag{13.4}$$

the *Madelung constant*.

From Eq. (13.3) one realizes that α has to be positive, in order to grant the aggregation of the ions to form a crystal. At the equilibrium interatomic distance

$$\left(\frac{dV}{dR}\right)_{R=R_e} = 0 = \frac{N\alpha e^2}{R_e^2} - \frac{Nz}{\rho}Be^{-R_e/\rho} \tag{13.5}$$

and then $(\rho\alpha e^2/zB) = R_e^2 e^{-R_e/\rho}$, thus giving for the total energy

$$V_T^{eq} = -\frac{N\alpha e^2}{R_e}\left[1 - \frac{\rho}{R_e}\right]. \tag{13.6}$$

The characteristic constant $\rho \ll R_e$ can be estimated from the crystal compressibility (see Problem 13.1) and usually turns out of the order of $0.1 R_e$. Thus Eq. (13.6) shows that the cohesive energy is of the order of the dissociation energy of the ideal molecule formed by positive and negative ions and that it is largely controlled by the Madelung constant.

The estimate of α is not trivial due to the slow convergence of the sum in Eq. (13.4). It has to be noticed that the series of the positive and of the negative terms, taken separately, diverge in 3D crystals. Numerical methods to grant fast convergence for α have been devised a long ago, based on the choice of reference regions where the monopole contribution vanishes (*Ewald* procedure). The remaining dipole or quadrupole contributions converge with increasing distance faster than the Coulomb terms.

Typical values for the Madelung constants are $\alpha = 1.7475$ for crystals with NaCl-type structure and $\alpha = 1.7626$ for crystals with CsCl-type structure. In the simple case of a chain with alternating positive and negative ions the evaluation of α is straightforward:

$$\alpha = 2\left[1 - \frac{1}{2} + \frac{1}{3} - \frac{1}{4} + ...\right] = 2ln2. \tag{13.7}$$

13.2.2 Lennard-Jones Interaction and Molecular Crystals

In order to describe the bonding mechanism in molecular crystals let us first derive the form of the attractive interaction among the atoms when no molecular-like mechanisms (as the ones described in Chaps. 8 and 9) are active. The mechanism we shall consider originates from *fluctuating electric dipoles*, first described as *Van der Waals interaction* and later on known in the quantum mechanical scenario as *London interaction*.

To derive the London interaction let us refer to two hydrogen atoms along the x direction:

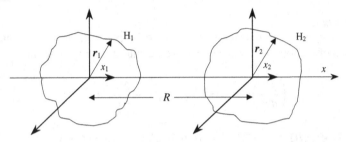

The distance R is larger than the one at which the bonding mechanisms leading to the Hydrogen molecule would become relevant (in other words R is a distance where the overlap, resonance or exchange integrals can be neglected; see Problem 8.6). Then the unperturbed wavefunction is

$$\phi^o(1,2) = \phi_{n_1, l_1}(\mathbf{r_1})\phi_{n_2, l_2}(\mathbf{r_2}), \tag{13.8}$$

with eigenvalue $E^o = E^o_{n1, l1} + E^o_{n2, l2}$. The perturbation Hamiltonian is the dipolar one

$$\mathcal{H}_d = \frac{e^2}{R^3}\left[\mathbf{r}_1 \cdot \mathbf{r}_2 - 3(\mathbf{r}_1 \cdot \hat{x})(\mathbf{r}_2 \cdot \hat{x})\right],$$

that is rewritten in the form (see Problem 8.6)

$$\mathcal{H}_d = -\frac{e^2}{R^3}\left[2x_1 x_2 - y_1 y_2 - z_1 z_2\right]. \tag{13.9}$$

From second order perturbation theory the ground-state energy turns out

$$E(R) = 2E^o_{1s} + <0|\mathcal{H}_d|0> + \sum_{k \neq 0} \frac{<0|\mathcal{H}_d|k><k|\mathcal{H}_d|0>}{E^o_0 - E^o_k} \tag{13.10}$$

$\mathcal{H}_d$ is an odd function and $<0|\mathcal{H}_d|0> = 0$.

By resorting to arguments already used in the derivation of the atomic polarizability (Sect. 4.2) and noticing that the denominator varies from $-e^2/a_o$ to $-3e^2/4a_o$, one can write

$$E(R) \simeq 2E_{1s}^o - \frac{a_o}{e^2}\left[\sum_k < 0|\mathcal{H}_d|k >< k|\mathcal{H}_d|0 >- < 0|\mathcal{H}_d|0 >< 0|\mathcal{H}_d|0 >\right]$$

$$= 2E_{1s}^o - \frac{a_o}{e^2} < 0|\mathcal{H}_d^2|0 >. \qquad (13.11)$$

Thus from Eq. (13.9),

$$E(R) \simeq 2E_{1s}^o - \frac{a_o}{e^2}\frac{e^4}{R^6}\left[4 < x_1^2 >< x_2^2 > + < y_1^2 >< y_2^2 > + < z_1^2 >< z_2^2 >\right].$$
$$(13.12)$$

For Hydrogen the expectation values of the square of the components x, y and z are $< r^2 > /3 = a_o^2$ and then

$$E(R) \simeq 2E_{1s}^o - \frac{6e^2a_o^2}{R^6}a_o^3, \qquad (13.13)$$

showing that an attractive interaction has arisen.

The London interaction can be depicted as related to the dipolar interaction between an instantaneous dipole in one atom and the one induced in the neighboring atom, thus explaining the role of the atomic polarizability $\alpha \propto a_o^3$ (see Sect. 4.2), as schematically described below

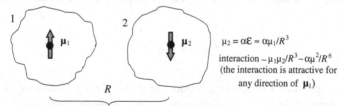

$\mu_2 = \alpha\mathcal{E} \approx \alpha\mu_1/R^3$

interaction $\sim \mu_1\mu_2/R^3 \sim \alpha\mu^2/R^6$
(the interaction is attractive for
any direction of μ_1)

The result in Eq. (13.13) can be generalized, leading to the assumption of an attractive potential energy of the form

$$V_{att} \simeq -\frac{e^2a_o^2}{R^6}\alpha, \qquad (13.14)$$

with α proper atomic polarizability. A short-range repulsive term given by $V_{rep} = B/R^{12}$ can be heuristically added.

The *Lennard-Jones potential* between two atoms collects the concepts described above and it reads

$$V_{ij}(R) = \varepsilon \left[\left(\frac{\sigma}{R_{ij}} \right)^{12} - 2 \left(\frac{\sigma}{R_{ij}} \right)^{6} \right], \tag{13.15}$$

where ε and σ are related to the repulsion coefficient B and to the atomic polarizability α.

Note that according to the form (Eq. 13.15) for the Lennard-Jones potential ε and σ are simply related to the shape of the interaction energy:

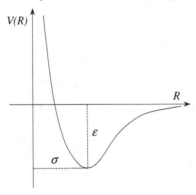

To evaluate the cohesive energy in molecular crystals one can proceed in a way similar to the one carried out in ionic crystals (Sect. 13.2.1). At variance with that case one can now limit the summation to the z first nearest neighbors for both the repulsive and the attractive terms. From the condition of minimum at $R = R_e$, one derives

$$V_T^{eq} = -\frac{Nz}{2R_e^6} e^2 a_o^2 \alpha. \tag{13.16}$$

The assumption of London interaction and of short-range repulsion as in Eq. (13.15) qualitatively justifies the cohesive energy in inert atoms crystals. In particular, through the dependence of the atomic polarizability from the third power of the "size" of the atom, Eq. (13.16) explains why the cohesive energy increases rapidly with the atomic number (see Fig. 13.1).

Problems

Problem 13.1 For ionic crystals assume that the short-range repulsive term in the interaction energy between two point-charge ions is of the form R^{-n}. Show that the cohesive energy is given by

$$E(R_e) = -\frac{N\alpha e^2}{R_e} \left(1 - \frac{1}{n} \right)$$

Fig. 13.1 Energy curves in crystals of inert atoms as a function of the interatomic distance

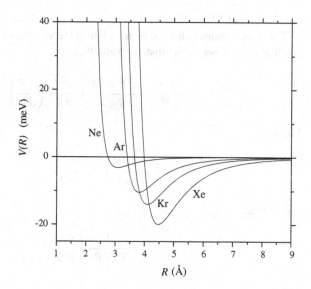

with α Madelung constant and R_e equilibrium nearest neighbor distance. Then, from the value of the bulk modulus $B = 2.4 \cdot 10^{11}$ dyne/ cm^2, estimate n for NaCl crystal. In KCl R_e is 3.14 Å and the cohesive energy (per molecule) is 7.13 eV. Estimate n.

Solution: Modifying Eq. (13.3) we write

$$E(R) = -N \left(\frac{\alpha e^2}{R} - A \frac{1}{R^n} \right)$$

where $A = \rho z$ (z number of first nearest neighbors and ρ a constant in the repulsive term ρ / R_{ij}^n). From

$$\left(\frac{dE(R)}{dR} \right)_{R=R_e} = 0$$

$$E(R_e) = -\frac{N\alpha e^2}{R_e} \left(1 - \frac{1}{n} \right).$$

The compressibility is defined (being the entropy constant)

$$k = -\frac{1}{V} \frac{dV}{dP}$$

and from $dE = -PdV$ the *bulk modulus* is

$$B = k^{-1} = V \frac{d^2 E}{dV^2}.$$

(see Problem 12.18).

For N molecules in the fcc Bravais lattice the volume of the crystal is $V = 2NR^3$ (with R nearest neighbors distance) and then

$$\frac{d^2E}{dV^2} = \frac{dE}{dR}\frac{d^2R}{dV^2} + \frac{d^2E}{dR^2}\left(\frac{dR}{dV}\right)^2 .$$

From

$$\left(\frac{dR}{dV}\right)^2 = \frac{1}{36N^2R^4}$$

one obtains

$$k^{-1} = \frac{1}{18NR_e}\left(\frac{d^2E}{dR^2}\right)_{R=R_e} .$$

Since

$$\frac{d^2E}{dR^2} = -N\left[\frac{2\alpha e^2}{R^3} - \frac{n(n+1)A}{R^{n+2}}\right],$$

$$k^{-1} = \frac{(n-1)\alpha e^2}{18R_e^4} .$$

For NaCl $\alpha = 1.747$ while $R_e = 2.82\,\text{Å}$, therefore $n \simeq 9.4$.
For KCl, from $E(R_e) = -7.13$ eV/molecule, one obtains $n = 9$.

Problem 13.2 In KBr the distance between the first nearest-neighbors is $R_e = 3.3$ Å, while the Madelung constant is $\alpha = 1.747$. The compressibility is found $k = 6.8 \cdot 10^{-12}$ cm^2/dyne. Evaluate the constant ρ in the Born-Mayer repulsive term and the cohesive energy.

Solution: From $V = 2NR^3$ and $dV = 6NR^2dR$, the pressure is

$$P = -\frac{1}{6R^2}\frac{dE}{dR} ,$$

where E is the cohesive energy per molecule. Then

$$\frac{dP}{dR} = -\frac{1}{6R^2}\left(\frac{d^2E}{dR^2}\right) + \frac{1}{3R^3}\frac{dE}{dR}$$

The second term being zero for $R = R_e$, the compressibility becomes

$$k = -\frac{3}{R}\frac{dR}{dP}.$$

Since

$$\frac{d^2E}{dR^2}\bigg|_{R=R_e} = \frac{18R_e}{k}$$

and

$$E = -\frac{\alpha e^2}{R}\left[1 - \frac{R\rho}{R_e^2}e^{-\frac{(R_e-R)}{\rho}}\right],$$

one finds

$$\frac{\rho}{R_e} = \left(2 + \frac{18R_e^4}{\alpha e^2 \kappa}\right)^{-1},$$

so that $R_e/\rho \simeq 9$, and $\rho \simeq 3 \cdot 10^{-9}$ cm. The cohesive energy per molecule turns out

$$E_C = |E| = \frac{\alpha e^2}{R_e}\left(1 - \frac{\rho}{R_e}\right) \simeq 6.8\,eV.$$

13.3 Electron States of Magnetic Ions in a Crystal Field

In a crystal the energy levels of partially filled d and f shells of transition metal and rare earth atoms are modified by the electric field generated by the neighboring atoms, yielding significant changes in the electronic and magnetic properties.

To account for the perturbative effect two approaches can be used: the *crystal field* (CF) approximation or the *ligand field theory*. In the first case the magnetic ion is assumed to be surrounded by point charges (with no covalency) which modify the electronic energies, in a way analogous to the Stark effect (Sect. 4.2). Thus one writes

$$\mathcal{H} = \mathcal{H}_{atom} + V_{CF}, \tag{13.17}$$

where

$$\mathcal{H}_{atom} = \sum_i \left(-\frac{\hbar^2}{2m}\nabla_i^2 - \frac{Ze^2}{r_i}\right) + \sum_{i>j}\frac{e^2}{r_{ij}} + \sum_i \xi_{nl}^{(i)}\mathbf{l}_i \cdot \mathbf{s}_i \tag{13.18}$$

The ligand field theory, at variance, takes into account the formation of covalent bonds with the neighboring atoms, within the molecular-orbital theory.

Let us discuss a few basic aspects of the electronic states for a magnetic ion within the CF approach. As regards the order of magnitude of the V_{CF} term one can remark the following:

(a) for $4d$ and $5d$ states usually one has $V_{CF} > \sum_{i>j} \frac{e^2}{r_{ij}} > \xi_{nl}$. In this *strong field limit* the CF yields splitting of the atomic levels of the order of 10^4 cm^{-1}.

(b) for $3d$ states one usually has $\sum_{i>j} \frac{e^2}{r_{ij}} \geq V_{CF} > \xi_{nl}$. In this case the splitting of the atomic levels due to the CF is of the order of $10^3 - 10^2$ cm^{-1}.

(c) for rare-earth atoms $\sum_{i>j} \frac{e^2}{r_{ij}} > \xi_{nl} > V_{CF}$, since the CF on the $4f$ electrons is sizeably shielded by the $5s$ and $5p$ electrons. Thus small CF splitting occurs, of the order of 1 cm^{-1}.

To understand qualitatively the role of the CF local symmetry in removing the d electron degeneracy, let us first consider the effect of point charges Ze placed at distances a from the reference ion along the x, y, z axes:

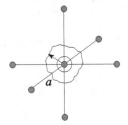

Then the perturbative potential, for instance from the charge at $(a, 0, 0)$ is

$$V_{CF} = -\frac{Ze^2}{|\mathbf{r} - a\mathbf{i}|} = -\frac{Ze^2}{\sqrt{(x-a)^2 + y^2 + z^2}} \equiv -\frac{Ze^2}{a} \frac{1}{\sqrt{1 + r^2/a^2 - 2x/a}},$$
(13.19)

where r is the nucleus-electron distance within the reference ion.

For $r \ll a$, by collecting the various terms and using[1]

$$(1+x)^{-1/2} = 1 - \frac{x}{2} + \frac{3x^2}{8} - \frac{5x^3}{16} + \frac{35x^4}{128} + \cdots$$

one writes

$$V_{CF} = -Ze^2 \left[\frac{6}{a} + \frac{35}{4a^5} \left(x^4 + y^4 + z^4 - \frac{3}{5}r^4 \right) + \cdots \right].$$
(13.20)

---

[1]
$$\frac{1}{|\mathbf{r} \pm a\mathbf{i}|} \simeq \frac{1}{a} \mp \frac{x}{a^2} - \frac{r^2}{2a^3} + \frac{3x^2}{2a^3} + \cdots$$

(see Problem 13.5).

More in general, the CF potential due to the surrounding ions, on a given ith electron is written

$$V_{CF}(\mathbf{r}_i) = -\sum_{k=1}^{N} \frac{Z_k e^2}{|\mathbf{R}_k - \mathbf{r}_i|}, \qquad (13.21)$$

with Z_k the charge of the ion at $\mathbf{R}_k$. Since $r_i \ll R_k$, the validity of the Laplace equation $\nabla^2 V(\mathbf{r}_i) = 0$ is safely assumed. Then the CF potential can be expanded in terms of Legendre polynomials P_l (see Problem 2.5):

$$V_{CF}(\mathbf{r}_i) = -e^2 \sum_{k=1}^{N} Z_k \sum_{l=0}^{\infty} \frac{r_i^l}{R_k^{(l+1)}} P_l(\cos \Omega_{ki}), \qquad (13.22)$$

with Ω_{ki} angle between $\mathbf{r}_i$ and $\mathbf{R}_k$. By expressing P_l in terms of spherical harmonics

$$P_l(\cos \Omega_{ki}) = \frac{4\pi}{(2l+1)} \sum_{m=-l}^{l} Y_{lm}(\theta_i, \phi_i) Y_{lm}^*(\theta_k, \phi_k), \qquad (13.23)$$

the CF Hamiltonian is written

$$\mathcal{H}_{CF} = \sum_{i=1}^{n} \sum_{l=0}^{\infty} \sum_{m=-l}^{l} A_l^m r_i^l Y_{lm}(\theta_i, \phi_i), \qquad (13.24)$$

with

$$A_l^m = \frac{-4\pi e^2}{(2l+1)} \sum_{k=1}^{N} \frac{Z_k Y_{lm}^*(\theta_k, \phi_k)}{R_k^{(l+1)}}, \qquad (13.25)$$

The coefficients A_l^m can be calculated once that the local coordination of the ion is known.

To give an example, let us consider the CF potential on one electron of a transition metal ion placed at the center of a regular octahedron formed by six negative charges Ze at distance R along the coordinate axes. In this case Eq. (13.24) reads

$$\mathcal{H}_{CF} = Ze^2 \left[\frac{6}{R} + \frac{7\sqrt{\pi} r^4}{3R^5} \left(Y_4^0 + \sqrt{\frac{5}{14}} (Y_4^4 + Y_4^{-4}) \right) \right] + \dots \qquad (13.26)$$

resembling Eq. (13.20).

Now one has to look for the effects of this perturbative hamiltonian on the degenerate d states. The electron wavefunction has to be of the form

$$\phi = c_0 \phi_0 + c_1 \phi_1 + c_{-1} \phi_{-1} + c_2 \phi_2 + c_{-2} \phi_{-2}, \qquad (13.27)$$

where $\phi_{0,\pm1,\pm2}$ are eigenfunctions of the unperturbed Hamiltonian.

One can notice that the matrix elements $< \phi_{0,\pm1,\pm2}|\mathcal{H}_{CF}|\phi_{0,\pm1,\pm2} >$ are all of the form nDq, with n an integer, $D = Ze^2/6R^5$ and $q \propto< r^4 >$.

The secular equation becomes

$$\begin{pmatrix} Dq - E & 0 & 0 & 0 & 5Dq \\ 0 & -4Dq - E & 0 & 0 & 0 \\ 0 & 0 & 6Dq - E & 0 & 0 \\ 0 & 0 & 0 & -4Dq - E & 0 \\ 5Dq & 0 & 0 & 0 & Dq - E \end{pmatrix} = 0 \qquad (13.28)$$

with solutions $E_1 = E_2 = 6Dq$ and $E_3 = E_4 = E_5 = -4Dq$, in correspondence to the eigenfunctions $\phi'_1 \equiv \phi_0 \equiv d_{z^2}$, $\phi'_2 = (1/\sqrt{2})(\phi_2 + \phi_{-2}) \equiv d_{x^2-y^2}$, $\phi'_3 = (1/\sqrt{2})(\phi_1+\phi_{-1}) \equiv d_{xz}$, $\phi'_4 = (-i/\sqrt{2})(\phi_1-\phi_{-1}) \equiv d_{yz}$ and $\phi'_5 = (-i/\sqrt{2})(\phi_2-\phi_{-2}) \equiv d_{xy}$.

The structure of the energy levels is shown in Fig. 13.2.

The core of high-temperature superconductors is an octahedron of oxygen atoms surrounding the Cu^{2+} $3d^9$ ion, yielding the splitting of the $3d$ levels depicted in Fig. 13.2 (it should be reminded that the CF levels for a single hole in the $3d$ sub-shell are equivalent to the ones for a single electron).

The case of one p electron in a perturbative CF due to ions in an octahedral symmetry is discussed in Problem 13.5, including the effect of an external magnetic field.

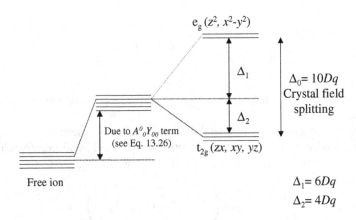

Fig. 13.2 Crystal field splitting of the $3d$ electron levels in regular octahedral coordination. The elongation of the octahedron along the z axis would cause the further splitting of the upper e_g levels

13.4 Simple Picture of the Electric Transport

Let us first recall a few introductory remarks based on the *Drude model*, basically classical considerations for a free electron gas, which help to grasp some aspects of electrical conductivity in solids.

In analogy to the molecular collisions in classical gases, for the electrons colliding with impurities or with the ions (oscillating around their equilibrium positions, see Chap. 14) one can define a *mean free path* λ. This is the average distance covered by an electron between two collisions, while it is moving with an average velocity $< v >$. This average velocity can be related to the Fermi energy E_F by referring to the average energy $< E > \simeq 3E_F/5$ (see Sect. 12.7.1): $< v >= \sqrt{< v^2 >} \sim \sqrt{E_F/m} \simeq 10^8$ cm/s.

An external electric field $\mathcal{E}$ modifies the random motions of the electrons in such a way that a charge flow opposite to the field arises, with a neat drift velocity v_d. The drift velocity is estimated as follows. After a collision a given electron experiences an acceleration $\mathbf{a} = e\mathcal{E}/m$, for an average time $\lambda/ < v >$. Then $v_d = a\lambda/ < v >= -e\mathcal{E}\lambda/m < v >$, which is usually much smaller than $< v >$. Then, indicating with n the electron density, the current density turns out

$$\mathbf{j} = -ne v_d = \frac{ne^2 \mathcal{E}\lambda}{m < v >}. \tag{13.29}$$

This equation corresponds to the Ohm law, where the resistivity is $\rho = \mathcal{E}/j$.

The *mobility* μ, defined by the ratio $|\mathbf{v}_d|/|\mathcal{E}|$, is thus given by $\mu = e\lambda/m < v >$ and the conductivity σ is

$$\sigma = ne\mu. \tag{13.30}$$

For totally filled bands the conductivity is zero, as it will be emphasized subsequently. When a band is *almost filled* an expression for the conductivity due to positive charges (*holes*) can be considered. A contribution to the conductivity analogous to Eq. (13.30) can then be written: $\sigma_h = n_h e_h \mu_h$.

It should be noticed that due to the opposite sign of their charges and of their drift velocities, both electron and hole conductivities contribute with the same sign to the electric transport.

In the *Drude* model for metallic conductivity all the free electrons contribute to the current, a situation in contradiction to the Pauli principle. In fact, the electron at energy well below E_F cannot acquire energy from the field, the states at higher energy being occupied. Furthermore the temperature dependence of the conductivity (which around room temperature goes as $\sigma \propto T^{-1}$) is not explicitly taken into account in Drude-like descriptions, the ions being considered immobile. Note that according to that simplified model the mean free path can increase to several lattice steps in the low temperature range (see Problem 13.3).

The quantum mechanical description of the current flow would require solving Schrödinger equation in the spatially periodic lattice potential in the presence of electric field. Here we shall limit to a semi-classical picture in order to better clarify the phenomenological concepts given above, taking into account the band structure and resorting to the wave-packet-like properties of the electrons.

In the semiclassical approach the motion of the electron (see Sects. 12.3 and 12.6) is based on the equation for the increase of energy δE in a time δt, due to the force associated with the electric field $\mathcal{E}$:

$$\delta E = -e\mathcal{E} \cdot \mathbf{v}\delta t. \tag{13.31}$$

Here $\mathbf{v}$ represents the group velocity of the Bloch wave-packet describing the electron:

$$\mathbf{v} = \nabla_\mathbf{k}\omega(\mathbf{k}) \equiv \frac{1}{\hbar}\nabla_\mathbf{k}E(\mathbf{k}). \tag{13.32}$$

It is recalled that in order to have particle properties, still retaining the required wave-like structure, an electron cannot have a precise definite momentum but must possess a range of $\mathbf{k}$ values.

From Eqs. (13.31) and (13.32) the equation of motion

$$\hbar\dot{\mathbf{k}} = -\mathcal{E}e \tag{13.33}$$

describes how the wave-vector and hence the state of the electron, changes. From Eqs. (13.32) and (13.33) the effective mass m^*, reflecting the effect of the crystal field included in $E(\mathbf{k})$, was obtained (Sect. 12.6). From the components of the acceleration

$$\dot{v}_\alpha = \frac{1}{\hbar}\frac{d}{dt}(\nabla_\mathbf{k}E)_\alpha = \frac{1}{\hbar}\sum_\beta \frac{\partial^2 E}{\partial k_\alpha \partial k_\beta}\dot{k}_\beta = \frac{1}{\hbar^2}\sum_\beta \frac{\partial^2 E}{\partial k_\alpha \partial k_\beta}(-e\mathcal{E}_\beta)$$

the components of the effective mass tensor turn out (see Eq. (12.24) and Problem 12.3)

$$(m^*)^{-1}_{\alpha\beta} = \frac{1}{\hbar^2}\frac{\partial^2 E(\mathbf{k})}{\partial k_\alpha \partial k_\beta}. \tag{13.34}$$

As already discussed at Sect. 12.6, the effective mass concept is useful to describe the effect of the lattice in regards of the response of the electrons to external forces. It has already been emphasized how the effective mass changes along a given band $E(\mathbf{k})$, so that the electrons can move along the direction of the electric field or along the opposite direction.

By extending Eq. (13.32) and considering that the density of **k** states is $Nv_c/8\pi^3$, the current density (Eq. (13.29)) can be written

$$\mathbf{j} = \frac{-e}{8\pi^3\hbar} \int_{BZ} \nabla_\mathbf{k} E_\mathbf{k} d\mathbf{k}, \tag{13.35}$$

where the integration is over all states occupied by electrons, within the Brillouin zone. For a fully occupied band the integral extends over all the BZ.

It must be remarked that for each electron with velocity $\mathbf{v}(\mathbf{k})$ there is another electron at $-\mathbf{k}$ for which

$$\mathbf{v}(-\mathbf{k}) = \frac{1}{\hbar}\nabla_\mathbf{k} E(-\mathbf{k}) = -\frac{1}{\hbar}\nabla_{-\mathbf{k}} E(-\mathbf{k}) = -\frac{1}{\hbar}\nabla_\mathbf{k} E(\mathbf{k}) = -\mathbf{v}(\mathbf{k}) \tag{13.36}$$

(since $E(\mathbf{k}) = E(-\mathbf{k})$, due to the inversion symmetry). Thus the current associated with a full band is *zero*, as it was anticipated. The crystal is an insulator, if no thermal excitation to the upper empty band is considered.

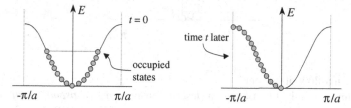

For a partially filled band, according to Eq. (13.33) the electric field redistributes the electrons, so that the distribution is no longer symmetric around $k = 0$. Therefore for a certain time interval there is no cancellation of the contributions to the drift and an electronic current flow along $-\mathcal{E}$ occurs, as sketched above in one-dimensional reciprocal space.

By extending to the band what has been derived for a single electron at Problem 12.4, one realizes that after some time the distribution in **k**-space changes. The states at positive **k** are refilled, as sketched below (for the moment, as in Problem 12.4, *no scattering* process is assumed to occur).

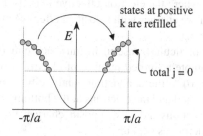

Then, at this moment the current flows due to the regions at positive and at negative **k**'s compensate each other. Later on a neat flow in opposite direction (see Eq. (13.36)) should occur. Therefore, as a whole, an oscillating current should be expected upon application of a constant electric field (the so-called *Bloch oscillations*, see Problem 12.4 for a single electron). However, we have to take into account the inelastic collisions of the electrons with impurities or oscillating ions. In a simple description one can imagine that after each collision the entire group of electrons is forced to re-take the equilibrium thermal distribution over the **k**-states. Then, for frequent collisions, only the evolution of the system in the first time interval mentioned above is practically effective. The net effect of the field can be thought to generate a stationary distribution skewed in the opposite direction of the field:

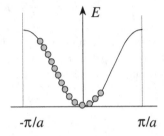

yielding a net flow of current.

For almost totally filled bands a description in terms of *pseudo-particles* (the *holes*) occupying the empty states can be given, as anticipated. In fact, the integral in Eq. (13.35) extends only over the occupied states. Therefore for the current density one can write

$$\mathbf{j} = \frac{-e}{8\pi^3}\left[\int_{BZ} \mathbf{v}(\mathbf{k})d\mathbf{k} - \int_{empty} \mathbf{v}(\mathbf{k})d\mathbf{k}\right] = +\frac{e}{8\pi^3}\int_{empty} \mathbf{v}(\mathbf{k})d\mathbf{k}. \qquad (13.37)$$

Thus the current has been formally transformed to a current of *positive* particles occupying *empty electron states*. To those quasi-particles Eqs. (13.31–13.35) and the related concepts do apply.

At the thermal equilibrium the holes are usually confined to the **k** states in the upper part of the band, where the electron effective mass is usually negative. Thus the holes behave as positive charges with a positive effective mass m_h^* moving along the electric field direction.

These concepts are particularly useful in intrinsic semiconductors, where the thermal excitations promote a limited number of electrons from the valence band (fully occupied at $T = 0$) to the conduction band (fully empty at $T = 0$). Since the holes in the valence band and the electrons at the bottom of the conduction band move along opposite directions and have opposite charges, the neat effect is that the electron and hole conductivities sum up.

Appendix 13.1 Magnetism from Itinerant Electrons

The magnetic properties associated with *localized magnetic moments*, therefore of crystals with magnetic ions, have been addressed at Chap. 4. At Sect. 12.7 and Problem 12.10 the paramagnetic susceptibility of the Fermi gas has been described.

The issue of the magnetic properties associated with an ensemble of delocalized electrons, with no interaction (Fermi gas) or in the presence of electron-electron interactions, is much more ample. In this Appendix we first recall the diamagnetism due to free electrons (*Landau diamagnetism*). Then some aspects of the magnetic properties of interacting delocalized electrons (*ferromagnetic or antiferromagnetic metals*) are addressed, in a simplified form.

The conduction electrons in metals are responsible of a *negative susceptibility*, associated with orbital motions under the action of external magnetic field. To account for this effect one has to refer to the generalized momentum operator (see Eq. (1.26)) $-i\hbar\nabla + (e/c)\mathbf{A}$, with $\mathbf{A} = (0, Hx, 0)$ (*second Landau gauge*),[2] for a magnetic field $\mathbf{H}$ along the z axes.

Then the Schrodinger equation takes the form

$$-\frac{\hbar^2}{2m}\left[\left(\frac{\partial}{\partial x}\right)^2 + \left(\frac{\partial}{\partial y} + \frac{ieHx}{\hbar c}\right)^2 + \left(\frac{\partial}{\partial z}\right)^2\right]\psi = E\psi \qquad (A.13.1.1)$$

Since $-i\hbar\nabla_{y,z}$ describe constants of motion with eigenvalues $\hbar k_{y,z}$ one can rewrite this equation in the form

$$\left[-\frac{\hbar^2}{2m}\frac{\partial^2}{\partial x^2} + \frac{1}{2}\frac{e^2 H^2}{mc^2}\left(x - \frac{\hbar k_y c}{eH}\right)^2 + \frac{\hbar^2 k_z^2}{2m}\right]\psi = E\psi \qquad (A.13.1.2)$$

where the first two terms represent the Hamiltonian for a displaced linear oscillator, with characteristic frequency

$$\omega_c = \frac{eH}{mc} = \frac{2\mu_B H}{\hbar} = 2\omega_L \qquad (A.13.1.3)$$

(ω_L Larmor frequency, see Problem 3.4). ω_c is the *cyclotron frequency*, while $x_o = \hbar c k_y/eH$ is the center of the oscillations.

Therefore, from Eq. (A.13.1.2) the eigenvalues turn out

$$E_{nk_z} = \frac{\hbar^2 k_z^2}{2m} + \left(n + \frac{1}{2}\right)\hbar\omega_c \qquad (A.13.1.4)$$

where the quantum number n labels the *Landau levels*.

---

[2]This gauge is translationally invariant along the y-axis, with eigenstates of the y-component of the momentum.

The one-electron eigenfunctions in the presence of the magnetic field are plane waves along one direction (dependent on the choice of the gauge for $\mathbf{A}$) multiplied by the wavefunctions for the harmonic oscillator.

The semiclassical view of the result given at Eq. (A.13.1.4) is that under the Lorenz force $\mathbf{F}_L = -(e/c)v_g \times \mathbf{H}$ (with $\mathbf{v}_g$ the group velocity) the evolution of thecrystal momentum $\hbar d\mathbf{k}/dt = \mathbf{F}_L$ induces a cyclotron rotational motion in the xy plane while the electron propagates along the z direction (see Problem 12.2).

It is noticed that each Landau level is degenerate, the degeneracy depending on the number of possible values for x_o. For a volume $V = L_x.L_y.L_z$, then $0 \leq x_0 \leq L_x$, while one has $0 \leq k_y \leq L_x eH/\hbar c \equiv k_y^{max}$. Therefore, k_y being quantized in steps $\Delta k_y = 2\pi/L_y$, the degeneracy of each Landau level, given by the number of oscillators with origin within the sample, is

$$N_L(H) = \frac{k_y^{max}}{\Delta k_y} = L_x L_y H \frac{e}{hc} = \frac{\Phi(H)}{\Phi_o}, \tag{A.13.1.5}$$

where $\Phi(H)$ is the flux of the magnetic field across the crystal and $\Phi_o = hc/e \simeq 4 \times 10^{-7}$ Gauss cm^2 is the *flux quantum*.[3]

It is observed that the degeneracy, the same for all the n levels, increases linearly with H. Hence, by increasing H one can vary the population of each level and eventually when H is very high (and for moderate electron densities) all electrons will occupy just the first $n = 0$ level. Accordingly, on increasing H different Landau levels will cross the Fermi energy.

By resorting to the results outlined above one can calculate the energy of the electrons $E(H)$ in presence of the field and then the magnetization. One can conveniently distinguish two regimes, for $k_B T$ large or small compared to $\hbar\omega_c$. For $k_B T \ll \hbar\omega_c$ an oscillatory behaviour of $E(H)$ is observed. The oscillations occur when the Landau level pass through the Fermi surface and cause changes in the energy of the conduction electrons, namely for

$$(n + \frac{1}{2})\hbar\omega_c = E_F, \tag{A.13.1.6}$$

Characteristic oscillations in the magnetization, known as *De Haas-Van Alphen oscillations* can be detected.

For $k_B T \gg \hbar\omega_c$ the discreteness of the Landau levels is no longer effective and the energy increases with H^2:

$$E(H) \propto \hbar\omega_c[\hbar\omega_c D(E_F)],$$

corresponding to an increase by $\hbar\omega_c$ of the energy for all the $\hbar\omega_c D(E_F)$ electrons in a Landau level ($D(E_F)$ density of states at the Fermi level, see Sect. 12.7.1).

---

[3]The flux quantum here is by a factor 2 larger than the superconducting fluxon $\Phi_{SC} = hc/2e$, since in the latter case a Cooper pair, of charge $2e$, is involved (see Chap. 18).

Therefore, the susceptibility turns out

$$\chi_L = -\frac{1}{12}\left(\frac{e\hbar}{mc}\right)^2 \frac{D(E_F)}{N v_c} = -\frac{1}{12\pi^2}\frac{e^2}{mc^2}k_F, \qquad (A.13.1.7)$$

k_F being the Fermi wave vector. From the Pauli susceptibility χ_P (see Problem 12.10) one can write

$$\chi_L = -\frac{1}{3}\chi_P. \qquad (A.13.1.8)$$

Modifications in χ_L (as well as in χ_P) have to be expected when the effective mass m^* of the electrons is different from m_e. For instance, when $m^* \ll m_e$ (as for example in bismuth, where $m^* \sim 0.01 m_e$) the metal can become diamagnetic. In fact, the total susceptibility for non-interacting delocalized electrons has to be written

$$\chi_{total} = \mu_B^2 D(E_F)\left[1 - \frac{1}{3}\left(\frac{m_e}{m^*}\right)^2\right] \equiv \chi_P\left[1 - \frac{1}{3}\left(\frac{m_e}{m^*}\right)^2\right]$$

For further insights on the behaviour of the Fermi gas in the presence of constant magnetic field, Chap. 15 in the book by *Grosso* and *Pastori Parravicini* should be read.

In transition metals, with partially occupied d bands, the electrons involved in the magnetic properties are itinerant, with relevant *many-body correlation* effects. The Fermi-gas picture for the conduction electrons is no longer adequate and significant modifications to the Pauli susceptibility have to be expected, including the possibility of the transition to an ordered state. In these cases one often speaks of *ferro* (or *antiferro)magnetic metals*. For example, an experimental evidence of a particular itinerant ferromagnetism is iron metal: the magnetic moment per atom is found around $2.2\mu_B$. This value cannot be justified in terms of localized moments on Fe^{2+} ion, in the $^5D_4$ state (see Sect. 3.2.3).

The simplest model to account for the correlation effects on the magnetic properties of itinerant electrons is the one due to *Stoner* and *Hubbard*. In this model the electron-electron Coulomb interaction is replaced by a constant repulsive energy U between electrons on the same site, with opposite spins according to Pauli principle. Then the total Hamiltonian is written

$$\mathcal{H} = \sum_k E(k)(n_{k,\uparrow} + n_{k,\downarrow}) + U\sum_m p_{m,\uparrow}p_{m,\downarrow} \qquad (A.13.1.9)$$

where the first term is the usual free electron kinetic Hamiltonian, while the second term describes the repulsive on-site interaction, with the sum running over all lattice sites.

The total magnetization can be derived in a way analogous to the one used for the Pauli susceptibility (Problem 12.10), by estimating the numbers of electrons with spin up and spin down, following the application of the magnetic field. For N electrons per cubic cm, in the conduction band of width larger than U, $N_\uparrow$ and $N_\downarrow$ are the numbers of electrons of spin up and spin down respectively. Then the energy for spin-up electrons turns out

$$E(\mathbf{k})_\uparrow = E(\mathbf{k}) + U n_\downarrow + \mu_B H \qquad (A.13.1.10)$$

while for electrons with spin-down

$$E(\mathbf{k})_\downarrow = E(\mathbf{k}) + U n_\uparrow - \mu_B H. \qquad (A.13.1.11)$$

where $n_{\uparrow,\downarrow} = N_{\uparrow,\downarrow}/N$.

The decrease of the energy of the spin-down band with respect to the spin-up band yields an increase in the population of spin-down electrons and a non-zero magnetization. Since (see again Problem 12.10) for $N_{\uparrow,\downarrow}$ one writes

$$N_\downarrow = \frac{1}{2} \int_{U n_\uparrow - \mu_B H}^{\infty} f(E) D(E - U n_\uparrow + \mu_B H) dE \simeq$$

$$\simeq \frac{1}{2} \int_0^{\infty} f(E) D(E) dE + \frac{1}{2}(\mu_B H - U n_\uparrow) D(E_F) \quad (A.13.1.12)$$

while

$$N_\uparrow \simeq \frac{1}{2} \int_0^{\infty} f(E) D(E) dE - \frac{1}{2}(\mu_B H + U n_\downarrow) D(E_F). \qquad (A.13.1.13)$$

The magnetization (per unit volume) becomes

$$M = \mu_B \frac{(N_\downarrow - N_\uparrow)}{V} \simeq \frac{\mu_B U D(E_F)}{2N}(N_\downarrow - N_\uparrow) + \mu_B^2 D(E_F) H \qquad (A.13.1.14)$$

(V the reference volume). Therefore the magnetic susceptibility becomes

$$\chi = \frac{M}{H} = \frac{\mu_B^2 D(E_F)}{1 - \frac{U D(E_F)}{2N}} = \frac{\chi_P}{1 - (U\chi_P/2\mu_B^2 N)}, \qquad (A.13.1.15)$$

with χ_P Pauli susceptibility (for bare electrons) and $D(E_F)$ the density of states per unit volume.

It is noted that when $U D(E_F)/2N \to 1$ (*Stoner criterium*) the susceptibility diverges and ferromagnetic order is attained.

Even if the Stoner condition is not fulfilled, Eq. (A.13.1.15) shows that the susceptibility is significantly modified with respect to the one for bare free-electrons. Equation (A.13.1.15) can be considered a particular case of Eq. (4.33), where the enhancement factor corresponds to the mean field acting on a particular electron due to the interaction with all the others. Stoner criterium rather well justifies the ferromagnetism in metals like Fe, Co and Ni, as well as the enhanced susceptibility (about 5 χ_P) measured in Pt and Pd metals.

Finally a few words are in order about the magnetic behaviour of itinerant electrons when the concentration n is reduced (*diluted electron fluid in the presence of electron-electron interaction*).

As shown in Problem 13.4 the Coulomb repulsive energy of the electrons goes as $< E_C > \propto e^2 n^{1/D}$ (D the dimensionality), while for the kinetic energy (for $T \to 0$) one has $< E > \propto n^{2/D}$. Thus the electron dilution causes a decrease of the average kinetic energy $< E >$ which is more rapid than the one for the average repulsion energy. Eventually, below $n_{3D} = 1.77 \times 10^{-1}/a_o^3$ and below $n_{2D} = 0.4/a_o^2$, when $< E_C >$ becomes dominant, a spontaneous "crystallization" could occur, in principle (*Wigner crystallization*).

Monte Carlo simulations predict a three-dimensional crystallization into the bcc lattice at densities below 2×10^{18} cm^{-3}, while at densities below 2×10^{20} cm^{-3} the Coulomb interaction should be strong enough to align all the spins, according to the Stoner criterium. Charge or spin ordering are hard to be experimentally tested, mainly because of the difficulty of the physical realization of the electron fluid at low density sufficiently free from impurities and/or defects.

Problems

Problem 13.3 Silver is a monovalent metal, with density 10.5 g/cm^3 and fcc structure. From the values of the resistivity at $T = 20$ K and $T = 295$ K given by $\rho_{20} = 3.8 \cdot 10^{-9}$ Ω cm and $\rho_{295} = 1.6 \cdot 10^{-6}$ Ω cm, estimate the mean free paths λ of the electrons.

Solution: The Fermi wavevector turns out $k_F = 1.2 \cdot 10^8$ cm^{-1} and the Fermi energy is $E_F = 64390$ K. The electron density is $n = 5.86 \times 10^{22}$ cm^{-3}.

From $\rho = m/ne^2\tau$, $\lambda = < v > \tau$ and $< v > \sim \sqrt{E_F/m}$ (see Sect. 13.4), one derives

$$\lambda = 3.6 \cdot 10^{-6} \text{ cm} \quad \text{at } 295 \text{ K} \quad \text{and} \quad \lambda = 1.53 \cdot 10^{-3} \text{ cm} \quad \text{at } 20 \text{ K}.$$

Problem 13.4 For three-dimensional and for two-dimensional metals, in the framework of the free-electron model and for $T \to 0$, evaluate the electron concentration n at which the average kinetic energy coincides with the average Coulomb repulsion (which can be assumed $U = e^2/d$, with d the average distance between electrons).

Solution: In 3D $d = 1/(4\pi n/3)^{1/3}$, while in 2D $d = 1/n^{1/2}$. Thus

$$U^{3D} = e^2 \left(\frac{4\pi}{3}\right)^{1/3} n^{1/3} \quad \text{and} \quad U^{2D} = e^2 n^{1/2}$$

The average kinetic energy per electron (for $T \to 0$) is $<E> = \int_0^{E_F} D(E) E \, dE$, with $D(E)^{3D} = (3/2) E^{1/2} / E_F^{3/2}$ and $D(E)^{2D} = 1/E_F$.
Then $<E>^{3D} = (3/5) E_F = (3\hbar^2/10m)(3\pi^2 n)^{2/3}$
and $<E>^{2D} = (1/2) E_F = \hbar^2 \pi n / 2m$.
The average kinetic energy coincides with the Coulomb repulsion for $n^{3D} = 1.77 \times 10^{-1}/a_0^3$ and $n^{2D} = 0.4/a_0^2$, with a_0 Bohr radius.

Problem 13.5 A magnetic field is applied on an atom with a single p electron in the crystal field at the octahedral symmetry (Sect. 13.3), with six charges Ze along the $\pm x$, $\pm y$, $\pm z$ axes. Show that without the distortion of the octahedron (namely $a = b$, with a the distance from the atom of the charges in the xy plane and b the one along the z axis) only a shift of the p levels would occur. Then consider the case $b \neq a$ and discuss the effect of the magnetic field (applied along the z axis) deriving the eigenvalues (neglect the spin magnetic moment).

Solution: By summing the potential due to the six charges, analogously to the case described at Sect. 13.3, for $r \ll a$ the crystal field perturbation turns out (see footnote 1 in this chapter)

$$V_{CF} = -Ze^2 \left\{ \left(\frac{1}{a^3} - \frac{1}{b^3} \right) r^2 + 3 \left(\frac{1}{b^3} - \frac{1}{a^3} \right) z^2 \right\} + \ldots \ldots = A(3z^2 - r^2) + const$$

where $A \neq 0$ only for $b \neq a$.
   From the unperturbed eigenfunctions the matrix elements of V_{CF} are

$$< \phi_{p_x} |V_{CF}| \phi_{p_x} > = A \int r^2 |\mathcal{R}(r)|^2 r^2 dr \int \sin^2 \theta \cos^2 \phi (3\cos^2 \theta - 1) \sin \theta \, d\theta \, d\phi$$

$$= -A <r^2> \frac{8\pi}{15} = < \phi_{p_y} |V_{CF}| \phi_{p_y} >$$

while

$$< \phi_{p_z} |V_{CF}| \phi_{p_z} > = A <r^2> \frac{16\pi}{15} .$$

In the absence of magnetic field the energy levels are

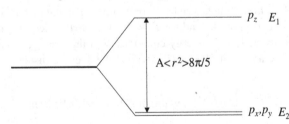

This effect can be interpreted in terms of quenching of angular momentum (see Problem 4.12). It can be observed that for orthorombic crystal symmetry, where the lowest degree ploynomial solution of the Laplace equation yields $V_{CF} = Ax^2 + By^2 - (A+B)z^2$, with A and B constants (with $A \neq B$), total quenching of the components of the angular momentum would occur.

For the electron in octahedral symmetry and in the presence of the field, the total perturbative Hamiltonian becomes $V_{CF} + \mu_B l_z H$.

The diagonal matrix elements of l_z in the basis of the unperturbed eigenfunctions are zero. In fact,

$$< \phi_{2p_x} |l_z| \phi_{2p_x} > = +i\hbar \int_0^\infty f(r)dr \int_0^\pi \sin^3\theta\, d\theta \int_0^{2\pi} \sin\phi \cos\phi\, d\phi = 0$$

(Problems 4.11 and 4.12) and, analogously,

$$< \phi_{2p_y} |l_z| \phi_{2p_y} > = < \phi_{2p_z} |l_z| \phi_{2p_z} > = 0.$$

The non-diagonal matrix elements are $< \phi_{2p_y} |l_z| \phi_{2p_x} > = i\hbar = - < \phi_{2p_x} |l_z| \phi_{2p_y} >$.

The secular equation becomes

$$\begin{vmatrix} E_0 - E & -i\mu_B H & 0 \\ i\mu_B H & E_0 - E & 0 \\ 0 & 0 & E_1 - E \end{vmatrix} = 0$$

yielding $E' = E_1$ and $E'' = E_0 \pm \mu_B H$, as sketched below

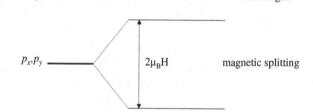

Specific References and Further Reading

1. M. Alonso and E.J. Finn, *Fundamental University Physics Vol.III- Quantum and Statistical Physics*, (Addison Wesley, 1973).
2. H. Ibach and H. Lüth, *Solid State Physics: an Introduction to Theory and Experiments*, Springer Verlag (1990).
3. G. Grosso and G. Pastori Parravicini, *Solid State Physics*, 2nd Edition, (Academic Press, 2013).
4. J.S. Blakemore, *Solid State Physics*, (W.B. Saunders Co., 1974).
5. N.W. Ashcroft and N.D. Mermin, *Solid State Physics*, (Holt, Rinehart and Winston, 1976).

6. G. Burns, *Solid State Physics*, (Academic Press Inc., 1985).
7. J.A. Cronin, D.F. Greenberg, V.L. Telegdi, *University of Chicago Graduate Problems in Physics*, (Addison-Wesley, 1967).
8. R. Fieschi e R. De Renzi, *Struttura della Materia*, (La Nuova Italia Scientifica, Roma, 1995).
9. C. Kittel, *Introduction to Solid State Physics*, 8th Edition, (J. Wiley and Sons, 2005).
10. H. Eyring, J. Walter and G.E. Kimball, *Quantum Chemistry*, (J. Wiley, New York, 1950).
11. L. Mihály and M.C. Martin, *Solid State Physics - Problems and Solutions*, (J. Wiley, 1996).
12. J.M. Ziman, *Principles of the Theory of Solids*, (Cambridge University Press, 1964).

Chapter 14
Vibrational Motions of the Ions and Thermal Effects

Topics

Elastic Waves in Crystals
Acoustic and Optical Branches
Debye and Einstein Models
Phonons
The Melting Temperature
Mössbauer Effect

14.1 Motions of the Ions in the Harmonic Approximation

Hereafter we shall afford the problem of the motions of the ions around their equilibrium positions in an ideal (disorder- and defect-free) crystal. The motions are called *lattice vibrations*. The Born-Oppenheimer separation and the adiabatic approximation (Sect. 7.1) will be implicit and the concepts involved in the description of the normal modes (Sect. 10.6) in the harmonic approximation will be used. In fact, the crystal cell will be considered as a molecular unit: its normal modes propagate along the crystal with a phase factor, in view of the spatial periodicity.

© Springer International Publishing Switzerland 2015
A. Rigamonti and P. Carretta, *Structure of Matter*,
UNITEXT for Physics, DOI 10.1007/978-3-319-17897-4_14

According to the definitions sketched below,

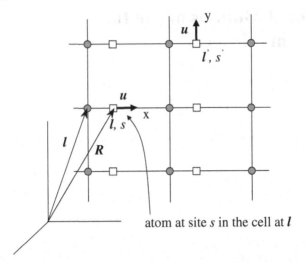

atom at site s in the cell at l

within the harmonic approximation the potential energy will be written

$$V_2 = \frac{1}{2} \sum_{l,s,\alpha} \sum_{l',s',\beta} \left(\frac{\partial^2 V}{\partial \alpha(l,s) \partial \beta(l',s')} \right)_o \mathbf{u}_\alpha(l,s) \mathbf{u}_\beta(l',s')$$

$$\equiv \sum_{l,s,\alpha} \sum_{l',s',\beta} \Phi^{(\alpha,\beta)}_{l,s,l',s'} \mathbf{u}_\alpha(l,s) \mathbf{u}_\beta(l',s'), \qquad (14.1)$$

where $\Phi^{(x,y)}_{l,s,l',s'}$ involves the force along the x direction on the ion at site s of the lth cell when the ion at site s' in the l' cell is displaced by the unit length along the y direction. From Eq. (14.1) the equations of motion turn out

$$m_s \frac{d^2 \mathbf{u}_{l,s}}{dt^2} = -\frac{\partial V_2}{\partial \mathbf{u}_{l,s}} = -\sum_{l',s'} \Phi_{l,s,l',s'} \mathbf{u}_{l',s'}, \qquad (14.2)$$

namely $3SN$ coupled equations (S number of atoms in each cell).

Recalling the normal modes in the molecules (Sect. 10.6) it is conceivable that due to the translational invariance, the motion of the atom at site s in a given cell differs only by a phase factor with respect to the one in another cell (this is the analogous of the Bloch orbital condition for the electron states). Therefore the displacement of the (l, s) atom along a given direction is written in terms of plane waves propagating the normal coordinates within a cell:

$$u^{(q)}_\alpha(l,s) = U_\alpha(s, \mathbf{q}) e^{i\mathbf{q} \cdot \mathbf{R}(l,s)} e^{-i\omega_q t}, \qquad (14.3)$$

where $\mathbf{q}$ are the wavevectors defined by the boundary conditions (the analogous of the electron wavevector $\mathbf{k}$, Sect. 12.4).

From Eqs. (14.2) and (14.3) for each $\mathbf{q}$, by taking $\mathbf{h} = \mathbf{l}_s - \mathbf{l}'_{s'}$, one has

$$m_s \omega_\mathbf{q}^2 U_\alpha(s, \mathbf{q}) = \sum_{\beta, s'} U_\beta(s', \mathbf{q}) M_{\alpha, \beta}(s, s', \mathbf{q}), \qquad (14.4)$$

where

$$M_{\alpha, \beta}(s, s', \mathbf{q}) \equiv \sum_\mathbf{h} \Phi^{(\alpha, \beta)}_{\mathbf{l}, s, \mathbf{l}', s'} e^{i\mathbf{q} \cdot \mathbf{h}} \qquad (14.5)$$

is the *dynamical matrix*, namely the Fourier transform of the elastic constants.

14.2 Branches and Dispersion Relations

For a given wave-vector Eq. (14.4) can be rewritten in the compact form

$$\omega^2 m \mathbf{U} = \mathbf{M} \mathbf{U} \qquad (14.6)$$

where $\mathbf{M}$ is a square matrix of $3S$ degree, m is a diagonal matrix and $\mathbf{U}$ is a column vector. As for the normal modes in molecules (see Eq. (10.53)) the condition for the existence of the normal coordinates is

$$|\mathbf{M} - \omega^2 m| = 0. \qquad (14.7)$$

For each wavevector $\mathbf{q}$ Eq. (14.7) yields $3S$ angular frequencies $\omega_{\mathbf{q}, j}^2$. Here j is a *branch index*. $3S - 3$ branches are called *optical* since, as it will appear at Sect. 14.3.2, they can be active in infrared spectroscopy, while 3 branches are called *acoustic*, since in the limit $\mathbf{q} \to 0$ the crystal must behave like an *elastic continuum*, where $\omega_\mathbf{q} = v_{sound} \mathbf{q}$. At variance, for the optical branches (see Sect. 14.3.2) for $q = 0$ one has $\omega_{\mathbf{q}, j} \neq 0$.

The $\mathbf{q}$-dependence of $\omega_{\mathbf{q}, j}$ is called *dispersion relation*. In analogy to the density of $\mathbf{k}$-states for the electrons (Sect. 12.5), one can define a density of $\mathbf{q}$ values in the reciprocal space: $D(\mathbf{q}) = N v_c / 8\pi^3$. One also defines the *vibrational spectrum* $D_j(\omega)$ for each branch, with the sum rule $\sum_{j=1}^{3S} \int D_j(\omega) d\omega = 3NS$.

In the next section illustrative examples of vibrational spectra will be given.

14.3 Models of Lattice Vibrations

In this section the classical vibrational motions of the ions within the harmonic approximation will be addressed for some model systems.

14.3.1 Monoatomic One-Dimensional Crystal

Let us refer to a linear chain of identical atoms, for simplicity by considering only the longitudinal motions along the chain direction:

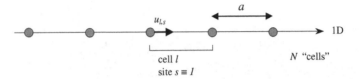

The equations of motions are of the form Eq. (14.2), the index s being redundant. One first selects in the reciprocal space a wavevector $q = n_1 2\pi/Na$, with $-N/2 \leq n_1 \leq N/2$. Then one writes the $u_{l,s}$ displacement as due to the superposition of the ones caused by the waves propagating along the chain, for each q (correspondent to Eq. (14.3)). From Eqs. (14.2) and (14.4) one writes

$$m_s \omega_{\mathbf{q}}^2 U(s, \mathbf{q}) = \sum_{s'} U(s', \mathbf{q}) M(s, s', \mathbf{q}),\qquad(14.8)$$

where

$$M(s, s', \mathbf{q}) \equiv \sum_{\mathbf{h}} \Phi_{l,s,l',s'} e^{i\mathbf{q}\cdot\mathbf{h}}\qquad(14.9)$$

is the *collective force constant*, representing the Fourier transform of the elastic constants. Equations (14.8) and (14.9) describe the propagation of the normal modes of the "cell" along the chain.

By limiting the interaction to the nearest neighbors,

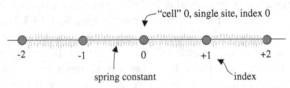

the equation of motion for the atom in the cell at the origin ($l = 0$) turns out

$$m\frac{d^2 u_0}{dt^2} = -2k u_0 + k u_1 + k u_{-1}\qquad(14.10)$$

implying $\Phi(0, 0) = 2k$ and $\Phi(\pm 1, 0) = -k$.

The dynamical matrix (Eq. (14.5)) is reduced to

$$M = \Phi(0,0) + \sum_{n=\pm 1} \Phi(n,0)e^{iqna}$$

and Eq. (14.8) takes the form

$$m\omega_q^2 U_q = (2k - 2k\cos(qa))U_q, \tag{14.11}$$

namely the one for a single *normal oscillator*, with an effective elastic constant taking into account the coupling to the nearest neighbors.

The solubility condition (Eq. (14.7)) corresponds to

$$\omega_q^2 = \frac{2k}{m}(1 - \cos(qa)), \tag{14.12}$$

yielding the dispersion relation

$$\omega_q = 2\sqrt{\frac{k}{m}}\sin(qa/2) \tag{14.13}$$

sketched below:

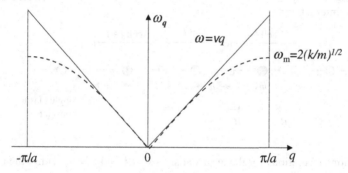

The vibrational spectrum, or density of states $D(\omega) = D(q)dq/d\omega$, with $D(q) = Na/2\pi$, turns out

$$D(\omega) = \left(2N/\pi\sqrt{\omega_m^2 - \omega^2}\right), \tag{14.14}$$

reported below

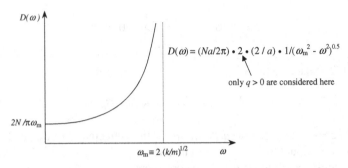

$$D(\omega) = (Na/2\pi) \cdot 2 \cdot (2/a) \cdot 1/(\omega_m^2 - \omega^2)^{0.5}$$

only $q > 0$ are considered here

The situation arising at the zone boundary, where $\omega_{q=\pi/a} \equiv \omega_m$, is equivalent to the one encountered at the critical points of the electronic states (see Sect. 12.5).

14.3.2 Diatomic One-Dimensional Crystal

For a chain with two atoms per unit cell, with mass m_1 and m_2 ($m_1 > m_2$), again considering the longitudinal modes and assuming a single elastic constant and nearest neighbour interactions,

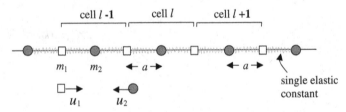

the equations of motions for the atoms at sites $s = 1$ and $s = 2$, within the lth cell, are

$$m_1 \frac{d^2 u_{l,1}}{dt^2} = -2k u_{l,1} + k u_{l,2} + k u_{l-1,2}$$

$$m_2 \frac{d^2 u_{l,2}}{dt^2} = -2k u_{l,2} + k u_{l,1} + k u_{l+1,1} \qquad (14.15)$$

Again resorting to solutions of the form

$$u(l, 1) = U_1 e^{iq2la} e^{-i\omega_q t}$$

and

$$u(l, 2) = U_2 e^{iq(a+2la)} e^{-i\omega_q t}$$

(the index q in $U_{1,2}$ is dropped here), one has

$$\left(\frac{2k}{m_1} - \omega^2\right) U_1 - \frac{k}{m_1}(e^{iqa} + e^{-iqa})U_2 = 0$$

$$-\frac{k}{m_2}(e^{iqa} + e^{-iqa})U_1 + \left(\frac{2k}{m_2} - \omega^2\right) U_2 = 0. \qquad (14.16)$$

The dynamical matrix is

$$M = \begin{pmatrix} 2k & -k(e^{iqa} + e^{-iqa}) \\ -k(e^{-iqa} + e^{iqa}) & 2k \end{pmatrix}$$

and the solubility condition

$$\begin{pmatrix} 2k - m_1\omega^2 & -2k\cos(qa) \\ -2k\cos(qa) & 2k - m_2\omega^2 \end{pmatrix} = 0$$

leads to

$$\omega_q^2 = k\left(\frac{1}{m_2} + \frac{1}{m_1}\right) \pm k\left[\left(\frac{1}{m_2} + \frac{1}{m_1}\right)^2 - \frac{4}{m_1 m_2}\sin^2(qa)\right]^{\frac{1}{2}}. \qquad (14.17)$$

The dispersion relations are shown in Fig. 14.1, with μ reduced mass.

At the boundaries of the Brillouin zone ($q = \pm\pi/2a$) the frequencies of the acoustic and optical modes are $\omega^A = \sqrt{2k/m_1}$ and $\omega^O = \sqrt{2k/m_2}$, respectively.

It is noted that when $m_1 = m_2$ the two frequencies coincide, the gap at the zone boundary vanishes: the situation of the monoatomic chain is restored, once that the length of the lattice cell becomes a instead of $2a$.

For a given wavevector one can obtain the atomic displacements induced by each normal mode. For instance, by choosing $q = 0$ for the acoustic branch one derives $U_A(0, 1) = U_A(0, 2)$, the same displacement for the two atoms, corresponding to

Fig. 14.1 Frequencies of the acoustic (A) and optical (O) longitudinal modes in one-dimensional diatomic crystal, according to Eq. (14.17)

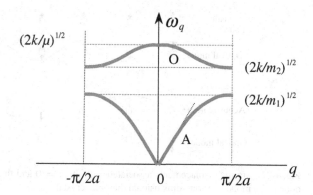

the translation of all the crystal. For the optical mode, again for $q = 0$ one has $m_1 U_O(0, 1) = -m_2 U_O(0, 2)$, keeping fixed the center of mass. As for the diatomic molecule (see Sect. 10.6) the difference of the two displacements corresponds to the normal coordinate.

In a similar way one can derive the displacements associated with the zone boundary wavevectors (Fig. 14.2, where also the transverse modes are schematized).

From the dispersion relations (Eq. (14.17)) the vibrational spectra reported in Fig. 14.3 are derived.

Up to now only longitudinal modes have been considered. To describe the transverse vibrations the elastic constants for the displacements perpendicular to the chain should be considered. In this way, for a given wave-vector, 3 vibrational branches would be obtained for the monoatomic chain and 6 branches for the diatomic one, at longitudinal (L) and transverse (T) optical and acoustic characters (see Fig. 14.2).

Finally one should observe that the interaction with electromagnetic waves requires the presence of *oscillating electric dipole* within the cell. To grant energy and momentum conservation, the absorption process should occur in correspondence to the *photon momentum* $q = \hbar\omega/c$, which for typical values of the frequencies

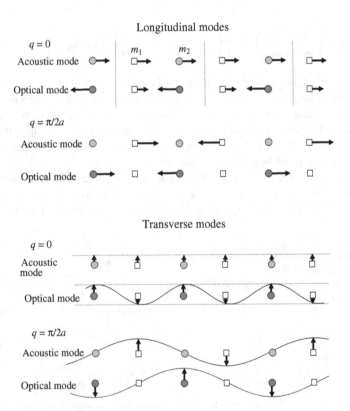

Fig. 14.2 Atomic displacements associated with the $q = 0$ and the $q = \pi/2a$ acoustic (A) and optical (O) modes, for one-dimensional diatomic crystal

Fig. 14.3 Vibrational spectra for the longitudinal acoustic (A) and optical (O) branches in one-dimensional diatomic crystal

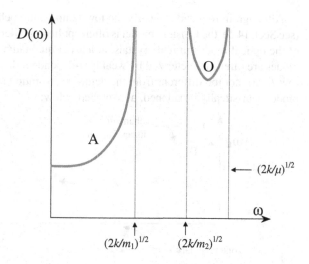

$(\omega \sim 10^{13} - 10^{14}$ rad s$^{-1})$ is much smaller than $h/2a$. For $q \to 0$, at the center of the Brillouin zone, the acoustic modes do not yield any dipole moment. Therefore only the optical branches, implying in general oscillating dipoles (as schematized in Fig. 14.2), can be active for the absorption of the electromagnetic radiation, similarly to the case described for the molecules.

14.3.3 Einstein and Debye Crystals

The phenomenological models due to Einstein and to Debye are rather well suited for the approximate description of specific properties related to the lattice vibrations in real crystals.

The *Einstein crystal* is assumed as an ensemble of independent atoms elastically connected to equilibrium positions. The interactions are somewhat reflected in a vibrational constant common to each oscillator, yielding a characteristic frequency ω_E. As regards the dispersion curves, one can think that for each $\mathbf{q}$ there is a three-fold degenerate mode at frequency ω_E. Thus, the vibrational spectrum could be schematized as below:

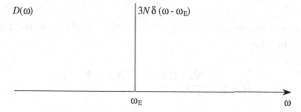

Although introduced to justify the low-temperature behavior of the specific heat (see Sect. 14.5), the Einstein model is often applied in order to describe the properties of the optical modes in real crystals, at least at qualitative level. In fact, the optical modes are often characterized by weakly q-dependent dispersion curves with a narrow $D(\omega)$, not too different from the delta-like vibrational spectrum of the Einstein model heuristically broadened, as sketched below:

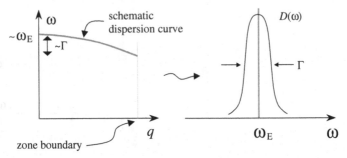

In the *Debye model* it is assumed that the vibrational properties are basically the ones of the *elastic* (and sometimes isotropic) *continuum*, with *ad hoc* conditions in order to take into account the discrete nature of any real crystal. In particular:

(i) the Debye model describes rather well the acoustic modes of any crystal, since for $\mathbf{q} \to 0$ the dispersion curves of the acoustic branches practically coincide with the ones of the continuum solid, the wavelength of the vibration being much larger than the lattice step.

(ii) the model cannot describe the vibrational contribution from optical modes.

(iii) one has to introduce a cutoff frequency ω_D in the spectrum in order to keep the number of modes limited to $3N$ (for N atoms).

(iv) only 3 branches have to be expected, with dispersion relations of the form $\omega_q^j = \mathbf{v}_{sound}^j \mathbf{q}$, where the sound velocity can refer to transverse or to longitudinal modes.

For a given branch, in the assumption of isotropy, the vibrational spectrum turns out

$$D_j(\omega) = \frac{Nv_c}{8\pi^3}d\mathbf{q} = \frac{Nv_c}{8\pi^3}4\pi q^2 dq = \frac{Nv_c}{8\pi^3}\frac{4\pi\omega^2}{v_j^3}. \qquad (14.18)$$

One can introduce an average velocity v and again in the isotropic case, $3/v^3 = 2/v_T^3 + 1/v_L^3$. Therefore

$$D(\omega) = \frac{Nv_c}{8\pi^3}\frac{12\pi\omega^2}{v^3} = \frac{Nv_c}{v^3}\frac{3}{2\pi^2}\omega^2, \qquad (14.19)$$

the typical vibrational spectrum characteristic of the continuum.

Now a cutoff frequency ω_D (known as *Debye frequency*) has to be introduced. The role of ω_D in the dispersion relation and in the vibrational spectrum $D(\omega)$ is illustrated below:

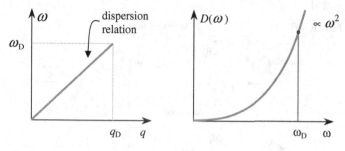

ω_D can be derived from the condition $\int D(\omega)d\omega = 3N$ or, equivalently, by evaluating the Debye radius q_D of the sphere in the reciprocal space which includes the N allowed wavevectors.

Thus $(Nv_c/8\pi^3)(4\pi q_D^3/3) = N$ and then

$$q_D = \left(\frac{6\pi^2}{v_c}\right)^{\frac{1}{3}} \tag{14.20}$$

and

$$\omega_D = vq_D = v\left(\frac{6\pi^2}{v_c}\right)^{\frac{1}{3}}. \tag{14.21}$$

In real crystals detailed descriptions of the vibrational modes are often difficult. One can recall the following. In the $\mathbf{q} \to 0$ limit one can refer to the conditions of the continuum and the acoustic branches along certain symmetry directions can be discussed in terms of effective elastic constants. These constants are usually derived from *ultrasound propagation* measurements.

The frequencies of the various branches can become equal in correspondence to certain wavevectors, implying *degeneracy*. Although the optical branches have non-zero frequency even for $q = 0$ they are not always optically active, since do not always imply oscillating electric dipoles. For instance, in diamond, although the optical modes cause the vibration of the two sublattices (see Sect. 11.3) against each other, no electric dipole is induced and no interaction with the electromagnetic waves can occur.

The dispersion curves are usually obtained by inelastic *neutron spectroscopy*. The schematic structure of a triple axes neutron spectrometer is reported below:

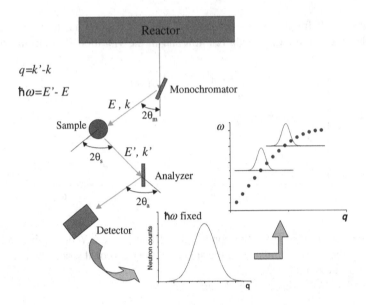

A suited description of the lattice vibrations, with theory and basic aspects of neutron spectroscopy, can be found in the report by *Cochran*.

14.4 Phonons

While discussing the normal modes in molecules (Sect. 10.6) it was shown how a non-normal Hamiltonian (in terms of local coordinates) could be transformed into a normal one by writing the local displacements as a superposition of excitations, each one associated to a normal oscillator. The collective normal coordinate was shown to be a linear combination of the local ones. The treatment given at Sect. 10.6 can be extended to the displacements of the atoms around their equilibrium positions in a crystal. Thus, returning to Eq. (14.3), for each branch (j) we write the displacement in the form

$$\mathbf{u} = \sum_{\mathbf{q}} \mathbf{U}_{\mathbf{q}} e^{i\mathbf{q} \cdot \mathbf{R}} e^{-i\omega_{\mathbf{q}} t} \tag{14.22}$$

Therefore the problem is reduced to the evaluation of the normal coordinates $Q_{\mathbf{q}}^{(j)}$ of the crystal cell, that one can build up from the amplitudes $\mathbf{U}_{\mathbf{q}}$ by including the masses and the normalization factors. The translational invariance of the crystal implies the propagation of the normal excitations of the cell with phase factor $e^{i\mathbf{q} \cdot \mathbf{R}}$.

Hence, one can start from Hamiltonians of the form $\mathcal{H} = \sum_{j} \mathcal{H}_{j}[Q^{j}(\mathbf{q})]$, for each wavevector $\mathbf{q}$ of a given branch j. By indicating with $\mathbf{Q}$ the group of the normal coordinates and with $\phi(\mathbf{Q})$ the related wavefunction, one expects

$$\phi(\mathbf{Q}) = \prod_{\mathbf{q},j} \phi_{\mathbf{q}}^{(j)} (Q^j(\mathbf{q})). \qquad (14.23)$$

In the harmonic approximation $\phi_{\mathbf{q}}^{(j)}$ is the eigenfunction of single normal oscillator, characterized by quantum number $n_j(\mathbf{q})$ and eigenvalues

$$E_{\mathbf{q}}^{(j)} = \hbar\omega_{\mathbf{q}}^{(j)} \left[1/2 + n_j(\mathbf{q}) \right].$$

The total energy is

$$E_T = \sum_j \sum_{\mathbf{q}} \left(n_j(\mathbf{q}) + \frac{1}{2} \right) \hbar\omega_{\mathbf{q}}^{(j)}. \qquad (14.24)$$

Therefore the vibrational state of the crystal is defined by the set of $3SN$ numbers $| \ldots, \ldots, n_j(\mathbf{q}), \ldots >$ that classify the eigenfunctions of the normal oscillators. At $T = 0$, the ground-state is labelled $|0, 0, 0 \ldots >$ and the wavefunction is the product of Gaussian functions (see Sect. 10.3.1).

At finite temperature one has to take into account the thermal excitations to excited states, for each normal oscillator. Two different approaches can be followed:

(A)—the *normal oscillators* are *distinguishable* and the numbers $n_j(\mathbf{q})$ select the stationary states for each of them. Then the Boltzmann statistics holds and for a given oscillator with characteristic frequency ν the average energy is

$$\overline{E} = \sum_v p_v E_v, \qquad (14.25)$$

with

$$p_v = \frac{e^{-E_v/k_B T}}{\sum_v e^{-E_v/k_B T}}$$

and

$$E_v = (v + 1/2)h\nu \quad v = 0, 1, 2, \ldots$$

For each normal mode the average energy $\overline{E}$ is found as shown at Problem 1.25 for photons (Planck derivation), here having to include the zero-point energy:

$$\overline{E} = h\nu \left(\frac{1}{2} + \frac{1}{e^{h\nu/k_B T} - 1} \right) \qquad (14.26)$$

The energy turns out the one for the quantum oscillator, provided that an *average excitation number*

$$< v > = \frac{1}{e^{\frac{h\nu}{k_B T}} - 1} \tag{14.27}$$

is introduced.

The total thermal energy of the crystal is obtained by summing Eq. (14.26) over the various modes, for each branch.

(B)—the crystal is considered as an assembly of *indistinguishable pseudo-particles*, each of energy $\hbar\omega_{\mathbf{q},j}$ and momentum $\hbar\mathbf{q} = (\hbar\omega_{\mathbf{q},j}/v_{j,\mathbf{q}})\hat{q}$. These quasi-particles are the quanta of the elastic field and are called *phonons* in analogy with the photons for the electromagnetic field.

Then the total energy has to be written

$$< \overline{E} > = \sum_{\mathbf{q},j} \left(\overline{n}_{\mathbf{q},j} + \frac{1}{2} \right) \hbar\omega_{\mathbf{q},j}, \tag{14.28}$$

where the average number of pseudo-particles is given by the Bose-Einstein statistics, i.e.

$$\overline{n}_{\mathbf{q},j} = \frac{1}{e^{\frac{\hbar\omega_{\mathbf{q},j}}{k_B T}} - 1}, \tag{14.29}$$

for a given branch j.

The two ways A and B to conceive the aspects of the lattice vibrations give equivalent final results, as it can be seen by comparing Eq. (14.26) (summed up to all the single oscillators) and Eq. (14.28). The derivation of some thermal properties (Sect. 14.5) will emphasize the equivalence of the two ways to describe the quantum aspects of the vibrational motions of the ions.

14.5 Thermal Properties Related to Lattice Vibrations

All the thermodynamical properties related to the vibrational motions can be derived from the total partition function $Z_{TOT} = \prod_{\mathbf{q},j} Z_{\mathbf{q},j}$, with

$$Z_{\mathbf{q},j} = \sum e^{\frac{-E(\mathbf{q},j)}{k_B T}} \tag{14.30}$$

where the sum is over all the energy levels, for each $\mathbf{q}$-dependent oscillator of each branch.

The thermal energy can be directly evaluated by resorting to the vibrational spectra $D(\omega)$, in the light of Eqs. (14.28) and (14.29), by writing

$$U = \int \hbar\omega \left(\frac{1}{2} + \frac{1}{e^{\frac{\hbar\omega}{k_B T}} - 1} \right) D(\omega) d\omega. \qquad (14.31)$$

For instance, for Einstein crystals where $D(\omega) = 3N\delta(\omega - \omega_E)$ one derives

$$U = 3N\hbar\omega_E \left[1/2 + 1/\left(e^{\frac{\hbar\omega_E}{k_B T}} - 1 \right) \right].$$

The molar ($N = N_A$) specific heat for $T \gg \Theta_E \equiv \hbar\omega_E/k_B$ (Θ_E often defined *Einstein temperature*) turns out $C_V \simeq 3R$. At variance with the classical results, for $T \ll \Theta_E$ one has

$$C_V \simeq 3R \left(\frac{\Theta_E}{T} \right)^2 e^{\frac{-\Theta_E}{T}} \qquad (14.32)$$

For Debye crystals, from Eq. (14.31) by resorting to Eq. (14.19), one writes

$$C_V = \frac{\partial}{\partial T} \left\{ \int_0^{\omega_D} D(\omega)\hbar\omega \frac{1}{e^{\frac{\hbar\omega}{k_B T}} - 1} d\omega \right\}$$

and then, for $N = N_A$

$$C_V = 9R \left(\frac{T}{\Theta_D} \right)^3 \int_0^{\Theta_D/T} \frac{z^4 e^z}{(e^z - 1)^2} dz, \qquad (14.33)$$

with $z = \hbar\omega/k_B T$.

For $T \gg \Theta_D$, with $\Theta_D \equiv \hbar\omega_D/k_B$ (known as *Debye temperature*), one again finds the classical result $C_V \to 3R$.

In the low temperature range ($\Theta_D/T \to \infty$) Eq. (14.33) yields $C_V \simeq (12\pi^4/5)R(T/\Theta_D)^3$. Equation (14.33) points out that the vibrational specific heat of Debye crystals is a universal function of the variable T/Θ_D. From Eq. (14.21) Θ_D can be written $\Theta_D = (\hbar v/k_B)(6\pi^2/v_c)^{1/3}$.

The temperature dependences of the molar specific heat in the framework of Einstein and Debye models are sketched below:

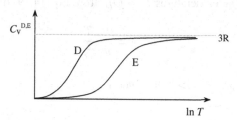

For $T \to 0$ the Debye specific heat C_V^D vanishes less rapidly than the Einstein C_V^E. The different behavior of C_V^D for $T \to 0$ originates from the fact that the vibrational spectrum in the Debye model includes oscillatory modes with energy separation of the order of $k_B T$, even at low temperature. On the contrary in the Einstein crystal in the low-temperature range one has $\hbar \omega_E \gg k_B T$.

In the Table below the Debye temperatures of some elements are reported.

Li 344	Be 1440	Debye Temperature θ_D (K) estimated from low-temperature (T<< θ_D) specific heat measurements											C diamond 2230			Ne 75
Na 158	Mg 400											Al 428	Si 645			Ar 92
K 91	Ca 230	Sc 360	Ti 420	V 380	Cr 630	Mn 410	Fe 470	Co 445	Ni 450	Cu 343	Zn 327	Ga 320	Ge 374	As 282	Se 90	Kr 72
Rb 56	Sr 147	Y 280	Zr 291	Nb 275	Mo 450		Ru 600	Rh 480	Pd 274	Ag 225	Cd 209	In 108	Sn 200	Sb 211	Te 153	Xe 64
Cs 38	Ba 110	La 142	Hf 252	Ta 240	W 400	Re 430	Os 500	Ir 420	Pt 240	Au 165	Hg 72	Tl 78.5	Pb 105	Bi 119		Rn 64

By resorting to the expression for the thermal energy in terms of the vibrational spectra, the mean square displacement of a given ion as a function of temperature can be directly derived. According to the extension of Eq. (14.3) to include all the normal excitations, the mean square vibrational amplitude of each atom around its equilibrium position is written

$$< |\mathbf{u}|^2 > = \sum_{\mathbf{q}, j} |\mathbf{U}_{\mathbf{q}, j}|^2. \qquad (14.34)$$

By recalling that for each oscillator the mean square displacement can be related to the average energy, $< u^2 > = < E > /(m\omega^2)$, then for a given branch j one can write $|\mathbf{U}_{\mathbf{q}}|^2 = < E_{\mathbf{q}} > /Nm\omega_{\mathbf{q}}^2$. Hence,

$$< u^2 > = \frac{1}{mN} \sum_{\mathbf{q}, j} \frac{< E_{\mathbf{q}, j} >}{\omega_{\mathbf{q}, j}^2} = \frac{\hbar}{mN} \int \left[\frac{1}{2} + \frac{1}{e^{\frac{\hbar \omega}{k_B T}} - 1} \right] \frac{D(\omega)}{\omega} d\omega. \quad (14.35)$$

For Debye crystals, at temperatures $T \gg \Theta_D$, from Eq. (14.19) one obtains

$$< u^2 > \simeq \frac{9 k_B T}{m \omega_D^2} \qquad (14.36)$$

and at low temperature $< u^2 > \simeq 9\hbar/4m\omega_D$.

It should be remarked that $< u^2 >$ controls the temperature dependence of the strength of the elastic component in scattering processes, through the *Debye-Waller* factor $e^{-4\pi <u^2>/\lambda^2}$, with λ wavelength of the radiation (see Sect. 14.6 for the derivation of this result).

According to the *Lindemann criterium* the crystal melts when the mean square displacement $< u^2 >$ reaches a certain fraction ξ of the square of the nearest neighbor distance R, $< u^2 > = \xi R^2$.

Empirically it can be devised that ξ is around 1.5×10^{-2} ($\sqrt{< u^2 >} \simeq 0.12R$). This criterium allows one to relate the melting temperature T_m to the Debye temperature. From Eq. (14.36)

$$T_m = \xi \Theta_D^2 \frac{m k_B R^2}{\hbar^2}. \tag{14.37}$$

Problems

Problem 14.1 Derive the vibrational entropy of a crystal in the low temperature range ($T \ll \Theta_D$).

Solution: From C_V^D (Eq. (14.33)) in the low temperature limit, by recalling that

$$S = \int_0^T \frac{C_V^D}{T} dT$$

the molar entropy is $S(T) = [12R\pi^4/(15\Theta_D^3)]T^3$. This result justifies the assumption for the lattice entropy used at Sect. 6.4. The contribution from optical modes can often be neglected.

Problem 14.2 Derive the vibrational contribution to the Helmoltz free energy and to the entropy in Einstein crystals.

Solution: For N oscillators the total partition function is $Z_T = Z^N$, with

$$Z = e^{-\hbar\omega_E/2k_BT} \sum_v e^{-\hbar\omega_E v/k_BT} = \frac{e^{-\hbar\omega_E/2k_BT}}{1 - e^{-\hbar\omega_E/k_BT}}$$

(remind that $\sum x^n = 1/(1-x)$, for $x < 1$).
Then the total free energy turns out

$$F = -Nk_BT lnZ = N \left\{ \frac{\hbar\omega_E}{2} + k_BT ln \left(1 - e^{-\hbar\omega_E/k_BT} \right) \right\}$$

and the entropy is

$$S = -\left(\frac{\partial F}{\partial T} \right)_V = -Nk_B \left\{ ln \left(1 - e^{-\hbar\omega_E/k_BT} \right) - \frac{\hbar\omega_E}{k_BT} \frac{1}{e^{\hbar\omega_E/k_BT} - 1} \right\}.$$

Problem 14.3 Evaluate the specific heat per unit volume for Ag crystal (fcc cell, lattice step $a = 4.07$ Å) at $T = 10$ K, within the Einstein model (the elastic constant can be taken $k = 10^5$ dyne/cm) and within the Debye model, assuming for the sound velocity $v \simeq 2 \times 10^5$ cm/s.

Solution: The Einstein frequency $\omega_E \simeq \sqrt{k/M_{Ag}}$, corresponds to the temperature $\Theta_E \simeq 170$ K. In the unit volume there are $n = 1/(N_A v_c)$ moles, with $v_c = a^3/4$ the volume of the primitive cell. Then, since $T = 10$ K $\ll \Theta_E$, from Eq. (14.32) one derives $C_V^E \simeq 280$ erg/K cm³.

The Debye frequency can be estimated from Eq. (14.21) and the corresponding Debye temperature turns out $\Theta_D \simeq 220$ K $\gg 10$ K. Then

$$C_V^D \simeq \frac{12\pi^4 k_B}{5 v_c} \left(\frac{T}{\Theta_D}\right)^3 \simeq 1.72 \times 10^5 \text{ erg/Kcm}^3.$$

Problem 14.4 Specific heat measurements in copper (fcc cell, lattice step $a = 3.6$ Å, sound velocity $v = 2.6 \times 10^5$ cm/s) show that C_V/T (in 10^{-4} Joule/mole K²) is linear when reported as a function of T^2, with extrapolated value (C_V/T) for $T \to 0$ given by about 7 and slope about 0.6. Estimate the Fermi temperature and the Debye temperature and the temperature at which the electronic and vibrational contributions to the specific heat are about the same (from the equations at Sects. 12.7.1 and 14.5) and compare the estimates with the experimental findings.

Solution: From the specific mass $\rho = 9.018$ g/cm³ the number of electrons per cm³ is found $n = 8.54 \cdot 10^{22}$ cm⁻³. From Eq. (12.28) $T_F = 7.8 \cdot 10^4$ K.

The Debye temperature, for the primitive cell of volume $v_c = a^3/4$, is

$\Theta_D = (\hbar v/k_B) \left(6\pi^2/v_c\right)^{1/3} = 323$ K.

From

$$\frac{\pi^2}{2} n k_B \frac{T^*}{T_F} = \frac{1}{v_c} k_B \frac{12\pi^4}{5} \left(\frac{T^*}{\theta_D}\right)^3,$$

(per unit volume) the temperature T^* at which the electronic and vibrational contributions are the same is obtained:

$$T^* = \sqrt{5 v_c n} (\Theta_D)^{3/2} / \left(\pi\sqrt{24 T_F}\right) \simeq 3 \text{ K}.$$

From the experimental data according to Eq. (12.31) for $NZ = N_A$
$\gamma = \pi^2 R/(2 T_F) = 7 \times 10^3$ erg/ mole K², one finds $T_F \simeq 5.8 \cdot 10^4$ K and from

$$C_V^D \simeq \frac{12\pi^4 R}{5} \left(\frac{T}{\Theta_D}\right)^3,$$

one derives $\theta_D \simeq 343$ K.

Problem 14.5 Write the zero-point vibrational energy of a crystal in the Debye model and derive the bulk modulus for $T \to 0$.

Solution: The zero-point energy is $E_0 = \frac{1}{2} \int_0^{\omega_D} \hbar\omega D(\omega) d\omega$ (Eq. (14.31)). From Eq. (14.19) one derives $E_0 = 9N\hbar\omega_D/8$.

At low temperature the bulk modulus is $(B \simeq V\partial^2 E_0/\partial V^2)$. Then, by writing ω_D in terms of the volume $V = N v_c$ one finds

$$B = \frac{1}{2}\frac{N}{V}\hbar\omega_D \equiv \frac{1}{2}\frac{1}{v_c}k_B\Theta_D.$$

14.6 The Mössbauer Effect

The recoil-free emission or absorption of γ-ray (for the first time experimentally noticed by Mössbauer in 1958) is strictly related to the vibrational properties of the crystals. Meantime it allows one to recall some aspects involving the interaction of radiation with matter.

Let us consider an atom, or a nucleus, ideally at rest, emitting a photon due to the transition between two electronic or nucleonic levels. At the photon energy $h\nu$ is associated the momentum $(h\nu/c)$. Then in order to grant the momentum conservation the atom has to recoil during the emission with kinetic energy $E_R = (h\nu/c)^2/2M$, with M the atomic mass. Because of the energy conservation the emission spectrum (from an assembly of many atoms) displays a Lorentzian shape,

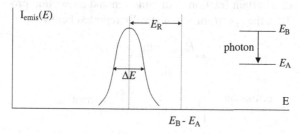

at least with the line broadening ΔE related to the *life-time* of the level (the inverse of the spontaneous emission probability, see Problem 1.24). Another source of broadening arises from the thermal motions of the atoms and the emission line usually takes a Gaussian shape, with width related to the distribution of the Doppler modulation in the emitted radiation (see Problem 1.30).

Let us suppose to try the *resonance absorption* of the same emitted photon from an equivalent atom (or nucleus). Again, by taking into account the energy and momentum conservation in the absorption process, the related spectrum must have an energy distribution of Gaussian shape, centered at $E = (E_B - E_A) + E_R$:

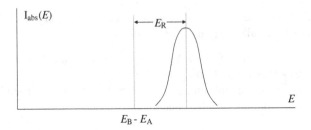

From the comparison of the emission and absorption spectra one realizes that the fraction of events that grant the resonance absorption is only the one corresponding to the energy range underlying the emission and absorption lines.

In atomic spectroscopy, where energy separations of the order of the eV are involved, the condition of resonant absorption is well verified. In fact, the recoil energy is $E_R \sim 10^{-8}$ eV, below the broadening $\Delta E \sim 10^{-7}$ eV typically associated with the life time of the excited state. At variance, when the emission and the absorption processes involve the γ-rays region, with energies around 100 keV, the recoil energy increase by a factor of the order of 10^{10}. Since the lifetime of the excited nuclear levels is of the same order of the one for electronic levels, only a limited number of resonance absorption processes can take place, for free nuclei.

In crystals, in principle, one could expect a *decrease* in the fraction of resonantly absorbed γ-rays upon cooling the source (or the absorber), due to the decrease of the broadening induced by thermal motions. Instead, an *increase* of such a fraction was actually detected by Mössbauer at low temperature. This phenomenon is due to the fact that in solids a certain fraction f of emission and absorption processes occurs *without recoil*. Thus the spectrum schematically reported below

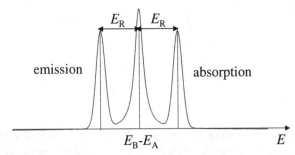

can be conceived, with a sizeable superposition of events around the energy difference $(E_B - E_A)$.

The momentum conservation is anyway granted, since the recoil energy goes to the whole crystal, with negligible subtraction of energy to the emitted or absorbed photons. The reason for the recoilless processes can be grasped by referring to the Einstein crystal, with energy $\hbar\omega_E$ larger than E_R. It is conceivable that when the quantum of elastic energy cannot be generated, then the crystal behaves *as rigid*.

Another interpretation (not involving the quantum character of the vibrational motions) is based on the classical consideration of the spectrum emitted by a source in motion. For a sinusoidal motion with frequency ω_S, the emitted spectrum has Fourier components at ω_i, $\omega_i \pm \omega_S$, ..., so that a component at the intrinsic frequency ω_i should remain.

The fraction f of recoilless processes can be evaluated by considering, in the framework of the time dependent perturbation theory used in Appendix 1.3, the emitting system as one nucleus imbedded in the crystal, looking for the transition probability between states having the same vibrational quantum numbers, while the nuclear state is changed. Since the long wave-length approximation cannot be retained, the perturbation operator reads $\sum_i \mathbf{A}_i \cdot \nabla_i$ (the sum is over all nucleons) (see Eq. (A.1.3.3)).

Let us refer to an initial state corresponding to the vibrational ground-state $|0, 0, 0, \ldots >$, by writing the amplitude of the time-dependent perturbative Hamiltonian $\sum_i e^{i\mathbf{k} \cdot \mathbf{R}_i}$. Expressing $\mathbf{R}_i$ in terms of the nucleon coordinates with respect to the center of mass, the effective perturbation term entering the probability amplitude $f^{1/2}$ is of the form $e^{i\mathbf{k} \cdot \mathbf{u}}$, with $\mathbf{u}$ the displacement of the atom from its lattice equilibrium position: $f^{1/2} \propto < 0, 0, 0 \ldots |e^{i\mathbf{k} \cdot \mathbf{u}}|0, 0, 0 \ldots >$.

The proportionality factor includes the matrix element of the variables and spins of the nucleons as well as the mechanism of the transition.

The vibrational ground-state (see Eq. (14.23)) for a given branch is $||0, 0, 0 \ldots > = \prod_{\mathbf{q}} e^{-Q_{\mathbf{q}}^2/4\Delta_{\mathbf{q}}^2}$. The displacement $\mathbf{u}$ can be written as a superposition of the normal modes coordinates: $\mathbf{u} = \sum_{\mathbf{q}} \alpha_{\mathbf{q}} Q_{\mathbf{q}}$ ($\alpha_{\mathbf{q}}$ normalizing factors which include the masses). Then, by referring to the component along the direction of the γ-rays, one writes

$$f^{1/2} \propto \int_{-\infty}^{+\infty} \prod_{\mathbf{q}} e^{\frac{-Q_{\mathbf{q}}^2}{2\Delta_{\mathbf{q}}^2}} e^{ik\alpha_{\mathbf{q}} Q_{\mathbf{q}}} dQ_{\mathbf{q}} \propto \prod_{\mathbf{q}} e^{\frac{-\alpha_{\mathbf{q}}^2 \Delta_{\mathbf{q}}^2 k^2}{2}} = e^{-\frac{1}{2} \sum_{\mathbf{q}} \alpha_{\mathbf{q}}^2 \Delta_{\mathbf{q}}^2 k^2}$$

The mean square displacement turns out

$$< 0, 0 \ldots |u_x^2|0, 0 \ldots > \equiv < 0, 0 \ldots | \sum_{\mathbf{q},\mathbf{q}'} \alpha_{\mathbf{q}} Q_{\mathbf{q}} \alpha_{\mathbf{q}'} Q_{\mathbf{q}'} |0, 0 \ldots > =$$

$$= \sum_{\mathbf{q}} \alpha_{\mathbf{q}}^2 < 0, 0 \ldots |Q_{\mathbf{q}}^2|0, 0 \ldots > = \sum_{\mathbf{q}} \alpha_{\mathbf{q}}^2 \Delta_{\mathbf{q}}^2$$

and then

$$f \propto e^{-k^2 <u_x^2>} = e^{-k^2 <\mathbf{u}^2>/3}.$$

Since for $k = 0$ one can set $f = 1$, one has

$$f = e^{-k^2 <\mathbf{u}^2>/3}. \tag{14.38}$$

For $T \to 0$ f depends from the particular transition involved in the emission process (through k^2) and from the spectrum of the crystal through the zero-point vibrational amplitude $< u^2(T = 0) >$.

The temperature dependence of f originates from the one for $< u^2 >$. f is also known as the *Debye-Waller* factor, since it controls the intensity of X-ray and neutron diffraction peaks. The Bragg reflections, in fact, do require elastic scattering and therefore recoilless absorption and re-emission.

By evaluating $< |\mathbf{u}|^2 >$ for the Debye crystal, for instance, (see Eq.(14.36)) for $T \ll \Theta_D$ one has

$$f = e^{-(3E_R/2k_B\Theta_D)}. \qquad (14.39)$$

The typical experimental setup for Mössbauer absorption spectroscopy is sketched below

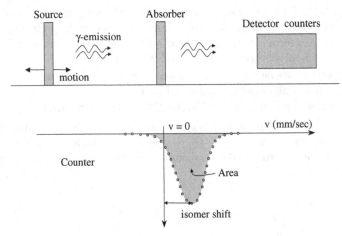

The source (or the absorber) is moved at the velocity v in order to sweep through the resonance condition. As a function of the velocity, one observes the Mössbauer absorption line, the area being proportional to the recoilless fraction f.

The shift with respect to the zero-velocity condition, *isomer shift*, is related to the finite volume of the emitting and absorbing nuclei (try to understand the shift by returning to Problems 1.6 and 5.23).

Since the motions do not affect the linewidth, the resolution of the Mössbauer line in principle depends only on the intrinsic lifetime of the level. Typically, for $\sim 100\,$keV γ-rays, a resolution around 10^{-14} can be achieved. Therefore, the Mössbauer spectroscopy can be used in solid state physics to investigate the magnetic and electric hyperfine splitting of the nuclear levels. It has been used also in order to detect subtle relativistic effects (see Problem 14.13).

Problems

Problem 14.6 Show that an approximate estimate of the Debye temperature in a monoatomic crystal can be obtained from the specific heat, by looking at the temperature at which $C_V \simeq 23 \cdot 10^7$ erg/mole K.

Solution: From Eqs. (14.19) and (14.31)

$$U = \int_0^{\omega_D} \frac{\hbar\omega}{e^{\frac{\hbar\omega}{k_B T}} - 1} D(\omega) d\omega = \frac{3}{2\pi^2} \frac{N v_c}{v^3} \int_0^{\omega_D} \frac{\hbar\omega^3}{e^{\frac{\hbar\omega}{k_B T}} - 1} d\omega,$$

(having neglected the zero-point energy which does not contribute to the thermal derivatives). v is the sound velocity (an average of the ones for longitudinal and transverse branches). From Eq. (14.33) the specific heat can be written

$$C_V = 9R \left[4 \left(\frac{T}{\theta_D} \right)^3 \int_0^{\frac{\theta_D}{T}} \frac{z^3}{e^z - 1} dz - \frac{\theta_D}{T} \frac{1}{e^{\frac{\theta_D}{T}} - 1} \right].$$

For $T = \theta_D$

$$C_V(T = \theta_D) \simeq 36R \left[\int_0^1 \frac{z^3}{e^z - 1} dz - \frac{1}{1.72} \right]$$

and then $C_V(T = \theta_D) \simeq 2.856R \simeq 23.74 \cdot 10^7$ erg/mole K.

Problem 14.7 In a 1D linear diatomic crystal of alternating Br^- and Li^+ ions and lattice step $a = 2$ Å, the sound velocity is $v = 2.7 \cdot 10^5$ cm/s. Derive the effective elastic constant for the sound propagation under the assumption used at Sect. 14.3.2. Estimate the gap between the acoustic and optical branches.

Solution: From Eq. (14.17), in the $q \to 0$ limit, the sound velocity turns out

$$v = \sqrt{\frac{2k}{m_1 + m_2}} a.$$

Then the elastic constant is

$$k = \frac{1}{2}(m_1 + m_2) \left(\frac{v}{a} \right)^2 \simeq 1.32 \times 10^4 \text{ dyne/cm}.$$

The gap covers the frequency range from $\omega_{min} = (2k/m_1)^{1/2}$ to $\omega_{max} = (2k/m_2)^{1/2}$, with

$$\omega_{min} = 0.14 \cdot 10^{14} \text{rad s}^{-1} \quad \text{and} \quad \omega_{max} = 0.47 \cdot 10^{14} \text{rad s}^{-1}.$$

Problem 14.8 For a cubic crystal, with lattice step a, show that within the Debye model and for $T \ll \Theta_D$, the most probable phonon energy is $\hbar\omega_p \simeq 1.6 k_B T$ and that the wavelength of the corresponding excitation is $\lambda_p \simeq a\Theta_D/T$.

Solution: In view of the analogy with photons (see Problem 1.25) the number of phonons with energy $\hbar\omega$ is given by

$$n(\omega) = D(\omega)/(e^{\hbar\omega/k_B T} - 1).$$

From Eq. (14.19) and from $dn(\omega)/d\omega = 0$, one finds

$$\frac{\hbar\omega_p}{k_B T} e^{\hbar\omega/k_B T} = 2(e^{\hbar\omega/k_B T} - 1)$$

and then $\hbar\omega_p/k_B T \simeq 1.6$.

Since $\lambda_p(\omega_p/2\pi) = v$, the average sound velocity, one has $\lambda_p \simeq 2\pi v\hbar/1.6 k_B T$. For cubic crystal $\Theta_D = (v\hbar/k_B a)(6\pi^2)^{1/3}$, and then $\lambda_p \simeq a\Theta_D/T$.

Problem 14.9 Show that in a Debye crystal at high temperature the thermal energy is larger than the classical one by a factor going as $1/T^2$.

Solution: From Eqs. (14.19), (14.21) and (14.31) the thermal energy is

$$U = 9 N k_B T \left(\frac{T}{\Theta_D}\right)^3 \int_0^{x_D} \left(\frac{x^3}{e^x - 1} + \frac{x^3}{2}\right) dx$$

with $x_D = \Theta_D/T$ and $x = \hbar\omega/k_B T$. For $x \to 0$, after series expansion of the integrand

$$\int_0^{x_D} \left(\frac{x^3}{e^x - 1} + \frac{x^3}{2}\right) dx \simeq \int_0^{x_D} \left(\frac{x^3}{x + \frac{x^2}{2} + \frac{x^3}{6} + \cdots} + \frac{x^3}{2}\right) dx \simeq$$

$$\simeq \int_0^{x_D} \left[x^2 \left(1 - \frac{x}{2} + \frac{x^2}{12} - \cdots\right) + \frac{x^3}{2}\right] dx \simeq \int_0^{x_D} x^2 \left(1 + \frac{x^2}{12} - \cdots\right) dx.$$

Note that the second term of the expansion cancels out the zero-point energy. Then one can write

$$U = 9 N k_B T \left(\frac{T}{\Theta_D}\right)^3 \left(\frac{1}{3} \left(\frac{\Theta_D}{T}\right)^3 + \frac{1}{60} \left(\frac{\Theta_D}{T}\right)^5 + \cdots\right).$$

The molar specific heat turns out

$$C_V \simeq 3R \left(1 - \frac{1}{20} \left(\frac{\Theta_D}{T}\right)^2 - \cdots\right).$$

Problem 14.10 In the figures below

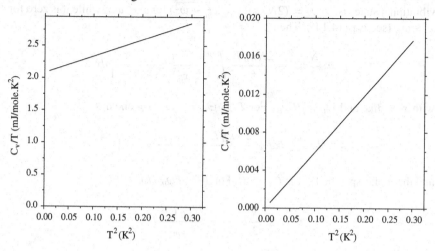

the low temperature specific heats of two crystals are reported. Are they metals or insulators? Estimate the Debye temperatures and the Fermi energy.

Solution: From $C_V/T = A + BT^2$, $A = R(\pi^2/3)D(E_F)k_B$ is the term associated with the free-electron contribution (see Problem 12.14 for N_A electrons), while $B = (12\pi^4/5)(R/\Theta_D^3)$ originates from the phonon contribution. Hence the figure on the left refers to a metal while the one on the right to an insulator ($A = 0$).

From the data on the left $A \simeq 2.1 \times 10^4$ erg/K^2mole one finds $E_F \simeq 1.7$ eV. From $B \simeq 2.6 \times 10^4$ erg/K^4mole, then $\Theta_D \simeq 90$ K. From the data on the right $B \simeq 590$ erg/K^4mole, yielding $\Theta_D \simeq 320$ K.

Problem 14.11 Derive the vibrational contribution to the specific heat for a monoatomic 1D crystal, at high and low temperatures, within the Debye and the Einstein approximations. Compare the results with the exact estimates obtained in the harmonic approximation and nearest-neighbor interactions (Sect. 14.3.1).

Solution: Within the Debye model the vibrational spectrum is $D(\omega) = Na/(2\pi v)$ and then according to Eq. (14.31)

$$U_D = \frac{N}{2}\hbar\omega + \frac{N}{\omega_D}\int_0^{\omega_D} \frac{\hbar\omega}{e^{\beta\hbar\omega} - 1}d\omega.$$

The molar specific heat turns out $C_V \simeq R$ for $T \gg \Theta_D = \hbar\omega_D/k_B$ and $C_V \simeq 2IR(T/\Theta_D)$ for $T \ll \Theta_D$, with $I = \int_0^\infty x/(e^x - 1)dx$.

Within the Einstein model $D(\omega) = N\delta(\omega - \omega_E)$ and results independent from the dimensionality are obtained (see Eq. (14.32)). One has $C_V \simeq R(\Theta_E/T)^2 exp(-\Theta_E/T)$ for $T \ll \Theta_E$ and $C_V \simeq R$ for $T \gg \Theta_E$.

In the harmonic approximation with nearest neighbors interactions the density of vibrational states is $D(\omega) = (2N/\pi)(1/\sqrt{\omega_m^2 - \omega^2})$ for $\omega \leq \omega_m$, while it is zero for $\omega > \omega_m$ (see Eq. (14.14)). Then

$$U = \frac{N}{2}\hbar\omega + \frac{2Nk_BT}{\pi} \int_0^{x_m} \frac{1}{\sqrt{x_m^2 - x^2}} \frac{x}{e^x - 1} dx$$

with $x = \beta\hbar\omega$ and $x_m = \beta\hbar\omega_m$. For $T \gg \Theta_m = \hbar\omega_m/k_B$ one has

$$U \simeq \frac{N}{2}\hbar\omega + \frac{2Nk_BT}{\pi} \left(\frac{\pi}{2} - \frac{x_m}{2} + \cdots\right)$$

and the molar specific heat is $C_V \simeq R$. For $T \ll \hbar\omega_m/k_B \equiv \Theta_m$

$$U \simeq \frac{N}{2}\hbar\omega + \frac{2N(k_BT)^2}{\pi\hbar\omega_m} I$$

so that

$$C_V \simeq \frac{4I}{\pi} R \frac{T}{\Theta_m}$$

showing that the Debye approximation yields the same low temperature behavior.

Problem 14.12 A diatomic crystal has two types of ions, one at spin $S = 1/2$ and $g = 2$ and one at $S = 0$. The Debye temperature is $\Theta_D = 200\,K$. Evaluate the entropy (per ion) at $T = 20\,K$ in zero external magnetic field and for magnetic field $H = 1\,kGauss$, for no interaction among the magnetic moments.

Solution: The vibrational entropy is

$$S_{vib} = \int_0^T \frac{C_V(T')}{T'} dT'$$

where for $T \ll \Theta_D$, neglecting the optical modes (see Problem 14.1)

$$C_V(T') = \frac{12\pi^4}{5} k_B \left(\frac{T'}{\theta_D}\right)^3 .$$

Then at $T' = 20\,K$

$$S_{vib} = k_B \frac{12\pi^4}{15} \left(\frac{T'}{\theta_D}\right)^3 = 0.078\,k_B.$$

The magnetic partition function is

$$Z_{mag} = \exp\left(-\frac{1}{2}y\right) + \exp\left(\frac{1}{2}y\right) \simeq 2 + \frac{y^2}{4}$$

with

$$y = \frac{\mu_B g H}{k_B T} \simeq \frac{0.9 \cdot 10^{-20}}{1.38 \cdot 10^{-16}} g \frac{H}{T} = 6.72 \cdot g \frac{H}{T} \cdot 10^{-5} \ll 1.$$

From

$$F = -k_B T \ln Z \quad \text{and} \quad S = -\frac{\partial F}{\partial T}$$

with

$$S_{mag}(T') \simeq k_B \left[\ln 2 - \frac{y^2}{4}\right] \simeq k_B ln2$$

one has

$$S = S_{vib} + \frac{1}{2}S_{mag} = k_B[0.078 + 0.34] = 0.42 k_B/\text{ion}.$$

Problem 14.13 The life time of the ^{57}Fe excited state decaying through γ emission at 14.4 keV is $\tau \simeq 1.4 \times 10^{-7}$ s (see Problems 1.24, 1.30 and 3.13). Estimate the height at which the γ-source should be placed with respect to an absorber at the ground level, in order to evidence the gravitational shift expected on the basis of Einstein theory.

Assume that a shift of 5 % of the natural linewidth of Mössbauer resonant absorption can be detected [in the real experiment by *Pound* and *Rebka* (Phys. Rev. Lett. 4, 337 (1960)) by using a particular experimental setup resolution of the order of $10^{-14} - 10^{-15}$ could be achieved, with a fractional full-width at half-height of the resonant Lorentzian absorption line of 1.13×10^{-12}]. Try to figure out why the source-absorber system has to be placed in a liquid He bath.

Solution: On falling from the height L the energy of the γ photon becomes

$$h\nu(0) = h\nu(L)\left[1 + \frac{gL}{c^2}\right]$$

where mgL/mc^2 can be read as the ratio of a gravitational potential energy mgL to the intrinsic energy mc^2 (mass independent and therefore valid also for photons).

The natural linewidth of the Mössbauer line is $2\hbar/\tau$. Therefore, to observe a 5 % variation

$$\frac{2\hbar}{20\tau} = h\nu(L)\frac{gL}{c^2}$$

and then

$$L = \frac{\hbar c^2}{10\,g\tau\,14.4\,\text{keV}} = 284\,\text{m}$$

(in the real experiment the height of the tower was about 10 times smaller!). Note that the natural linewidth, when sweeping with velocity v the absorber (or the source) corresponds to a velocity width

$$\Delta v = \frac{2\hbar c}{h\nu\tau} \simeq 0.2\,\text{mm/s}$$

(the actual full-width at half height in the experiment by Pound and Rebka was 0.43 mm/s).

A difference in the temperatures of the source and the absorber of 1 K could prevent the observation of the gravitational shift because of the temperature-dependent *second-order* Doppler shift resulting from lattice vibrations, since $< v^2 > \sim k_B T/M$. Low temperature increases the γ-recoilless fraction f.

Specific References and Further Reading

1. W. Cochran, *Lattice Vibrations*, Reports on Progress in Physics XXVI, 1 (1963).
2. A. Abragam, *L'effet Mossbauer et ses applications a l'etude des champs internes,* (Gordon and Breach, 1964).
3. N.W. Ashcroft and N.D. Mermin, *Solid State Physics*, (Holt, Rinehart and Winston, 1976).
4. F. Bassani e U.M. Grassano, *Fisica dello Stato Solido*, (Bollati Boringhieri, 2000).
5. G. Burns, *Solid State Physics*, (Academic Press Inc., 1985).
6. H.J. Goldsmid (Editor), *Problems in Solid State Physics*, (Pion Limited, London, 1972).
7. G. Grosso and G. Pastori Parravicini, *Solid State Physics*, 2nd Edition, (Academic Press, 2013).
8. H. Ibach and H. Lüth, *Solid State Physics: an Introduction to Theory and Experiments*, Springer Verlag (1990).
9. C. Kittel, *Introduction to Solid State Physics*, 8th Edition, (J. Wiley and Sons, 2005).
10. L. Mihály and M.C. Martin, *Solid State Physics - Problems and Solutions*, (J. Wiley, 1996).
11. J.M. Ziman, *Principles of the Theory of Solids*, (Cambridge University Press, 1964).

Chapter 15
Phase Diagrams, Response Functions and Fluctuations

Topics

Phenomenology of Phase Transitions
Critical Points and Thermodynamic Relationships
The Concept of Order Parameter
Free Energy at Phase Transitions
Homogeneous and Non-homogeneous Systems
Fluctuations and Their Time Dependence
Dynamical Structure Factor and Generalized Susceptibility
Experimental Techniques

During the last decades emerging scientific attention has been directed towards the subject of *phase transitions*, namely the dramatic changeover in the properties of a macroscopic system upon variation of a thermodynamic variable (in most cases the temperature). In this chapter and in the subsequent three chapters, we shall deal with the description of some phases of solid state matter and describe the basic aspects of the microscopic mechanisms that control the transition between different phases. The key concept and the role of the *order parameter* will be introduced, first in the framework of the thermodynamic description and then in the framework of specific theories dealing with the microscopic dynamics that drive the crossover from one phase to the other.

It will be shown how on approaching the *critical point*, marking the boundary between two phases at *second-order phase transitions*, the *fluctuations* around the average quantities exhibit strong enhancement. At the same time the metastable "heterophase droplets" of one phase into the other (reminiscent of the liquid droplets forming in the vapour phase when cooling along the critical isochore towards the vapour-liquid transition point) live longer and longer in time, thus implying *slowing-down of the fluctuations*.

A rather general description of phase transitions and of the critical dynamics dealing with the behaviour of thermodynamic functions, shall be given in the framework

© Springer International Publishing Switzerland 2015
A. Rigamonti and P. Carretta, *Structure of Matter*,
UNITEXT for Physics, DOI 10.1007/978-3-319-17897-4_15

of a *Landau-type statistical theory*, which can be considered to include more specific *mean field* approaches, such the *Weiss* theory for magnetic systems and the *Van der Waals* theory for fluids.

In developing those topics we will have the chance to recall, and at the same time to systematize, some issues and phenomena already mentioned in previous parts of the book, for instance clarifying the concepts of *instability* at a critical point or the meaning of the fluctuations and their role in forcing the transition from one phase to the other.

15.1 Phase Diagrams, Thermodynamic Responses and Critical Points: Introductory Remarks

Condensed matter is conventionally considered to be found in solid, liquid and gaseous states, also known as phases. In the solid state, a variety of other "phases" can indeed be defined, regarding electric properties (*dielectrics, ferroelectrics*), magnetic properties (*paramagnets, ferro* or *antiferromagnets*), transport properties (*insulators, semiconductors, metals, superconductors*), crystal structures (*polymorphs*, ordered or disordered *binary alloys*). Furthermore in condensed matter other particular phenomena accompanying phase transformations are known, such as *superfluidity*, the *incommensurate phases*, the transitions in *disordered crystals* and in *martensitic materials*, the roughening of the surfaces or the *collapse transition* in gels. Other phases can be obtained through artificial lattices produced by optical confinement as, for example, the *Bose-Einstein condensate*.

The phenomena occurring in the neighbourhood of the crossover from one phase to the other, when one physical variable (e.g. the temperature, pressure or external electric or magnetic fields) drives the system in a different state, have attracted strong scientific interest. *Instability* phenomena, *criticality* of some parameters or *critical divergences* of some quantities are detected at the transition, accompanied by *enhancement* and *slowing down* of the fluctuations, as we shall discuss in subsequent sections. In spite of the relevant microscopic differences in these systems and phenomena, the phase transitions can be discussed within a common theoretical framework, by means of rather general approaches that justify the large variety of the related effects.

The phase diagrams divide the planes defined by certain electric, magnetic or thermodynamic variables in areas which refer to the existence of a given state. The typical phase diagram of gas-liquid-solid matter, as the water, is shown in Fig. 15.1. In the phase diagram the vaporization curve is conveniently defined as the *coexistence line* and it ends at the *critical point*, where the *critical pressure* P_c and *critical temperature* T_c are defined. For instance, for water the critical temperature is $T_c = 647$ K and P_c is 218 atm, in correspondence to a *critical density* $\rho_c = 0.329$ g/cm^3. It should be remarked that for water the line dividing the solid and liquid states has

Fig. 15.1 The water pressure-temperature phase diagram

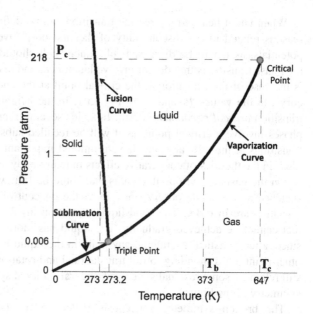

negative derivative along the fusion curve (see Fig. 15.1). This is a very special case in condensed matter (see Problem 15.1).

When on increasing temperature the solid melts the temperature stays constant for the time of complete melting. Similarly, when heat is provided and the liquid is crossing to vapour, along the vaporization process the temperature remains constant. *Melting* and *vaporization* are phase transitions for which one must provide heat along the process, meanwhile the temperature is constant, namely the *latent heat*[1] For these transitions, by moving along pathways parallel to the temperature axis in the figure above, the crossover from one phase to the other is *discontinuous*, both for increasing or for decreasing temperature.[2] Other changes of states requiring latent heat are the ones related to *polymorphism in crystals*.

---

[1]The *melting latent heat* is the one required to transform a gram of solid into liquid (for ice 79.7 cal/g). To push the liquid into the vapour phase the *latent heat of vaporization* L_{vap} is required (for water 539.6 cal/g = 40.7 kJ/mol). When at the boiling temperature T_b water is transformed in steam, a jump in the entropy occurs and correspondently a spike in the heat capacity is observed. From the definition $L_{vap} = (\Delta Q)_{rev} = T_b \Delta S$, the discountinuous jump in the entropy turns out $\Delta S = 110$ J/K.mol, that corresponds to 13.1 R (namely $L_{vap} = 13.1 R T_b$) in approximate agreement with the empirical *Trouton rule* $L_{vap} \simeq 10 R T_b$. This rule can be approximatively derived from the comparison of the liquid and the vapour densities, by resorting to the definition of entropy in terms of the number of microstates available for a given macrostate (see Problem 15.2).

[2]A certain asymmetry is related to the delay in the solidification process for a liquid, when the meta-stable *under-cooled liquid* is produced, while no analogous thermal hysteresis is detected in the melting process of the solid. On the other hand, the meta-stable state of *superheated liquid* is known to be possible.

When latent heat is required the transition can be defined to be *first order*, for reasons related to the discontinuity of the first derivative of the thermodynamic potentials, as it will be discussed subsequently. It should be remarked that there are phase transitions that do not involve latent heat and *occur with continuity*. This is the case of the transition at the critical point at the end of the vapour pressure curve, at the values P_c and T_c in Fig. 15.1. In the neighbourhood of this type of transition *critical phenomena* occur. In solids several transitions between different phases involve a critical point, as it will be recalled subsequently. The studies of transitions at the critical points have promoted important developments and novel ideas in the theories of cooperative effects in many-body systems.

On the basis of the possible paths that could be followed in the phase diagram sketched above, some observations about the concept of *broken symmetry* at the transition are in order. The solid-liquid transition involves a *change of symmetry* that cannot be achieved gradually. The liquid has more symmetry than the solid since, after statistical averaging, any point in the liquid is exactly the same as any other point, corresponding to full translational and rotational symmetry. The solid, still retaining some residual symmetry (the translational symmetry) has lost the full symmetry of the liquid.

The broken symmetry concept is better understood at the *paramagnetic-ferromagnetic phase transition*. Above T_c complete rotational symmetry is present, since the magnetic moments can point in any direction. Instead, below the ordering temperature (see Sect. 4.4) a unique direction for all the spins is chosen, thus breaking the rotational symmetry present in the disordered phase.

The path from gas to liquid and *viceversa* does not involve discontinuous symmetry breaking but still implies a *discontinuous change* of the density and there is not a critical point. By moving along a line which goes around the critical point any discontinuous change would be avoided. Thus there is no change of symmetry and the gas and the liquid differ only in their densities. Only at the end of the coexistence line (the vaporization curve) between liquid and gas a critical point is present.

By cooling a gas along the *critical isochore* the critical point is reached at the temperature T_c and a continuous change of density starts to occur on further reducing the temperature. In fact, another diagram known from elementary physics is the one dealing with the isotherms of a real gas (or *Van der Waals isotherms*), often accompanied by the *Clapeyron equation* for the derivative of the pressure as a function of temperature in terms of the specific volumes (inverse of densities), as sketched in Fig. 15.2.

In Fig. 15.3 the analogies between the transition at the critical point of a fluid and the ones occurring in ferroelectrics and in magnetic systems are illustrated.

Pressure, electric and magnetic fields are called *fields*. The volume (or the inverse density ρ^{-1}), the electric polarization $\mathcal{P}$ and the magnetization M are the *densities*, thermodynamically conjugated to the corresponding fields. The densities go gradually to zero on approaching the critical points by moving along the coexistence

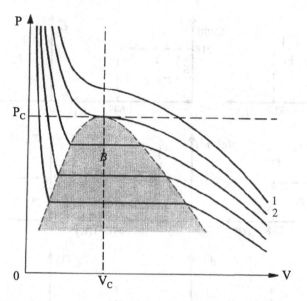

Fig. 15.2 The pressure-volume phase diagram for a fluid. The *solid lines* are isothermal curves. Line 1 is at $T \gg T_c$ and line 2 at $T = T_c$. In region B one has liquid and vapour. The *Clapeyron equation* is $(dP/dT) = \Delta S/\Delta V = L/T(V_2 - V_1)$, with L the latent heat and $V_{1,2}$ the specific volumes. For water at 100 °C, one has $(dP/dT) = 27\,\text{mm Hg/K}$. Since at the liquid-ice transition $V_1 > V_2$, (dP/dT) turns out to be negative, as shown in Figs. 15.1 and 15.3 (see Problem 15.1)

curves and thus they can be assumed as the *thermodynamic order parameter*[3] ρ, M or $\mathcal{P}$.

Other systems displaying critical points are the *binary alloys*. For example in *CuZn* the Cu and Zn atoms are randomly distributed on the lattice sites of two interpenetrating cubic lattices in the high temperature disordered phase (above $T_c \simeq 640\,°C$). The percent of atoms on a given sub-lattice are $n_{Cu} = n_{Zn} = 1/2$. In the totally ordered phase (for temperature close to zero) the bcc structure has $n_{Cu} = 1$ and $n_{Zn} = 0$ in a sub-lattice and $n_{Cu} = 0$ and $n_{Zn} = 1$ in the other one. The order parameter $(n_{Cu} - n_{Zn})$ is zero in the disordered phase and 1 in the totally ordered phase. No latent heat is required at the transition.

Other phase transitions we shall deal with occur in *antiferromagnets* and in *superconductors*. Finally it is mentioned that the Helium *superfluid transition*, around 2.16 K, is known as λ *transition*, in view of the temperature behaviour of the specific heat, resembling this Greek letter. In this case the ordering does not involve

---

[3]The thermodynamic order parameter is usually related to the local order parameter, namely the statistical average (see Eq. (15.1)) of a given microscopic variable (for instance the expectation values of spin components for magnetic systems) through a sum, with proper phase factors. In the presence of inhomogeneities, with spatial variation of the local order parameter, as it is the case when thermal fluctuations are taken into account (Sect. 15.3), the definition of the local order parameter requires an averaging process over a certain number of lattice sites (*coarse grain average*).

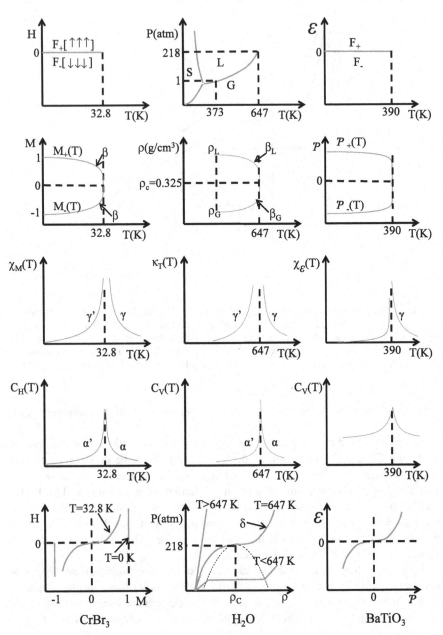

Fig. 15.3 Sketches which illustrate the analogies among different kind of phase transitions: from *left* to *right* the ferromagnetic transition in *CrBr₃*, the vapour-liquid transition in water and the ferroelectric transition in *BaTiO₃*. In the *first line* the phase diagrams are shown. The *second line* reports the behaviours of the order parameters while in the *third* and *fourth lines* the behaviours of some response functions are sketched. The *last line*, at the *bottom*, reports alternative representations of the phase diagrams. P, $\mathcal{E}$ and H are the *fields*; $V = \rho^{-1}$, $\mathcal{P}$ and M the *densities*. The density going to zero on moving along the coexistence curve defines the thermodynamic order parameter (see text)

the positions of the atoms but rather it occurs in the momentum space. Below 2.2 K Helium suddenly acquires unusual flow properties, with no viscosity, as the consequence of the "order" of the atoms in the state of zero momentum.

Now a few recalls of thermodynamic-statistical relations are in order. The statistical thermal average of an observable $\mathcal{O}$ is written

$$< \mathcal{O} > = \frac{\sum_n e^{-\beta E_n} < n|\mathcal{O}|n >}{\sum_n e^{-\beta E_n} < n|n >} \equiv \frac{Tr\{\rho \mathcal{O}\}}{Tr\{\rho\}} \tag{15.1}$$

where $< n|\mathcal{O}|n >$ are the eigenvalues of the observable while $\rho = exp(-\beta \mathcal{H})$ is the *density matrix* and $Z = Tr\{\rho\}$ the *partition function* (see Problems 4.10 and 4.18).

The *Gibbs free energy* is

$$G = F + xX = U - TS + xX \tag{15.2}$$

where U, the *internal energy*, is the statistical thermal average of the Hamiltonian, i.e. $U = < \mathcal{H} >$, while here X indicates the fields and x the conjugated densities (which are extensive variables, as the *entropy S*).

The *Helmholtz free energy* $F = U - TS$ (see Problem 6.4) is

$$F = -k_B T \ln Z \tag{15.3}$$

The Gibbs free energy $G = G(X, T)$, with $xX = PV$, is the relevant thermodynamic potential for a system at given T and X. At the equilibrium, coexistence phases have the same chemical potential and $dG_{total} = 0$. The Helmholtz free energy $F = U - TS$ is the thermodynamic potential in terms of T and V. The latent heat corresponds to the discontinuity in the first derivative of G and in particular in the entropy and the density. The discontinuities in the specific heat (see Fig. 15.3) and in the compressibility correspond to the discontinuities in the second derivatives of G. The discontinuity in the third derivative of G with respect to T implies the divergence of the specific heat at constant pressure for $T \rightarrow T_c$ and it accompanies the λ *transition*.

The *order parameters* are the first derivatives of free energy with respect to the conjugated variable:

$$\rho^{-1} = -(\partial G/\partial P)_T \tag{15.4}$$
$$M(H, T) = -(\partial G/\partial H)_T \tag{15.5}$$
$$\mathcal{P}(\mathcal{E}, T) = -(\partial G/\partial \mathcal{E})_T \tag{15.6}$$

In view of the *fluctuation-dissipation relationships* (see Problems 4.18 and 6.12 and Eq. (6.14)) one can figure out that the divergences in the response functions must correspond to an enhancement of the fluctuations, as it will be addressed subsequently. In fact, it is reminded that the specific heat $C_V = [\partial U/\partial T] = -T[\partial^2 F/\partial T^2]$ can be related to the energy fluctuations (see Problem 6.9):

$$C_V = \frac{-1}{T^2} \frac{\partial}{\partial \beta} \left(\frac{Tr \mathcal{H} e^{-\beta \mathcal{H}}}{Tr e^{-\beta \mathcal{H}}} \right) = \frac{1}{k_B T^2} \left(<E^2> - <E>^2 \right), \qquad (15.7)$$

while for the susceptibility $\chi = (<\Delta M^2> /k_B T)$ (see Eq. (6.14)) and for the isothermal compressibility of a fluid $\kappa_T / \kappa_0 = <\Delta N^2> / <N>$, with N the number of particles (e.g. molecules) and κ_0 the compressibility for the non-interacting particles (see Problem 15.5).

According to the criterion envisaged by *Ehrenfest, first order phase transitions* are the ones with discontinuity in the first derivatives of the free energy, such as the entropy, yielding a latent heat and a change of volume as observed in the melting and solidification of a fluid. *Second order phase transitions* show discontinuities in the second derivatives of the thermodynamic potentials as most of the magnetic, ferroelectric and superconductive phase transitions.[4]

The *response functions* are defined as the derivatives of the densities with respect to the fields:

$$\kappa_T = -\left(\frac{1}{V} \right) (\partial V / \partial P)_T = -\left(\frac{1}{V} \right) (\partial^2 G / \partial P^2)_T \qquad (15.8)$$

$$\chi_T = (\partial M / \partial H)_T = -\left(\frac{1}{V} \right) (\partial^2 G / \partial H^2)_T \qquad (15.9)$$

$$C_P = T(\partial S / \partial T)_P = -T(\partial^2 G / \partial T^2)_P \qquad (15.10)$$

$$C_V = -T(\partial^2 F / \partial T^2)_V \qquad (15.11)$$

Then one can again observe that at second order phase transition the discontinuities only involve the response functions and not the order parameters.[5]

By referring to Fig. 15.3 one notices that the order parameter and the response functions in the proximity of T_c follow power laws, characterized by critical indexes α, β, γ, etc. The index is defined as the limit, for *reduced temperature* $\epsilon = (T - T_c)/T_c$ going to zero, of the ratio $ln f(\epsilon)/ln \epsilon$, for a given "critical function" f. For example, in regards of the order parameter, β defines the slope of the magnetization or of the polarization for temperature approaching T_c from below. γ and γ' define the slope of the response functions on approaching the critical points.

---

[4]*Fisher* defines the second order phase transition as continue (discontinuities occurring at higher order, as in the compressibility or in the specific heat). For *structural* and *ferroelectric* phase transitions in crystals Landau defines as first order transition the one with sudden variation of the order parameter at T_c and with equilibrium of two phases without symmetry constrains at this temperature. At variance, according to the Landau criterion are second order the transitions with symmetry constrains (the point group below T_c is a subgroup of the one above T_c) and for them an expansion of the free energy in terms of powers of the order parameter is allowed (see our generalization at Sect. 15.2).

[5]It is remarked that most second order transitions in practice display some discontinuity in the order parameter at T_c. When the discontinuity is small one speaks of *quasi second-order transitions* and the same theoretical framework is used, possibly with renormalization of some quantities, for instance the transition temperature.

The critical indexes depend on two main factors: the *system dimensionality D* (namely if the interactions involve one, two or three dimensions) and the *dimensionality of the critical variable d*. To illustrate the meaning of this last quantity let us refer to an insulating magnetic system where the critical variables corresponds to the spins localized at the lattice sites. If the spin is fully isotropic, namely its x, y and z components are equivalent, the system is called an *Heisenberg system* ($d = 3$). If the z component is zero and the x and y components are equivalent one says it is an *XY system* ($d = 2$). If only the z component is different from zero one says it is an *Ising system* ($d = 1$). For a magnetic system the dimensionality d is thus determined by the *magnetic anisotropy*, typically driven by the spin-orbit interaction. Once D and d are defined the critical exponents are determined (see Table 15.1), no matter which is the nature of the interaction causing the phase transition. In order to clarify those relevant quantities involved in phase transitions here we are anticipating some issues to be dealt with at Chap. 17.

Accordingly, different systems as, for example, magnetic ones and fluids, can be characterized by the *same critical behaviour* if D and d are the same, namely if they belong to the same *universality class*. As it is shown in Table 15.1 in some cases (e.g. $D = 2$ and $d = 3$) the critical exponents are not defined. This is due to the fact that for certain universality classes no phase transition at a finite temperature is permitted. Indeed, as demonstrated by the *Mermin-Wagner theorem*, for an Heisenberg system no phase transition at a finite temperature is allowed for $D < 3$. In Table 15.2 the occurrence or not of a phase transition for the different universality classes is summarized. A special case is represented by the two-dimensional XY systems which are characterized by anomalies in the response functions but the order parameter is non-zero below T_c just in the case of finite size systems.

We conclude this introductory section with some observations on the thermodynamic description of magnetic systems. In general terms that description can be obtained from the usual thermodynamics of a gas by replacing $-V$ by the magneti-

Table 15.1 The values of the critical exponent β for different dimensionalities of the lattice D and of the critical variable d (see also Sect. 17.1)

D/d	3 (Heisenberg)	2 (XY)	1 (Ising)
3	0.345	0.33	0.3125
2	–	0.235 for finite systems	0.125
1	–	–	–

Table 15.2 The occurrence ($\odot$) or the absence ($\times$) of the phase transition at finite temperature is reported for different dimensionalities of the lattice D and of the critical variable d

D/d	3 (Heisenberg)	2 (XY)	1 (Ising)
3	$\odot$	$\odot$	$\odot$
2	$\times$	$\otimes$	$\odot$
1	$\times$	$\times$	$\times$

the $D = 2$ XY system, where at the *Berezinskii-Kosterlitz-Thouless* transition a divergence in the correlation length occurs while the order parameter remains zero for an infinitely large system zation M and the pressure P with the field H. However some more comments are in order.

Let us refer to a solenoid of length L and with N turns of area A, wound around a magnetic sample. The field created by the current i is $H = 4\pi ni/c$ ($n = N/L$ number of turns per unit length). The elementary work required to increase the current from i to $i + di$ is given by

$$dW = -\mathcal{E} \cdot \mathbf{i}dt = -(NAi/c)dB = NA\mathbf{H} \cdot \mathbf{dB}/4\pi n, \qquad (15.12)$$

having used for the electromotive force $\mathcal{E} = (N/c)d\Phi/dt = (NA/c)dB/dt$. Since $\mathbf{B} = \mathbf{H} + 4\pi\mathbf{M}$, one has

$$dW = V\left[\mathbf{H} \cdot \mathbf{dM} + \mathbf{H} \cdot \mathbf{dH}/4\pi\right]; \qquad (15.13)$$

the first term is the magnetic work done on the sample while the second term is the work, independent from the presence of the sample, required to create the energy density $(H^2/8\pi)$ in the empty coil. Then the first law of thermodynamics, expressed per unit volume, becomes

$$dU = TdS + PdV + \mathbf{H} \cdot \mathbf{dM} + d(H^2/8\pi), \qquad (15.14)$$

namely

$$dU \simeq TdS + \mathbf{H} \cdot \mathbf{dM}, \qquad (15.15)$$

having considered that $dV \simeq 0$ and neglecting terms not involving the sample. $\mathbf{H} \cdot \mathbf{dM}$ is the work on the sample equivalent to $-PdV$ in a fluid.

The internal energy could be defined in terms of S and M but usually it is more convenient to work with the external field and the temperature as variables. Then, by taking into account Eq. (15.15), one writes the thermodynamic potentials $G(T, H)$ or $F(T, M)$ as

$$dG = dU - SdT - TdS - \mathbf{M} \cdot \mathbf{dH} - \mathbf{H} \cdot \mathbf{dM} = -SdT - \mathbf{M} \cdot \mathbf{dH}. \qquad (15.16)$$

F and G are related by the *Legendre transformation*

$$G(T, H) = U - TS - \mathbf{H} \cdot \mathbf{M}(H) = F[T, \mathbf{M}(H)] - \mathbf{H} \cdot \mathbf{M}(H), \qquad (15.17)$$

F is the thermodynamic potential most used when T and M are constant. In fact, when the induced magnetization is small G and F practically coincide. This equivalence is commonly used in superconductors around the transition temperature T_c (where on the other hand the term $H^2/8\pi$ has to be added to F in order to account for the magnetic energy in the volume, see Sect. 18.9).

15.2 Free Energy for Homogeneous Systems

The phenomenology recalled in the previous section evidences that phase transitions are collective phenomena occurring in many-body systems with inter-particle interactions. Once that a local "force" (an external force or even an "internal" force related to the *fluctuations around the equilibrium value*) initiates to induce a certain order in a given system of variables, the interaction can propagate the order to distances larger than the interaction range. Thus, without any external action, as a consequence of the fluctuations a collective response can occur, with the spontaneous transition to the "ordered" state. In general terms, when a phase transition is driven at a finite temperature, one can look at the free energy $F = U - TS$ and observe that the interaction energy U favours order, the entropy S favours disorder and T is the tilting factor in the balance.

A rather general description of the phase transitions and of the underlying dynamics involving some microscopic variables, can be given according to a *Landau-type statistical approach*, which includes more specific theories such as the *Weiss mean field theory* for magnetic systems and the *Van der Waals theory* for fluids. This approach is suited for second order or slightly first-order transitions where one observes anomalous behaviours (such as divergences or tendencies to zero) of the *response functions*, the derivatives of the thermodynamic densities (magnetization, particle density, entropy) with respect to the conjugate fields (magnetic or electric fields, pressure, temperature). It will be shown that those behaviours reflect the onset of correlation and slowing down of the fluctuations around the equilibrium values, driven by some microscopic *critical dynamics*.

In the framework of Landau-type statistical theory one defines a generalized site-dependent order parameter $m(\mathbf{r})$ having $\mathbf{h}$ as thermodynamically conjugated field (we shall use local scalar quantities for simplicity). An appropriate average $< m >$ of $m(\mathbf{r})$ is defined as the *thermodynamic density* introduced at Sect. 15.1, going gradually to zero on moving along the coexistence curve towards the critical point. From a microscopic point of view there is a certain correspondence of $< m >$ to a local critical variable (e.g. magnetization corresponding to the expectation values of local spins).

The general principle is to expand the free energy density $f[m(\mathbf{r}, T)]$ in powers of $m(\mathbf{r})$, with coefficients that depend on $(T - T_c)$, the total free energy being $F = \int f[m(\mathbf{r}, T)]d\mathbf{r}$. By taking into account that f must increase when a gradient of m occurs (and that ∇m must be involved to the second power for isotropic systems), one writes[6]

$$f[m(r, T)] = f_0(T) + \alpha_L(T)m^2(r) + \frac{\beta_L(T)}{2}m^4(r) + \gamma_L(T)|\nabla m(r)|^2. \quad (15.18)$$

---

[6]The cost in energy when the order parameter varies in the space can be derived in the form involving $\nabla \mathbf{m}(\mathbf{r})$ (Eq. (15.18)) from the limit for $\mathbf{h} \to 0$ of the expression $\sum_{\mathbf{h}}[m(\mathbf{r} + \mathbf{h}) - m(\mathbf{r})]^2/|\mathbf{h}|^2$ (limit of the continuum condition).

In this expansion the odd terms are not present because the form of the forces cannot change on reversing the sign of m. The term ∇m has the role of damping out the spatial variation in m (for $\gamma_L(T) > 0$), f having a minimum for m constant.[7] The expansion does not converge but it is customary to suppose that this involves the coefficients of the terms not explicitly considered.

It is reasonable to expand $\alpha_L(T)$ and $\beta_L(T)$ in series of $(T - T_c)$. Then, for a homogeneous system $|\nabla m(r)| = 0$ and by assuming $\beta_L(T) = b_0$, temperature-independent, and $\alpha_L(T) = a_0(T - T_c)$, Eq. (15.18) is rewritten

$$f[m, T] = f_0(T) + a_0(T - T_c)m^2 + \frac{b_0}{2}m^4. \qquad (15.19)$$

For $T > T_c$ the minimum of f as a function of m is at $m = 0$. For $T < T_c$ the minimum is at the temperature-dependent equilibrium value

$$m_e = \left[-\frac{a_0}{b_0}(T - T_c) \right]^{1/2} = m_0(-\varepsilon)^{1/2}, \qquad (15.20)$$

where $\varepsilon = (T - T_c)/T_c$ (see sketch below).

Therefore the critical exponent describing the temperature dependence of the order parameter on cooling below T_c is $\beta = 1/2$, often indicated as the *classical* or *mean-field exponent*.

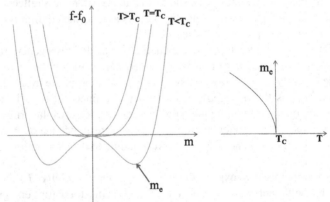

From the equation of state yielding $m = m(T, H)$ (as for instance the equation for the magnetization in the Weiss mean field theory, see Sect. 4.4, or for the inverse density $\rho^{-1} = V$ for the isotherms of real gases) one can find the response of m to the conjugate field, namely the generalized susceptibility $\chi = dm/dh$. According to the mean field derivation at Sect. 4.4 (see also Sect. 15.1), around T_c for a ferromagnet one has $\chi \propto |(T - T_c)|^{-1}$, namely a *classical critical exponent* $\gamma = 1$.

---

[7]The condition for stability are $df/dm = 0$ and $df^2/dm^2 > 0$. In order for f to have a true minimum at $m = 0$ for $T > T_c$ the term in m^3 has to be eliminated.

From the free energy density as in Eq. (15.18), the specific entropy and the specific heat (the response function to thermal stimulus) can be directly derived (examples are given at Chap. 6).

It should be anticipated that for the superconducting transition the role of the free energy expansion requires further discussion. In fact, in that case the order parameter $\Psi(\mathbf{r}) = \sqrt{n_C} exp(i\theta(\mathbf{r}))$ is complex (with n_C the density of *Cooper pairs* and $\theta(\mathbf{r})$ the phase) and it cannot be considered site-independent, as we shall discuss at Sect. 18.9.

15.3 Non Homogeneous Systems and Fluctuations

The assumption of homogeneity for m fails in the neighbourhood of a phase transition, where the fluctuations around the equilibrium value play a key role, the curvature of f as a function of m going to zero for $T \rightarrow T_c$, according to Eq. (15.19). One could say that it is just the enhancement of the fluctuations for $T \rightarrow T_c$ from above that drives the transition to the low-temperature phase.

Statistical mechanics yields for the probability in a subsystem of the fluctuation of m around the mean value, at constant temperature, the expression

$$W(m) = Ae^{S_T(m)/k_B}, \tag{15.21}$$

where S_T is the entropy of the total system, which is characterized by a peak around the average $< m >$ and by a Gaussian distribution (see Problem 6.1). The local fluctuation at constant temperature implies an increase of the free energy

$$\delta F(m(\mathbf{r})) = \delta U - T\delta S. \tag{15.22}$$

Since the total system remains in equilibrium $\delta U_T = 0$ and then $\delta S_T = -\delta F/T$. Thus

$$W[m(\mathbf{r}) - < m(\mathbf{r}) >] \propto e^{-\Delta F/k_B T}. \tag{15.23}$$

where here $\Delta F = F[m(\mathbf{r})] - F[< m(\mathbf{r}) >]$.

When the transition is approached, the free energy being a slowly varying function of m (see the sketchy behaviour above), large fluctuations occur. The mean square value $< m^2 >$ is related to the response function χ by the *fluctuation-dissipation relationship*. Therefore the divergence of χ at the critical point, for a second-order phase transition, evidences the divergence of the fluctuations, as already mentioned.

Furthermore, as it is known from the *variational principle* in thermodynamics, when a fluctuation occurs the system tends to return to the equilibrium according to the law $\partial F/\partial t \leq 0$ (or the analogous one in term of G for T and fields constant).

Then, in the presence of fluctuations around the equilibrium value $< m >$ (that is zero above T_c) one has to go back to Eq. (15.18) by considering $\gamma_L \neq 0$, often assumed temperature independent.[8] The term in m^4 is neglected in the so called *Gaussian approximation*, or *first-order fluctuation correction*.[9] Furthermore one has to expand $m(\mathbf{r})$ in Fourier series

$$m(\mathbf{r}) = \sum_{\mathbf{q}} m_{\mathbf{q}} e^{-i\mathbf{q}\cdot\mathbf{r}} \tag{15.24}$$

and then

$$\frac{\Delta F}{V} = \frac{\int (f - f_0) dV}{V} = \sum_{\mathbf{q}} |m_{\mathbf{q}}|^2 \alpha_L(T) \left(1 + \frac{\gamma_L q^2}{\alpha_L(T)}\right). \tag{15.25}$$

One has to take the thermodynamic average of $|m_{\mathbf{q}}|^2$ over all the possible values of the fluctuating order parameter, in accordance to Eqs. (15.23) and (15.25):

$$< |m_{\mathbf{q}}|^2 > = \frac{\int |m_{\mathbf{q}}|^2 e^{-\Delta F/k_B T} dm_{\mathbf{q}}}{\int e^{-\Delta F/k_B T} dm_{\mathbf{q}}}, \tag{15.26}$$

namely

$$< |m_{\mathbf{q}}|^2 > = \frac{k_B T}{2V\alpha_L(T)\left(1 + \frac{\gamma_L q^2}{\alpha_L(T)}\right)}. \tag{15.27}$$

In the light of the form for $\alpha_L(T) = a_0(T - T_c)$ (see Eqs. (15.19) and (15.20)) this expression is rewritten

$$< |m_{\mathbf{q}}|^2 > = \frac{k_B T}{2V T_c a_0 \left(\varepsilon + \frac{\gamma_L q^2}{a_0 T_c}\right)} \propto \left(\varepsilon + \frac{q^2}{k_0^2}\right)^{-1}, \tag{15.28}$$

with k_0 a constant wave vector. For uniform fluctuations (i.e. $\mathbf{q} = 0$) this equation predicts the divergence of $< |m_{\mathbf{q}}|^2 >$ as for the static homogeneous susceptibility (at $\mathbf{q} = 0$ and $\omega = 0$), with critical exponent $\gamma = 1$. This is the phenomenon of the *enhancement of the fluctuations* on approaching the transition. Here $\mathbf{q} = 0$ can be defined as the *critical wave vector*. Notice that for antiferromagnets the critical

---

[8]When the system is non-homogeneous the free energy as in Eq. (15.18) is no more a local function of the magnetization, instead becoming a *functional* of the field $m(r)$, in the form of a proper integral of the energy density. For the concept of functional and its role at the superconductive transition, where inhomogeneities are crucial, see Sect. 18.9.1.

[9]This approximation breaks down in the so-called *critical region*. In most systems this region corresponds to a very narrow temperature range. Only in particular cases one cannot neglect the term in m^4, as for instance in restricted dimensionality or for the superconducting transition in small-size grains, where the critical region is expanded. It is remarked that the *range of the interactions* (long or short range) also controls the width of the critical region.

wave-vector $\mathbf{q_c}$ is not $q = 0$ but, for example in a planar antiferromagnet on a square lattice, $\mathbf{q_{AF}} = (\pi/a, \pi/a)$ since $m(\mathbf{r})$ has a periodicity which is twice the unit cell. Accordingly the Fourier components of $|m_\mathbf{q}|^2$ will be peaked around $\mathbf{q_{AF}}$ and Eq. (15.28) still applies provided that the wave vectors origin is shifted to $\mathbf{q_{AF}}$ (see Appendix 15.1). This aspect will be clarified at Chap. 17.

Equation (15.28) is often written in the form

$$< |m_\mathbf{q}|^2 > = \frac{< |m_{\mathbf{q}=0}|^2 >}{\left(1 + q^2 \xi(T)^2\right)}, \qquad (15.29)$$

where

$$\xi(T) = \sqrt{\frac{\gamma_L}{\alpha_L(T)}} = \frac{1}{k_0 \varepsilon^\nu} = \xi_0 \varepsilon^{-1/2}, \qquad (15.30)$$

is the *correlation length*, diverging with the *critical exponent* $\nu = 1/2$, for $T \to T_c$.[10] $k(T) = k_0 \varepsilon^\nu$ is called the *inverse correlation length*.

The role of ξ as a correlation length can be better understood by looking at the spatial correlation function for $m(\mathbf{r})$. By referring for simplicity to the $T > T_c$ temperature range, where $< m(r) > = 0$, the correlation function is

$$g(\mathbf{r}, \mathbf{r}') = < m(\mathbf{r})m(\mathbf{r}') > = \sum_\mathbf{q} < |m_\mathbf{q}|^2 > e^{-i\mathbf{q} \cdot (\mathbf{r} - \mathbf{r}')}, \qquad (15.31)$$

For $\mathbf{R} = (\mathbf{r} - \mathbf{r}')$, from Eq. (15.28) and through the usual transformation $\sum_\mathbf{q} \to (V/8\pi^3) \int d\mathbf{q}$ one obtains

$$g(\mathbf{R}) \propto \frac{k_B T}{R} e^{-Rk_0(T-T_c)^{1/2}/T_c^{1/2}} \propto \frac{k_B T}{R} e^{-R/\xi(T)}, \qquad (15.32)$$

showing that $\xi(T)$ is indeed a measure of the correlation. For $T \gg T_c$ one has a fast decay of the correlations, while at T_c, where $\xi(T)$ diverges (note that $\gamma_L > 0$), $g(\mathbf{R}) \propto (k_B T_c/R)$. Thus the onset of correlated fluctuations and their enhancement for T approaching T_c from above, has been deduced. Below T_c, where the fluctuations occur around $< m(r) > \neq 0$ and $\alpha = 2a_0(T_c - T)$, analogous divergences are obtained for $< |m_{\mathbf{q_c}}|^2 >$ and for the correlation length, with the same critical exponents as above T_c.

In Problem 15.6 the divergence of the correlation length and the relationship to the response function are illustrated for a fluid.

----

[10] As recalled in Table 15.1 in most cases the critical exponents are different from the "classical" values derived in mean field theories. In particular the difference might be relevant when the temperature range of the critical region is expanded, when the interactions are short-range or when the dimensionality of the system is reduced.

15.4 Time Dependence of the Fluctuations

So far we did not have to take into account the time-dependence of the fluctuations.
However the life time of the fluctuations, the corresponding microscopic dynamics
of the local critical variables (spin-components or local density for fluids, to which
the macroscopic order parameter can be connected) have a crucial role around the
phase transitions.

In the framework of Landau-type statistical theory, the time-dependence of $m(\mathbf{r}, t)$
around the equilibrium value can be discussed by starting from an equation based
on the thermodynamics of the irreversible processes. For the collective components
and for $\mathbf{q}$ around the critical wave vector, the following heuristic equation

$$M\frac{d^2m_{\mathbf{q}}}{dt^2} + \Omega_q\frac{dm_{\mathbf{q}}}{dt} + k_{\mathbf{q}}m_{\mathbf{q}} = 0, \tag{15.33}$$

typical of damped harmonic oscillators, will be assumed.[11]

In Eq. (15.33) the effective elastic constant $k_{\mathbf{q}}$ must be of the form $k_{\mathbf{q}=0} \propto |T - T_c|$,
since it is related to the curvature of the free energy as a function of m. The generalized
mass M can be assumed temperature independent.

From the Eq. (15.33) for the collective generalized order parameter $m_{\mathbf{q}}$ two ways to
obtain the correlation and the slowing down of the fluctuations are possible. One way
is to look for the response of the system to an external, wave-dependent and oscillating
field, thus deriving the response function, the generalized susceptibility $\chi(\mathbf{q}, \omega)$.
Then, by means of the fluctuation-dissipation theorem, the power spectrum of the
correlation function for the collective component $< m_{\mathbf{q}}(0)m_{-\mathbf{q}}(t) >$ is obtained.
This power spectrum, which is the q and ω transform of the correlation function
$g(\mathbf{R}, t)$ (including the time t in Eq. (15.31)) is known as the *dynamical structure
factor* (DSF):

$$S(\mathbf{q}, \omega) = \int e^{-i(\omega t - \mathbf{q}.\mathbf{R})} g(\mathbf{R}, t) d\mathbf{R}dt.$$

Another method to describe the fluctuations consists in constructing directly
the correlation function from Eq. (15.33). As it is shown in Problem 15.7, or by

---

[11]This equation can be justified by extension of the thermodynamics of the irreversible processes,
in which for the order parameter one writes $dm/dt = -c(\partial f/\partial m)$, meaning that the speed to
approach the equilibrium (after a variation induced by the fluctuation) is proportional to the restoring
force. Then, from $f = a_0|T - T_c|m^2$ (Eq. (15.19)) for $\delta m = m- < m >$, one would have
$\delta m(t) = \delta m(0)exp(-t/\tau)$, that for $\tau \propto (T - T_c)^{-1}$, describes the *classical Landau-Khalatnikov*
slowing down of the fluctuations. With respect to the usual relaxation function, in Eq. (15.33) we
have added an *inertial term*, with generalized mass M. This term completes the relaxation function
yielding a form which allows an underdamped motion. In particular, this term is required in order
to describe structural transitions, while it can be neglected for magnetic transitions where the local
critical variable can be identified with spin operator.

resorting to the susceptibility of damped harmonic oscillator (see Appendix 15.1 and Eq. (15.46)), one derives

$$S(\mathbf{q}, \omega) \propto \frac{4\Omega_q \omega_q^2 / M}{(\omega_q^2 - \omega^2)^2 + (4\Omega_q^2 \omega_q^2 / M)}, \tag{15.34}$$

namely in the form of two resonant peaks centered at $-\omega_q$ and $+\omega_q$ with width Ω_q / M.

For any extent of the damping, the solution of Eq. (15.33) can be written in the form

$$m_q(t) = exp(-i\widetilde{\omega}t), \tag{15.35}$$

$\widetilde{\omega}$ being a complex frequency with imaginary part $\Omega_q / 2M$ and real part $\sqrt{[4Mk_q - \Omega_q^2]}/2M$. If T^* is defined as the temperature at which for a given wavevector $\Omega_q^2 = 4Mk_q$, then for $T < T^*$ both poles lie on the imaginary axis and the frequency $\omega = -ik_{q=0}/\Omega_{q=0}$ moves towards the origin of the complex plane. Accordingly the transition is approached with this frequency going as $(T - T_c)$.

Thus Eq. (15.35) describes the slowing-down of the fluctuations for $T \to T_c^+$. In terms of the DSF it can be said that above a given temperature T^* one has temperature-dependent resonant peaks of constant width while for $T < T^*$ the slowing-down is rather described by a *diffusive-type central peak*, centered at $\omega = 0$, its width decreasing while the transition is approached. Equivalently, an effective correlation time $\tau_{q=0} = i[\omega_{q=0}]^{-1}$ goes to infinity for $T \to T_c^+$.

As already mentioned, the fluctuation-dissipation theorem relates the spectrum of the fluctuations to the response to a dynamic perturbation and in its more general form it can be written as

$$S(\mathbf{q}, \omega) = \frac{2\hbar\chi''(\mathbf{q}, \omega)}{1 - e^{-\hbar\omega/k_B T}}, \tag{15.36}$$

$\chi''(\mathbf{q}, \omega)$ being the dissipative imaginary part of the generalized susceptibility. Notice that the static structure factor is

$$S(\mathbf{q}) = \frac{1}{2\pi} \int S(\mathbf{q}, \omega) d\omega, \tag{15.37}$$

and then by using the *Kramers-Kronig relations* (see Problem 15.4) one writes

$$S(\mathbf{q}) = k_B T \chi(\mathbf{q}, 0), \tag{15.38}$$

Now one faces the problem to express the $\mathbf{q}$-dependence of the static and the dynamical quantities involved in the above equations, that are correct only in the limit of $\mathbf{q}$ close to the *critical wave-vector*, for which the thermodynamic approach is valid. For temperature around T_c where the fluctuations at the critical wave vector are dominant, for the amplitude of the fluctuations (and then for the static structure factor) one can adopt the form given in Eq. (15.29) in term of the correlation length.

The $\mathbf{q}$-dependence of the frequencies involved in the time-dependent phenomena is a complicated issue, in essence dealing with the dynamics of cooperative effects. Several tentative theories have been developed, all with limitations and possible criticism. One could use an expansion of the time-dependent correlation function in series of t, the coefficients being the equal-time correlation functions related to the lowest moments of the spectral function. Then approximants are often used, the procedure being essentially the attempt to guess the long-time behaviour of $g(\mathbf{r}, t)$ from the short-time behaviour. Other theoretical lines are based on the *Mori continuous fraction* approximation or on *Monte Carlo* computer simulations.

Here we will illustrate a phenomenological approach based on *scaling concepts*. We start by writing the spectral density for the dynamical part of the DSF in the form

$$\mathcal{J}_\mathbf{q}(\omega) = \frac{S(\mathbf{q}, \omega)}{S(\mathbf{q})} = \frac{2\pi}{\Gamma_\mathbf{q}} G(\omega/\Gamma_\mathbf{q}), \tag{15.39}$$

where $\Gamma_\mathbf{q}(T)$ is a characteristic frequency and $G(\omega/\Gamma_\mathbf{q})$ is a well behaved function which depends on the frequency only through the ratio $(\omega/\Gamma_\mathbf{q})$ and has width and area of the order of unity. The detailed form of the function G has to be determined experimentally or according to theoretical models.

Then, both the amplitude of the fluctuations and their decay rates $\Gamma_\mathbf{q}$ are scaled in their $\mathbf{q}$-dependence in term of the correlation length.

With a slight generalization (see Eq. (15.29)) we shall write

$$< |m_q|^2 > = k_B T \chi(\mathbf{q}, 0) \propto \xi^{2-\eta} F_s(q\xi), \tag{15.40}$$

with a correction term η that is negligible for lattice dimensions $D \geq 2$. The simplest form for F_s is

$$F_s(q\xi) = \frac{1}{1 + (q\xi)^2}, \tag{15.41}$$

as suggested by Eq. (15.29).[12]

For the decay rate the analogous equation is

$$\Gamma_\mathbf{q}(T) = \Gamma_{\mathbf{q}=0} G_s(q\xi) = \Gamma_{\mathbf{q}=0}[1 + (q\xi)^2] \tag{15.42}$$

in its simplest form, namely with $G_s = F_s^{-1}$. According to the arguments given above the critical decay rate, the one corresponding to the critical wave vector $\mathbf{q}_c$, must decrease towards zero for $T \rightarrow T_c$. It can be assumed of the form

---

[12]This form for the static part of the DSF corresponds to the *Orstein-Zernike expansion* of the correlation function in the theory for the critical opalescence at the vapour-liquid transition.

$$\Gamma_{\mathbf{q}_c}(T) \propto \xi^{-z} \tag{15.43}$$

where z is the *dynamical critical exponent*.[13]

It is noted that our scaling relationships correspond to scale the generalized susceptibility in the form

$$\chi(\mathbf{q}, \omega) = \chi^0 \xi^z F'(q\xi, \omega/\xi^z) \tag{15.44}$$

χ^0 being the *single particle response function*. In other words $\chi(\mathbf{q}, \omega)$ can be expressed as a function independent of the reduced temperature ε provided that length and frequencies are rescaled by appropriate powers of ε. It can be remarked that if one assumes $z\nu = \gamma = 1$ then the so-called *thermodynamic slowing-down condition* is attained: on approaching T_c from above the size of the "islands" where the correlation is effective (correlation length) increases at the same rate of their life-time.

15.5 Generalized Dynamical Susceptibility and Experimental Probes for Critical Dynamics

As mentioned in the illustration of the phenomenological aspects in Sect. 15.1, in the neighborhood of a critical point complex cooperative motions occur, that one can call *critical dynamics*. This dynamic is usually "anomalous" in comparison to the ordinary excitations of the system, as for example the phonons, the spin waves, the rotational and diffusion motions in fluids. The critical dynamic represents the microscopic correspondence of the enhancement and slowing down of the fluctuations, described at Sects. 15.3 and 15.4 in general terms. One could say that is just the critical microscopic dynamic that drives the phase transition, by inducing marked spatial and temporal correlations, controlling the transport coefficients and the responses to external perturbations.

The modern theories on phase transitions represent the link between the thermodynamic-statistical description and the microscopic phenomena or, in other words, between the anomalous critical dynamics and the response functions or between single-particle response function and the cooperative response for the strongly correlated systems.

In Appendix 15.1 it is described a simple model to derive the linear response of a system, in the equilibrium state and within the mean field approximation, to a time-dependent perturbation in the presence of fluctuations. It is shown that the generalized dynamical susceptibility $\chi(\mathbf{q}, \omega)$ is related to the single particle susceptibility $\chi^0(\omega)$ according to a relation that generalizes the one in Eq. (4.34):

---

[13]In the dynamical scaling theory z also controls the q-dependence of $\Gamma_{\mathbf{q}}$ at T_c, i.e. $\Gamma_{\mathbf{q}} \sim Aq^z$; in other words the exponent expressing the q-dependence of $\Gamma_{\mathbf{q}}$ at T_c is the same exponent that expresses the dependence of $\Gamma_{\mathbf{q}}$ above T_c. For details see the book by *Stanley*.

$$\chi(\mathbf{q}, \omega) = \frac{\chi^0(\omega)}{1 - I_{\mathbf{q}} \chi^0(\omega)}, \tag{15.45}$$

where $I_{\mathbf{q}}$ is the Fourier transform of the interaction. In the light of this equation the generalized susceptibility for model systems can be obtained. Notice that the static susceptibility derived within the *Stoner-Hubbard model* in Appendix 13.1, accounting for the effect of electronic correlations in a metal, is a particular case of the more general Eq. (15.45).

For instance, for a one-dimensional damped oscillator, for which the single particle equation in the presence of external stimulus of amplitude ef_0 is written

$$m \left(\frac{d^2 x}{dt^2} + \gamma \frac{dx}{dt} + \omega_0^2 x \right) = ef_0 e^{-i\omega t}. \tag{15.46}$$

The stationary solution $x(t) = x_0 exp(-i\omega t)$ yields the well known response

$$\chi^0(\omega) = \frac{(e/m)}{\omega_0^2 - \omega^2 - i\omega\gamma} = \frac{\chi^0(0)\omega_0^2}{\omega_0^2 - \omega^2 - i\omega\gamma} \tag{15.47}$$

(with ω_0 proper frequency and γ damping factor), the collective susceptibility becomes

$$\chi(\mathbf{q}, \omega) = \frac{\chi^0(0)\omega_0^2}{\omega_{\mathbf{q}}^2 - \omega^2 - i\omega\gamma}, \tag{15.48}$$

having written

$$\omega_{\mathbf{q}}^2 = \omega_0(1 - I_{\mathbf{q}} \chi^0(0)). \tag{15.49}$$

Thus one sees that the static susceptibility at the critical wave vector diverges when $\omega_{\mathbf{q}_c}$ tends to zero, in correspondence to $I_{\mathbf{q}_c} \chi^0(0) \rightarrow 1$, in agreement with the deduction at Sect. 15.3 (see an illustration in Fig. 15.4).

For a Debye-type relaxor the single particle susceptibility is

$$\chi^0(\omega) = \frac{(e/m)}{\omega_0^2 - i\omega\gamma} = \frac{\chi^0(0)}{1 - i\omega\tau_D}, \tag{15.50}$$

where $\tau_D = \gamma/\omega_0^2$ is the *Debye relaxation time*. In an analogous way, when the homogeneous static susceptibility at $\mathbf{q} = 0$ goes to infinity, a generalized frequency defined as $\omega_{\mathbf{q}} = i\tau_{\mathbf{q}}^{-1}$ goes to zero. Equivalently, the collective relaxation time at the critical wave vector goes to infinity. More details regarding the above description, including the spectrum of the fluctuations, are given in Appendix 15.1.

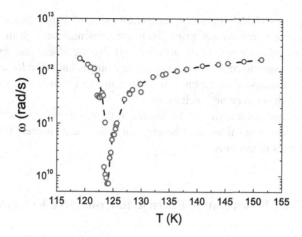

Fig. 15.4 Temperature dependence of the frequency of the $q = 0$ phonon mode close to the paraelectric-ferroelectric transition of tris-sarcosine calcium chloride, as derived from dielectric measurements (see R. Mackeviciute et al., J.Phys. Condens. Matter 25, 212201 (2013)). To understand how the frequency is obtained from the real and the imaginary part of the dielectric constant see Chap. 16

Now we briefly mention the main experimental approaches that can be used to investigate the critical dynamics, giving access to the *generalized susceptibility* $\chi(\mathbf{q}, \omega)$ or equivalently to the *DSF*. These experiments, essentially of spectroscopic character, may differ in the wave vector $\mathbf{q}$ and in the frequency ω of the stimulus and thus probe different characteristics of the response of the system under study.

The cross section of *inelastic neutron scattering* (for some detail see Sect. 14.3) is often considered the most complete tool. In fact, in principle one can select the wave vector $\mathbf{q}$ and the energy $\hbar\omega$ and thus the whole spectrum of the excitations can be probed. With appropriate experimental arrangements also the character of the interaction (e.g. magnetic or elastic) can be selected. In practice, some restrictions are obviously present: the frequency can typically be chosen in the range 10^{10}–10^{14} Hz while geometrical constrains imply limitations to the resolution in $\mathbf{q}$-space. Furthermore the sample usually needs to be a rather large one ($\simeq 1\,\text{cm}^3$).

Dielectric measurements with conventional audio, RF bridges, Q-meters or waveguides directly can provide the real and the imaginary part of the electric susceptibility, in a wide frequency range that typically spans from zero up to 10^{10} Hz. However, since the response to a uniform field is detected, only the $q = 0$ wave-vector is probed. Analogous performances have the conventional equipments for susceptibility in regards of the magnetic properties, as the apparatus using double coils or the modern ones based on SQUID units (extremely sensitive but usually limited to a frequency range up to 10^6 Hz, see Chap. 18).

The *light scattering* is particularly useful in fluids, by means of the homodyne or heterodyne spectroscopy of the Rayleigh diffusion: even the very low frequency range is explored but the wave vectors are usually limited to values much smaller that the ones characteristic of the excitations of the systems. *Synchrotron radiation* can provide more insights.

Radio frequency (*NMR*) and *microwave* (*EPR*) spectroscopies based on nuclear or electron magnetic resonances respectively (in particular through the relaxation processes) or the *Muon spin relaxation* (μSR) and also the *Mössbauer spectroscopy* (see Sect. 14.6) span the corresponding frequency ranges and have *local character*. Therefore they correspond to responses involving an average over wide ranges of the wave vectors of the elementary excitations.

Some more illustration of the results derived by those experimental techniques for the studies of phase transitions shall be given in subsequent chapters when dealing with special topics or systems.

Appendix 15.1 From Single Particle to Collective Response

The main lines of a general procedure which allows one to derive the response of a collective system (in the mean-field equilibrium state) to a time-dependent external perturbation are sketched in the following. The condition of linear response to both the external force and to the internal fluctuations shall be assumed. Furthermore interaction of bi-linear character will be considered while the local potential is somewhat arbitrary. The relevant advantage of this model is to *include the fluctuations in the framework of the mean field approximation*, where they are usually neglected. Thus the simplicity of the mean field approach is retained in spite of the inclusion of the fluctuations.[14]

The system is perturbed by a time-dependent perturbation $\Delta F e^{-i\omega t}$, which modifies the *local critical variable* v_l according to

$$v'_l(t) = \Delta v_l e^{-i\omega t} + v_l(t)$$

Weak perturbations and linear responses $\Delta v_l = \chi^0(\omega)\Delta F_l$ are assumed, while $v_l(t)$ represents the spontaneous "local dynamics". Finally the interactions among the local variables are described through a bilinear Hamiltonian

$$\mathcal{H} = \frac{1}{2}\sum_{ll'} \mathbf{I}_{ll'} v_l v_{l'},$$

as for the *Heisenberg exchange* or for the *harmonic lattice vibrations*, for example. Then v_l will not experience just the external perturbation ΔF but also the feedback $\sum_{l'} \mathbf{I}_{ll'} v_{l'}(t)$. Hence, v_l will be globally perturbed by a term

$$\Delta F_l(t) = \Delta F e^{-i\omega t} + \sum_{l'}\mathbf{I}_{ll'}\Delta v_{l'}e^{-i\omega t} + \sum_{l'}\mathbf{I}_{ll'}v_{l'}(t)$$

---

[14]For details on the basic assumptions see *H. Thomas* in the book edited by K.A. Müller and A. Rigamonti.

where the last term describes the interactions involving the spontaneous fluctuations, which are present also in the absence of external stimulus. Within the mean field approximation one writes

$$\Delta v_l = \chi^0(\omega) \left[\sum_{l'} \mathbf{I}_{ll'} \Delta v_{l'} + \Delta F_l \right],$$

where $\chi^0(\omega)$ is the single-particle "bare" susceptibility. By resorting to the normal coordinates $\Delta v_{\mathbf{q}}$ and expanding the stimulus in its Fourier components

$$\Delta v_l = \frac{1}{\sqrt{N}} \sum_{\mathbf{q}} \Delta v_{\mathbf{q}} e^{i\mathbf{q}\cdot\mathbf{R}_l}$$

$$\Delta F_l = \frac{1}{\sqrt{N}} \sum_{\mathbf{q}} \Delta F_{\mathbf{q}} e^{i\mathbf{q}\cdot\mathbf{R}_l}$$

from the previous equation one has

$$\sum_{\mathbf{q}} \Delta v_{\mathbf{q}} e^{i\mathbf{q}\cdot\mathbf{R}_l} = \chi^0(\omega) \left[\sum_{l'} \mathbf{I}_{ll'} \sum_{\mathbf{q}} \Delta v_{\mathbf{q}} e^{i\mathbf{q}\cdot\mathbf{R}_{l'}} + \sum_{\mathbf{q}} \Delta F_{\mathbf{q}} e^{i\mathbf{q}\cdot\mathbf{R}_l} \right],$$

which is rewritten in the form

$$\sum_{\mathbf{q}} \Delta v_{\mathbf{q}} = \chi^0(\omega) \left[\sum_{\mathbf{q}} \sum_{l'} \mathbf{I}_{ll'} \Delta v_{\mathbf{q}} e^{i\mathbf{q}\cdot(\mathbf{R}_{l'}-\mathbf{R}_l)} + \sum_{\mathbf{q}} \Delta F_{\mathbf{q}} \right].$$

Thus for each $\mathbf{q}$ one has

$$\Delta v_{\mathbf{q}} = \chi^0(\omega) \left[\sum_{l'} \mathbf{I}_{ll'} \Delta v_{\mathbf{q}} e^{i\mathbf{q}\cdot(\mathbf{R}_{l'}-\mathbf{R}_l)} + \Delta F_{\mathbf{q}} \right]$$
$$= \chi^0(\omega) \left[\mathbf{I}_{\mathbf{q}} \Delta v_{\mathbf{q}} + \Delta F_{\mathbf{q}} \right],$$

where $\mathbf{I}_{\mathbf{q}} = \sum_{l'} \mathbf{I}_{ll'} e^{i\mathbf{q}\cdot(\mathbf{R}_{l'}-\mathbf{R}_l)}$ is the *Fourier transform* of the spatially varying interaction. The collective susceptibility is

$$\chi(\mathbf{q}, \omega) = \frac{\Delta v_{\mathbf{q}}}{\Delta F_{\mathbf{q}}},$$

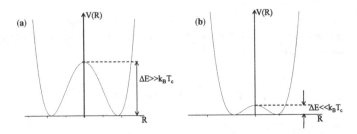

Fig. 15.5 Sketchy forms of local strongly anharmonic potential (a) and of quasi-harmonic potential (b). **a** can describe order-disorder, relaxational-type systems while **b** quasi-harmonic vibrational lattices, displacive type systems (see Chap. 16)

and then

$$\chi(\mathbf{q}, \omega) = \frac{\chi^0(\omega)}{1 - \chi^0(\omega)I_{\mathbf{q}}}. \tag{A.15.1.1}$$

One observes that at T_c, for a given wavevector $\mathbf{q_c}$, so that $\chi^0(0)I_{\mathbf{q_c}} \to 1$, a divergence of the static susceptibility $\chi(\mathbf{q_c}, 0)$ occurs, implying the onset of a long range order with a modulation of the local variable determined by $\mathbf{q_c}$.

Let us refer to a system that can be described by a *local quasi-harmonic potential* of the form (b) sketched in Fig. 15.5.

In the case of no interaction the susceptibility would be the one of damped oscillator

$$\chi^0(\omega) = \chi^0(0)\frac{\omega_0^2}{\omega_0^2 - \omega^2 - i\gamma\omega},$$

with ω_0 fundamental frequency and γ the damping parameter. The frequency ω_0 can be considered slightly temperature dependent, for instance in the form $\omega_0 \propto (a + bT)$, $a \gg bT$ (quasi-harmonic approximation, see Problem 15.3). As a consequence of the interactions, from Eq. (A.15.1.1), one writes

$$\chi(\mathbf{q}, \omega) = \frac{1}{\frac{1}{\chi^0(\omega)} - I_{\mathbf{q}}} = \frac{\omega_0^2\chi^0(0)}{\omega_0^2 - \omega^2 - i\gamma\omega - I_{\mathbf{q}}\chi^0(0)\omega_0^2},$$

which preserves the same form of the susceptibility, with a frequency of the mode at wavevector $\mathbf{q}$ renormalized as

$$\omega_{\mathbf{q}}^2 = \omega_0^2(1 - I_{\mathbf{q}}\chi^0(0)). \tag{A.15.1.2}$$

In fact

$$\chi(\mathbf{q}, \omega) = \frac{\omega_0^2\chi^0(0)}{\omega_{\mathbf{q}}^2 - \omega^2 - i\gamma\omega} = \frac{\omega_{\mathbf{q}}^2\chi(\mathbf{q}, 0)}{\omega_{\mathbf{q}}^2 - \omega^2 - i\gamma\omega}$$

and

$$\frac{\chi(\mathbf{q}, 0)}{\chi^0(0)} = \frac{\omega_0^2}{\omega_{\mathbf{q}}^2}.$$

When for $T \to T_c$ $\mathbf{I_q}\chi^0(0) \to 1$, $\omega_{\mathbf{q}} \to 0$.

For particles in a *double-well strongly anharmonic potential* (see (a) in the sketch above) (which with some modifications allows one to describe the parallel suscepti-bility of magnetic systems, see Chap. 17) the single particle susceptibility is given by Eq. (15.50) (*relaxational behaviour*) and the relaxation time τ is usually temperature dependent in the form $\tau \simeq \tau_0 exp(\Delta E / k_B T)$. Then, from

$$\chi(\mathbf{q}, \omega, T) = \frac{\chi^0(0, T)}{1 - \mathbf{I_q}\chi^0(0, T) - i\omega\tau},$$

a frequency $\omega_{\mathbf{q}} = i/\tau_{\mathbf{q}}$ goes to zero when according to Eq. (A.15.1.1) $\chi(\mathbf{q}, 0, T)$ goes to infinity and

$$\tau_{\mathbf{q}} = \frac{\tau}{1 - \mathbf{I_q}\chi^0(0, T)}$$

also diverges (here the T−dependence is explicitly indicated).

Summarizing, it has been derived that for a given wave vector $\mathbf{q}$, for which as a function of the temperature $\mathbf{I_q}\chi^0(0, T)$ approaches unity, a *critical temperature* $T_c(\mathbf{q})$ exists so that $\chi(\mathbf{q}, 0, T) \to \infty$ and $\omega_{\mathbf{q}}$ *moves towards zero* (see sketch below). Correspondingly the phase at $< v_l >= 0$ becomes unstable against the generalized mode for which $T_c(\mathbf{q})$ takes the largest value.

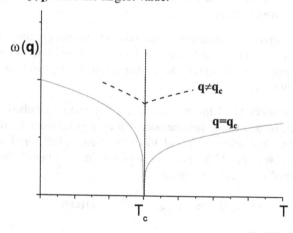

The temperature dependence of the *soft mode* corresponds to the slowing down discussed at Sect. 15.4 in a different scenario. The detailed temperature dependence of the soft mode would need the knowledge of the microscopic Hamiltonian.

Since for fluctuations decoupled from the response the fluctuation-dissipation theorem holds, the spectrum of the collective fluctuations is written

$$\int_{-\infty}^{+\infty} < v_{\mathbf{q}}(0)v_{\mathbf{q}}(t) > e^{-i\omega t} dt = \frac{-2k_B T}{\omega} \chi''(\mathbf{q}, \omega, T).$$

The mean square amplitude of the fluctuations is directly obtained:

$$< |v_{\mathbf{q}}^2| > = \frac{1}{2\pi} \int \frac{-2k_B T}{\omega} \chi''(\mathbf{q}, \omega, T) d\omega = k_B T \chi(\mathbf{q}, 0, T)$$

At $T = T_c$ the amplitude of the fluctuations for the collective component at $\mathbf{q}_c$ diverges. In other words, on cooling towards T_c larger and larger fluctuations become correlated over longer times until at the instability limit one has long-range fluctuations at zero frequency.[15]

Problems

Problem 15.1 In the light of the Clapeyron equation discuss the slope of the function P versus T of the solid -liquid line for water in comparison to the one for all the other systems (see Figs. 15.1 and 15.3), explaining why ice is required in order to allow skating rather than using a floor of any solid material.

Solution: The particular behaviour of ice versus water is related to the fact that ice has a specific volume reduced with respect to water. Thus $V_2 < V_1$ (return to Fig. 15.2 caption) and from the Clapeyron equation $[dP/dT] < 0$. The pressure due to the sharp blade of the skater melts the surface of the ice, thus allowing a smooth sliding.

Another aspect related to the particular behaviour of water is the fact that rivers and lakes freeze from the top.

Problem 15.2 Derive an approximate expression of the entropy jump at the boiling process of water by resorting to the definition of entropy in terms of the number of microstates corresponding to a given macrostate (estimate inspired from the book by *Blundell* and *Blundell*).

Solution: The number Ω of microstates (states that can be labelled by the wave vector $\mathbf{k}$) can be related to the volume occupied by a given quantity, in analogy to the cases of the normal modes of the radiation (see Problem 1.25) or of the electronic states in crystals (see Sect. 12.5). For a mole of vapour or of liquid, the ratio of the specific number of states can be written

$$\Omega_{vap}/\Omega_{liq} = (V_{vap}/V_{liq})^{N_A} \sim (1000)^{N_A},$$

---

[15]Once again it should be remarked that the breakdown of the mean field approximation is expected when entering the critical region, where the fluctuations become strong. Correspondingly also the decoupling of the fluctuations from the responses and the linear response approximation can be expected to loose part of their validity.

having assumed a ratio between vapour and liquid densities around 10^3. Then, from $S = k_B ln\Omega$, the jump of the entropy turns out

$$\Delta S = \Delta(k_B ln\Omega) \sim k_B ln(1000)^{N_A} \sim 7R.$$

The *Trouton rule* (see footnote 1) is usually written $L_{vap} \simeq 10RT_b$. A remarkable violation is found for Helium, where $L_{vap}/RT_b = 2.4$, an indication of quantum effects.

Problem 15.3 From the equation of motion of a single normal mode $Q(t)$ with an effective elastic constant linearly temperature dependent, by assuming a local electric field proportional to the polarization and then to Q, show that instability can occur, with a frequency approaching zero (return to Sect. 10.6). This problem somewhat anticipates some issues to be described at Chap. 16 (see Problem 16.5).

Solution: The equation of motion is

$$\mu\ddot{Q}(t) + (a + bT)Q(t) = qE_{loc} = \alpha Q(t),$$

(which pertains to a transverse optical mode, see Sect. 16.3). μ is a reduced mass. Thus

$$\mu\ddot{Q}(t) + b[T - (\alpha - a)/b]Q(t) = 0$$

and the effective frequency becomes $\Omega^2(T) = (b/\mu)[T - T_c]$ with $T_c = (\alpha - a)/b$. In order to have instability T_c must be positive.

In crystals the electric polarization associated with a transverse optical mode at zero wave vector is the sum of an ionic term NqQ (q the ionic charge) plus the term due to the local field $E_{loc} = 4\pi P/3$. In general a small correction to the short range elastic constant occurs (this justifies why at Sect. 14.3.2 it has been neglected). Transition of displacive character to a ferroelectric phase might occur when the proportionality factor between E_{loc} and P is larger than the Lorentz factor $4\pi/3$ (see Sect. 16.2).

Problem 15.4 Starting from the fluctuation-dissipation relationship and by resorting to *Kramers-Kronig relationships* derive the expression for the paramagnetic static uniform susceptibility in the high temperature limit ($T \gg J$, the exchange interaction).

Solution: The Kramers-Kronig relationships are

$$\chi'(\mathbf{q}, \omega) - \chi'(\mathbf{q}, \infty) = \frac{1}{\pi}P\int \frac{\chi''(\mathbf{q}, \omega')}{\omega' - \omega}d\omega'$$

$$\chi''(\mathbf{q}, \omega) = -\frac{1}{\pi}P\int \frac{\chi'(q, \omega')}{\omega' - \omega}d\omega'$$

(for remind see Sect. 16.1).

From the first equation by expressing the imaginary part of the spin susceptibility in terms of the DSF (see Sect. 15.4 and Eq. (15.36)), one notices that for $\omega \to 0$

$$
\chi'(0, 0) = \frac{1}{\pi} P \int \frac{\chi''(0, \omega')}{\omega'} d\omega'
$$

$$
= \frac{1}{\pi} P \int d\omega' \omega' S_{\alpha\alpha}(0, \omega') \frac{e^{\beta \hbar \omega} - 1}{2 e^{\beta \hbar \omega}}, \text{ with } \beta = \frac{1}{k_B T}
$$

For $\beta \hbar \omega \ll 1$ one can write

$$
\chi'(0, 0) = \frac{1}{\hbar} \int \frac{1}{k_B T} \frac{\hbar \omega d\omega}{\omega} \int dt e^{i\omega t} < S_0^\alpha(t) S_0^\alpha(0) > =
$$

$$
= \frac{1}{k_B T} \int dt \delta(t = 0) < S_0^\alpha(t) S_0^\alpha(0) > =
$$

$$
= \frac{1}{k_B T} < \left| \sum_i S_i^\alpha \right|^2 >
$$

For $k_B T \gg J$ the spins are uncorrelated and one has

$$
\chi'(0, 0) = \sum_i \frac{< |S_i^\alpha|^2 >}{k_B T} = \frac{S(S + 1)}{3 k_B T},
$$

having assumed isotropic spin fluctuations (return to Problem 4.10).

Problem 15.5 Show that the isothermal compressibility of a fluid is proportional to the fluctuation of the mean square number of particles.

Solution: In a grand canonical ensemble the fluctuation in the total number of particles N is

$$
< (N - < N >)^2 > = \frac{1}{Z} \sum_{N=0}^{\infty} \frac{1}{N! h^{3N}} \int d\mathbf{r} \, d\mathbf{p} \, N^2 e^{-\beta U_N} e^{\beta \mu N} -
$$

$$
- \left[\frac{1}{Z} \sum_{N=0}^{\infty} \frac{1}{N! h^{3N}} \int d\mathbf{r} d\mathbf{p} \, N e^{-\beta U_N} e^{\beta \mu N} \right]^2
$$

$$
= (k_B T)^2 \frac{\partial^2 \ln Z}{\partial \mu^2}.
$$

Since $PV/k_BT = \ln Z$ (see the book by *Amit* and *Verbin*).

$$< (N- < N >)^2 >= (k_BT)^2 \left[\frac{\partial^2 PV/k_BT}{\partial \mu^2}\right]_{T,V} = k_BTV \left[\frac{\partial^2 P}{\partial \mu^2}\right]_{T,V}$$

Now $\left[\frac{\partial P}{\partial \mu}\right]_{T,V} = \frac{<N>}{V} = n$, so that

$$< (N- < N >)^2 > = k_BTV \left[\frac{\partial(\frac{<N>}{V})}{\partial \mu}\right]_{T,V} = -\frac{< N > k_BTV}{V^2}\left[\frac{\partial V}{\partial \mu}\right]_{T,N}.$$

By recalling that $K_T = -\frac{1}{V}\left[\frac{\partial V}{\partial \mu}\right]_{T,N}$

$$< (N- < N >)^2 > = \frac{< N >^2 k_BT}{V} K_T = < N > nk_BT K_T.$$

In ideal gas $PV = Nk_BT$ and then

$$K_T^0 = -\frac{1}{V}\frac{\partial(Nk_BT/P)}{\partial P} = \frac{1}{V}\frac{1}{P^2}Nk_BT = \frac{V}{N}\frac{1}{k_BT} = \frac{1}{nk_BT}$$

and finally

$$\frac{K_T}{K_T^0} = \frac{< (N- < N >)^2 >}{< N >}.$$

For detailed description see the books by *Stanley* and by *Amit* and *Verbin*.

Problem 15.6 Derive the relationship between the isothermal compressibility and the density correlation function in fluids. Then, show that the intensity of the radiation scattered with momentum **q** is directly proportional to the static structure factor.

Solution: Since

$$< (N- < N >)^2 > = < \int d\mathbf{r}(n(\mathbf{r})- < n(\mathbf{r}) >) \int d\mathbf{r}' \left(n(\mathbf{r}')- < n(\mathbf{r}') >\right) >$$

$$= \int\int d\mathbf{r}d\mathbf{r}'g(\mathbf{r} - \mathbf{r}') = V \int g(\mathbf{R})d\mathbf{R},$$

while for a fluid $g(\mathbf{r}' - \mathbf{r}) = < (n(\mathbf{r}')- < n >)(n(\mathbf{r}) - < n >) >$, according to Problem 15.5

$$\frac{K_T}{K_T^0} = \frac{1}{n}\int g(\mathbf{r})d\mathbf{r}.$$

The behaviour of the correlation function can be studied by means of radiation scattering techniques. The intensity of the scattered radiation, after having exchanged a wavevector $\mathbf{q}$ with the fluid, is

$$I(\mathbf{q}) = \left< \left| \sum_j a_j(\mathbf{q}) \right|^2 \right>,$$

where $a_j(\mathbf{q}) = ae^{-i\mathbf{q}\cdot\mathbf{r_j}}$ is the amplitude of the radiation scattered at wave vector $\mathbf{q}$ by the molecule at site j. Then $I(\mathbf{q}) = a^2 \left< \left| \sum_{j=1}^{N} e^{-i\mathbf{q}\cdot\mathbf{r_j}} \right|^2 \right>$, in the case of uncorrelated molecules leading to $I^0(\mathbf{q}) = Na^2$. Thus

$$\frac{I(\mathbf{q})}{I^0(\mathbf{q})} = \frac{1}{N} \left< \sum_{i,j} e^{-i\mathbf{q}\cdot(\mathbf{r_i}-\mathbf{r_j})} \right>$$

$$= \frac{1}{N} \int\int d\mathbf{r}d\mathbf{r}' \left< \sum_{i,j} \delta(\mathbf{r}-\mathbf{r_i})\delta(\mathbf{r}'-\mathbf{r_j})e^{-i\mathbf{q}\cdot(\mathbf{r}-\mathbf{r}')} \right>$$

$$= \frac{1}{N} \int\int d\mathbf{r}d\mathbf{r}'e^{-i\mathbf{q}\cdot(\mathbf{r}-\mathbf{r}')} \left< \underbrace{\sum_i \delta(\mathbf{r}-\mathbf{r_i})}_{n(r)} \underbrace{\sum_j \delta(\mathbf{r}'-\mathbf{r_j})}_{n(r')} \right>$$

$$= \frac{1}{N} \int\int d\mathbf{r}d\mathbf{r}'e^{-i\mathbf{q}\cdot(\mathbf{r}-\mathbf{r}')} \left[g(\mathbf{r}-\mathbf{r}') + n^2 \right].$$

Namely

$$\frac{I(\mathbf{q})}{I^0(\mathbf{q})} = \frac{1}{n} \int d\mathbf{R}g(\mathbf{R})e^{i\mathbf{q}\cdot\mathbf{R}} + \frac{V^2}{N}n^2\delta(\mathbf{q}).$$

The first term, the Fourier transform of the correlation function at wavevector $\mathbf{q}$, is the static structure factor $S(\mathbf{q})$ which gives the amplitude of the collective modes at wave-vector $\mathbf{q}$

$$S(\mathbf{q}) = \int d\mathbf{R}g(\mathbf{R})e^{i\mathbf{q}\cdot\mathbf{R}} \propto < |m_\mathbf{q}|^2 > \propto (1+q^2\xi^2)^{-1}$$

(see Eq. (15.40)). The compressibility, being the response function to a uniform $q = 0$ perturbation, is directly proportional to $S(q = 0)$.

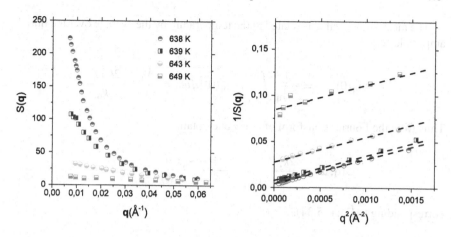

As illustrative example, the neutron scattering intensity is reported above as a function of the wavevector q exchanged with D_2O on approaching the critical temperature $T_c \simeq 637$ K (left). The enhancement of the fluctuations at $q = 0$ is detected. The inverse intensity is reported as a function of q^2 in the plot on the right and a linear trend is observed, as expected. The intercept is proportional to the inverse correlation length. The decrease in the intercept on approaching T_c evidences the divergence in the correlation length.

Problem 15.7 From the equation of motion for underdamped oscillator, derive the dynamical structure factor.

Solution: For underdamped normal oscillators, i.e. such that $\Omega_{\mathbf{q}}^2 \ll 4k_{\mathbf{q}}M$, the solution of Eq. (15.33) is

$$m_{\mathbf{q}}(t) = e^{-\frac{\Omega_{\mathbf{q}} t}{2M}}\left(m_{\mathbf{q}}(0)\cos\omega_{\mathbf{q}}t + \left[\frac{\partial m_{\mathbf{q}}(t)}{\partial t} + \frac{m_{\mathbf{q}}(0)\Omega_{\mathbf{q}}}{2M}\right]\frac{\sin\omega_{\mathbf{q}}t}{\omega_{\mathbf{q}}}\right),$$

with

$$\omega_{\mathbf{q}} = \left(\frac{k_{\mathbf{q}}}{M} - \frac{\Omega_{\mathbf{q}}^2}{4M^2}\right)^{1/2}.$$

yielding for the correlation function

$$g_{\mathbf{q}}(t) = <|m_{\mathbf{q}}(0)|^2> e^{-\Omega_{\mathbf{q}}t/2M}\left(\cos\omega_{\mathbf{q}}t + \left[\frac{\Omega_{\mathbf{q}}}{2M\omega_{\mathbf{q}}}\right]\sin\omega_{\mathbf{q}}t\right).$$

Due to the deterministic character of the motion the statistical ensemble average for the correlation function only involves the initial condition, i.e. the value of $m_{\mathbf{q}}(0)$. In the equation above one can set $\partial m_{\mathbf{q}}(0)/\partial t = 0$ since the recovery of the collective order parameter towards equilibrium, after a fluctuation, initiates with zero velocity.

The mean square value is related to the temperature by the average over the initial amplitude $A_{\mathbf{q}}$:

$$< |m_{\mathbf{q}}(0)|^2 > = \frac{1}{2A_{\mathbf{q}}} \int_{-A_{\mathbf{q}}}^{+A_{\mathbf{q}}} m_{\mathbf{q}}(0)^2 dm_{\mathbf{q}} = \frac{A_{\mathbf{q}}^2}{3} = \frac{2k_B T}{3k_{\mathbf{q}}}.$$

Thus from the Fourier transform of $g_{\mathbf{q}}(t)$ one obtains

$$S(\mathbf{q}, \omega) = \frac{2k_B T}{3k_{\mathbf{q}}} \frac{4\Omega_{\mathbf{q}}\omega_{\mathbf{q}}^2/M}{(\omega_{\mathbf{q}}^2 - \omega^2)^2 + (4\Omega_{\mathbf{q}}^2\omega_{\mathbf{q}}^2/M)}.$$

corresponding to Eq. (15.34).

Specific References and Further Reading

1. H. Stanley, *Introduction to Phase Transitions and Critical Phenomena*, (Oxford University Press, Oxford, 1971).
2. H. Thomas *in Local Properties at Phase Transitions*, Eds. K.A. Müller and A. Rigamonti, (North-Holland Publishing Company, Amsterdam, 1976).
3. D.J. Amit and Y. Verbin, *Statistical Physics - An Introductory Course*, (World Scientific, 1999).
4. S.J. Blundell and E.K. Blundell, *Concepts in Thermal Physics, 2nd Edition*, (Oxford University Press, Oxford, 2010).
5. D.L. Goodstein, *States of Matter*, (Dover Publications Inc., 1985).
6. R. Blinc and B. Zeks, *Soft Modes in Ferroelectrics and Antiferroelectrics*, (North-Holland Publishing Company, Amsterdam, 1974).
7. M. Gitterman and V. Halperin, *Phase Transitions*, (World Scientific, 2013).
8. L.D. Landau and E.M. Lifshitz, *Statistical Physics*, (Pergamon Press, Oxford, 1959).

Chapter 16
Dielectrics and Paraelectric-Ferroelectric Phase Transitions

Topics

Local Electric Fields in Solids
Dielectric Relaxation
Clausius-Mossotti Relation
Instability of Crystal Lattice and Optical Modes for Displacive Ferroelectrics
Lyddane-Sachs-Teller Relation
Critical Dynamics for Quasi-Harmonic Lattice
Order-Disorder Ferroelectrics and Pseudo-Spin Dynamics

16.1 Dielectric Properties of Crystals. Generalities

In the linear approximation the dielectric displacement $\mathbf{D} = \mathcal{E} + 4\pi\mathbf{P}$ and the electric field $\mathcal{E}$ are connected by a second order tensor ε, the *dielectric function*, invariant under the point-group symmetry operation of the crystal. We shall assume for simplicity scalar ε and dielectric susceptibility χ (in $\mathbf{P}/\mathcal{E} = \chi$ and $\varepsilon = 1 + 4\pi\chi$).

The frequency dependence of the *complex* dielectric constant can be determined from the decay function $g(t)$ describing the gradual decrease of $\mathbf{D}$ when the field is suddenly turned off:

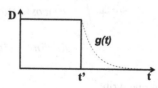

© Springer International Publishing Switzerland 2015
A. Rigamonti and P. Carretta, *Structure of Matter*,
UNITEXT for Physics, DOI 10.1007/978-3-319-17897-4_16

In fact, let us suppose to apply an electric field pulse for a time interval $d\tau$:

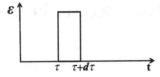

One has an immediate response $\mathbf{D_{imm}}$ which defines the high frequency dielectric constant

$$\mathbf{D_{imm}} = \varepsilon(\infty)\mathcal{E}(\tau). \tag{16.1}$$

Furthermore, in view of the tendency of $\mathbf{D}$ to lag in phase behind $\mathcal{E}$, some polarization persists for $t > \tau + d\tau$. Thus, if one imagines to apply a varying field divided into a series of pulses, in the assumption that the same decay function holds for each element, at the time t the total displacement resulting from a sequence of field pulses is

$$\mathbf{D}(t) = \mathbf{D_{imm}} + \mathbf{D_{ret}} = \varepsilon(\infty)\mathcal{E}(t) + \int_{-\infty}^{t} \mathcal{E}(\tau)g(t-\tau)d\tau. \tag{16.2}$$

For periodically varying field $\mathcal{E}(t) = \mathcal{E}_0 cos(\omega t)$ (that one can assume to persist for time much longer than a characteristic time τ_c over which $g(\tau)$ vanishes), $\mathbf{D}$ will be periodic in t, although in general dephased with respect to $\mathcal{E}(t)$, namely

$$\mathbf{D}(t) = \mathbf{D}_0 cos(\omega t - \delta).$$

By introducing the real and imaginary part of the ω-dependent dielectric constant

$$\mathbf{D}(t) = \mathcal{E}_0 \left[\varepsilon'(\omega)cos(\omega t) + \varepsilon''(\omega)sin(\omega t) \right],$$

as it is shown in Problem 16.2, from Eq. (16.2) one can obtain

$$\varepsilon'(\omega) = \varepsilon(\infty) + \int_0^\infty g(t)cos(\omega t)dt \tag{16.3}$$

$$\varepsilon''(\omega) = \int_0^\infty g(t)sin(\omega t)dt. \tag{16.4}$$

In a single complex equation one writes

$$\varepsilon(\omega) = \varepsilon(\infty) + \int_0^\infty g(t)e^{-i\omega t}dt, \tag{16.5}$$

where the Fourier transform at the frequency ω of the decay function appears.

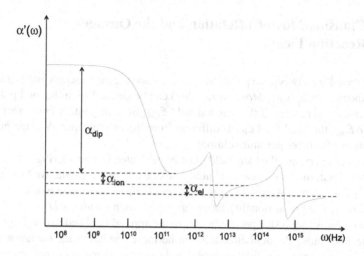

Fig. 16.1 Schematic form of the polarizabilities as a function of the frequency

ε' and ε'' being derived from the same function cannot be independent. In fact, by resorting to Fourier transformation (see Problem 16.3) one obtains

$$\varepsilon'(\omega) - \varepsilon(\infty) = \frac{2}{\pi} \mathcal{P} \int_0^\infty \omega' \frac{\varepsilon''(\omega')}{\omega'^2 - \omega^2} d\omega'$$

$$\varepsilon''(\omega) = \frac{-2\omega}{\pi} \mathcal{P} \int_0^\infty \frac{\varepsilon'(\omega') - \varepsilon(\infty)}{\omega'^2 - \omega^2} d\omega'. \tag{16.6}$$

which are the *Kramers-Kronig relations* (see Sect. 15.4 and Problem 15.4). $\mathcal{P}$ means the integral principal value.

It should be noted that, according the above relationships, from the measurements of ε'' over a broad frequency range ε' can be derived. For example, for a delta-like absorption peak of the form $\varepsilon''(\omega) \sim \delta(\omega - \omega_0)$, from Eq. (16.6) a static dielectric constant $\varepsilon'(0)$ inversely proportional to ω_0 is expected.

From a phenomenological point of view the typical frequency dependence of the dielectric constant (or of the *electric polarizability* $\alpha(\omega)$) can be sketched as in Fig. 16.1.

In the frequency range up to about 10^8 Hz the *dipolar polarizability* $\alpha_{dip}(\omega)$ may be related to permanent dipoles and its decrease above a frequency of that order is related to the inability of these dipoles to follow the field. Usually *audio and RF bridges* or *Q-meters* are used for dielectric measures. The *ionic polarization* (involving $\alpha_{ionic}(\omega)$) is related to field-induced dipoles in the crystal cell. In the correspondent frequency range ($10^{11} - 10^{13}$ Hz) the typical device to measure the polarizability is the *infrared spectrometer*. The *electronic polarizability* (see Stark effect at Sect. 4.2 for $\alpha_{electronic}(0)$) involves very high frequency, where only the electronic clouds can respond to the field. Special techniques, belonging to the realm of the optical spectroscopies of solids, are used to measure the dielectric constant.

16.2 Clausius-Mossotti Relation and the Onsager Reaction Field

The relationship between the polarizabilities α's and the macroscopic dielectric constant (known as *Clausius -Mossotti relation*) can be derived from the total polarization **P** expressed in term of the external field $\mathcal{E}_{ext}$, by taking into account that while $\mathbf{P} = N\alpha\mathcal{E}_{loc}$, the local field $\mathcal{E}_{loc}$ is different from the external one (N is the number of atoms or of dipoles per unit volume).

In dielectric crystals the local field must be evaluated by considering various contributions. First, the total "external" field (the one entering into Maxwell equations) is $\mathcal{E}_{ext} + \mathcal{E}_{P}$, where $\mathcal{E}_{P} = -D_{P}\mathbf{P}$ is the *depolarization field*, due to the outer surface charges. The proportionality factor (*depolarization factor*) is $D_{P} = 4\pi/3$ for a sphere, $D_{P} = 4\pi$ for a slab perpendicular to the applied field and $D_{P} = 0$ for a slab with the plane along the field direction. Furthermore one has to add the *Lorentz field* $\mathcal{E}_{L}$ due to the charges in a fictitious spherical cavity centered at the reference atom. From elementary electrostatics one derives that $\mathcal{E}_{L} = (4\pi/3)\mathbf{P}$.

Finally there would be the field due to the dipoles inside that fictitious cavity. This is averaged to zero for cubic lattice or for an isotropic distribution of the dipoles and we shall not consider it. Thus we write the local field in the Lorentz form[1]:

$$\mathcal{E}_{loc} = \mathcal{E}_{ext} + \frac{4\pi}{3}\mathbf{P}. \qquad (16.7)$$

By referring to the geometry of samples in form of a slab within the plates of a capacitor,[2] one can use for the local field the Lorentz expression (Eq. (16.7)). Then the electric polarization becomes

$$\mathbf{P} = N\alpha\left[\mathcal{E}_{ext} + \frac{4\pi}{3}\mathbf{P}\right], \qquad (16.8)$$

---

[1] The *Lorentz form* for the local field is not totally appropriate when there are permanent dipoles inside the reference cavity. In that case the *Onsager reaction field* (measuring the dis-alignment of the dipoles) has to be taken into account. The Onsager field has the relevant result of modifying the relationship between the single particle and the collective response functions. Instead of Eq. (15.45) one has to write

$$\chi(\mathbf{q}, \omega) = \chi_0(\omega)/\left[1 - \mathbf{I}_{\mathbf{q}}\chi_0(\omega) + \lambda(\mathbf{q})\right]$$

and the correction factor $\lambda(\mathbf{q})$ allows one to preserve the validity of the *fluctuation-dissipation theorem* in the framework of the mean field approximation (see Appendix 15.1 and Sect. 15.4.).

[2] The contributions $-4\pi P$ (from the depolarization) and the one $+4\pi P$, resulting from the plates of the condenser in the usual experimental set-up for the measure of the dielectric constant (usually carried out from the comparison of the capacity in the presence and in the absence of the specimen) compensate each other and do not appear in the local field (for illustration see the book by *Kittel*).

and since $\mathbf{D} = \varepsilon(\mathcal{E}_{\text{ext}} + 4\pi\mathbf{P})$, the *Clausius-Mossotti* relation is found:

$$\frac{\mathbf{P}}{\mathcal{E}_{\text{ext}}} = \frac{N\alpha}{1 - \frac{4\pi}{3}N\alpha} = \frac{\varepsilon - 1}{4\pi}$$

or

$$\varepsilon = \frac{1 + \frac{8\pi}{3}N\alpha}{1 - \frac{4\pi}{3}N\alpha}. \tag{16.9}$$

From this relation it is noted that when $N\alpha \to 3/4\pi$, then $\varepsilon \to \infty$. Therefore one can make the qualitative prediction of finite polarization in zero field, or the *ferroelectric catastrophe*. For permanent dipoles this consideration (that for instance would predict a ferroelectric transition in water, see Problem 16.4) does not hold. In fact in Eq. (16.8) one has to add the *Onsager field*: the dipoles are mis-aligned and the local field is close to the external one.

16.3 Dielectric Response for Model Systems

For the *orientational polarization* due to permanent dipoles (namely the dipolar polarizability, see Fig. 16.1) a crude approach could be to rely on second-order perturbation theory, as described at Sect. 10.2.4, in the assumption of purely rotational states for the permanent dipoles in a crystal. Then

$$P = N\mu_e < cos\theta >= L(x), \tag{16.10}$$

where θ is the angle formed by the permanent dipole μ_e and the electric field and $L(x)$ is the Langevin function for the variable $x = \mu_e\mathcal{E}_{ext}/k_BT$. For $x \to 0$, a rather standard experimental condition, the orientational contribution to the electric susceptibility would be

$$\chi = N\frac{\mu_e^2}{3k_BT}.$$

The assumption to neglect the field due to the dipoles nearby the reference one may be justified in view of the reaction field, which in practice almost cancels the Lorentz field.

A more realistic description for the contribution of permanent dipoles to the *orientational polarization*, in particular for its frequency dependence, is based on the model of local double-well potential (see Appendix 15.1), which in the presence of the external electric field can be sketched in the form

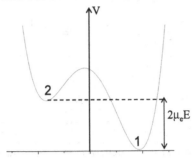

The polarization is $P = (N_1 - N_2)\mu_e$, with $N_{1,2}$ statistical populations on the levels, while $N_1/N_2 = exp(2\mu_e E/k_B T)$ and $N_1 + N_2 = N$.

For periodically oscillating field the susceptibility can be derived from the equations at Sect. 16.1 once that a certain *decay function* $g(t)$ is assumed. Following the theory due to *Debye* and by recalling that $g(t)$ must be of the form ε/t with normalization to unit, one can assume

$$g(t) = \frac{\varepsilon'(0) - \varepsilon(\infty)}{\tau} e^{-t/\tau}. \qquad (16.11)$$

On physical grounds the correlation time τ is often written in the form

$$\tau = \tau_0 e^{\Delta E/k_B T}, \qquad (16.12)$$

with τ_0 slightly T-dependent and typically of the order of $10^{-11} - 10^{-13}$ s. Then

$$\varepsilon(\omega) = \varepsilon(\infty) + \frac{\varepsilon(0) - \varepsilon(\infty)}{1 - i\omega\tau}, \qquad (16.13)$$

as it has been assumed in general terms in Appendix 15.1 for overdamped harmonic oscillators. It should be remarked that in most cases a single correlation time is not appropriate and one has to refer to a distribution of τs (see Appendix 16.2).

For the *ionic dielectric response*, related to the charge displacements within the crystal cell, we remind that the static polarizability can be written (see Problem 10.16)

$$\alpha_{ion} = \frac{(fe)^2}{\mu\omega_0^2} \quad \text{with} \quad fe\mathcal{E}_{loc} = \mu\omega_0^2 \Delta x, \qquad (16.14)$$

where fe is an effective electric charge of the ions, μ the reduced mass and $\mu\omega_0^2$ the elastic constant. In spite of the crudeness of this assumption, still for ionic crystals the order of magnitude of the static dielectric constant is rather well accounted for. For instance, by assuming an atomic-like polarizability around $\alpha \simeq 10^{-24}$ cm^3, from the *Clausius-Mossotti* relation, for a number density of ions around 10^{22} cm^{-3}, one estimates static dielectric constants in the range 3–5. The experimental value for NaCl is $\varepsilon(0) = 5.6$, and it includes the high frequency contribution expected around $\varepsilon(\infty) \simeq 2.2$.

To derive the frequency dependent contribution for α_{ion} one can assume a dielectric response typical of underdamped harmonic oscillators, namely of the form $exp(-\gamma t)cos(\omega_0 t)$. This assumption might be appropriate for optical modes (see Sect. 14.3) at wave vector close to zero (the homogeneous electric field in the dielectric measurements evidently corresponds to $q = 0$, see Sect. 15.2). Then, consistently with Eq. (16.5) and according to Eq. (15.48) one has

$$\varepsilon(\omega) = \varepsilon(\infty) + \frac{(\varepsilon(0) - \varepsilon(\infty))\omega_0^2}{(\omega_0^2 - \omega^2) - i\gamma\omega}. \tag{16.15}$$

Thus, the frequency dependence of the ionic contribution to the dielectric constant takes the form qualitatively sketched at Sect. 16.1 for α_{ion}.

According to *Kramers-Kronig relations* (Eq. 16.6) the absorption peak in $\varepsilon''(\omega)$ implies

$$\varepsilon(0) = \varepsilon(\infty) + \frac{2}{\pi} \mathcal{P} \int_0^\infty \frac{\varepsilon''(\omega')}{\omega'} d\omega'.$$

Now we show that the fundamental frequency ω_0 of the underdamped oscillator appearing in Eq. (16.15) must correspond to an *transverse optical (TO) mode* of the lattice vibrations. From simple pictorial models for the $q = 0$ vibration of longitudinal and transverse character in an ionic crystal one deduces that $\omega_{TO} \ll \omega_{LO}$. In fact, for the $q = 0$ longitudinal modes planes of ions move back and forth and the front and rear surfaces have sheets of positive and negative charges, so that $\mathcal{E}_{loc} = -4\pi P + (4\pi/3)P = -(8\pi/3)P$ (see sketch below).

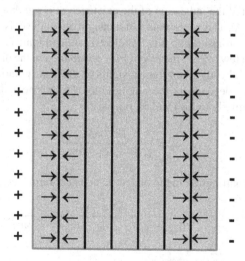

In other words the bulk polarization, i.e. the local field (originated from long range interactions and therefore not considered at Sect. 14.3), tends to resist to the deformation. At variance, for transverse optical modes around $q = 0$, no macroscopic polarization is generated and the local field $(4\pi/3)P$ acts as a feedback mechanism that enhances the distortion associated with the mode.

Thus, being necessarily $\omega_{TO} \ll \omega_{LO}$ one can expect $\varepsilon'(\omega \gg \omega_{TO}) \simeq 0$ and for frequencies much larger than ω_{TO} one writes

$$\varepsilon'(\omega) - \varepsilon(\infty) \simeq \frac{2}{\pi}\mathcal{P}\int_0^\infty \omega' \frac{\varepsilon''(\omega')}{\omega'^2 - \omega_{LO}^2}d\omega' \simeq \frac{2}{\pi}\mathcal{P}\int_0^\infty \frac{\omega_{TO}^2}{\omega'}\frac{\varepsilon''(\omega')}{\omega_{TO}^2 - \omega_{LO}^2}d\omega' \quad \text{or}$$

$$-\varepsilon(\infty) = \left(\frac{\omega_{TO}^2}{\omega_{TO}^2 - \omega_{LO}^2}\right)(\varepsilon(0) - \varepsilon(\infty)), \tag{16.16}$$

$\varepsilon''(\omega)$ being peaked around ω_{TO}.

By collecting $\varepsilon(0)/\varepsilon(\infty)$ along this approximate derivation one obtains the relevant result

$$\frac{\varepsilon(0)}{\varepsilon(\infty)} \simeq \frac{\omega_{LO}^2}{\omega_{TO}^2}, \tag{16.17}$$

known as *Lyddane-Sachs-Teller* (LST) relation.[3] It is noted that according to the LST relation the divergence of the static dielectric constant at the ferroelectric transition (see Sect. 15.1 and Appendix 15.1) implies

$$\omega_{TO}^2 \propto (T - T_c), \tag{16.18}$$

that can be considered the indication of the existence of a *soft mode*, the transverse optical mode at $q = 0$. This represents a first insight about the existence of a *critical dynamics* driving the transition to the ferroelectric state in ionic crystals.

Now we devote a few lines to the *electronic contribution* to the dielectric constant. At Sect. 4.2 we have already given some insights on the static polarizability $\alpha(0)$. As regards the frequency dependence the simplest model is the one of damped harmonic oscillator. Thus, similarly to Eq. (16.15), one writes

$$\alpha(\omega) = \frac{(e^2/m)}{(\omega_e^2 - \omega^2) - i\gamma\omega}$$

and the electronic contribution to the dielectric constant turns out

$$\varepsilon(\omega) \simeq 1 + \frac{(Ne^2/m)}{(\omega_e^2 - \omega^2) - i\gamma\omega - (e^2 4\pi N/3m)}, \tag{16.19}$$

---

[3] For a purely relaxational oscillator (over-damped mode) the analogous of Eq. (16.17) would involve the *Debye relaxation times* at constant field and at constant polarization (see Sect. 15.4).

where the last term at the denominator can be neglected when the local field is close to the external one (see Eqs. (16.8) and (16.9)). ω_e is typically in the ultraviolet range. The behaviour of $\varepsilon(\omega)$ as a function of frequency is similar to the one for the ionic contribution, with the shift of the central frequency from ω_0 to the characteristic electronic frequency ω_e.

16.4 The Ferroelectric Transition in the Mean Field Scenario

Ferroelectrics are polar crystals in which the polarization can be reversed by an electric field. The reason of this behaviour is that the polar phase is a non-polar one only slightly distorted. According to the phenomenological pictures given at Sect. 15.1 the dielectric response, the elastic and the electro-optical properties display divergences around the transition temperature.

Below the transition temperature the raise of the spontaneous polarization (the order parameter) is observed, often with a small discontinuity. It should be remarked that the dielectric constants around the transition are several orders of magnitude larger than the usual ones. In the ferroelectric phase the polarization as a function of the electric field shows an hysteretic behaviour similar to the one in ferromagnets (see Fig. 16.2).

When permanent dipoles are involved in the ferroelectric transition one usually speaks of *order-disorder ferroelectrics*. When in the paraelectric phase no permanent dipoles are present then the transition involves ionic displacements from the equilibrium positions of the ions above T_c and one speaks of *displacive ferroelectrics*. Some crystals exhibiting mixed effects are known.

A list of a few examples for both types of ferroelectric crystals, with data for the spontaneous polarization, is given in Table 16.1.

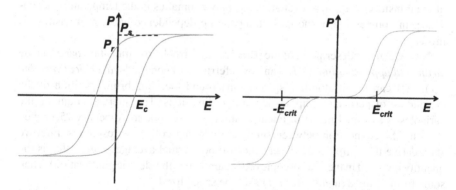

Fig. 16.2 Schematic behaviour of the polarization as a function of the external field in a ferroelectric crystal (*left*) and in anti-ferroelectric crystal (*right*). For antiferroelectrics a strong field is usually require to drive the crystal in the condition to display a plot similar to the one in the ferroelectric state. To understand this effect it may help to look at Sect. 17.3 for the similar case of ferromagnets

Table 16.1 The values of the critical temperature T_c and of the polarization P_s (at the temperature on the right column) for a group of ferroelectric compounds

	$T_c(K)$	P_s (μC cm^{-2})	At the Temperature (K)
Order-disorder ferroelectrics			
CsH_2AsO_4	143	–	–
HCl	98	1.2	83
KH_2AsO_4	92	5	78
KH_2PO_4 (KDP)	123	4.75	96
$NaNO_2$	436	8	373
RbH_2AsO_4	110	–	–
RbH_2PO_4	147	5.6	90
TGS	323	2.8	293
Displacive ferrolectrics			
$BaTiO_3$	393	26	296
$KNbO_3$	712	30	523
$LiNbO_3$	1483	71	296
$PbTiO_3$	763	>50	300

Illustrations of the vibration mode driving the transition for the typical displacive ferroelectric *BaTiO₃* (see Fig. 11.3) and for the typical order-disorder ferroelectric *NaNO₂*, in terms of the electric dipole NO₂⁻, are provided below. A sketch of the correspondent local potential is recalled (see Appendix 15.1).

As already mentioned, the *enhancement* and *slowing down of the fluctuations* described in general terms at Chap. 15 have microscopic correspondence in the *critical dynamics* of some *local variables*. The critical dynamics can involve the *optical soft mode* of vibrational character, as pointed out at Sect. 16.3, or low-frequency *relaxational* modes, sometimes reported as *re-orientational modes*. In the vicinity of the transition both these excitations display anomalies in the temperature dependence, in comparison to the usual temperature dependence when no transition is involved.

Now we give a description of the paraelectric-ferroelectric transition based on the *mean field approximation (MFA)* and by referring to a generic *local critical variable* $v(t)$. $v(t)$ can be an atomic displacement around the equilibrium position or the orientation of permanent dipoles along two directions. The interaction among the variables is assumed in the bi-linear form, as it was done in Appendix 15.1 when deriving the connection between single particle and collective responses. First we shall derive the temperature dependence of the thermal average $< v >$, that is the quantity involved in the macroscopic order parameter (the electric polarization). Then some form of the dynamics for $v(t)$ shall be specialized.

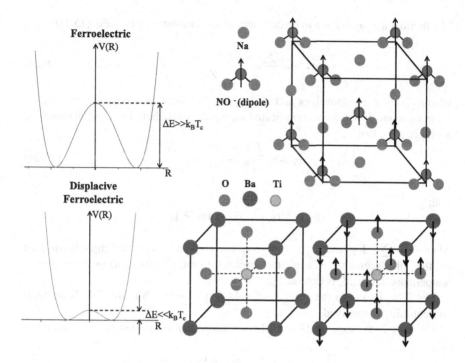

The model Hamiltonian we shall refer to is

$$\sum_l \left(\frac{\Pi_l^2}{2}\right) + V(v_1, v_2, v_3, \ldots v_N) = \sum_l \left[\frac{\Pi_l^2}{2} + V_l\right] + \text{interaction terms},$$

(16.20)

while the rest of the crystal is considered embedded in the thermal bath. Π_l is the conjugate momentum of v_l. The local potential V_l for each single particle is a *quasi-harmonic* one for *displacive transitions* or a *double-well potential* for *order-disorder transitions* (see sketch above). The interaction is written as bilinear coupling

$$\mathcal{H} = -\frac{1}{2}\sum_{l,l'} \mathbf{I}_{l,l'} v_l v_{l'},$$

(16.21)

that can approximately describe, in a simplified form, dipolar interactions, *Ising* or *Heisenberg* type Hamiltonians. In the presence of the external field $\mathcal{E}$ the term $-\sum_l v_l \mathcal{E}_l$ is added to the Hamiltonian in Eq. (16.20). As already emphasized in other circumstances, *MFA* consists in singling out a particular v_l while all other $v_{l'}$ are replaced by their thermal average. Thus a single-particle Hamiltonian is written in the form

$$\mathcal{H} = \frac{\Pi_l^2}{2} + V(v_l) - \mathcal{E} v_l - \sum_{l'} \mathbf{I}_{l'} v_l < v_{l'} > .$$

(16.22)

The thermal averages have to be determined self-consistently (see Eq. (15.1)):

$$< v_l >= \frac{\sum_i < i|v_l|i > e^{-E_i/k_B T}}{Z}, \tag{16.23}$$

where E_i are the eigenvalues and Z the *partition function*. When the eigenvalues are not known, as usual for complicated local potentials, then the classical ensemble averages are taken:

$$< v_l >= \frac{\int_{-\infty}^{\infty} v_l e^{-E_l/k_B T} dv_l}{\int_{-\infty}^{\infty} e^{-E_l/k_B T} dv_l}, \tag{16.24}$$

with

$$E_l = V(v_l) - \mathcal{E}v_l - Iv_l < v_{l'} >$$

where $\mathbf{I} = \sum_{l'} \mathbf{I}_{l,l'}$ and $< v_{l'} >=< v_l >=< v_l >_{\mathcal{E}}$. Thus an implicit equation for the "displacement" $< v >$ (to be related to the polarization) as a function of temperature and field, is obtained.

For $T > T_c$ and in the absence of field $< v >= 0$, while for $\mathcal{E} \neq 0$ the local susceptibility is involved.

For $T < T_c$ the Eqs. 16.23 or 16.24 can have a temperature-dependent solution

$$< v >_{\mathcal{E}=0} \neq 0,$$

namely a spontaneous polarization arises as a consequence of the interactions (for a preliminary illustration in magnetic systems return to Sect. 4.4). The equation

$$< v >_{0,T} =< v >_{0,0} f[V(v_l), T], \tag{16.25}$$

corresponds to the temperature dependence of the order parameter. In Problem 16.1 it is shown how from Eq. (16.25), applied to Ising-like local variable corresponding to spin $1/2$, in the absence of interactions, the *Brillouin function* for $< v >$ is obtained, with a *Curie-like* law for the susceptibility $\chi_0 =< v > /\mathcal{E}$.

By accounting for the interactions, in the MF scenario the total field is

$$\mathcal{E}_{\text{tot}} = \mathbf{I} < v >_{\mathcal{E}} + \mathcal{E}, \tag{16.26}$$

and then

$$< v >_{\mathcal{E},T} = v B\left(\frac{v(\mathbf{I} < v >_{\mathcal{E}} + \mathcal{E})}{k_B T}\right), \tag{16.27}$$

which specializes Eq. (16.25) by showing that the *Brillouin function* B now includes the mean field.

For zero external field the implicit equation for $< v >_T \equiv < v >$ becomes

$$< v >= v_0 B\left(\frac{Iv < v >}{k_B T}\right) \tag{16.28}$$

which can be rewritten

$$\frac{<v>}{v} = tanh\left[\left(\frac{Iv^2}{k_BT}\right)\left(\frac{<v>}{v}\right)\right] \tag{16.29}$$

or $yz = tanh(y)$, with $z = (k_BT/Iv^2)$ and $y = (<v>/vz)$. A graphical solution can be found by looking for the intersection of the curve $tanh(y)$ versus y with a series of straight lines yz versus y, for various z.

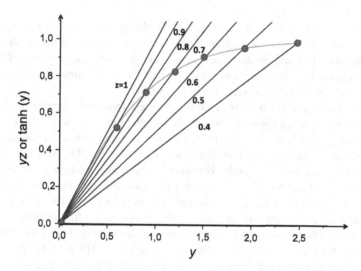

As it appears from the sketch above a solution is obtained for $z \leq 1$, namely for $k_BT \leq Iv^2$, in other words for $T \leq T_c$, with *critical temperature* $T_c = (Iv^2/k_B)$.

The temperature behaviour of $<v>$ in the vicinity of the transition to the ferroelectric phase can be found from Eq. (16.28) by expanding the Brillouin function, since around T_c one can assume $<v> \ll v$. Then, from the expansion of $<v>$ $/v = B[(<v>/v)/(T/T_c)]$ one finds

$$<v> \propto \left(1 - \frac{T}{T_c}\right)^{1/2}. \tag{16.30}$$

Thus, the *critical exponent* $\beta = 1/2$ for the order parameter is found, typical MFA result, as already anticipated at Sect. 15.2. On the other hand, in more sophisticated descriptions which take into account critical effects, the critical exponent β for three-dimensional lattices turns out $\beta \simeq 1/3$. Furthermore, often a small discontinuity in the temperature dependence of the polarization is present below T_c, indicating *quasi-second order transition*.

To derive the susceptibility the presence of the external field must be considered and one writes

$$<v>_{\mathcal{E},T} = vB\left(\left(\frac{v}{k_BT}\right)(I<v>+\mathcal{E})\right). \tag{16.31}$$

Again, by expanding the Brillouin function

$$< v > \simeq \left(\frac{v^2}{k_B T} \right) (\mathbf{I} < v > + \mathcal{E})$$

or

$$< v > \propto \frac{(\mathcal{E} v^2 / k_B T)}{\frac{T}{T_c} - 1} \tag{16.32}$$

and then $\chi = (< v > / \mathcal{E}) \propto [(T - T_c)/T_c]^{-1}$ yielding the *critical exponent* $\gamma = 1$.

It can be noticed that below the transition $< v >$ tends to saturate toward the value v_0 and therefore the response decreases because of the biasing due to the polarization, in a way equivalent to the presence of a strong field (see for instance Fig. 4.6), a field that here is of internal origin (see plots in Fig. 15.3).

For *order-disorder ferroelectrics*, in the simplest form with permanent dipoles having two eigenvalues corresponding to the orientations of the dipole in the local double-well potential, $< v >$ has to be thought as the statistical or the quantum average of the pseudo-spin variable taking the values $+1$ or -1. The description given above evidently yields MFA results. However in some cases a relevant improvement is necessary. One has to take into account the possibility of *quantum tunnelling* between the two wells of the local potential. This is particularly true in the case, as for instance the *KDP-type ferroelectric* family, where the Hydrogen atom can take the right and left positions in the local potential. The dramatic change of the transition temperature from $T_c = 123$ K for ordinary KDP and $T_c = 213$ K for the deuterated crystal, indicates that some other term must be added to the Hamiltonian 16.20 in order to describe in a proper way the mass dependence.

This type of description can be carried out in the framework of the *pseudo-spin formalism* by associating to the dipole a fictitious spin 1/2 operator and introducing the possibility of quantum tunnelling with a rate Γ between the two orientations by means of the spin matrix S_x for the transverse component, namely the *tunnelling operator*. The Hamiltonian 16.20 is therefore written

$$\mathcal{H} = -2\Gamma\hbar \sum_i S_x^i - \sum_{i,j} {}' I_{ij} S_z^i S_z^j, \tag{16.33}$$

where

$$S_x = \frac{1}{2} \begin{pmatrix} 0 & 1 \\ 1 & 0 \end{pmatrix} \quad \text{and} \quad S_z = \frac{1}{2} \begin{pmatrix} 1 & 0 \\ 0 & -1 \end{pmatrix}$$

are the spin operators. The bilinear interaction Hamiltonian is the Ising-like one (involving in most cases the dipolar coupling among the dipoles, see Problems 5.2 and 6.5 for the magnetic case).

The MFA leads to

$$\mathcal{H} = \sum_i \left[-2\Gamma \hbar S_x^i - S_z^i I_0 < S_z > \right] \equiv \sum_i \mathcal{H}_i, \qquad (16.34)$$

with $I_0 = \sum_j I_{ij}$.

Thus the single-particle Hamiltonian $\mathcal{H}_i$ is the one of a pseudo-spin in a pseudo-field. In the paraelectric phase $< S_z > = 0$ and the pseudo-spin precesses around H_x (see sketch below).

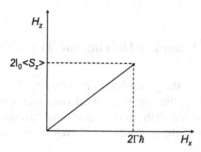

The expectation value reads (see Sects. 4.4 and 6.1)

$$< S_x > = \frac{Tr[S_x e^{-\mathcal{H}_i/k_B T}]}{Tr[e^{-\mathcal{H}_i/k_B T}]} = tanh\left(\frac{\Gamma \hbar}{k_B T}\right). \qquad (16.35)$$

Below T_c $< S_z > \neq 0$ and the polarization $P = 2\mu_e < S_z >$ arises.

The diagonalization of the pseudo-spin Hamiltonian is obtained by the transformation

$$S_x = S_\perp cos\theta + S_\parallel sin\theta \qquad S_z = -S_\perp sin\theta + S_\parallel cos\theta,$$

with $cos\theta = H_z/H_{eff}$ and $sin\theta = H_x/H_{eff}$ $(H_{eff} = [H_x^2 + H_z^2]^{1/2})$.

The single spin Hamiltonian becomes $\mathcal{H}_i = -S_\parallel H_{eff}$, with $< S_\perp > = 0$ while

$$< S_\parallel > = \frac{Tr[S_\parallel e^{-\mathcal{H}_i/k_B T}]}{Tr[e^{-\mathcal{H}_i/k_B T}]} = \frac{1}{2} tanh\left(\frac{S_\parallel H_{eff}}{2k_B T}\right). \qquad (16.36)$$

Then

$$< S_z > = \left(\frac{H_z}{H_{eff}}\right) < S_\parallel > \qquad (16.37)$$

$$= \frac{I_0 < S_z >}{[(2\Gamma \hbar)^2 + (I_0 < S_z >)^2]^{1/2}} \cdot tanh\left(\frac{[(2\Gamma \hbar)^2 + (I_0 < S_z >)^2]^{1/2}}{2k_B T}\right),$$

which is the self-consistent equation corresponding to Eq. (16.29) and pertaining to the rise of the polarization. By singling out the temperature at which the crossover from $< S_z > = 0$ to $< S_z > \neq 0$ occurs, in a way analogous to what done for Eq. (16.29), one has

$$\frac{2\Gamma\hbar}{I_0} = tanh\left(\frac{\Gamma\hbar}{k_B T_c}\right), \tag{16.38}$$

which sets the condition for the occurrence of the *ferroelectricity*: $I_0 > 2\Gamma\hbar$.

One sees that the *driving interaction* is the *local field* (favouring the polar state) while the *restoring force* is the *tunnelling*, which favors the state of equal populations in the two wells of the local potential.

16.5 The Critical Dynamics Driving the Transition

In *displacive ferroelectrics* the general concept regarding the microscopic dynamics leading to the transition, is the *soft mode*, as already mentioned. The stability of a crystal structure requires that all the normal modes have frequency *real* and *positive*, while as a consequence of the softening of a given mode the condition of "*lattice instability*" is approached.

In the framework of the pseudo-spin formalism for *order-disorder ferroelectrics* the critical dynamics can be described, in the MFA approximation,[4] by resorting to Heisenberg equations for the time-dependence of the spin operators involved in the Hamiltonian 16.33. In the assumption that the tunnelling term in the Hamiltonian is neglected one could also describe the slowing down of a collective pseudo-spin component by resorting to an extension to the dipole lattice of the *Glauber model*, which forces the dynamics in the Ising-like Hamiltonian by considering the role of the thermal bath. In Appendix 16.1 the main lines of the above mentioned descriptions are recalled.

Here we give some more insights on the critical dynamics for displacive ferroelectrics. We shall refer to a cubic crystal, with ions at charge $\pm Q$, in the simplifying assumption that the negative ions have infinite mass and zero electronic polarizability. Thus the equations for the motion of the positive ion of mass M (single particle per cell and negative ions not involved) and for the polarization are written

$$M\ddot{x} = -kx + Q\mathcal{E}_{loc}, \tag{16.39}$$

$$P = \frac{Qx}{V_c} + \alpha_e \mathcal{E}_{loc},$$

with α_e the electronic polarizability (see Fig. 16.1).

---

[4]The mean field approximation extended to time-dependent phenomena corresponds to the evaluation of the commutator $[S_i, \mathcal{H}]$ in the Heisenberg equation with the substitution of the *density matrix* of the N-body system with the product of the single-spin density matrix. It is usually called *random phase approximation* (RPA)(see Appendix 16.1).

In the light of the derivation at Sect. 16.3, the local field is

$$\mathcal{E}_{loc} = \mathcal{E}_{ext} + \frac{4\pi}{3}\mathbf{P} \quad \text{for transverse waves}, \tag{16.40}$$

$$\mathcal{E}_{loc} = \mathcal{E}_{ext} + \frac{4\pi}{3}\mathbf{P} - 4\pi\mathbf{P} \quad \text{for longitudinal waves},$$

with the correction related to the depolarization field. For vibrational modes at $q = 0$, in the absence of external field, from the equation for motion and polarization one obtains

$$-\omega_T^2 x = -\frac{k}{M} + \frac{\beta_{ion}}{1 - \beta_{elec}} \quad \text{for transverse waves}$$

$$-\omega_L^2 x = -\frac{k}{M} - \frac{2\beta_{ion}}{1 + 2\beta_{elec}} \quad \text{for longitudinal waves}, \tag{16.41}$$

where $\beta_{ion} = 4\pi Q^2 / 3 M V_c$ and $\beta_{elec} = 4\pi\alpha_e / V_c$.
Therefore two results are addressed:
(i) for the transverse optical mode (at zero wave vector)

$$\omega_T^2 \to 0 \quad \text{when} \quad (k/M) \to \beta_{ion}/(1 - \beta_{elec}), \tag{16.42}$$

(ii) for the dielectric constant

$$\frac{\varepsilon(\infty)}{\varepsilon(0)} = \frac{1 - \beta_{elec} - \beta_{ion}}{1 + 2(\beta_{elec} - \beta_{ion})} \cdot \frac{1 + \beta_{elec}}{1 - \beta_{elec}} = \frac{\omega_T^2}{\omega_L^2}, \tag{16.43}$$

namely the *LST relation* already anticipated (see Eq. (16.17)).

One should remark that the condition of $\omega_T^2 \to 0$ (Eq. (16.42)) is unlikely in most ionic crystals. Returning, for instance, to the case of NaCl order of magnitude estimates yield

$$\frac{k}{M} \simeq 2\frac{\beta_{ion}}{1 - \beta_{elec}}$$

and therefore no instability can be expected. The effect of the local field generated in the vibration motions is to reduce by a factor around 30% the frequency ω_T with respect to the value controlled by the short range elastic constant. At variance, in ferroelectric crystals the elastic and the electric terms are of the same order of magnitude and some temperature dependence of the form $k = (a + bT)$, as assumed in the *quasi-harmonic approximation*, leads to

$$\omega_T^2 \propto a(T - T_0) \quad \text{and} \quad \varepsilon(0) = \frac{C}{T - T_0},$$

(see Problem 15.3).

The generalized electric susceptibility is written

$$\chi(\mathbf{q}, \omega) = \frac{C}{\omega^2(\mathbf{q}) - \omega^2 - i2\gamma\omega}$$

where C is a constant characteristic of the crystal. The $\mathbf{q}$-dependence of the vibrational frequency can be taken

$$\omega^2(\mathbf{q}) = \alpha(T - T_0) + \delta q^2$$

as the result of the expansion of the interaction term $I(\mathbf{q})$ (see Appendix 6.1, Eq. (A.16.1.11)). It is remarked that $T_c(\mathbf{q}) < T_c(\mathbf{q} = 0)$, namely the instability temperature is the highest for the wave-vector corresponding to the homogeneous ($\mathbf{q} = 0$) polarization in the ferroelectric phase.

Appendix 16.1 Pseudo-Spin Dynamics for Order-Disorder Ferroelectrics

In the following the response of an assembly of interacting dipoles μ_e to a small and time-dependent external field is described in the framework of the *mean field approximation*, by resorting to the *pseudo-spin formalism*.

In the light of Eq. (16.33) one starts from the Hamiltonian

$$\mathcal{H} = -2\Gamma\hbar \sum_i S_x^i - \sum_{i,j}{}' I_{ij} S_z^i S_z^j - 2\mu_e \sum_i H_i(t) S_z^i, \qquad (A.16.1.1)$$

H_i being the local field. The statistical average of the spin operators is time-dependent according to the equation

$$\frac{d < \mathbf{S}^i > (t)}{dt} = -\frac{i}{\hbar}[< \mathbf{S}^i >, \mathcal{H}]_t \qquad (A.16.1.2)$$

The MFA extended to time dependent phenomena (the so called *RPA*, random phase approximation) corresponds to evaluate the commutator by substituting the *density matrix* of the N-body system with the product of single-spin density matrices (note 4 in the present chapter). In turn, this is equivalent to substitute the products as $< S_\alpha^i S_\beta^j >$ (with $i \neq j$ and $\alpha, \beta = x, y, z$) with the products of the expectation values of the type $< S_\alpha^i > < S_\beta^j >$. Thus the equations of motions can be written in terms of single particle, becoming

$$\frac{d < \mathbf{S}^i > (t)}{dt} = < \mathbf{S}^i > (t) \times \mathbf{H_i}(t) \qquad (A.16.1.3)$$

where the average field is

$$\mathbf{H_i}(t) = -\frac{\partial < \mathcal{H} > (t)}{\partial < \mathbf{S}^i > (t)},$$

$< \mathcal{H} > (t)$ being the expectation value of the Hamiltonian in MFA.

In the light of the description given at Sect. 16.4, Eq. (A.16.1.3) corresponds to the single spin precession around an effective instantaneous, time-dependent mean field.

In the assumption of linear response to the external field, $< \mathbf{S}^i > (t)$ is the sum of the expectation value plus $\delta < \mathbf{S}^i > exp(i\omega t)$, corresponding to the deviation due to the field

$$\mathbf{H} = \mathbf{H}_i + \delta\mathbf{H}_i e^{i\omega t}$$

From Eq. (A.16.1.3), by taking into account only the terms linear in $\delta < \mathbf{S}^i >$ and in $\delta\mathbf{H_i}$, one has

$$i\omega\delta < \mathbf{S}^i >= \delta < \mathbf{S}^i > \times \mathbf{H_i} + < \mathbf{S}^i > \times \delta\mathbf{H_i} \qquad (A.16.1.4)$$

$< \mathbf{S}^i > \times \mathbf{H_i}$ being zero, the mean value of the spin operator being along $\mathbf{H_i}$. The excitations are the deviations from the time-independent values and the eigenfrequencies are obtained from Eq. (A.16.1.4) for zero external field.

By taking into account that the average molecular field is

$$H_i^x = 2\Gamma\hbar, \quad H_i^y = 0, \quad H_i^z = \sum_j {}' I_{ij} < S_z^j >_t,$$

the equation of motions for the spin deviations turn out

$$i\omega\delta < S_x^i >= \left(\sum_j I_{ij} < S_z^j >\right)\delta < S_y^i >$$

$$i\omega\delta < S_y^i >= 2\Gamma\hbar\delta < S_z^i > - \sum_j \left[I_{ij} < S_z^j > \delta < S_x^i > -I_{ij} < S_x^i > \delta < S_z^j >\right]$$

$$\qquad (A.16.1.5)$$

$$i\omega\delta < S_z^i >= -2\Gamma\hbar\delta < S_y^i >,$$

namely $3N$ coupled linear equations. Then we turn to the collective components

$$< \delta S_{\mathbf{q}} >= \sum_i \delta < \mathbf{S}^i > exp[-i\mathbf{q} \cdot \mathbf{R_i}].$$

In the paraelectric phase, being $< S_z >= 0$, for a given wave vector $\mathbf{q}$ one has

$$i\omega\delta < S_{\mathbf{q}}^x >= 0$$

$$i\omega\delta < S_{\mathbf{q}}^y >= 2\Gamma\hbar\delta < S_{\mathbf{q}}^z > -I_{\mathbf{q}} < S_{\mathbf{q}}^x > \delta < S_{\mathbf{q}}^z > \qquad (A.16.1.6)$$

$$i\omega\delta < S_{\mathbf{q}}^z >= -2\Gamma\hbar\delta < S_{\mathbf{q}}^y >$$

with $I_{\mathbf{q}} = \sum_{i,j} I_{ij} exp[-i\mathbf{q} \cdot (\mathbf{R_i} - \mathbf{R_j})]$.

As usual, the excitation frequencies are given by the secular equation

$$\begin{pmatrix} i\omega & 0 & 0 \\ 0 & i\omega & -2\Gamma\hbar + I_{\mathbf{q}} < S_{\mathbf{q}}^x > \\ 0 & 2\Gamma\hbar & i\omega \end{pmatrix} = 0. \qquad (A.16.1.7)$$

One of the eigenfrequencies pertains to the longitudinal motion along the molecular field (along S_x) while two frequencies describe the precessional motion of the pseudo-spin around the mean field.

For temperature well above the transition temperature the only excitation involves the tunnelling frequency $\omega_{2,3}^2(\mathbf{q}) = 4\Gamma^2$. On decreasing temperature one has

$$\omega_{2,3}^2(\mathbf{q}) = \left(\frac{I_0 < S_z >}{\hbar}\right)^2 + 2\Gamma\left(2\Gamma - \frac{I_{\mathbf{q}} < S_x >}{\hbar}\right), \qquad (A.16.1.8)$$

the first term being zero for $T > T_c$, while $< S_x >= tanh(\Gamma\hbar/k_BT)$ (see Eq. (16.35)).

Equation (A.16.1.8) evidences the slowing down of the frequencies and the *instability limit* for $q = 0$ in correspondence to the maximum value of the *Fourier transform of the interaction*, the one for $q = 0$. On approaching T_c, for $q = 0$ one can expand the above equation in power of $(T - T_c)$, then writing

$$\omega_{2,3}^2(q = 0) = \left(\frac{\partial \omega_{2,3}^2(q = 0)}{\partial T}\right)_{T=T_c} (T - T_c) = \frac{\Gamma^2 I_{q=0}}{k_B T_c^2 cosh^2\left(\frac{\Gamma\hbar}{k_B T_c}\right)}\left(\frac{T - T_c}{T_c}\right) \equiv$$

$$\equiv A_{q_c=0}\left(\frac{T - T_c}{T_c}\right), \qquad (A.16.1.9)$$

indicating a *dynamical critical exponent* $\gamma = 1$, as expected.

According to Eq. (A.16.1.9), the pictorial behaviour (derived from Eq. (A.16.1.5) without assuming $< S_z > = 0$) of the frequencies are sketched below (see also Chap. 15)

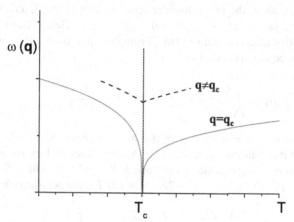

As regards the response to the external field, by resorting to Sect. 15.4 and adding at hand a damping factor, the *generalized susceptibility* is written

$$\chi''(\mathbf{q}, \omega) \propto \frac{N \mu_e^2}{(\omega(\mathbf{q})^2 - \omega^2)^2 + \gamma(\mathbf{q})^2 \omega^2}, \qquad (A.16.1.10)$$

The modes have resonant character for small damping while turn to relaxational modes for strong damping.

Two comments are in order. The tunnelling integral being expected mass-dependent, one can realize why the deuteration in KDP increases the transition temperature from 123 K to 213 K.

For the cubic lattice of dipoles at distance a, from

$$\mathbf{I_q} = 2I[\cos(q_x a) + \cos(q_y a) + \cos(q_z a)],$$

by expanding $\mathbf{I_q}$ and $\omega_{2,3}^2(\mathbf{q})$ around $q = 0$ Eq. (A.16.1.9) becomes

$$\omega_{2,3}^2(\mathbf{q}) = A_{q=0} \frac{T - T_c}{T_c} + 4\Gamma^2 \frac{a^2 q^2}{6}, \qquad (A.16.1.11)$$

a dispersion relation for the pseudo-spin excitations consistent with the general form of the q-dependence discussed at Sect. 15.4.

It is also reminded that when $\mathbf{I_q}$ is maximum in correspondence to a zone-boundary wave vector $\mathbf{Q_{BZ}}$ then the transition involves the crossover to an *antiferroelectric* phase, the order parameter is the sublattice polarization and the crystal cell doubles below T_c.

Finally we just mention that when the maximum of the Fourier transform of the interaction is neither at $q = 0$ nor at $\mathbf{Q_{BZ}}$, the transition can involve an *incommensurate phase*, in which the order parameter of the critical variable (e.g. the expectation value of the pseudo-spin or the lattice displacement of an atom) is not commensurate with the underlying lattice.

A few words about the order-disorder ferroelectrics below T_c can be added. In the ferroelectric phase both $< S_x >$ and $< S_z >$ are different from zero and the fictitious effective field is in the xz plane. From the equations for the spin deviations and the secular equations one can obtain

$$\omega_{2,3}^2(\mathbf{q}) = \left(\frac{I_0 < S_z >}{\hbar} \right)^2 + 4\Gamma^2 \left(1 - \frac{I_\mathbf{q}}{I_0} \right). \qquad (A.16.1.12)$$

The critical mode is at $q = 0$ and $\omega_{2,3}(q = 0) \propto I_0 < S_z >$. The temperature dependence of the polarization related to $< S_z >$ follows. In fact, the effective field $(H_x + H_z)$ below T_c implies the eigenvalues $\pm W = \pm [\Gamma^2 + (I_0 < S_z >)^2]^{1/2}$.

The partition function being $Z = 2cosh(W/k_B T)$, the polarization is

$$P = \frac{1}{k_B T} \frac{\partial \ln Z}{\partial H} = \frac{I_0 < S_z >}{2W} tanh \left(\frac{W}{k_B T} \right) \propto$$

$$\propto tanh \left[\frac{[\Gamma^2 + (I_0 < S_z >)^2)]^{1/2}}{k_B T} \right].$$

For exhaustive presentation of the issues related to Sect. 16.4 and particularly to this Appendix, the book by *Lines* and *Glass* or the one by *Blinc* and *Zeks* are suggested.

Appendix 16.2 Distribution of Correlation Times and Effects around the Transition

For a system of permanent dipoles with single-particle response of Debye character Eq. (16.13) holds and then

$$\varepsilon'(\omega) = \varepsilon(\infty) + \frac{\varepsilon(0) - \varepsilon(\infty)}{1 + \omega^2 \tau^2} \qquad (A.16.2.1)$$

$$\varepsilon''(\omega) = \frac{(\varepsilon(0) - \varepsilon(\infty))\omega\tau}{1 + \omega^2 \tau^2}$$

where $\tau(T) = \tau_0 exp(\Delta E / k_B T)$ is the *relaxation time*.

In *mono-dispersive* crystals, where a single correlation time occurs, these equations can be used for the MFA dielectric response measured at $q = 0$, with (see Appendix 15.1)

$$\tau_p = \tau_{q=0} = \frac{\tau(T)}{1 - I_{q=0}\chi^0} \qquad (A.16.2.2)$$

Real dielectrics are hardly mono-dispersive and rather exhibit a distribution of $\tau's$. that has to be taken into account, particularly around the transition from the disordered to the ordered phase. Empirical account of dielectric dispersion and absorption measurements in *poly-dispersive systems* can be given by using for the dielectric constant the following relations:

$$\varepsilon'(\omega) = [\varepsilon'(0) - \varepsilon(\infty)]\frac{1 + bZ}{1 + 2bZ + Z^2}$$

$$\varepsilon''(\omega) = \varepsilon'(\omega)\frac{aZ}{1 + bZ} \qquad (A.16.2.3)$$

where $a = sin(\pi B/2)$, $b = cos(\pi B/2)$ and $Z = (\omega\tau_p)^B$, while B measures the width of the distribution of the relaxation times, with $0 < B \leq 1$.

These equations are known as *Cole-Cole relationships* and result from the integration over τ of Eq. (A.16.2.1) with a distribution function of the form

$$y(\tau) = \frac{1}{\pi}\frac{sin(B\pi)}{x^B + x^{-B} + 2cos(B\pi)} \qquad (A.16.2.4)$$

with $x = \tau/\tau_p$, τ_p being the correlation time measured with the homogeneous electric field (see Eq. (A.16.2.2)), with the critical behaviour

$$\tau_p \propto \left(\frac{T - T_c}{T_c}\right)^{-\Delta},$$

for $T \rightarrow T_c^+$. The *critical exponent* is $\Delta = \gamma = 1$, in the MFA. For $B = 1$ in Eq. (A.16.2.3) one again obtains the Debye relations, consistent with the MFA susceptibility at $q = 0$.

For $B \neq 1$, by plotting $\varepsilon''(\omega)$ versus $\varepsilon'(\omega)$ and then $[\varepsilon''/(\varepsilon'a - \varepsilon''b)]^{1/B}$, one can extract $\varepsilon(0)$, B and τ_p. For illustration and for comparison with a 3D MFA system (return to Fig. 15.4), see the plots reported below for a 2D system with short-range interactions ($SnCl_2.2H_2O$).

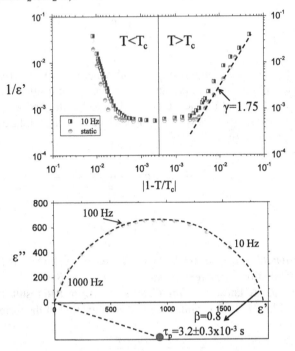

Illustrative plots of the data around the transition temperature $T_c = 219.45$ K in a ferroelectric crystal characterized by planar structure of dipoles (data from Mognaschi, Rigamonti and Menafra, Phys. Rev. B 14, 2005 (1976)). It is noted that the MFA value $\gamma = 1$ is clearly ruled out.

Problems

Problem 16.1 By applying Eq. (16.25) to permanent electric dipoles with *Ising-like pseudo-spin variable* v of dichotomic character, show that the *Brillouin function* for $< v >$ (in terms of $x = v\mathcal{E}/k_B T$) and the *Curie-like* law are obtained, in the assumption of no interactions.

Solution: For no interaction one rewrites

$$< v >= \frac{v\exp(v\mathcal{E}/k_B T) + (-v)\exp(-v\mathcal{E}/k_B T)}{Z}$$

and from the expansion of the Brillouin function

$$\chi^0 = \frac{< v >}{\mathcal{E}} = \frac{v^2}{k_B T}.$$

Problem 16.2 By starting from Eq. (16.2) and considering two frequency dependent in-phase and out-of phase dielectric constants, derive Eqs. (16.3) and (16.4).

Solution: By changing the variable from τ to $t' = t - \tau$ in Eq. (16.2), one writes

$$\mathbf{D}(t) = \varepsilon(\infty)\mathcal{E}_0 cos(\omega t) + \mathcal{E}_0 \int_{-\infty}^{t} cos(\omega \tau)g(t - \tau)d\tau =$$

$$= \varepsilon(\infty)\mathcal{E}_0 cos(\omega t) + \mathcal{E}_0 \int_{0}^{\infty} cos(\omega(t - t'))g(t')dt'.$$

Thus

$$\mathbf{D}(t) = \varepsilon(\infty)\mathcal{E}_0 cos(\omega t) + \mathcal{E}_0 \int_{0}^{\infty} g(t')[cos(\omega t)cos(\omega t') + sin(\omega t)sin(\omega t')]dt' =$$

$$= \mathcal{E}_0 cos(\omega t)\left[\varepsilon(\infty) + \int_{0}^{\infty} cos(\omega t')g(t')dt'\right] + \mathcal{E}_0 sin(\omega t)\int_{0}^{\infty} sin(\omega t')g(t')dt'.$$

Then, from

$$\mathbf{D}(t) = \mathcal{E}_0\left[\varepsilon'(\omega)cos(\omega t) + \varepsilon''(\omega)sin(\omega t)\right]$$

Equations (16.3) and (16.4) follow.

Problem 16.3 Derive the relationship between $\varepsilon'(\omega)$ and $\varepsilon''(\omega)$.

Solution: By Fourier transformation of Eqs. (16.3) and (16.4)

$$g(x) = \frac{2}{\pi} \int [\varepsilon'(\omega) - \varepsilon(\infty)]cos(\omega'x)d\omega'$$

$$g(x) = \frac{2}{\pi} \int \varepsilon''(\omega)sin(\omega'x)d\omega',$$

Equations (16.6) follow.

Problem 16.4 In the light of Eq. (16.9) show that without the *Onsager reaction field* one would predict the *ferroelectric catastrophe* in water.

Solution: One has $\alpha = \mu_e^2/3k_BT$ and from Eqs. (16.7) and (16.9) for $\mu_e^{H_2O} = 1.87$ Debye

$$\varepsilon = 1 + \frac{4\pi N\mu_e^2}{3k_B(T - T_c)}$$

with $T_c = (4\pi N\mu_e^2/9k_B) \simeq 1200$ K.

Problem 16.5 From the equation of motion of a single ion in the lattice under short-range elastic constant (temperature dependent in the form $k_{sh} + bT$) and a long range electrostatic force $k_{el}x$, derive the susceptibility and the temperature at which the frequency goes to zero, with lattice instability.

Solution: The equation of motion is

$$m\ddot{x} + \gamma\dot{x} + (k_{sh} - k_{el} + bT)x = qE_0 e^{i\omega t}.$$

The polarization is Nqx. Then, by solving this equation

$$\chi(\omega) = \frac{Nq}{m(\omega_0^2 - \omega^2 + i\gamma\omega)}$$

(see Eq. (15.47)) and

$$\omega_0^2 = \frac{b}{m}\left(T + \frac{k_{sh} - k_{el}}{b}\right).$$

Thus $\omega_0 \to 0$ for $T \to T_c$, with $T_c = -(k_{sh} - k_{el})/b$ (return to Problem 15.3).

Problem 16.6 By referring to the vibrational motion of a diatomic chain and by considering anharmonic terms of the form $(V_4/4!)\sum_i(x_{i+1} - x_i)^4$, show that the frequency of the $q = 0$ *optical mode* goes to zero at the critical temperature $T_c = 2|k_1|k_2/(V_4 k_B)$, with k_1 and k_2 elastic constants for nearest neighbours and next nearest neighbours.

Solution: The equations of motion for optical and acoustic modes in the absence of anharmonic terms follow from

$$m\ddot{x}_i = -2(k_1 + k_2)x_i + k_1(x_{i+1} + x_{i-1}) + k_2(x_{i+2} + x_{i-2}),$$

yielding $\omega_{ac}^2(q)$ and $\omega_{op}^2(q)$ (see Eq. (14.15)).

With the anharmonic contribution, by writing the Hamiltonian in terms of the normal coordinates and averaging the acoustic ones, one derives

$$\omega_{op}^2(q) = \left[\frac{V_4 k_B T}{2mk_2} - \frac{4|k_1|}{m}\right]\cos^2\left(\frac{qa}{2}\right) + \left(\frac{4k_2}{m}\right)\sin^2(qa).$$

Thus, for $T = T_c = 2|k_1|k_2/V_4 k_B$ the lattice instability occurs.

Specific References and Further Reading

1. C. Kittel, *Introduction to Solid State Physics*, 8th Edition, (J. Wiley and Sons, 2005).
2. M.E. Lines and A.M. Glass, *Principles and Applications of Ferroelectrics and Related Materials*, (Clarendon Press, Oxford, 1977).
3. R. Blinc and B. Zeks, *Soft Modes in Ferroelectrics and Antiferroelectrics*, (North-Holland Publishing Company, Amsterdam, 1974).
4. N.W. Ashcroft and N.D. Mermin, *Solid State Physics*, (Holt, Rinehart and Winston, 1976).
5. F. Bassani e U.M. Grassano, *Fisica dello Stato Solido*, (Bollati Boringhieri, 2000).
6. J.S. Blakemore, *Solid State Physics*, (W.B. Saunders Co., 1974).
7. H.J. Goldsmid (Editor), *Problems in Solid State Physics*, (Pion Limited, London, 1972).
8. G. Grosso and G. Pastori Parravicini, *Solid State Physics*, 2nd Edition, (Academic Press, 2013).
9. H. Ibach and H. Lüth, *Solid State Physics: an Introduction to Theory and Experiments*, (Springer Verlag, 1990).
10. Y.-K. Lim (Editor), *Problems and Solutions on Thermodynamic and Statistical Mechanics*, (World Scientific, Singapore, 2012).
11. L. Mihály and M.C. Martin, *Solid State Physics - Problems and Solutions*, (John Wiley, 1996).
12. H. Thomas in Local Properties at Phase Transitions, Eds. K.A. Müller and A. Rigamonti, (North-Holland Publishing Company, Amsterdam, 1976).
13. H. Stanley, *Introduction to Phase Transitions and Critical Phenomena*, (Oxford University Press, Oxford, 1971).
14. J.M. Ziman, *Principles of the Theory of Solids*, (Cambridge University Press, 1964).

Chapter 17
Magnetic Orders and Magnetic Phase Transitions

Topics

Electronic Correlations and the Hubbard Hamiltonian
The Magnetic Phase Transition within the Mean-Field Approximation
Ferromagnets and Antiferromagnets
Ordered Magnetic Systems and Magnons
Scaling Arguments
Dimensionality Effects
Superparamagnetism, Spin Glasses and Magnetic Frustration
Two-dimensional Quantum Heisenberg Antiferromagnet

17.1 Introductory Aspects on Electronic Correlation

In previous sections (Chaps. 4 and 6) the cases of isolated magnetic moments, with the related phenomena of atomic diamagnetism and paramagnetism, have been addressed. Some of the experimental techniques for the study of their properties (magnetic resonances, muon and Mossabuer spectroscopies, neutron scattering) have been mentioned. In general terms, the scenario of localized and weakly interacting magnetic moments can be defined as the *local moments representation*.

Another model scenario previously described along the book is the one of delocalized electrons in the Fermi gas, particularly for the paramagnetism and diamagnetism of itinerant electrons in metals (see Sect. 12.7.1, Problems 12.10 and 12.19, Appendix 13.1).

In the framework of the local moments representation the magnetization $M(J, H)$ has been derived (Sect. 4.4). A way to take into account the interactions among the magnetic moments has been indicated by resorting to the mean field approximation (MFA), writing for the total magnetic field (see Eq. (4.33)) experienced by a given magnetic moment $H = H_{ext} + \lambda M$. The MFA susceptibility

© Springer International Publishing Switzerland 2015
A. Rigamonti and P. Carretta, *Structure of Matter*,
UNITEXT for Physics, DOI 10.1007/978-3-319-17897-4_17

$$\chi = \frac{\chi^0}{1 - \lambda\chi^0} \tag{17.1}$$

was derived, with χ^0 the bare susceptibility for non-interacting moments (Eq. 4.34). The transition from the disordered to the ordered state is thus conceivable, with a transition temperature $T_c = N\mu_J^2\lambda/3k_B$. One could argue that in Eq. (17.1) the parameter λ hides *exchange* and *correlation effects*, to be addressed subsequently.

In the scenario of *delocalized electrons*, the Pauli and Landau susceptibilities have been derived and it has been observed how by decreasing the electron concentration the average kinetic energy of the electrons decreases faster than the Coulomb repulsive interaction U (see Appendix 13.1).

Experimental evidences indicate that one has to go beyond those simple descriptions and that the detailed role of interactions/correlations has to be considered. Among others one can mention: (*i*) in 3d electron metals the saturation magnetic moment per atom is not an integer number of Bohr magnetons; (*ii*) in alloys the average magnetic moment per site is not linearly dependent on the concentration of the magnetic atoms and sometimes magnetization is found in alloys of non-magnetic atoms; (*iii*) the pressure usually decreases the saturation magnetization, while in principle it should be rather insensitive to the small variations in the interatomic distances; (*iv*) first order transitions, with discontinuous jumps of the magnetization can be driven by an external magnetic field. Thus one can realize that it is necessary to go beyond the mean field approach and the *one-electron approximation*.

A simple way to account for interaction and correlation effects has been mentioned through the introduction in Eq. (17.1) of Stoner correction $\lambda = U/2\mu_B^2 N$, a term which yields a susceptibility enhancement and then a drive towards the ordered state (Eq. A.13.1.15). That result has been derived starting from the *Stoner-Hubbard* Hamiltonian (Eq. A.13.1.9). The *many-body Hubbard* Hamiltonian has been constructed in the space of single-electron *Wannier functions*, namely wave functions centered at a given lattice site and obtained as a sum over all the states **k** of a band of normalized Bloch functions (Eq. 12.41). The properties of these functions are such that the overlap integral between adjacent sites i and j is exactly zero. Thus the assumption in Eq. (12.45) is no longer required and the tight-binding approach extends its validity. The more general expression of the many-particle Hamiltonian devised by *Hubbard* in order to account for correlation effects, for a single band, is

$$\mathcal{H} = \sum_{i,j,\sigma} t_{ij} c_{i\sigma}^\dagger c_{j\sigma} + \frac{1}{2} \sum_{i,j,i',j',\sigma,\sigma'} <ij|V|i'j'> c_{i\sigma}^\dagger c_{j\sigma}^\dagger c_{i'\sigma'} c_{j'\sigma'}, \tag{17.2}$$

where c are fermionic annihilation or creation $(c^\dagger)$ operators of an electron at a given site and σ and σ' labels the spin up or down states. t_{ij} represents the *hopping integral* for a single electron between sites i and j which corresponds to the one derived in the tight-binding approximation in the band theory. The term $<ij|V|i'j'> = <ij|e^2/r_{ij}|i'j'>$ in Eq. (17.2) is the matrix element of the two electrons interaction.

In many systems one can assume $t_{ij} \neq 0$ for i and j nearest neighbours and $t_{ij} = 0$ otherwise. The interaction term is more conveniently treated by making the strong assumption that only on-site Coulomb repulsion is at work: $< ij|V|i'j' > = U$ for $i = j = i' = j'$ and $< ij|V|i'j' > = 0$ otherwise. Under these approximations Hubbard Hamiltonian takes the form

$$\mathcal{H} = \sum_{<i,j>\sigma} t_{ij} c_{i\sigma}^{\dagger} c_{j\sigma} + U \sum_i \hat{n}_{i\uparrow} \hat{n}_{i\downarrow}, \qquad (17.3)$$

where $\hat{n}_{i,\sigma}$ are the number operators $\hat{n}_{i,\sigma} = c_{i,\sigma}^{\dagger} c_{i,\sigma}$ and the sum runs over the lattice sites while $< i, j >$ means that it is limited to the nearest neighbour pairs.

The first term of the Hubbard Hamiltonian is sometimes represented in the **k**-space as $\sum_{\mathbf{k},\sigma} E_{\mathbf{k}} a_{\mathbf{k},\sigma}^{\dagger} a_{\mathbf{k},\sigma}$, where $a_{\mathbf{k},\sigma}$ indicate fermionic operators for the creation and the destruction of the elementary particles at wave vector **k** and spin σ. $E_{\mathbf{k}}$ is the electron band dispersion of width W, derived within a tight-binding approximation. In 3d metals one has typical values for W around $4\,\text{eV}$ while the repulsion energy U is of the order of $1–3\,\text{eV}$.[1]

The Hamiltonian in Eq. (17.3) has the remarkable role to allow one to bridge the gap between localized ($U > W$) and delocalized electrons ($W > U$) scenarios: now the positions and the motion of all the electrons are correlated, since they interact each other inducing forces among them during the motion.

Unfortunately the Hubbard Hamiltonian, in spite of its apparent simplicity, can be solved analytically for any value of the ratio U/W only in one dimension and for a single band. Many solutions in the limits $U \gg W$ or $W \gg U$ have been reported. For $U/W \ll 1$ the *Hartree-Fock approach* is usually employed, corresponding to a first-order perturbation in U/W. When the density functional method (Sect. 3.4 and Chap. 9) includes the spin, one deals with the *spin-density functional* theory. The energy of the ground state, written as a functional of the electron density and of the spin polarization, is then minimized. The *exchange-correlation term* (not exactly known) is thought to include all the many body effects. Nevertheless, from numerical approaches good results for the band structure and for the spin density are usually achieved.

The opposite $U \gg W$ limit has to be considered in several other cases. In particular when from band calculation one would predict a metallic state while the localized electrons condition is actually attained. An example of such a situation is found in many transition metal oxides, as La_2CuO_4, which is the parent of high T_c superconductors. In fact, La_2CuO_4 is an insulating crystal while neglecting the correlation one would predict a metal (see Appendix 17.1). In the $U \gg W$ limit the *hopping term* can be considered as a perturbation (to the second order) of the purely repulsive term of Eq. (17.3). In that limit, for a *half-filled band* (namely one electron per site),

---

[1]When W is of the same order of U upon slight variation of certain interaction parameters the transition from the insulated to the metal compound is possible. This is the *Mott transition*. The Hubbard model is possibly the simplest way to show how the electronic interaction can produce novel magnetic states in solids.

the Hubbard Hamiltonian yields an effective interaction which has the *Heisenberg form*

$$\mathcal{H}_{eff} = \sum_{i \neq j} \frac{|t_{ij}|^2}{U} \left(4\mathbf{S_i} \cdot \mathbf{S_j} - \frac{1}{2} \right), \qquad (17.4)$$

with *exchange coupling* given by $4|t_{ij}|^2/U$ (see Problem 17.4). It is noted that in this case, the coupling being *essentially positive*, an antiferromagnetic state is induced by the correlations (see Sect. 4.4 and Appendix 17.1). This is what happens in La_2CuO_4.

For non-half filled bands the Heisenberg-like Hamiltonian has to be replaced by the $t - J$ *Hamiltonian*, where instead of U the exchange interaction J is present.

Similarly to Eq. (17.3), several other Hamiltonians have been introduced as starting points in order to describe the magnetic properties of strongly correlated electrons systems. They are generally defined *model spin systems*. A large variety of models with analytical or numerical solutions is known. From the Heisenberg nearest-neighbours Hamiltonian

$$\mathcal{H}_{eff} = - \sum_{i,j} J_{ij} \mathbf{S_i} \cdot \mathbf{S_j}, \qquad (17.5)$$

the spin models are characterized on the basis of the dimensionality d (generally 1, 2, 3) of the order parameter (here the spin $\mathbf{S}$) and of the dimensionality D of the lattice (see Sect. 15.1). Analytical solutions for the partition function, so that all the thermodynamical quantities can be derived, are known for: (i) $d = 1, D = 1$ and $J > 0$ (namely positive exchange constant for nearest neighbours), known as *1D Ising model*, with ferromagnetic ground state occurring only at $T = 0$ (namely no transition to an ordered state at finite temperature) as for all $D = 1$ cases; (ii) for $d = 1$ and $D = 2$ (*Onsager model*); (iii) for all the cases at $D = 4$ or more, since the MFA solutions turn out to be valid; (iv) for all the cases where the range of the interaction can be considered infinite (again the MFA solutions are valid); (v) for the case of $d \to \infty$ and any D (the so called *spherical model*); (vi) for $d = 2$ (*XY model*) where no phase transition is known to occur below $D = 2$ (see Table 15.2).

No exact solutions are known for the most interesting case, namely $D = 3$ and limited range of the interaction parameter J. A general semi-empirical method to attack the problems of the possible transition to ordered states and of the critical behaviour for spin models is the one based on *scaling arguments* and on the concept of *universality* (see Sect. 15.1 and Appendix 17.2 for some insights).

Here, we only recall the description of the transition from the high temperature paramagnetic phase to the low temperature ordered state by resorting to the mean field approximation. The description of phase transition in the thermodynamic framework and the relationships between the single particle and collective responses have been developed (Sect. 16.2, Appendix 15.1) in a way to allow direct transposition to the case of spin variables. In particular it is reminded that within the MFA the total magnetic field in paramagnets is written

$$\mathbf{H} = \mathbf{H_{ext}} + I < \mathbf{S} >_{H_{ext}} \equiv \mathbf{H_{ext}} + \lambda \mathbf{M}, \qquad (17.6)$$

where the energy scale I rather than being related to the dipolar interaction (as at Sect. 16.4) has to be related to the exchange integral. The magnetization is

$$M(H_{ext}, T) = M_0 B_S \left[\mathbf{H_{ext}} + \lambda \mathbf{M}(H_{ext} \mu_B g S / k_B T) \right], \qquad (17.7)$$

with B_S the *Brillouin function* and M_0 the saturation magnetization.

In the absence of external field the implicit equation for the spontaneous magnetization is derived

$$M(0, T) = M_0 B_S \left[\lambda \mathbf{M}(0, T) \right] \propto < S_z > (T), \qquad (17.8)$$

which, for the case of $S = 1/2$ is written (see Problem 4.18)

$$M(0, T) = M_0 tanh \left(\frac{\lambda M(0, T)}{k_B T} \right), \qquad (17.9)$$

yielding for $T \to T_c^-$, i.e. $M(0, T) \ll M_0$,

$$M(0, T) \propto < S_z > (T) \propto \left(1 - \frac{T}{T_c} \right)^{1/2}, \qquad (17.10)$$

corresponding to the *critical exponent* $\beta = 1/2$ (see comment at Sect. 15.3). For the response $\chi(\mathbf{q}, 0)$ see Sects. 4.4 and 16.4.

In subsequent sections of this chapter we shall clarify how the magnetic interaction, in most cases the exchange interaction, originates for atoms at different lattice positions. Then ordered states will be described and the elementary excitations in ordered systems, called *magnons*, presented. It is observed that magnetism in solids has a very large variety of aspects. We shall only mention the *superparamagnetism*, *spin-glasses* and the *magnetic frustration* . Magnetic systems are also ideal cases for the theoretical treatments of exact statistical models (such as Ising and Heisenberg models) and in order to emphasize the dimensionality effects.

Some more specific topics shall be discussed in the Appendices of the present chapter: magnetic scaling and the phase diagram of the two-dimensional $S = 1/2$ Heisenberg antiferromagnet (2DQHAF), that besides being a very interesting system for quantum magnetism, is the father of the *high temperature superconductors* (to be described at Chap. 18). It will also give the chance to illustrate in some detail the correlation effects and how they can be taken into account in an apparently simple way by means of the Hubbard Hamiltonian.

17.2 Mechanisms of Exchange Interaction

In solids, in most cases the overlap between the wave functions centered at different atomic sites is small, because of their short range character. Therefore the exchange mechanism described at Sect. 2.2 for the two electrons in He atom cannot be simply transferred in the attempt to justify the magnitude of the exchange interaction leading to the I or J terms in the Hamiltonians introduced in the previous section. As it can be argued from the discussion given at Sect. 17.1, the exchange mechanism cannot be thought as resulting from the overlap of localized electrons but rather one has to include the electron itinerancy as well as correlation effects.

Illustrative examples, in this respect, are offered by *transition metal oxides*, systems to which the Hubbard model was initially (1963) applied. Let us refer to *MnO*, which involves Mn^{2+} ions with the $3d^5$ subshell implying $J = S = 5/2$, according to the Hund rules described at Sect. 3.2.3. The value of the magnetic moment experimentally determined from susceptibility measurements is around $5.9\mu_B$. MnO is an *antiferromagnet* (AF) below $T_N = 116$ K, with the localized magnetic moments of Mn^{2+} nearest neighbours, connected by Oxygen atoms, pointing along opposite directions. In other terms, the antiferromagnetic coupling via the Oxygen lowers the energy of the whole magnetic structure. In a simple picture one can justify the occurrence of the AF state by involving in the exchange mechanism also the 2p electrons of O. In fact, the decrease in the kinetic energy induced by the electron delocalization along the Mn-O-Mn unit, and the strong overlap between 3d Mn and 2p O orbitals (thanks to Pauli principle) yields a neat AF coupling. A sketch of this *superexchange mechanism* is shown in Fig. 17.1.

Fig. 17.1 Sketch of the superexchange mechanism yielding the MnO antiferromagnetic structure (partial view)

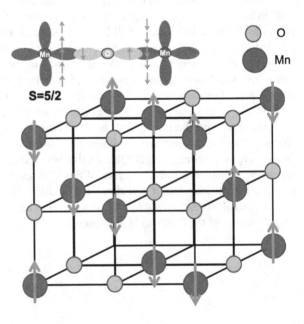

One can think that the ordering process is related to a magnetic Hamiltonian of the type in Eq. (17.4). Although the presence of five 3d electrons of the Mn^{2+} ion and of two 2p electrons of the Oxygen make things more complicated, the basic aspects of the superexchange mechanism can be described through the simple form given in Eq. (17.5). Similar situation occurs in the planar array of Cu^{2+} and Oxygen ions in AF La_2CuO_4, as we shall discuss in Appendix 17.1. In general the superexchange coupling is related to the overlap between atomic orbitals and hence it significantly depends on the angle formed by the paramagnetic ions (Cu^{2+}, for example) and the diamagnetic ion (O^{2-}). In fact, one can notice a change by orders of magnitude in the superexchange coupling upon varying that angle from π to $\pi/2$ (Fig. 17.2).

There is also another mechanism of indirect magnetic coupling between localized moments: the one mediated by delocalized electrons. This *indirect* (or *itinerant*) *exchange* in metals is known as *RKKY interaction* (from *Ruderman, Kittel, Kasuya* and *Yoshida*, the discoverers). This interaction, mediated by the Fermi sea, is a long range one, with an oscillating behaviour with the distance. Thus, depending on the distance between the magnetic ions, the exchange coupling can change from ferromagnetic to antiferromagnetic. This behaviour is at the basis of the *spin-glass state* found in metals doped with paramagnetic ions, as $Cu_{1-x}Mn_x$ and it is the relevant coupling in many intermetallic systems formed by paramagnetic lanthanide ions.

In order to derive the form of the RKKY coupling let us consider two paramagnetic ions, say A and B, coupled with the itinerant electrons of the metal via a local contact-like interaction J

$$\mathcal{H}_{e-A} = -J \sum_i \mathbf{S_A} \cdot \mathbf{s}_i \delta(\mathbf{r_A} - \mathbf{r_i}), \qquad (17.11)$$

with $\mathbf{S_A}$ the localized spin and $\mathbf{s}_i$ the ith itinerant electron spin. Accordingly the itinerant electrons will experience a local effective field $H_{eff} = -JS_A/g\mu_B$ and the electron gas will acquire a neat spin polarization

Fig. 17.2 Superexchange coupling J versus the Cu-O-Cu angle in a series of copper oxides

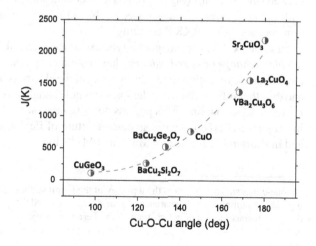

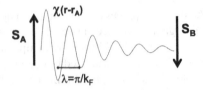

Fig. 17.3 Schematic view of the RKKY coupling between two localized spins through the polarization of the itinerant electrons via non-local susceptibility $\chi_P(\mathbf{r} - \mathbf{r_A})$

$$< s(\mathbf{r} - \mathbf{r_A}) > = -\chi_P(\mathbf{r} - \mathbf{r_A})\frac{\mathbf{H_{eff}}}{g\mu_B} = \frac{1}{V}\sum_q e^{i\mathbf{q}\cdot(\mathbf{r}-\mathbf{r_A})}\chi_P(\mathbf{q})\mathbf{S_A}\frac{J}{g^2\mu_B^2}, \quad (17.12)$$

where $\chi_P(\mathbf{q})$ is the static $\mathbf{q}$-dependent Pauli susceptibility (see Problem 17.3).

For three-dimensional Fermi gas one obtains

$$\frac{1}{V}\sum_q e^{i\mathbf{q}\cdot\mathbf{r'}}\chi_P(\mathbf{q}) = \frac{3g^2\mu_B^2}{8E_F}\frac{N}{V}\frac{k_F^3}{16\pi}\left[\frac{sin(2k_Fr') - 2k_Frcos(k_Fr')}{(k_Fr')^4}\right], \quad (17.13)$$

$(\mathbf{r'} = \mathbf{r} - \mathbf{r_A})$ which evidences how the electron gas spin polarization oscillates with a periodicity determined by Fermi wavevector k_F and changing sign upon varying the distance from the localized ion A. The polarized itinerant electrons will in turn interact with the localized spin $\mathbf{S_B}$, so that at the end an effective coupling between $\mathbf{S_A}$ and $\mathbf{S_B}$ described by the Hamiltonian

$$\mathcal{H}_{AB} = -\frac{J^2}{g^2\mu_B^2}\frac{1}{V}\sum_q e^{i\mathbf{q}\cdot(\mathbf{r_A}-\mathbf{r_B})}\chi_P(\mathbf{q})\mathbf{S_A}\cdot\mathbf{S_B} \quad (17.14)$$

occurs. Thus, the exchange coupling between $\mathbf{S_A}$ and $\mathbf{S_B}$ oscillating with the distance between the two spins (Fig. 17.3). It is noticed that since $\chi_P(q)$ changes significantly with the lattice dimensionality (see Problem 17.3 and return to Problem 12.7)[2] the same occurs for the RKKY coupling.

In some oxides a ferromagnetic-type exchange mechanism could occur, known as *double exchange* or *mixed valency,* due to the fact that the magnetic ion can exist in more than one oxidation state. The hopping of the electrons between one ionic species and the other is favoured when the spin does not change its orientation, yielding a neat ferromagnetic coupling. This process describes the indirect ferromagnetic exchange in *magnetite* (Fe_3O_4), having an equal mixture of Fe^{2+} ($3d^6$) and Fe^{3+} ($3d^5$) ions, and in *manganites* formed by Mn^{3+} and Mn^{4+} ions.

---

[2]In one-dimensional conductors the topology of the Fermi surface yields a divergence of $\chi_P(q)$ at $q = 2k_F$, resulting in the appearance of *spin density wave* (SDW) phases, an ordered state where the spontaneous local spin polarization of the Fermi gas varies with a periodicity determined by $2k_F$.

Finally we mention that in a way similar to the superexchange mechanism the spin-orbit interaction can lead to an *anisotropic exchange interaction* via the exchange process of the excited state of an ion with the ground state of the nearby ion. This process involves an interaction Hamiltonian of the form

$$\mathcal{H}_{DM} = \mathbf{D} \cdot \mathbf{S_i} \times \mathbf{S_j} \tag{17.15}$$

as derived by *Dzyaloshinsky* and *Moriya*, with a vectorial term **D** which is zero only when the crystal field has inversion symmetry. This coupling mechanism is rather common in antiferromagnets and since it tends to induce a canted ferromagnetic structure in the spins, one often speaks of *weak ferromagnetism*.

17.3 Antiferromagnetism, Ferrimagnetism and Spin Glasses

Again by referring to the generic Hamiltonian (Eq. 17.5), when the exchange interaction J_{ij} is negative the ordered state is antiferromagnetic, with nearby spins along opposite direction. Illustrative example is provided by MnF_2 crystal. Below about 67 K the structure of MnF_2 can be seen as resulting from two interpenetrating *sublattices* of cubic structure, each one being characterized by ferromagnetic order, while the sublattices have magnetic moments along opposite directions. No net magnetization is obviously present for the whole crystal and only a rather small discontinuity is usually observed in the susceptibility at the ordering temperature, known as *Néel temperature* T_N. As for the case of the ferromagnetic transition, anomalies in the specific heat and in the expansion coefficient are detected around T_N.

A simple picture of the AF state and of the phase transition can be provided in the two sub-lattices model within the mean field approximation. By limiting the range of the interactions to the first and to the second nearest neighbours[3] the internal field at the A site is written

$$\mathbf{H_{intA}} = -\lambda_{AB}\mathbf{M_B} - \lambda_{AA}\mathbf{M_A} \tag{17.16}$$

where λ_{AB} is *positive* in view of the negative sign of the exchange integral which grants an antiferromagnetic order, while λ_{AA}, which is expected much smaller than λ_{AB}, in principle could be positive or negative. Equivalently, at the sites of the B sublattice one has

$$\mathbf{H_{intB}} = -\lambda_{BA}\mathbf{M_A} - \lambda_{BB}\mathbf{M_B} \tag{17.17}$$

When the atoms in the two sublattice are the same, as in the example of the MnF_2 crystal, one can set $\lambda_{AB} = \lambda_{BA} = \lambda$ and $\lambda_{AA} = \lambda_{BB} = \alpha$. Then, in the presence of the external field, by extending Eq. (17.7), we write for the magnetization of the A sublattice

---

[3]It should be remarked that only for the bcc crystal one can refer to a model of first n.n. of type A and second n.n. of type B. This separation would not be possible, for instance, for the fcc crystal structure, where four sublattices have to be considered.

$$\mathbf{M_A} = \frac{1}{2}N\mu_J B_J\left[\frac{\mu_J H_A}{k_B T}\right], \tag{17.18}$$

with $\mathbf{H_A} = \mathbf{H_{ext}} - \lambda\mathbf{M_B} - \alpha\mathbf{M_A}$ (N is the number of atoms per unit volume).
Analogous expression holds for $\mathbf{M_B}$. Since Eq. (17.18) is strictly similar to Eq. (17.7),
the derivation of the susceptibility and of the phase transition for an antiferromagnet
follows the same procedure as for ferromagnets (see Chap. 4). For weak external
magnetic field, one can develop the Brillouin functions for $\mathbf{M_A}$ and for $\mathbf{M_B}$ and

$$\mathbf{M_A} = \frac{C}{2T}\left[\mathbf{H_{ext}} - \lambda\mathbf{M_B} - \alpha\mathbf{M_A}\right], \tag{17.19}$$

$$\mathbf{M_B} = \frac{C}{2T}\left[\mathbf{H_{ext}} - \lambda\mathbf{M_A} - \alpha\mathbf{M_B}\right], \tag{17.20}$$

with $C = N\mu_J^2/3k_B$ the Curie constant. In the paramagnetic region the internal and
the external fields are parallel and then

$$\chi = \frac{M_A + M_B}{H_{ext}} = \frac{C}{2T H_{ext}}\left[2H_{ext} - (\lambda+\alpha)(M_A + M_B)\right] = \frac{C}{T+\Theta}, \tag{17.21}$$

with $\Theta = (C/2)(\lambda+\alpha)$ *Curie-Weiss temperature*. The transition from the disordered
to the ordered state for each sublattice can be expected at a temperature where for
$H_{ext} = 0$ the homogeneous system of Eqs. (17.19) and (17.20) has non-zero solution
for M_A and M_B. By setting to zero the determinant of the coefficients for M_A and
M_B one finds $(2T/C) + \alpha = \lambda$ and then the Néel temperature turns out[4]

$$T_N = \frac{C}{2}(\lambda - \alpha) \tag{17.22}$$

and $(T_N/\Theta) = (\lambda - \alpha)/(\lambda + \alpha)$, showing that when the second n.n. interaction is
neglected $T_N = \Theta$. At $T = T_N$ one has $\mathbf{M_A} = -\mathbf{M_B}$, namely opposite directions of
the sublattice magnetizations arise.

For $T < T_N$ the response to the external field H_{ext} can be evaluated in the same
way as for the ferromagnet, by expanding the Brillouin function as for Eq. (17.9). If
the field is applied along the magnetization direction the susceptibility (the response
to a homogeneous field, therefore at $q = 0$ and at $\omega = 0$, see Sect. 15.3) is obtained
expanding the Brillouin functions in term of the variables

$$x_A = \frac{\mu_J}{k_B T}\left[H_{ext} + \lambda M_B - \alpha M_A\right] \tag{17.23}$$

---

[4]It can be observed that on increasing α the magnetic moments belonging to a given sublattice tends
to align in the AF configuration and a magnetic structure of more than two sublattices should be
considered.

and

$$x_B = \frac{\mu_J}{k_B T}\left[-H_{ext} + \lambda M_A - \alpha M_B\right], \qquad (17.24)$$

around the value x_0 of x_A and x_B derived for $H_{ext} = 0$. Thus, one obtains (see Problem 17.1)

$$\chi_{\|}(T) = N\mu_J^2|\frac{dB_J}{dx}|_{x=x_0}\left[k_B T + \frac{\alpha+\lambda}{2}N\mu_J^2|\frac{dB_J}{dx}|_{x=x_0}\right]^{-1}, \qquad (17.25)$$

which for $T \to 0$ yields $\chi_{\|} = 0$. For field perpendicular to the magnetization, for $T < T_N$, the *spin canting* mechanism yields an effective magnetization along the field (Problem 17.8) and

$$\chi_{\perp}(T) = \frac{1}{\lambda}, \qquad (17.26)$$

namely a temperature independent susceptibility directly related to the exchange coupling (and α-independent) (see Fig. 17.4).

For powder or polycrystalline sample, by averaging over the angle formed by the sublattice magnetization and the magnetic field, the isotropic susceptibility becomes

$$\chi_{meas} = \frac{1}{3}\chi_{\|} + \frac{2}{3}\chi_{\perp}, \qquad (17.27)$$

so that for $T \to 0$ $\chi_{meas} \to (2/3)\chi_{\perp}$ (see Fig. 17.4).

In several real cases $T_N \neq \Theta$, as it is shown in Table 17.1 for some typical antiferromagnets.

Fig. 17.4 Susceptibility of $S = 5/2$ antiferromagnet with $T_N = 67$ K. Below T_N the behaviour of χ for field parallel and perpendicular to the sublattice magnetization is shown, as well as for powder samples

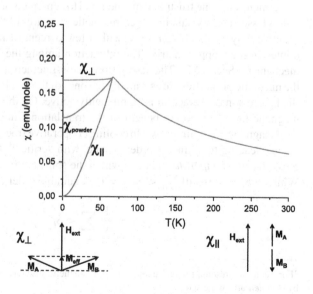

Table 17.1 Néel temperature
and Curie-Weiss temperature
for selected antiferromagnetic
compounds

	$T_N(K)$	$\Theta\,(K)$
MnO	122	610
MnF_2	67	80
FeO	198	570
NiO	515	~2000
CoO	291	330
FeF_2	85	117

It is noted that in antiferromagnets the order parameter for the transition (see Sect. 15.1) is the sub-lattice magnetization or the amplitude of the staggered magnetization, being equivalent to the difference $\mathbf{M_A} - \mathbf{M_B}$ of the two sub-lattice magnetizations.

In the description of the antiferromagnetic state given above we have assumed that the two sublattices have the same type of magnetic ions and equal magnetizations in modulus. When in the sublattices there are different magnetic ions (or there are non-equivalent sublattices for crystallographic reasons) then a net magnetization arises below T_N. These type of magnetic compounds, typically the ferrites, are called *ferrimagnets*. A representative case is magnetite (Fe_3O_4), typical example of semiconducting ferrite family (the majority being insulators) where two types of iron magnetic ions are present, involving the double exchange process.

A large number of other magnetic structures are known. When nearest neighbours and next-nearest neighbours exchange couplings are competing *helical phases* may appear with an order parameter characterized by a period of modulation of several lattice steps or which can even be incommensurate.[5]

We shall only mention a group of systems called *spin glasses* (while at Sect. 17.5 some attention to the frustrated magnets and to superparamagnetism will be devoted). Spin glasses are crystals in which magnetic ions are diluted in a *random way*. An example may be the *MnCu* alloy, with a few percent of magnetic manganese ions diluted in the Copper matrix. The interaction among the Mn^{2+} ions is via RKKY mechanism (Sect. 17.2). The description of non-interacting magnetic ions (for which the magnetic properties are essentially the ones of the insulated atoms) has to be modified since some interaction is significantly active. On the other hand, the localized magnetic moments do not benefit of the translational invariance adopted for standard magnetic configurations. On cooling cooperative freezing of the spin excitations occurs, leading to a kind of ordered state, with "critical" temperature T_g (the *spin-glass freezing temperature*) below which the spins are frozen in random directions. While $<\mathbf{S_i}> \neq 0$ still $\sum_i <\mathbf{S_i}> = 0$. A suitable order parameter can be written

$$q_{EA} = \frac{1}{N}\sum_i < S_i >^2_{T,J}, \tag{17.28}$$

---

[5]For a concise presentation of the many ordered magnetic structures the Chaps. 5 and 8 in the book by *Blundell* are suggested.

which is different from zero for $T < T_g$ (*Edward-Anderson order parameter*). The local "critical" variable is first thermally averaged and subsequently averaged over the distribution of the interactions. $\sqrt{q_{EA}}$ becomes an indication of the "frozen" local moment per lattice site. The occurrence of a real phase transition in the thermodynamical sense can be debated. In fact, no long range preferential orientation occurs and furthermore T_g (as detected for instance by the peak in the susceptibility) might be found different according to the different time scales of the experiment, as a result of a wide distribution of correlation times for the local spins. In spin glass the randomness in the distances implies a distribution of the interaction so that even the sign of the coupling constant is not defined in a definite way. Thus the so-called *magnetic frustration* is present and a single ion cannot take a precise direction with respect to the ones nearby (see Sect. 17.5). There is not a defined ground state but rather a large number of different ground states.

17.4 The Excitations in the Ordered States

At zero temperature ferromagnets (FE) or antiferromagnets (AF) are perfectly ordered and neglecting the quantum, not-thermally driven, fluctuations (see Appendix 17.1) no excitations are present. On increasing temperature the order is somewhat disrupted by magnetic excitations, called *magnons*, similar to the phonons in the crystal lattice. A relevant fact is that for isotropic FE or AF even a vanishing thermal energy is effective in creating excitations. At variance, in the presence of magnetic anisotropy an energy gap in the excitations is found. Usually FE's and AF's differ in the energy dependence on the wavevector (the dispersion relations, see at Sect. 14.3 for the phonons): for $q \rightarrow 0$ the energy of the magnons is parabolic in the wave vectors for FE's while it is linear for AF's.

In the following the magnon excitations in an isotropic FE, in the low temperature range are described. For high temperature and for the transition from the paramagnetic to the ordered state it is reminded that the spin dynamics driving the transition is rather of diffusive character, poorly quantized, as it has been illustrated in general terms at Chap. 15. A simplified semiclassical approach (resembling the pseudo-spin dynamics described for order-disorder ferroelectrics) is given in the following, by referring to a one-dimensional lattice of spins interacting with the nearest neighbours only.

From the Heisenberg Hamiltonian $\mathcal{H}$ (Eq. 17.5) for ferromagnets the time dependence of the expectation value of a given spin operator is

$$
\begin{aligned}
\frac{d < \mathbf{S}_j >}{dt} &= \frac{1}{i\hbar} < [\mathbf{S}_j, \mathcal{H}] > \\
&= -\frac{2J}{i\hbar} < [\mathbf{S}_j, \ldots + \mathbf{S}_{j-1} \cdot \mathbf{S}_j + \mathbf{S}_j \cdot \mathbf{S}_{j+1} + \ldots] > \\
&= -\frac{2J}{i\hbar} < [\mathbf{S}_j, \mathbf{S}_{j-1} \cdot \mathbf{S}_j] + [\mathbf{S}_j, \mathbf{S}_j \cdot \mathbf{S}_{j+1}] > \\
&= \frac{2J}{\hbar} < \mathbf{S}_j \times (\mathbf{S}_{j-1} + \mathbf{S}_{j+1}) > .
\end{aligned}
\tag{17.29}
$$

At low temperatures for the low energy excitations in the transverse components
$S_j^{x,y} \ll S$ and by considering the spins as classical vectors, one writes

$$\frac{dS_j^x}{dt} \simeq \frac{2JS}{\hbar}(2S_j^y - S_{j-1}^y - S_{j+1}^y)$$

$$\frac{dS_j^y}{dt} \simeq -\frac{2JS}{\hbar}(2S_j^x - S_{j-1}^x - S_{j+1}^x)$$

$$\frac{dS_j^z}{dt} \simeq 0. \qquad (17.30)$$

As discussed at Sect. 14.2, one looks for normal mode solutions, namely $S_j^x = A \, exp[i(qja - \omega t)]$ and $S_j^y = B \, exp[i(qja - \omega t)]$. Then one derives $A = iB$ and the dispersion relation for the *spin waves* turns out

$$w(q) = \frac{4JS}{\hbar}[1 - cos(qa)], \qquad (17.31)$$

and for $q \to 0$ $\hbar \omega(q) \simeq 2Ja^2q^2$, as anticipated.

In a way similar to what carried out at Sect. 14.5, the excitations being bosons as the phonons, the number of magnons at a given temperature is written

$$n_m = \int_0^\infty \frac{D(\omega)}{e^{(\hbar\omega/k_BT)} - 1}d\omega. \qquad (17.32)$$

Since the dispersion curve is quadratic in the wave vector as the one for free electrons, one immediately realizes (see Chap. 12) that in three dimensions the density of states $D(\omega) \propto \omega^{1/2}$ and by setting $z = \hbar\omega/k_BT$ one rewrites

$$n_m = \left(\frac{k_BT}{\hbar}\right)^{3/2} \int_0^\infty \frac{z^{1/2}}{e^z - 1}dz \propto T^{3/2}. \qquad (17.33)$$

The reduction of the magnetization upon thermal excitation is proportional to the number of the magnons being excited. Then the *Bloch $T^{3/2}$ law* follows:

$$\frac{M(0) - M(T)}{M(0)} \propto T^{3/2},$$

while the thermodynamic energy associated to the spin waves is

$$U = \int_0^\infty \frac{\hbar\omega D(\omega)}{e^{(\hbar\omega/k_BT)} - 1}d\omega \propto T^{5/2}. \qquad (17.34)$$

Therefore the magnons contribution to the specific heat in a ferromagnet is proportional to $T^{3/2}$.

In the schematic plot given below, the temperature behaviour of the magnetization (in an ideal single domain of a ferromagnetic crystal) is reported in the temperature range up to the transition to the paramagnetic phase, the critical behaviour in the vicinity of T_c being deduced as described at Chap. 15 or in the MFA as in Eq. (17.10).

It can be remarked that the integral in Eq. (17.32) diverges for lattice dimensionalities $d = 2$ and $d = 1$. Therefore, in correspondence to the infinity in the number of magnons being excited by an infinitesimal increase of temperature the magnetization is zero at any finite temperature (provided that the magnetic compound is of Heisenberg isotropic character). This observation corresponds to the *Mermin-Wagner-Berezinskii* theorem: no long range order is possible for $d \leq 2$ in the presence of continuous symmetry. At variance, for anisotropic systems in $d = 2$ this is no longer the case (see Problem 17.5).

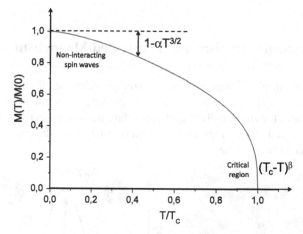

At this point it is appropriate to remark that the considerations carried out until now refer to the ideal case of a single magnetic crystal, namely to an ideally infinite array of ordered spins. In reality such a configuration is not stable, the energy due to the dipolar interaction (disregarded in the above, being much smaller than the short-range exchange interaction) does not favours such an extended order. Therefore the magnetic compounds are arranged in mesoscopic regions (*domains*) each having the spin-orientational properties considered until now, but differing each other in the direction of the spontaneous magnetization. Thus a macroscopic magnetic compound does not exhibit, in general, a spontaneous magnetization. The complex domain structure is energetically favourable with respect to an ordered state with magnetization pointing along the same direction over an infinite range of distances.

The domains are separated each other by the *domain walls*, regions of the sample where rapid variations in the direction of the spontaneous magnetization occur. The width of the domain walls depends on the ratio between the magnetic anisotropy and the exchange coupling (see Problem 17.6), being progressively reduced for increasing magnetic anisotropy.

The domain structure is responsible of the magnetization curve

$$\mathbf{B} = \mathbf{H_{ext}} + 4\pi\mathbf{M}$$

versus $\mathbf{H_{ext}}$ in a ferromagnet, the classical loop being associated with the domains motions, with the possibility of the remnant field B for $H_{ext} = 0$ when the domains alignment is locked. We will not discuss these aspects pertaining to macroscopic magnetism.[6]

Finally we remind that in regards of the local field at a given site in a magnetic compound, all the considerations given at Sect. 16.1 for the electric field can be transferred to the magnetic field, including in particular the issue related to the depolarization factors.

17.5 Superparamagnetism and Frustrated Magnetism

If a certain amount of ferromagnetic nano-sized particles (see the figure below) is diluted in a non-magnetic matrix and the average distance is much larger than the interaction range, for sufficiently high temperature the compound will behave like a paramagnet with a "local" magnetic moment resulting from the coupling of many elementary magnetic moments.

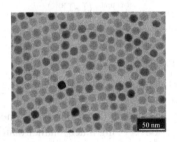 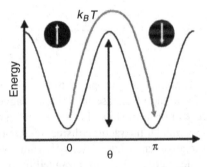

The temperature excitations in superparamagnets are somewhat similar to the ones already discussed for permanent electric moments in a double well potential (see sketch above). In particular one can define a relaxation time as in Appendix 15.1, with a barrier E, of the form

$$\tau = \tau_0 e^{(E/k_B T)} \tag{17.35}$$

and τ_0 is typically of the order of 10^{-9} s.

---

[6]Chapter 33 of the book by *Ashcroft* and *Mermin* can be advised in this respect.

The slowing down, on cooling, of the reorientational fluctuations of the effective magnetic moment of the nanoscopic particles implies that below a certain temperature T_g (that might depend on the time scale of the experiment) the superparamagnet appears ordered, with each particles locked into one of the two minima of the local potential. It should be noted that the barrier E is related to the volume of the particle, to the magneto-crystalline anisotropy and/or to the shape anisotropy related to the demagnetization factors.

The difference between spin glasses and superparamagnets is basically in the range of the interaction. While in superparamagnets the particles are essentially independent, in the spin glass there are sizeable interaction and the transition to the quasi-ordered state is a cooperative effect (the transition to the ordered state being possibly frustrated by the disorder in the positions of the elementary magnetic moments).

Frustrated spin systems are characterized by a highly degenerate ground state which prevents to reach a unique state of minimal energy. The frustration is related to the impossibility in fulfilling, at the same time a given configurational interaction among different pairs of magnetic moments. There are several sources of possible frustration in spin systems. The simplest case is when a particular geometry of the magnetic lattice is responsible for it. The prototype situation of this type is the triangular lattice, as sketched below:

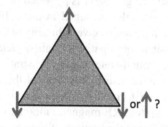

The coupling is assumed through n.n. Heisenberg-like Hamiltonian, with J negative: two spins try to arrange anti-parallel to each other and the third remains frustrated. Another way to understand the presence of a two-fold degenerate ground-state is by considering that classically the energy of the triangular spin configuration can be minimized once adjacent spins form an angle of $2\pi/3$. That condition can be attained either for positive *chirality* of the three spins or for negative chirality:

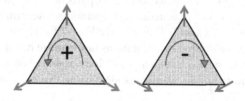

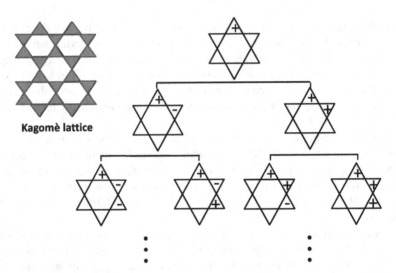

Fig. 17.5 The Kagomé lattice (*top left*). On the *right* the degeneracy arising from the possible chiralities is evidenced

Another example of frustrated geometry is provided by the *Kagomé lattice*, characterized by corner sharing triangles arranged in a star-like fashion, as shown in Fig. 17.5. Starting from a triangle with a given chirality one realizes that the adjacent triangle can have either positive or negative chirality, yielding a neat macroscopic degeneracy of the magnetic ground-state, namely frustration.

Among the three-dimensional frustrated systems the one which has recently attracted more attention is that formed by paramagnetic rare-earth (RE) ions on a pyrochlore lattice. In this system RE magnetic moments characterized by a large uniaxial magnetic anisotropy are at the vertices of regular tetrahedra. The anisotropy usually favours the orientation of the moments along the direction connecting the vertex with the center of the tetrahedron. In the presence of dipolar interaction among the RE magnetic moments of the same order of magnitude of the exchange coupling among nearest neighbour spins the energy of each tetrahedron is minimized if the ice-rule is obeyed: two spins point inwards and two outwards (see plot).

For each tetrahedron there are 6 different configurations that comply to this rule and since there are 4 spins per tetrahedron, for the pyrochlore lattice with N tetrahedra there will be $W = (6/4)^N$ degenerate configurations. Accordingly the $T \rightarrow 0$ entropy $S = k_B ln W = N k_B ln(3/2)$ arises (see Fig. 17.6). These systems are called *spin-ices* since the possible spin configurations inside the tetrahedron are equivalent to those characterizing the displacement of H^+ ions around oxygen in ice. In fact, the low-temperature entropy of the spin-ice is identical to that calculated in the thirties by *Pauling* for ice.

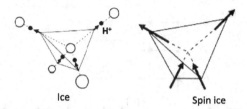

Ice   Spin ice

Owing to the large magnetic anisotropy the Hamiltonian describing the spin-ice is Ising-like

$$\mathcal{H} = J_{eff} \sum_{i,j} {}' \sigma_i \sigma_j$$

with $\sigma_i = \pm 1$ and $J_{eff} = J_{nn} + D_{nn}$ is an effective exchange coupling including both n.n. exchange (J_{nn}) and dipolar (D_{nn}) couplings. Depending on the ratio between those two latter quantities the spin-ice or an antiferromagnetic ground-state can be stabilized.

As already mentioned another possible source for frustration is the disorder of the magnetic lattice, as in the spin glass system $Cu_{1-x}Mn_x$, with dilution of the paramagnetic ion Mn^{2+} in the Copper metallic matrix. The *RKKY interaction* is responsible for frustration because of the change in the sign with the distance. Accordingly, for random distribution of Mn^{2+}, a given ion can interact either through ferromagnetic or antiferromagnetic couplings with its nearest neighbours and these competing interactions will generate frustration.

Finally frustration may arise from the geometry of the competing interactions. An emblematic case is represented by the so-called $\mathbf{J_1 - J_2}$ configuration on a *square lattice*, where the nearest neighbour exchange interaction J_1 competes with the next-nearest neighbour exchange interaction J_2 provided that the latter one is of antiferromagnetic character (as in the sketch below). For $J_2 \to 0$ the ground state is the standard Néel antiferromagnetic order. In the opposite limit $J_2/J_1 \to \infty$ one can view the system as formed by two interpenetrating Néel orders which can freely rotate one with respect to the other by an angle θ suggesting an infinitely degenerate ground-state.

Fig. 17.6 Temperature dependence of the magnetic entropy in $Ho_2Sn_2O_7$, a typical spin ice compound

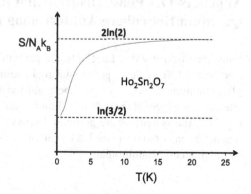

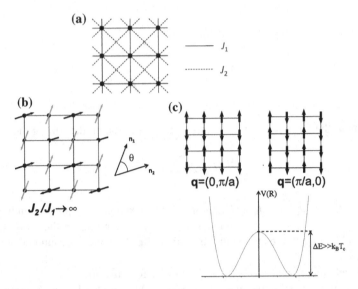

However, for $S = 1/2$, at any finite J_1 value, the exact solutions of the $J_1 - J_2$ Hamiltonian

$$\mathcal{H} = -J_1 \sum_{<i,j>} {}' \mathbf{S_i} \cdot \mathbf{S_j} - J_2 \sum_{\ll i,j \gg} {}' \mathbf{S_i} \cdot \mathbf{S_j},$$

the first sum running over all n.n. and the second one over n.n.n. spins, yields a two-fold degenerate ground-state, characterized by wave-vector $\mathbf{q} = (\pi/a, 0)$ or $\mathbf{q} = (0, \pi/a)$. The two degenerate ground-states sketched above are separated by an energy barrier ΔE related to J_1, J_2 and ξ, the in-plane correlation length (see Chap. 15 and Appendix 17.1). When $k_B T < \Delta E$ the system collapses in one of the two ground states through an order-disorder Ising transition analogous to the one described in ferroelectrics at Chap. 16. For $|J_2/J_1| \simeq 0.5$ frustrations is so strong that no magnetic order sets in.

Appendix 17.1 Phase Diagram and Related Effects in 2D Quantum Heisenberg Antiferromagnets (2DQHAF)

Since the discovery that La_2CuO_4, the parent of high temperature superconductors (see Sect. 18.8), is the experimental realization of the model for two-dimensional (2D) quantum ($S = 1/2$) Heisenberg antiferromagnet (2DQHAF), a great deal of interest was triggered towards low-dimensional quantum magnetism. As sketched in Fig. 17.7, the system we are going to discuss is basically a planar array of $S = 1/2$ magnetic ions onto a square lattice, in antiferromagnetic interaction and described by the magnetic Hamiltonian

Fig. 17.7 Sketch of a planar antiferromagnet with weak inter-planes interaction $J_\perp$

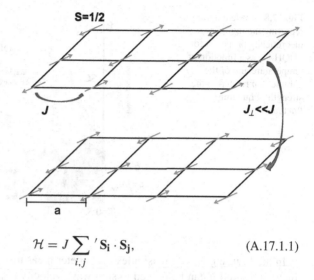

$$\mathcal{H} = J \sum_{i,j} {}' \mathbf{S_i} \cdot \mathbf{S_j},$$ (A.17.1.1)

with $J > 0$ and summation limited to the nearest neighbors spins.

$S = 1/2$ characterizes the quantum condition and Heisenberg character means the absence of single ion anisotropy.

We shall devote our attention to a variety of aspects involving static and dynamical properties of that 2D array: (i) the temperature dependence of the in-plane magnetic correlation length ξ_{2D} entering in the equal-time correlation function $< \mathbf{S_i}(0) \cdot \mathbf{S_j}(0) >$; (ii) the critical spin dynamics driving the system towards the long-range ordered state (at $T = 0$ in pure 2DQHAF in the absence of interplanar interaction $J_\perp$); (iii) the validity of the dynamical scaling, where ξ_{2D} controls the relaxation rate Γ of the order parameter according to a law of the form $\Gamma \propto \xi_{2D}^{-z}$, with critical exponent z (see Sect. 15.1).

We shall also comment on the modifications induced by *spin dilution* (or *spin doping*) namely when part of the $S = 1/2$ magnetic ions are substituted by non-magnetic $S = 0$ ions, as well as mentioning the effects related to *charge doping*, namely the injection (for instance by hetero-valent substitutions) of $S = 1/2$ holes, thus creating local singlets which can itinerate onto the plane, locally destroying the magnetic order and inducing novel spin excitations. These aspects are of particular interest in the vicinity of the percolation thresholds, where the AF order is about to be hampered at any finite temperature. This can be considered a situation similar to a *quantum critical point (QCP)*, where no more the temperature but rather the Hamiltonian parameters can drive the transition.

In Fig. 17.8 the phase diagram for the system sketched in Fig. 17.7 is shown. This diagram results from a variety of experimental studies in synergistic interplay with theoretical descriptions, that will not be recalled in detail.[7] We only present it and define some characteristic parameters.

---

[7]For an introduction and an exhaustive review of the studies of the magnetic properties of 2DQHAF's, see *Johnston*.

Fig. 17.8 Phase diagram
reporting the regimes
theoretically proposed for
2DQHAF, as a function of
temperature and of the
parameter g related to the
strength of quantum
fluctuations

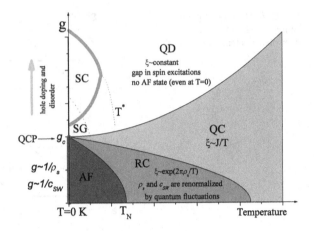

In the Figure g is a dimensionless parameter measuring the strength of quantum
fluctuations and it can be related to spin wave velocity c_{sw} and to the spin stiffness
ρ_s:

$$g = \frac{\hbar c_{sw}\sqrt{2\pi}}{k_B \rho_s a} \qquad (A.17.1.2)$$

(*a* lattice parameter, see Fig. 17.7). The *spin stiffness* ρ_s measures the increase in the
ground state energy for rotation of the magnetization of the two sublattices by an
angle θ ($\Delta E = \rho_s k_B \theta^2/2$). It can be written $\rho_s \propto (c_{sw}^2/\chi_\perp) \chi_\perp$ being the transverse
spin susceptibility. The parameter g is expected to *increase upon doping* and *disorder*.

It has been customary to map the Hamiltonian (A.17.1.1) onto the so-called *non-
linear σ model*, that for $T \to 0$ is the simplest continuum model with the same
symmetry and the same spectrum of excitations. In this way the diagram in Fig. 17.8
has emerged.

Below a given value g_c the ground state, at $T = 0$, is the Néel AF state, which
extends to finite temperature T_N because of the interplane interaction $J_\perp \ll J$.

The *percolation threshold* for the AF state at $T = 0$ as a function of g is at g_c.
For $g < g_c$, upon increasing temperature, above T_N, one enters in the *renormalized
classical (RC) regime*. Here the effect of the quantum fluctuations is to renormalize
ρ_s and c_{sw} with respect to the mean field values for the "classical" 2D Heisenberg
paramagnet. Thus the in-plane correlation length goes as

$$\xi_{2D} \propto e^{2\pi\rho_s/T}. \qquad (A.17.1.3)$$

Weakly damped *spin waves* exist for wave vectors $q \geq \xi^{-1}$ while for longer wave
lengths only diffusive spin excitations of hydrodynamic character are present (see
Sect. 15.4). For $T \geq J/2 \simeq 2\rho_s$ instead of entering into the classical limit (that
would be reached for $T \gg J$) the planar QHAF should cross to the *quantum critical
(QC) regime*. In this phase, typical of 2D and 1D quantum magnetic systems, the

only energy scale is set by temperature and $\xi \propto J/T$. On increasing g, according to proposals still under debate (in the scenario of quantum phase transitions theories) the increase in quantum fluctuations can inhibit an ordered state even at $T = 0$. The system is then in the *quantum disordered (QD) regime*, the correlation length being short and temperature independent. In the spectrum of spin excitations a gap of the order of hc_{sw}/ξ_{2D} opens up.

The somewhat speculative phase diagram illustrated in Fig. 17.8 is still debated. In particular the validity of the non-linear σ model is not entirely accepted for large T and/or for large g regions. Furthermore, there is not clear evidence of crossovers from RC to QC or to QD regimes. Also the real nature of the low-energy excitations remains an open question. The *cluster-spin-glass (SG)* phase is the one for $g > g_c$ in which the experiments indicate the presence of mesoscopic "islands" of AF character separated by domains walls, with effective magnetic moments undergoing collective spin freezing, without long range order even at temperature close to zero.

Above a certain amount of charge doping (e.g. hole injection as in $La_{2-y}Sr_yCuO_4$, see Sect. 18.8), as discovered by *Müller* and *Bednorz*, the systems become superconductors (*SC phase*), with the so-called *underdoped* and *overdoped regimes* characterized by a transition temperatures $T_c < T_{max}$ (the one pertaining to the optimal doping).

In SC underdoped phases a gap in the spin excitations at the AF wave vector $\mathbf{q_{AF}} = (\pi/a, \pi/a)$ has been experimentally observed to arise at a given temperature. The spin-gap (and charge pseudo-gap) region has possibly to be related to superconducting fluctuations (see Sect. 18.11) of "anomalous" character or to AF fluctuations locally creating a "tendency" towards a mesoscopic *Mott insulator*. Exotic excitations of various nature have been considered to occur in the regions of high g's. We shall not go into detail involving these aspects, which are still under debate and less settled than the ones for low g, namely for the doped non-superconducting 2DQHAF.

Summarizing conclusions that can be drawn from the studies in pure 2DQHAF are the following: (i) the absolute value and the temperature dependence of the in-plane magnetic correlation length follows rather well the theoretical expression given by A.17.1.3; (ii) in La_2CuO_4 and in similar 2DQHAF the RC regime appears to hold, up to temperature of the order of $1.5\,J$; (iii) no evidence of crossover to QC or QD regimes has been clearly observed.

Some more quantitative comments can be given about the effect of spin dilution. As already mentioned, in La_2CuO_4 spin dilution is obtained by $S = 0$ Zn^{2+} (or Mg^{2+}) for $S = 1/2$ Cu^{2+} substitutions. While in $La_{2-y}Sr_yCuO_4$ the Néel temperature drops very fast with the Sr content, analogous effect but at much lower rate, is driven by the spin dilution (see Fig. 17.9).

In the limit of weak doping, the dilution model should hold. The dilution model modifies the Hamiltonian A.17.1.1 simply by considering the probability that a given site is spin-empty:

$$\mathcal{H} = J \sum_{i,j}{}' p_i \mathbf{S_i} \cdot p_j \mathbf{S_j} = J(0)(1-x)^2 \sum_{i,j}{}' \mathbf{S_i} \cdot \mathbf{S_j}. \qquad (A.17.1.4)$$

Fig. 17.9 Doping
dependence of the Néel
temperature in diluted
2DQHAF
$(La_2Cu_{1-x}(Zn,Mg)_xO_4)$
and in hole-doped 2DQHAF
$(La_{2-x}Sr_xCuO_4)$, from a
variety of measurements

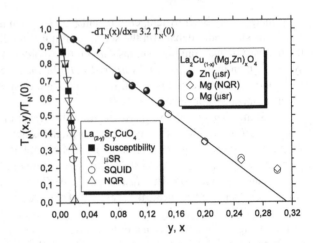

Fig. 17.9 Doping dependence of the Néel temperature in diluted 2DQHAF $(La_2Cu_{1-x}(Zn,Mg)_xO_4)$ and in hole-doped 2DQHAF $(La_{2-x}Sr_xCuO_4)$, from a variety of measurements

Then the spin stiffness should depend on doping according to $\rho_s(x) = \rho_s(0)(1-x)^2$, the correlation length becoming (see Eq. A.17.1.4)

$$\xi_{2D}(x, T) \simeq \xi_{2D}(0, T)e^{-(2-x)x1.15J(0)/T}.$$

An indication for the value of the correlation length at T_N can be obtained from the mean field argument:

$$\xi_{2D}^2(x, T_N)J_\perp(x) = T_N(x)$$

Then for the correlation length one can write

$$\xi_{2D}(x, T_N) \simeq \xi_{2D}(0, T_N)\frac{(1 - 4x)^{1/2}}{1 - x}$$

In Fig. 17.10 the *doping dependence of the spin stiffness*[8] is compared with the prediction of the dilution model. As it could be expected, the dilution model is reasonably well obeyed for light doping while for x amount of the non-magnetic ions larger than about 0.1 it evidently fails.

It should be remarked that although in the strong dilution regime the reduction of the spin stiffness dramatically departs from the one predicted by the dilution model, still the transition to the AF state occurs when the correlation length reaches an in-plane value around 150 lattice steps, as in pure or lightly doped systems.

Another quantity of interest for the quantum effects in disordered 2DHQAF is the zero-temperature staggered magnetic moment $< \mu(x, T \to 0) >$ along the local quantization axis, in other words the dependence of the sub-lattice magnetization on spin dilution. The staggered magnetic moment is different from the classical $S = 1/2$ value because of the *quantum fluctuations*, which in turn are expected to increase with spin dilution. The quantity

---

[8]Derived also from measurements of NQR relaxation rates, that we shall not report.

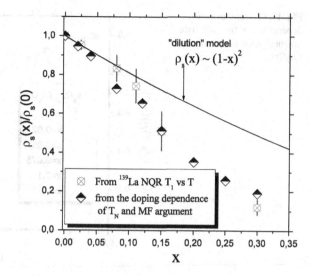

Fig. 17.10 Spin stiffness $\rho_s(x)$ in spin diluted La$_2$CuO$_4$ and comparison with the dependence expected within the dilution model

$$R(x, T = 0) = \frac{< \mu(x, 0) >}{< \mu(0, 0) >}$$

has been obtained to a good accuracy from the magnetic perturbation due to the local hyperfine field on ^{139}La NQR spectra, from μSR precessional frequencies and from neutron scattering also close to the percolation threshold of Zn-Mg doped La$_2$CuO$_4$ (see Fig. 17.11).

While the classical doping dependence (for $S \to \infty$) as well as the one predicted by the quantum non-linear σ model are not supported by the experimental findings, the data indicate a doping dependence of the form $R = (x_c - x)^\beta$, with critical exponent $\beta = 0.45$, close to the behaviour deduced from spin wave theory.

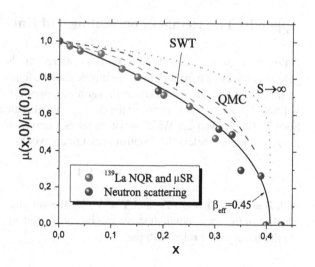

Fig. 17.11 The zero temperature normalized sublattice magnetization in spin diluted La$_2$CuO$_4$. Comparison with *spin-wave theory* (SWT), *Quantum Monte Carlo* (QMC) and classical spin ($S \to \infty$) theoretical behaviour is presented

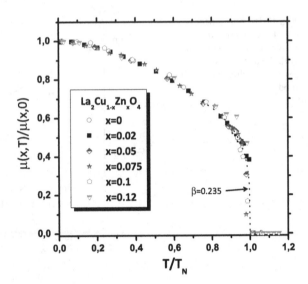

Fig. 17.12 Temperature dependence of the sublattice magnetization in spin diluted La_2CuO_4

As regards the temperature dependence of R, it appears that both in the light doping regime as well as for strong dilution a universal law of the form

$$R(x = const, T) \propto [T_N(x) - T]^{\beta}$$

holds, with a small critical exponent β that appears to be around 0.2 for light doping while on approaching the percolation it increases to 0.3 (Fig. 17.12).

Details and further experimental data can be found in the article by Rigamonti, Carretta and Papinutto in *Novel NMR and EPR Techniques*, J. Dolinsek, M. Vilfan and S. Zumer (Eds.) Springer (2006).

Appendix 17.2 Remarks on Scaling and Universality

An example of scaling and universality is offered by the equation of state of real gases. As it is known from elementary physics, all the fluids obey to the same equation of state once that the thermodynamic variables, pressure (P), temperature (T) and volume (V), are scaled in terms of the correspondent critical variables P_c and T_c (see Sect. 15.1). The Van der Waals equation takes a *universal form*, independent on the microscopic parameters that control pressure and volume:

$$\mathbf{P} = f(\mathbf{T}, \mathbf{V}),$$

with $\mathbf{P} = P/P_c$, $\mathbf{T} = T/T_c$ (and $\mathbf{V} = V/V_c$, the volume being the inverse density). The law of the correspondent states can be considered an example of *universality* in the thermodynamic relationships.

It is conceivable that also at the phase transitions of the second order, when on approaching the critical temperature the correlation length tends to infinity, the thermodynamic behaviour becomes independent on the detailed parameters of the short range interaction. Then some kind of universal laws can be written, disregarding irrelevant terms and/or variables. This is also the root of the so-called *renormalization group theory* (RG) which has been devised by *Wilson* in order to determine the partition function or the critical exponents very close to T_c by means of recursion relations.

To give some remarks about the property of scale invariance and an example of class of universality of the thermodynamic potentials, let us return to the free energy for homogeneous system (Eq. 15.19):

$$f[m, T] = f_0(T) + a_0(T - T_c)m^2 + \frac{1}{2}b_0 m^4 + \cdots \qquad (A.17.2.1)$$

From $h = (\partial f/\partial m)_T$ the equation of state involving the field h and the magnetization m around the transition is written

$$h = 2a_0 T_c \varepsilon m + 2b_0 m^3, \qquad (A.17.2.2)$$

with $\varepsilon = (T - T_c)/T_c$, namely in the form

$$h = q \left[2 \left(\frac{m}{p} \right) + 4 \left(\frac{m}{p} \right)^3 \right], \qquad (A.17.2.3)$$

having defined new factors $p = (2a_0 T_c \varepsilon/b_0)^{1/2}$ and $q = (a_0 T_c \varepsilon)^{3/2}/(b_0/2)^{1/2}$.
From Eq. (A.17.2.3) it is noted that

$$\left(\frac{m}{p} \right) = g_\pm(h/q) \qquad (A.17.2.4)$$

where $g_\pm$ is universal function (g_- corresponding to $\varepsilon < 0$, g_+ to $\varepsilon > 0$), namely the same function for any system (provided with the same symmetry properties). In other words, when the magnetization is scaled by $\varepsilon^{1/2}$ it is no more a function of temperature and field separately but it becomes a function only of the ratio $h/\varepsilon^{3/2}$. Then the equation of state A.17.2.3 involves a single independent variable instead of two, ε and h. In a similar way it is possible to rewrite the free energy A.17.2.1 in terms of (h/q) and (m/p) to get a universal function, with $f[m, T]$ scaled by $pq \propto \varepsilon^2$ in terms of the variable $(h/\varepsilon^{3/2})$.

The possibility to rescale the independent variables of a function in a way to decrease their number is a characteristic of the homogeneous functions. By following a conjecture by Widom we reformulate the scaling properties sketched above by writing the equation of state in the form

$$h = m\psi(\varepsilon, m^{1/\beta}) \qquad (A.17.2.5)$$

where ψ is homogeneous function of degree γ, meaning $\psi(\lambda x, \lambda y) = \lambda^\gamma \psi(x, y)$. Therefore, if one knows the value of the function at the point x_0, y_0 and the degree of homogeneity γ as well, then the function is known everywhere. This is the condition of scale invariance analogous to the one described by the equation of correspondent states: the equation of state does not change (it has the same functional form) when the thermodynamic variables are scaled by any given quantity λ to a certain exponent. Thus, Eq. A.17.2.5 indeed takes universal form, being invariant under the scale transformations

$$\varepsilon \to \lambda\varepsilon, \quad m \to \lambda^{1/\beta}m \quad \text{and} \quad h \to \lambda^{\gamma+1/\beta}h.$$

In this way the critical behaviour appears to be controlled by two parameters only, β and γ. Again in analogy to the law of correspondent states, according to the scaling hypothesis (which has been supported by more advanced theories as the renormalization group, by solutions of exact models and particularly by experiments) one can speculate that for a certain class of materials (for instance having the same dimensionality D of the lattice and d of the order parameter, see Sect. 17.1) universal behaviour occurs, with a minimum number of free parameters. In particular, all the critical exponents should be related to the two, β and γ, defined above.

Now we are going to show that β and γ indeed are critical exponents, namely the same phenomenological exponents introduced at Sect. 15.1 and already derived in the MFA scenarios for ferroelectrics and for magnetic systems. In fact, the spontaneous magnetization m_s is solution of the implicit equation

$$h(\varepsilon, m_s) = 0$$

(return for similarities at Sect. 16.4 and Eqs. (17.6)–(17.10)). Then from Eq. (A.17.2.5) we set

$$\psi(\varepsilon, m_s^{1/\beta}) = 0$$

ψ being homogeneous, i.e. $\psi(\varepsilon, m^{1/\beta}) = \lambda^{-\gamma}\psi(\lambda\varepsilon, \lambda m^{1/\beta})$, so that $\psi(\lambda\varepsilon, \lambda m^{1/\beta}) = 0$. The solution being a function of ε only, one can write

$$m_s^{1/\beta} = g(\varepsilon)$$

and in analogous way for the scaled equation $\lambda m_s^{1/\beta} = g(\lambda\varepsilon)$. Therefore g must be homogeneous of degree 1 and thus

$$m_s \propto (\varepsilon)^\beta.$$

Analogous demonstration can be carried out for the critical exponent γ, that controls the temperature dependence of the isothermal susceptibility for evanescent field. From Eqs. A.17.2.3 and A.17.2.5, the derivative of h with respect to m yields the inverse of the susceptibility:

$$\chi^{-1}(h \to 0) = \left(\frac{\partial h}{\partial m}\right)_T = \left[\psi(\varepsilon, m^{1/\beta}) + m\left(\frac{\partial \psi(\varepsilon, m^{1/\beta})}{\partial m}\right)\right]_{h\to 0} \quad \text{(A.17.2.6)}$$

For $T > T_c$ $m = 0$ and then

$$\chi^{-1}(h \to 0) = \psi(\varepsilon, 0) = \lambda^{-\gamma}\psi(\lambda\varepsilon, 0).$$

By setting the scaling factor $\lambda = \varepsilon^{-1}$ one derives

$$\chi^{-1}(h \to 0) = \varepsilon^{\gamma}\psi(1, 0)$$

and since $\psi(1, 0)$ is temperature-independent, γ is the *critical exponent* for χ.

Below T_c it can be proved, by starting from Eq. (A.17.2.6), that the same critical exponent controls the temperature dependence of χ for $T \to T_c^-$.

A comprehensive presentation of the scaling theory and of the related consequences on the critical behaviour at the phase transitions (in particular in regards of the relationships among the various critical exponents) can be found in the book by *Stanley*.

Problems

Problem 17.1 Derive Eq. (17.25).

Solution: The expansion of the Brillouin function in the mean field approximation reads

$$M_A = \frac{N\mu_J}{2}\left[B_J(x_0) + \frac{\mu_J}{k_B T}\left(H_{ext} + \lambda(M_B - M_0) - \alpha(M_A - M_0)\right)|\frac{dB_J}{dx}|_{x_0}\right]$$

and in analogous way

$$M_B = \frac{N\mu_J}{2}\left[B_J(x_0) + \frac{\mu_J}{k_B T}\left(-H_{ext} + \lambda(M_A - M_0) - \alpha(M_B - M_0)\right)|\frac{dB_J}{dx}|_{x_0}\right]$$

with M_0 the magnetization in zero field. Being the magnetization $M = (M_A - M_B)$, the susceptibility (M/H_{ext}) turns out as in Eq. (17.25).

Problem 17.2 By means of quantum mechanical procedure re-derive Eq. (17.31).

Solution: The ground state of a ferromagnet consists of all the spins ($S = 1/2$) along the z-direction and the Heisenberg Hamiltonian reads

$$\mathcal{H} = -2J\sum_i\left[S_z^i S_z^{i+1} + \frac{1}{2}\left(S_+^i S_-^{i+1} + S_-^i S_+^{i+1}\right)\right].$$

The eigenvalue for the ground state is $-NS^2J$. When an excitation arises a given spin at the site j is flipped and as a consequence the total spin of the system is changed by $1/2 - (-1/2) = 1$. By applying the Hamiltonian one has

$$\mathcal{H}|j> = 2\Big[(-NS^2J + 2SJ)|j> -SJ|j+1> -SJ|j-1>\Big],$$

showing that it is not an eigenstate.

The Hamiltonian can be diagonalized by looking for plane wave solutions

$$|p> = \frac{1}{\sqrt{N}} \sum_j |j> e^{i\mathbf{q}\cdot\mathbf{R}_j}$$

the total spin of the state $|p>$ being $(NS-1)$. Thus $\mathcal{H}|p> = E(p)|p>$ with $E(p) = -2NS^2J + 4JS[1 - cos(qa)]$, implying energy of the excitation as in Eq. (17.31).

The present treatment follows the one by *Blundell*. A comprehensive description of the spin waves in terms of response to the time-dependent and space-dependent external field can be found at Chap. 6 of the book by *White*.

Problem 17.3 Derive the q-dependent static magnetic susceptibility $\chi(\mathbf{q}, 0)$ for delocalized, non-interacting electrons (the Fermi gas) in lattice dimensions $D = 3, 2$ and 1. In this latter case comment how the divergence of $\chi(q, 0)$ at $q = 2k_F$ implies a spin density wave instability (*Kohn anomaly*).

Solution: $\chi(\mathbf{q}, 0)$ is the response function to a static spatially varying magnetic field yielding a perturbation $\mathcal{H}_P = g\mu_B\mathbf{S}\cdot\mathbf{H}cos(\mathbf{q}\cdot\mathbf{r})$. Since the eigenstates of the Fermi gas Hamiltonian are $|\mathbf{k}> = exp(i\mathbf{k}\cdot\mathbf{r})$, one immediately realizes that for any $q \neq 0$ $\mathcal{H}_P$ does not give any first order correction. In fact, the cosine term yields non-zero matrix elements of the form $< \mathbf{k}|\mathcal{H}_P|\mathbf{k}\pm\mathbf{q} >$, connecting an initial state $|\mathbf{k}>$ with a final state $|\mathbf{k}\pm\mathbf{q}>$. Then, from second order perturbation theory, by weighting for the probability that the initial state is occupied with Fermi-Dirac distribution function $f_\mathbf{k}$ and that the final state must be empty, one derives

$$\Delta E = -\frac{\mu_B^2 H^2}{2}\sum_\mathbf{k}\left[\frac{f_\mathbf{k}(1-f_{\mathbf{k}+\mathbf{q}})}{E_{\mathbf{k}+\mathbf{q}} - E_\mathbf{k}} + \frac{f_\mathbf{k}(1-f_{\mathbf{k}-\mathbf{q}})}{E_{\mathbf{k}-\mathbf{q}} - E_\mathbf{k}}\right],$$

with $E_\mathbf{k}$ the free-electron dispersion curve. Since the sum runs over all $|\mathbf{k}>$ states it is possible to replace $\mathbf{k}$ with $\mathbf{k}+\mathbf{q}$ in the second term and taking the second derivative of the energy with respect to the magnetic field one derives

$$\chi(\mathbf{q}, 0) = \mu_B^2 \sum_\mathbf{k} \frac{f_\mathbf{k} - f_{\mathbf{k}+\mathbf{q}}}{E_{\mathbf{k}+\mathbf{q}} - E_\mathbf{k}}$$

(for the derivation of Eq. (17.13) see the book by White). The sum depends on the lattice dimensionality. The above equation shows that one should expect a large contribution to $\chi(\mathbf{q}, 0)$ at those wave-vectors $\mathbf{q}$ connecting a large number of initial filled states with quasi-degenerate empty final states. This happens if $\mathbf{q}$ connects large portions of the Fermi surface or, in other terms, if $\mathbf{q}$ is a *nesting wave-vector*. In particular, this occurs in quasi-1D systems for $q = 2k_F$, where a divergence is present, as shown below.

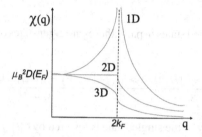

Problem 17.4 Starting from the Hubbard Hamiltonian (Eq. 17.3) show that for half-filled band in the $U \gg t$ limit one derives an effective antiferromagnetic exchange coupling as in Eq. (17.4).

Solution: For $U \gg t$ the hopping term $\mathcal{H}_t$ of the Hamiltonian in Eq. (17.3) is a perturbation of the term $\mathcal{H}_U$ involving U. The energies of the triplet and singlet configurations for two electrons on neighbouring sites i and j are

$$|S = 1, M_S = 1 > = | +_i +_j >, \quad |1, -1 > = | -_i -_j >,$$

$$|1, 0 > = \frac{1}{\sqrt{2}}\left(| +_i -_j > + | -_i +_j > \right)$$

and

$$|0, 0 > = \frac{1}{\sqrt{2}}\left(| +_i -_j > - | -_i +_j > \right).$$

These four states are characterized by single occupancy and are eigenstates of $\mathcal{H}_U$ with eigenvalue $E_n^0 = 0$. Now consider the effect of $\mathcal{H}_t$ on these four states. Since the electron hopping driven by $\mathcal{H}_t$ leads to double occupancy it will generate states orthogonal to these four states and no first order correction. At the second order, i.e.

$$\Delta E_n^{(2)} = \sum_{m \neq n} \frac{| < m|\mathcal{H}_t|n > |^2}{E_n^0 - E_m^0},$$

$\mathcal{H}_t$ applied to $|1, \pm 1 >$ triplet states yields zero correction since it would lead to double occupancy of a given site with same spin orientation. On the other hand

$$\mathcal{H}_t| +_i -_j > = t(| +_i -_i > + | +_j -_j >), \text{ while}$$

$$\mathcal{H}_t| -i +j > = -t(| +i -i > +| +j -j >),$$

where the change of sign is associated with the different order of the spin orientations. By taking into account of the sign reversal one realizes that $< m|\mathcal{H}_t|1, 0 > = 0$, thus none of the triplet states show second order correction. For the singlet $|0, 0 >$ state one has

$$\mathcal{H}_t|0, 0 > = \frac{2t}{\sqrt{2}}(| +i -i > +| +j -j >).$$

Since the doubly occupied states in parentheses are eigenstates of $\mathcal{H}_U$ with eigenvalue U it turns out

$$\Delta E_{|0,0>}^{(2)} = \frac{| < +i -i |\mathcal{H}_t|0, 0 > |^2}{0 - U} + \frac{| < +j -j |\mathcal{H}_t|0, 0 > |^2}{0 - U} = -\frac{4|t|^2}{U},$$

showing that the energy of the singlet state is lowered by $4|t|^2/U$ with respect to the one of the triplet state. So, for $U \gg t$ Eq. (17.4) holds.

Problem 17.5 From the dispersion curve for the magnons in isotropic (Heisenberg) ferromagnet (Eq. 17.31) show that no magnetic order can be present in two-dimensions (2D), while in the presence of single ion anisotropy magnetic order can exist.

Solution: From Eq. (17.32) the number of magnons excited at a given temperature is given by

$$n_m = \int_0^\infty \frac{D(\omega)}{e^{\hbar\omega/k_B T} - 1} d\omega,$$

with $D(\omega)$ the density of states. The dispersion relation for a ferromagnet in the $q \to 0$ limit is quadratic in q, in 2D $D(\omega)$ is constant. Hence

$$n_m \propto \frac{k_B T}{\hbar} \int_0^\infty \frac{1}{e^x - 1} dx$$

with $x = \hbar\omega/k_B T$. For $x \to 0$ the integrand is $1/x$, giving rise to a logarithmic divergence in the number of excited magnons and then to the disruption of the magnetization. On the other hand, if anisotropy is present there will be a minimum energy cost in exciting a spin wave and an energy gap Δ in the dispersion curve arises. Then the previous equation can be written

$$n_m \propto \frac{k_B T}{\hbar} \int_{\Delta/k_B T}^\infty \frac{1}{e^x - 1} dx,$$

the logarithmic divergence is truncated and a magnetic order can set in.

Problem 17.6 Consider a domain wall in a ferromagnet and show that the size is related to the magnetic anisotropy and to the exchange coupling.

Solution: Consider a π rotation of the magnetic moments across the domain wall. Then, for a domain wall of N lattice steps, with $N \gg 1$, the cost in exchange energy is

$$E_{exc} = N J S(S+1) \left(\frac{\pi}{N} \right)^2.$$

The cost for the anisotropy energy is given by $E_{an} = K_{an} N$, with K_{an} a phenomenological constant accounting for the magnetic anisotropy. From the derivative of $E_{exc} + E_{an}$ with respect to N one derives

$$N = \left(\frac{\pi^2 J S(S+1)}{K_{an}} \right)^2.$$

Thus the thickness of the domain wall increases with J and decreases with the magnetic anisotropy.

Problem 17.7 For a one-dimensional antiferromagnet with elastic coupling among the nearest neighbour magnetic ions, show that the spatial dependence of the exchange coupling favours a lattice distortion.

Solution: By including in the Heisenberg Hamiltonian an elastic coupling characterized by a constant k_{el}, for small displacements from the equilibrium configuration, the energy variation for a pair of adjacent spins turns out

$$\Delta E = \frac{1}{2} k_{el} \Delta x^2 - \left(\frac{\partial J}{\partial x} \right)_0 S(S+1) \Delta x.$$

ΔE is minimized for

$$\Delta x = \frac{\left(\frac{\partial J}{\partial x} \right)_0 S(S+1)}{k_{el}},$$

namely for any finite change of the exchange coupling with the distance. In quasi-one-dimensional antiferromagnets (e.g. $CuGeO_3$) a structural dimerization is present at the so-called Spin-Peierls transition.

Problem 17.8 Show that in antiferromagnet the static uniform susceptibility for magnetic field perpendicular to the sublattice magnetization is given by Eq. (17.26).

Solution: The magnetic field yields the canting of the sublattice magnetizations as shown below (return to Fig. 17.4):

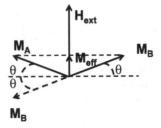

One writes the energy as

$$E = \lambda \mathbf{M}_A \cdot \mathbf{M}_B + \alpha \mathbf{M}_B \cdot \mathbf{M}_B + \alpha \mathbf{M}_A \cdot \mathbf{M}_A - \mathbf{H}_{ext} \cdot (\mathbf{M}_A + \mathbf{M}_B)$$

with λ, $\alpha > 0$. For small θ values, by indicating with M the absolute value of the sublattice magnetizations, one has

$$E = -\lambda M^2 \left[1 - \frac{1}{2}(2\theta)^2 \right] + 2\alpha M^2 - 2M H_{ext}\theta,$$

which is minimized for $\theta = H_{ext}/(2\lambda M)$, α-independent. The transverse susceptibility is $\chi_\perp = M_{eff}/H_{ext} = 2M\theta/H_{ext}$ and Eq. (17.26) follows.

Specific References and Further Reading

1. S.J. Blundell, *Magnetism in Condensed Matter*, (Oxford Master Series in Condensed Matter Physics, Oxford U.P., 2001).
2. N.W. Ashcroft and N.D. Mermin, *Solid State Physics*, (Holt, Rinehart and Winston, 1976).
3. D.C. Johnston, *Handbook of Magnetic Materials Vol.10*, Ed. K.H.J.Buschow, Chapter 10, (Elsevier, 1997).
4. H. Stanley, *Introduction to Phase Transitions and Critical Phenomena*, (Oxford University Press, Oxford, 1971).
5. M. White, *Quantum Theory of Magnetism*, (McGraw-Hill, 1970).
6. G. Burns, *Solid State Physics*, (Academic Press Inc., 1985).
7. J.M.D. Coey, *Magnetism and Magnetic Materials*, Cambridge University Press, Cambridge (2009).
8. D.L. Goodstein, *States of Matter*, (Dover Publications Inc., 1985).
9. G. Grosso and G. Pastori Parravicini, *Solid State Physics*, 2nd Edition, (Academic Press, 2013).
10. A.P. Guimares, *Magnetism and Magnetic Resonance in Solids*, (J. Wiley and Sons, 1998).
11. H. Ibach and H. Lüth, *Solid State Physics: an Introduction to Theory and Experiments*, (Springer Verlag, 1990).
12. C. Kittel, *Introduction to Solid State Physics*, 8th Edition, (J. Wiley and Sons, 2005).
13. K.A. Müller and A. Rigamonti (Editors), *Local Properties at Phase Transitions*, (North-Holland Publishing Company, Amsterdam, 1976).

Chapter 18
Superconductors, the Superconductive Phase Transition and Fluctuations

Topics

Phenomenology and Main Experimental Aspects
Cooper Pairs
Special Meaning of the Superconductive Wave Function
Meissner Effect and London Penetration Length
Flux Quantization in Rings
Josephson Junction and SQUID
Type II Superconductors
High Temperature Superconductors
Ginzburg-Landau Theory
Flux Lines
Superconductive Transition, Fluctuations and Diamagnetic Effects
Nanoparticles, Zero-Dimensional Condition and Critical Region

18.1 Historical Overview and Phenomenological Aspects

In the following the main steps in the study of the superconductive state along the twentieth century are schematically collected.

The first detection of the sudden drop to zero-resistance on cooling Mercury[1] to about 4 K goes back to *Kamerlingh-Onnes*, in 1911. A few years later the super-

---

[1]The use of Hg, a liquid metal at room temperature, was related to the possibility to perform a series of distillation processes at relatively low temperature, since at that time a role of the impurities in levelling to a finite value the resistivity in metals at very low temperatures was suspected. Kamerlingh-Onnes first claimed that the drop to zero of the resistance supported his point of view, that without the impurities the resistance had to tend to zero on cooling towards zero temperature. Then he corrected himself, observing that in reality a new state, that he called superconductivity, was induced on crossing the critical temperature of about 4.2 K.

© Springer International Publishing Switzerland 2015
A. Rigamonti and P. Carretta, *Structure of Matter*,
UNITEXT for Physics, DOI 10.1007/978-3-319-17897-4_18

conductivity was found in other metals (Sn at the temperature $T_c = 3.2\,\text{K}$, Pb at $T_c = 7.2\,\text{K}$ and Nb at $T_c = 9.1\,\text{K}$). In 1933 the *Meissner effect* (repulsion of a magnetic field from the sample on crossing the superconducting transition temperature T_c), was discovered. While T_c was found to increase up to about $20\,\text{K}$ in intermetallic alloys,[2] the *London brothers* (1935) were able to formulate a first quantitative description of the properties of the superconducting state. For reasons to be pointed out subsequently, the *London theory*, based on the existence of superfluid electrons, had substantial successes in spite of the fact that no quantum mechanics was used and no knowledge of the pairing mechanism was at hand.

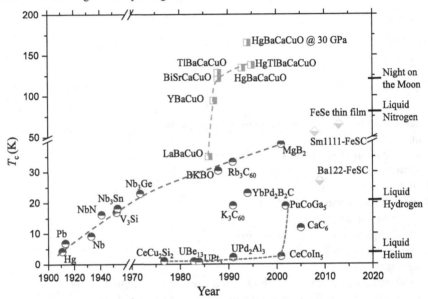

The first steps towards the microscopic theory of superconductivity, around the years 1950, can be attributed to *Frölich*, for the observation of the *isotope effect* (namely the modification of T_c upon isotopic substitution) and to *Cooper*, for the concept of novel pseudo-particles, the *Cooper pairs*, resulting from the pairing in the singlet state ($S = 0$) of two electrons. Meantime a powerful semi-phenomenological theory was developed by *Ginzburg* and *Landau*, theory which later on turned out to have deep microscopic roots. At the same time, that approach allowed *Abrikosov* to formulate a predictive theory on the existence of a novel type of superconductivity, namely the *type II superconductors*, implying high critical fields and critical currents.

In 1955 *Bardeen*, *Cooper* and *Schrieffer* published the famous *BCS theory*, a comprehensive quantum-mechanical description of the superconducting state. In the early sixties *Josephson* predicted the possibility that the Cooper pairs could jump

---

[2]The increase of the superconducting temperature and of the critical fields and critical currents in the alloys is related to the reduction of the coherence length accompanying the decrease of the *mean free path of the electrons*. The alloying process transforms the metallic superconductors from first-type to second type (Sects. 18.7 and 18.9).

across an insulating layer, thus extending the superconductivity to a system known as *Josephson junction*, extremely sensitive to magnetic field. That discovery initiated the field of the *superconductive electronics* and of a variety of electromagnetic devices with extraordinary properties.

Meantime the critical temperature T_c, evidently a crucial parameter for the applications, was increasing to a slow rate (see sketch above): up to about 1980 the record was around 23 K, for Nb-based alloys. An unexpected and sudden jump in T_c happened in 1986, when *Bednorz* and *Muller* discovered the superconductivity in some Lanthanum-Barium *Copper oxides* (nowadays the family is known as *cuprates*, or high temperature superconductors, *HTcSC*) that not only were poorly conducting at room temperature but also included a magnetic ion (Cu^{2+}, at $S = 1/2$), a feature that was previously thought to be detrimental for superconductivity. With proper substitution of some atoms (thus increasing the internal chemical pressure) just one year later *Chu* was able to overcome the liquid nitrogen temperature with the Yttrium-Barium-Copper oxide (nowadays known with the acronyms *YBCO*). In other *Hg based cuprates* the transition temperature can be increased up to 164 K, under a pressure of about 30 GPa. Other compounds that can be considered HTcSC are MgB_2 ($T_c = 39$ K, where on the other hand the pairing mechanism basically is of the form of the conventional BCS scenario) or the *Fe-based superconductors* (discovered in 2008) which, similarly to the cuprates, are characterized by a layered structure and where the pairing mechanism is still under debate (as substantially it remains for the cuprate family as well).

It is remarkable that several crystals with a poor room temperature conductivity do attain the superconducting state while very good conductors, such as Cu, Au or Ag (as well as the magnetic metals Fe or Ni) do not exhibit superconducting transition down to very low temperature.

The most striking phenomenological aspect of the transition from the ordinary to the superconducting state probably is the sudden drop to zero[3] of the resistivity. In reality other dramatic phenomena occur around T_c, for instance the anomalous behaviour of the electronic contribution to the specific heat (see Sect. 12.7) as shown in Fig. 18.1. It should be remarked that in the temperature range involved in this figure the contribution to the specific heat from lattice vibrations (see Sect. 14.5) can be neglected.

The superconducting transition (which in the absence of external magnetic field is a *second order one*, with no latent heat, see Sects. 15.1 and 15.2) implies the crossover to a more ordered state. The behaviour of the specific heat below T_c is typical of electronic systems with gap (see Sect. 12.7.2):

---

[3]The resistance does actually drop to zero, provided that competing processes like *flux flow of the vortices* are properly avoided. From precise measures by means of NMR spectra (see Sect. 6.3) of the magnetic field due to superconducting current launched in a solenoid, it can be proved that no appreciable variation would occur at least for a period of about 10^5 years. However the resistance is not zero for alternate current and in the presence of external magnetic field.

Fig. 18.1 Sketchy
behaviour of the specific heat
in the superconducting state
of *Aluminium*

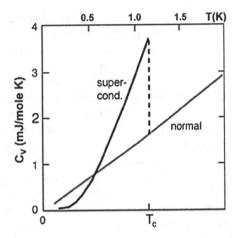

Fig. 18.2 Sketchy
behaviour of the temperature
dependence of the critical
field in *Lead*, approximately
of the form
$H_c(T) = H_c(0)[1 - (T/T_c)^2]$

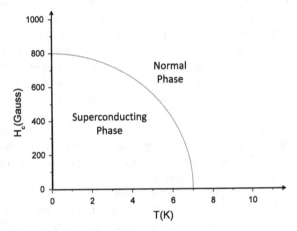

$$C_p \propto exp[-\Delta/k_B T],$$

with Δ turning out numerically around $k_B T_c$.

It is noted that a magnetic field can suppress the superconductivity (as indicated in Fig. 18.1, straight line). Analogous effect has an electric current, the *critical current* (in classical metals of the order of some hundreds of Amp per square mm, evidently by a mechanism similar to the one related to the external magnetic field).[4] A sketch of the temperature behaviour of the magnetic *critical field* for lead is given in Fig. 18.2.

---

[4]The maximum current in a wire can be related to the critical field. Approximately one has $I_{max} \simeq 5RH_c$, in Amp., the radius R of the wire being in cm and H_c in Oe.

The superconducting state is preserved in the presence of electromagnetic radiation, provided that the frequency is below a certain threshold. The response to the electromagnetic radiation involves the frequency dependence of the conductivity, which in turns controls, for instance, the reflection from a bulk sample or the transmission through thin films of the superconductor. Therefore, when an energy gap Δ in the electronic excitation spectrum opens up below T_c, the high frequency conductivity becomes remarkably different from the one in the normal state: no electronic absorption is observed, provided that $\hbar\omega \leq \Delta$. On the contrary, above this threshold the behaviour of the conductivity is the one for the normal state.

For temperature well below T_c the threshold value of the electromagnetic frequency turns out of the order of $k_B T_c/\hbar \sim 10^{12}$ rad s^{-1}, in the range of microwave or far infrared frequencies. At the optical frequencies the behaviour of the superconductor becomes the same as in the normal state.

The interplay of superconductivity with associated *total conductivity* in regards of the magnetic response, in particular the *perfect diamagnetism*, is worthy of further comment.

Let us refer to an ideal metal that becomes a perfect conductor below a given temperature, in the absence of any external field H_{ext}. Below that temperature the magnetic field is applied: inside the material B stays zero, since the induced currents cause a field that according to the Lenz law is against the one applied. Therefore one has *total screening*. One could think that the total diamagnetism, the *Meissner effect*, could just be due to the occurrence of zero resistivity ρ. In fact in this case the electric field $\mathcal{E}$ is *necessarily zero* (to avoid the catastrophic increase in the electron velocity in view of the relation[5] $\mathcal{E} = \rho\mathbf{j}$, for finite current density $\mathbf{j}$). Then, from

$$rot\mathcal{E} = -\frac{1}{c}\frac{\partial \mathbf{B}}{\partial t}, \tag{18.1}$$

one infers that the magnetic field $\mathbf{B}$ inside the sample *cannot change*. Therefore in a perfect conductor at zero electric resistance below the so-defined T_c the application of an external field cannot induce $B \neq 0$. However, since $\mathbf{B}$ inside the sample cannot change in a perfect conductor one could have two possible states[6] below T_c: one at

---

[5]This equation has to be taken with a certain care since in its general form it is valid only in stationary condition, while a variation of the magnetic field launches a transitory current. On the other hand the currents launched in the transient circulate in a narrow outer sheet of the material and the electric field in the bulk has to be considered zero.

[6]Thus in the condition of total conductivity only, one would not have a single thermodynamic state and the equilibrium thermodynamics could not be applied. This is another relevant difference between the total conductor and the superconductor. In the superconductor the magnetic field is anyway repelled, irrespective of the order of the operations, cooling in the magnetic field or applying the field below T_c (the field to be considered is smaller than the critical field).

$B = 0$ and the second at $B \neq 0$, according to the order of the cooling process and of the field application, as schematically described below:

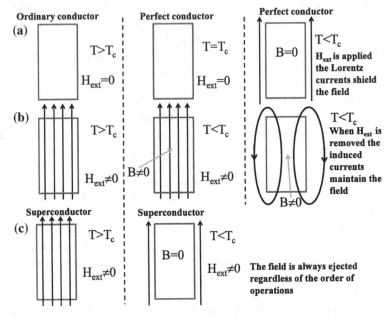

Therefore the superconductor *auto-generates* the currents exactly of the strength required in order to expel the field H_{ext}, either on crossing T_c on cooling or by applying the field below T_c.

Since $B = \mu H_{ext} = [1 + 4\pi\chi]H_{ext}$, the $B = 0$ condition implies

$$\chi = -\frac{1}{4\pi},\qquad(18.2)$$

and the superconductor can be defined as a *superdiamagnet*, with the maximum diamagnetic susceptibility[7] χ and no dependence from the magnetic history. It is reminded that according to the derivations at Chaps. 4 and 13, the volume diamagnetic susceptibility in ordinary insulating or metallic compounds is around 10^{-6}, a factor 10^5 smaller than in the superconducting state. The electromagnetic properties of the superconductor can be derived in the framework of phenomenological descriptions based on the equilibrium thermodynamics, as carried out by *Heinz* and *Fritz London*. The *London equations* (to be discussed a t Sect. 18.3) seemed to indicate that all the

---

[7]Such a strong value of the diamagnetic susceptibility makes possible the phenomenon of the *diamagnetic levitation* of large masses. It can be remarked since now that for strongly type II superconductors, as the *HTcSC cuprates*, a similar impressive phenomenon is the *diamagnetic suspension*: the superconductor follows a magnet lifted up, thus appearing magically suspended.

current carriers (in 1935 they were thought the superconducting electrons) should be found in the same macroscopic quantum state, evidently in contrast with the distribution (Sect. 12.7.1) expected according to the Fermi-Dirac statistics.

18.2 Microscopic Properties of the Superconducting State

As recalled in the historical overview, significant progresses towards the understanding of the microscopic properties of the conventional superconductors (metals and alloys) go back to the early 1950. The discovery of the isotope effect (emphasizing the *role* of *lattice vibrations*) and the understanding of the pairing mechanisms between two electrons (for which the electron-phonon interaction acts as glue for the Cooper pseudo-particles) paved the way to the BCS theory, that can be considered one of the most impressive solutions of many-body problems in condensed matter. Being more interested to the aspects of the superconducting phase transition in terms of the Landau-type statistical description, we shall not go into the details of the BCS theory, for which one could read, among the many texts dealing with solid state theory, the comprehensive treatments achieved by *Grosso* and *Pastori Parravicini* or by *Annett*. Rather we shall first (Sect. 18.2.1) clarify how two electrons can pair in a singlet state as a consequence of an *arbitrarily weak* attractive interaction related to the polarization of the lattice (the second electron "follows" the first one through the polarization tail that this latter leaves in the positive ions of the crystal[8]). Then in the state of correlated electrons each Cooper pair has the momentum of any other, irrespective of the momenta $\hbar\mathbf{k}$ of the individual electrons. This pair, at total crystal momentum $k = 0$ and total spin $S = 0$, then a boson, is the key to overcome the problem of the Pauli principle, since it grants the possibility of condensation in a common state, the BCS state. Then (Sect. 18.2.2) the main properties of this state shall be recalled. Finally the meaning of the wave function of all the Cooper pairs and the related physical consequences will be emphasized (Sect. 18.2.3).

18.2.1 The Cooper Pair

Let us consider the Bloch-type wave function of two independent electrons (see Chap. 12)

$$\phi(\mathbf{k}_1, \mathbf{k}_2) = \phi(\mathbf{k}_1)\phi(\mathbf{k}_2), \tag{18.3}$$

---

[8]One could observe that an electron, after having caused a local shrinking of the positive lattice, leaves the place at a speed of the order of the Fermi velocity while the distortion evaporates in much longer times, the process involving the sound velocity.

neglecting for the moment the spin part. After a scattering process involving a quantum vibrational excitation according to the sketch below

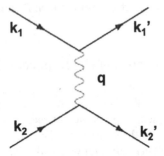

the wave function becomes $\phi(\mathbf{k_1'})\phi(\mathbf{k_2'})$ and the total momentum $\mathbf{b}$ is conserved:

$$\mathbf{k_1'} + \mathbf{k_2'} = \mathbf{k_1} + \mathbf{k_2} = \mathbf{b}.$$

The energies E and E' of the initial and final states cannot differ more than by the typical phonon energy $\hbar\omega_D$, ω_D being the Debye frequency (see Sect. 14.3.3). Furthermore the available states must be around the Fermi level E_F, namely must have $k \simeq k_F$. On the other hand, from $E_k = \hbar^2 k^2 / 2m$, it turns out that the change in energy implies a change in momentum of about

$$\delta k = |\mathbf{k_1} - \mathbf{k_1'}| \simeq \frac{m\omega_D}{\hbar k_F} \ll k_F. \tag{18.4}$$

As sketched in Fig. 18.3 the number of electrons satisfying the above conditions is *maximized when* $\mathbf{b} = 0$. Therefore the best situation to decrease the energy occurs when

$$\mathbf{k_2} = -\mathbf{k_1} \quad \text{so that} \quad \mathbf{b} = 0.$$

Thus one can write the two-particles wave function as

$$\psi(\mathbf{r_1} - \mathbf{r_2}) \equiv \psi(\mathbf{r}) = \frac{1}{V_{crystal}} exp[i\mathbf{k} \cdot (\mathbf{r_1} - \mathbf{r_2})].$$

Fig. 18.3 The volume measuring the states available is about $V = 2\pi k_F sin\theta (\delta k)^2$, maximized when $\mathbf{b} = 0$, in this case becoming $V = 4\pi k_F^2 \delta k$

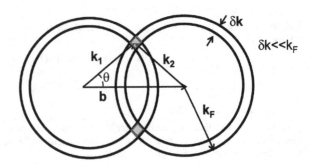

For the Schrodinger equation

$$\left[-\frac{\hbar^2}{2m}(\nabla_1^2 + \nabla_2^2) + V(\mathbf{r}_1 - \mathbf{r}_2)\right]\psi(\mathbf{r}) = (2E_F + \varepsilon)\psi(\mathbf{r}) \qquad (18.5)$$

and by expanding the wavefunction in the $\mathbf{k}$-space in the form $\psi(\mathbf{r}) = \sum_{\mathbf{k}} g_{\mathbf{k}} exp(i\mathbf{k} \cdot \mathbf{r})$, from Eq. (18.5) it can be shown (see Problem 18.2) that the pair is bounded, i.e. $\varepsilon < 0$, the new state thus having energy less than $2E_F$. The total spin of the pair will depend on the symmetry of the orbital wave function. In many cases for repulsive interactions (e.g. e^2/r_{12} in the Helium atom, Sect. 2.3) the lowest energy state is the triplet. One argues that since here the electron-electron interaction is attractive, the *singlet state* is favoured. Thus the complete, spin part included, wave function for the pair is written

$$\psi(\mathbf{r}) = \phi_\uparrow(\mathbf{k})\phi_\downarrow(-\mathbf{k}), \qquad (18.6)$$

with two electrons having energy below $2E_F$ and opposite momenta and spin. Beyond this simple picture for the pair, the more rigorous description accounting for the correct antisymmetry of the wave function and taking into account the matrix elements for the electron-phonon-electron interaction, allows one to obtain the decrease of the energy in the form

$$\Delta E \simeq -2k_B\Theta_D exp[-2/D(E_F)V], \qquad (18.7)$$

$D(E)$ being the density of states, V the interaction energy for two electrons around the Fermi surface (see Sect. 12.7.1) and Θ_D the Debye temperature (Sect. 14.5) (see Problem 18.2).

18.2.2 Some Properties of the Superconducting State

As a consequence of the pairs formation the Fermi gas is unstable towards an attractive potential, even very weak. Therefore, according to the general concepts described at Chap. 15, a phase transition has to occur. The superconductive state involves the pairing of a sizeable part of about 10^{22} electrons per cubic centimeter. The pairs at $S = 0$ obey the Bose-Einstein statistics and all can set in a single quantum state, while single electrons are continuously scattered in different single-electron states in the momentum range δk (see Eq. (18.4)). The BCS theory describes a variety of properties of the superconducting state resulting from the transition driven by pair condensation. We are going to summarize a few of these properties.

Somewhat consistent with Eq. (18.7) the transition temperature turns out

$$T_c = \frac{1.14\hbar\omega_D}{k_B} exp[-1/D(E_F)V] \simeq \theta_D exp[-1/\lambda], \qquad (18.8)$$

Fig. 18.4 Density of states in a superconductor in comparison to the one in an ordinary metal, at zero temperature and at finite temperature. The dashed area indicates $D(E)$ times the electron distribution function. In the superconducting state, on measuring energies from E_F, one can write to a good approximation that $D(E) = D(0)E/\sqrt{E^2 - \Delta^2}$, for $|E| \geq \Delta$

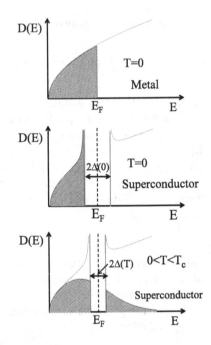

where we have introduced the *coupling factor* $\lambda = D(E_F)V$, which usually falls in the range 0.1–0.5. Therefore T_c is expected a small fraction of the Debye temperature. The isotope effect is evidently implicit in this term.

In the sketch in Fig. 18.4 the density of states in the superconductor is compared to the one in the metallic state (see Sect. 12.7), at zero and at finite temperature (in zero external magnetic field). The dashed areas correspond to occupied states. 2Δ is the lowest energy required in order to carry electrons from the superconducting to the ordinary state. The theory provides the temperature dependence of the gap $\Delta(T)$ through a self-consistent implicit equation that can easily be solved in two limits. In the neighbourhood of the transition temperature one has

$$\Delta(T) \simeq \Delta(0)\left(1 - \frac{T}{T_c}\right)^{1/2}, \tag{18.9}$$

while for $T \to 0$

$$\Delta(0) = 2\hbar\omega_D exp[-1/\lambda].$$

Thus, from Eq. (18.8) one can write

$$2\Delta(0) = 3.5 k_B T_c. \tag{18.10}$$

This relevant relation is rather well obeyed for *weak coupling* (namely systems at small V) while it breaks down for strong electron-phonon interaction. The extension of the BCS theory to the *strong coupling* situation,[9] carried out primarily by *Eliashberg* and by *Mc Millan*, leads to rather good agreement with the experimental findings for a variety of properties of most superconducting metals. The behaviour of the critical field and of the specific heat, as well as the Meissner effect, are well explained in microscopic terms. The fact that no electric resistance is involved in the transport can be crudely justified in terms of scattering process of the pair (Problem 18.1).

It should be remarked that the density of pairs in the BCS picture is expected fairly large. In fact, the complete superconducting wave function is the product of single-pairs wave functions of the form (18.6) and the correlation range between the two electrons of the pair is much larger than the lattice constant.[10] Therefore in the "region" of a pair a large number of centers of mass of other pairs do fall (see Problem 18.3).

18.2.3 The Particular Meaning of the Superconducting Wave Function

The large pairs density n_c in the same quantum state allows one to give a particular meaning to the superconducting wave function ψ_{SC}. In fact, because of this situation $\psi_{SC}^* \psi_{SC}$ becomes the real number of pairs per unit volume and the charge density in the superconducting state can be written $-2e\psi_{SC}^* \psi_{SC}$.

From the probability of presence $P(\mathbf{r}, t) = \psi_{SC}^*(\mathbf{r}, t)\psi_{SC}(\mathbf{r}, t)$ and the condition of charge continuity, one writes

$$\frac{\partial P}{\partial t} = \psi_{SC}^* \frac{\partial \psi_{SC}}{\partial t} + \psi_{SC} \frac{\partial \psi_{SC}^*}{\partial t} = -div\mathbf{j}, \tag{18.11}$$

$\mathbf{j}$ being the current density. On the other hand the Schrodinger equation for a particle of mass m^* and charge q in the presence of an electromagnetic field described by a

---

[9]The relationship between transition temperature T_c and the Debye temperature to a good approximation is provided by the semi-empirical formula due to McMillan

$$T_c = \frac{\Theta_D}{1.45} exp\left(\frac{-(\lambda^* + 1)}{[\lambda^* - \mu^*(1 + 0.62\lambda^*)]}\right),$$

where the electron-phonon coupling parameter λ^* typically varies in the range 0–2 while the repulsive Coulombic parameter μ^* is between 0.1 and 0.2. λ^* and μ^* are usually determined by means of electron *tunneling experiments*.

[10]At Sect. 18.9, in describing the *Ginzburg-Landau theory* in the general framework of phase transitions, it will be shown that the "correlation range" can be considered the equivalent of the *coherence length* introduced at Sect. 15.2. It is remarked that in most cases the distance over which the correlation is effective is limited to an almost temperature-independent *mean free path of the electrons*. The Cooper pair coherence length is smaller than the Ginzburg-Landau coherence length that we shall introduce at Sect. 18.9. Only when the *mean free path of the electrons* can be assumed infinite the two coherence lengths (Cooper and Ginzburg-Landau) coincide.

potential vector **A** reads

$$\frac{1}{2m^*}\left(-i\hbar\nabla - \frac{q\mathbf{A}}{c}\right)^2 \psi_{SC} + q\varphi\psi_{SC} = i\hbar\frac{\partial\psi_{SC}}{\partial t}, \qquad (18.12)$$

with $div\mathbf{A} = 0$ in the Coulomb gauge, $\varphi = 0$ and $-i\hbar\nabla = \mathbf{p}$ (see note 3 at Sect. 1.6). By comparing this equation with Eq. (18.11) one obtains for $\partial P/\partial t$ the divergence of a function, which according to Eq. (18.11), implies a current density due to the pairs given by

$$\mathbf{j} = \frac{1}{2m^*}\left[\left(\left[\mathbf{p} - \frac{q\mathbf{A}}{c}\right]\psi_{SC}\right)^* \psi_{SC} + \psi_{SC}^*\left(\left[\mathbf{p} - \frac{q\mathbf{A}}{c}\right]\psi_{SC}\right)\right]. \qquad (18.13)$$

For reasons again related to the large number of particles in the same state this expression for the current can be used to describe a charge current of macroscopic character. Therefore in the superconducting state both the charge density and the electrical current are directly connected to the wave functions of the pairs, which take a physical meaning of macroscopic, quasi-classical character, somewhat resembling the situation pertaining to the electromagnetic field in terms of the photons, other non-interacting bosonic particles.

Then the most general form for the wave function shall be written

$$\psi_{SC} = \sqrt{n_c(\mathbf{r})}exp[i\theta(\mathbf{r})], \qquad (18.14)$$

where one has the square root of the number of pairs per unit volume $n_c = |\psi_{SC}|^2$ and the general phase factor involving $\theta(\mathbf{r})$.[11]

In the superconducting state even the phase factor acquires a precise physical meaning. In fact, from Eq. (18.14) and by resorting to Eq. (18.13), one obtains

$$\mathbf{j} = \left[-\hbar\nabla\theta + \frac{2e\mathbf{A}}{c}\right]\frac{2en_c}{m^*} = e^*n_c v_{SC}, \qquad (18.15)$$

showing that the phase θ is involved through its gradient in part of the current or equivalently in the velocity $v_{SC} = [\hbar\nabla\theta + (e^*\mathbf{A}/c)]/m^*$ of the pair. One should remark that is just the "coincidence" of quantum parameters with classical charge and current that explains why in spite of the classical approach the London brothers have been able to derive results quantitatively correct (see Sect. 18.3).

---

[11]One could argue that this form of the wavefunction, being based on the condition of very large correlation range, could break down in superconductors where the range of the interaction in the pair (namely the correlation length) is a few lattice steps, as for instance in the high temperature superconductors (Sect. 18.8). In reality, the *Ginzburg-Landau theory*, valid both in the BCS superconductors at very large coherence length as well as in high T_c superconductors, is based on the inspiring guess that the density of superconducting carriers (the order parameter in the scenario of the phase transition, see Sect. 15.2) can be written as the modulus square of some effective wave function.

18.3 London Theory and the Flux Expulsion

The valuable equations that the London brothers derived before any quantum description of the superconducting state were based on some heuristic assumptions. When a certain fraction of electrons become superconducting the current density has to be written

$$\mathbf{j} = -n_e e \mathbf{v},$$

where $\mathbf{v}$ is an average velocity and n_e the density of these electrons. A *transient* electric field $\mathcal{E}$ (required in order to have $\partial \mathbf{B}/\partial t \neq 0$, according Eq. (18.1)) launches the supercurrent through

$$-e\mathcal{E} = m\frac{\partial \mathbf{v}}{\partial t}.$$

Then

$$\frac{d\mathbf{j}}{dt} = n_e \frac{e^2}{m}\mathcal{E}$$

and from Maxwell equation (see Eq. (18.1))

$$rot\frac{d\mathbf{j}}{dt} = n_e \frac{e^2}{m} rot\mathcal{E} = -n_e \frac{e^2}{mc}\frac{\partial \mathbf{B}}{\partial t}. \tag{18.16}$$

Here $\mathbf{B}$ is due to the external field and to the superconducting current. Therefore

$$\frac{\partial}{\partial t}\left[rot\mathbf{j} + n_e \frac{e^2}{mc}\mathbf{B}\right] = 0. \tag{18.17}$$

This equation in itself holds for any metal, but cannot explain the Meissner effect (Sect. 18.1). In addition to any arbitrary solution of Eq. (18.17) with $\mathbf{j}$ and $\mathbf{B}$ constant, in order to have $\mathbf{B} = 0$ from this equation London argued that the *argument* of the *derivative* had to be zero. Thus Eq. (18.15) was anticipated. In fact, for a superconducting block at the equilibrium one must have

$$div\mathbf{j} = 0 \quad and \quad div\mathbf{A} = 0$$

(*London gauge* or *Landau* or *Coulomb gauge*). In Eq. (18.15) the phase is constant and one derives

$$\mathbf{j} = -\frac{n_e e^2}{mc}\mathbf{A} \quad or \quad rot\mathbf{j} = -\frac{n_e e^2}{mc}\mathbf{B}, \tag{18.18}$$

known as *London equation* (here $m^* = m$ and $e^* = e$ according to the London's scenario).

From $rot\mathbf{B} = (4\pi/c)\mathbf{j}$, being $rot(rot\mathbf{A}) = \nabla div\mathbf{A} - \nabla^2\mathbf{A}$, and

$$\nabla^2\mathbf{A} = -\frac{4\pi}{c}\mathbf{j} \qquad (18.19)$$

one obtains

$$\nabla^2\mathbf{A} = -\frac{\mathbf{A}}{\lambda_L^2} \quad \text{or} \quad \nabla^2\mathbf{B} = -\frac{\mathbf{B}}{\lambda_L^2} \quad \text{or} \quad \nabla^2\mathbf{j} = -\frac{\mathbf{j}}{\lambda_L^2}. \qquad (18.20)$$

Furthermore

$$\frac{\partial\mathbf{j}}{\partial t} = \mathcal{E}\frac{n_e e^2}{m}$$

and

$$\lambda_L^2 = \frac{m^* c^2}{4\pi n_c (2e)^2}, \qquad (18.21)$$

where here we take into account that pairs are the real charge carriers. It is noted that since $m^* = 2m$, $n_c = n_e/2$ and $e^* = 2e$ the London length λ_L could not prove that the carriers are pairs.

As a consequence, while the field $\mathcal{E}$ does not penetrate in the superconducting metal at all, the vector potential $\mathbf{A}$ and therefore the magnetic field $\mathbf{B}$ *exponentially decreases from the surface towards the interior of the material*, as sketched below (see Problem 18.4):

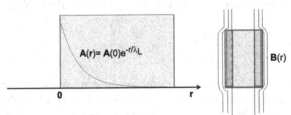

The *penetration length* λ_L turns out of the order of 10^3 Å for typical values of the electron density in metals, increasing on decreasing n_c. For its temperature dependence we mention that one should write

$$\lambda_L(T) = \lambda_L(0)\left[1 - \left(\frac{T}{T_c}\right)^4\right]^{-1/2}. \qquad (18.22)$$

Thus one observes that the supercurrents are running in a narrow sheet of the order of λ_L, screening the external magnetic field so that $\mathbf{B} = 0$ in the bulk of the material.

It is noted that the density of charge carriers is only about 10^{-4} the conduction electrons density, around the transition temperature. Furthermore λ_L is affected by impurities and to a certain extent by the external magnetic field.

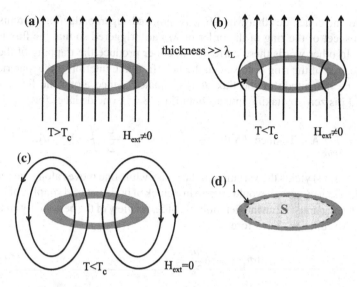

Fig. 18.5 Trapping of the external magnetic field in a ring. **a** H_{ext} is applied for $T > T_c$; **b** the metallic ring is FC below T_c; **c** the field is removed. Part **d** of the figure shows the line for the electric field circulation (well inside the ring) and the surfaces we are referring to

For the temperature dependence of the penetration depth we shall see at Sect. 18.9.5 how it is derived from the Ginzburg-Landau theory in a form consistent with Eq. (18.22).

18.4 Flux Quantization in Rings

Rather than to a block of bulk superconducting material now we shall refer to a ring having thickness much larger than λ_L, in the presence of magnetic field H_{ext} applied above T_c. Then the ring is cooled down in the presence of the field (*FC*) and when in the superconducting state, the *Meissner expulsion* having occurred, the field is removed (Fig. 18.5a–c).

When the external field is removed the flux through the surface defined by the ring cannot change since $\partial\Phi/\partial t = 0$, as it is inferred from the circulation of the electric field $\mathcal{E}$ along the line internal to the ring itself[12]:

$$\frac{\partial\Phi}{\partial t} = -\frac{\partial}{\partial t}\int_S \mathbf{B}\cdot ds = c\int rot\mathcal{E}\cdot d\mathbf{l} = 0. \qquad (18.23)$$

---

[12]See note for Eq. (18.1). The currents launched in the transient circulate in a narrow sheet close to the surface of the ring, while the electric field is zero along the internal line.

Therefore one concludes that when H_{ext} is removed superconducting currents (that flow in a sheet of the ring of the order of λ_L) are triggered so that the flux is kept constant. In other words these currents can either produce the trapping of the field inside the area of the ring or to screen the field from the interior of the material.

Now the magnitude of the flux Φ is evaluated. Since along the line l in the Fig. 18.5d **j** is zero, by taking into account Eq. (18.15), one deduces for

$$\Phi = \int_{circ} \mathbf{A} \cdot d\mathbf{l} \;\; \text{since}\;\; \hbar \nabla \theta = -2e\frac{\mathbf{A}}{c}\;,\;\; \Phi = \frac{\hbar c}{-2e} \int_{circ} \nabla \theta \cdot d\mathbf{l}. \qquad (18.24)$$

Equation (18.24) yields the variation of the phase while one moves along the ring and it is not necessarily zero (it would be zero in a block of bulk material simply connected, **A** being zero). θ must be an integer multiple of 2π, in view of the periodicity condition for the wave function. Therefore

$$|\Phi| = 2\pi n \frac{\hbar c}{2e} = n\frac{hc}{2e} = n\Phi_0. \qquad (18.25)$$

Namely the flux trapped by the ring is an integer number n of the *fluxon* Φ_0 given by $\Phi_0 = 2.07 \times 10^{-7}$ Gauss·cm^2.

This quantum phenomenon, which has been proved by experiments, is a further support for the formation of Cooper pairs. In fact, the flux quantum in the superconductor differs by a factor 2 from that derived in a metal (hc/e) when *Landau levels* form (see Appendix 13.1). One should also remark that having deduced that result from the circulation of the potential vector along a line of macroscopic length, this implies that the superconducting wave function maintains its coherence all along the ring (in the experiments of diameter around 10^{-2}–10^{-1} cm).

Finally it is anticipated that the flux penetration plays a role of particular relevance in type II superconductors, as we shall discuss subsequently (Sects. 18.7 and 18.10).

18.5 The Josephson Junction

The phase of a superconductor can be compared with the one in an analogous compound nearby, causing the occurrence of very interesting effects predicted by *Josephson* in 1962. Basically, the *Josephson effect* consists in the fact that the super-current can tunnel across an insulating layer in between two superconductors A and B, without loosing the coherence of the wave function and without dissipation. The insulating layer is usually the oxide (O) of a metal, of thickness around some tens of Å. The amount of supercurrent is related to the difference in the phases θ_A and θ_B. This current flow is usually known as *continuous Josephson effect* and it is extremely sensitive to an external magnetic field, as it will be shown. On the other hand it is also possible to induce in the junction SC$_A$-O-SC$_B$ an alternating current by applying a continuous voltage to the insulating layer. This *alternating Josephson effect* is currently used in a variety of application (the *superconductive electronics*) including, for instance, radiation detectors, manipulation of high frequency signal or for fast-switching devices as the ones used in computers and in metrology.

We shall only derive the basic principles of the Josephson junction by resorting to the time evolution of a two-state system (see Appendix 1.2).[13]

ψ_A and ψ_B being the wave functions of A and B (superconducting compounds that shall be assumed of the same nature), the coupling between the two is granted by a *tunnelling integral* K, a constant characterizing the junction A-O-B, as sketched below:

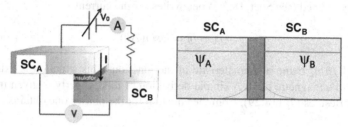

As for any two-states system the equations for the wave functions can be written

$$i\hbar\frac{\partial\psi_A}{\partial t} = E_A\psi_A + K\psi_B$$

$$i\hbar\frac{\partial\psi_B}{\partial t} = E_B\psi_B + K\psi_A, \tag{18.26}$$

where E_A and E_B are the energies for the ground state of the two superconductors. $E_A = E_B$ in the absence of external electric field, while for a voltage V applied at the junction $E_A - E_B = -2\,e V$. With a shift in the reference for the energy one writes

$$i\hbar\frac{\partial\psi_A}{\partial t} = e V\psi_A + K\psi_B$$

$$i\hbar\frac{\partial\psi_B}{\partial t} = -e V\psi_B + K\psi_A. \tag{18.27}$$

For ψ_A and ψ_B the expression given in Eq. (18.14) holds. By equating the real and the imaginary parts, four equations are obtained, for the time dependencies of the densities of the Cooper pairs and for the phases. From the pair densities

$$\hbar\frac{\partial n_A}{\partial t} = 2K(n_A n_B)^{1/2}sin(\theta_B - \theta_A)$$

$$\hbar\frac{\partial n_B}{\partial t} = -2K(n_A n_B)^{1/2}sin(\theta_B - \theta_A)$$

---

[13] Such an approach as been devised by Feynman and it is often reported as the two-states *Feynman model*.

the current is written

$$j_A = 2e\frac{\partial n_A}{\partial t} = 2e\frac{2K}{\hbar}(n_A n_B)^{1/2}sin(\Delta\theta)$$

$$\text{or}\ \ j_B = 2e\frac{\partial n_B}{\partial t} = -2e\frac{2K}{\hbar}(n_A n_B)^{1/2}sin(\Delta\theta) \tag{18.28}$$

with $\Delta\theta = (\theta_B - \theta_A)$. By assuming n_A and n_B constant[14] and equal to the n_c previously defined (see Sect. 18.2.3) one writes for the current

$$j = j_A - j_B = j_0 sin(\Delta\theta), \tag{18.29}$$

$j_0 = (4eK/\hbar)n_c$ being a characteristic of the junction. This situation is extremely sensitive to a magnetic field. A simple derivation of this property is given in Problem 18.6. Besides Eq. (18.29), from the equation for the phases one obtains

$$\frac{\partial\theta_A}{\partial t} = -\frac{K}{\hbar}cos(\theta_B - \theta_A) - \frac{eV}{\hbar}$$

$$\frac{\partial\theta_B}{\partial t} = -\frac{K}{\hbar}cos(\theta_B - \theta_A) + \frac{eV}{\hbar}.$$

Being $(\partial\Delta\theta/\partial t) = (\partial\theta_B/\partial t) - (\partial\theta_A/\partial t) = 2eV/\hbar$, then

$$\Delta\theta(t) = \Delta\theta(0) + \frac{2e}{\hbar}\int_0^t V(t')dt'. \tag{18.30}$$

Equations (18.29) and (18.30) represent the essence of the Josephson effect.

By referring to the current-voltage curve sketched below, one can comment the following.

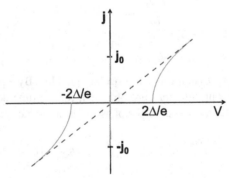

---

[14]n_A and n_B are constant while a non-zero charge flows because the source acts in order to feed the junction of further pairs. To take into consideration also the currents provided by the source would not modify the current related to the tunnelling of the insulating layer, which is the one we are evaluating here (see the book by Feynmann).

For voltage $V < |2\Delta/e|$ the current is *apparently* missing. In reality a current at very high frequency is running, undetected by the meter (see text and Eq. (18.31)). For $V \gg |2\Delta/e|$ the system behaves as an ordinary resistor.

For $V = 0$ the phase difference is constant although not necessarily zero and from Eq. (18.29) one can expect a current through the junction, up to j_0, with zero voltage between SC_A and SC_B. Therefore one has *supercurrent through the insulator*. While in a block of superconducting material the current requires a gradient of the phase (see Eq. (18.15)) through the junction the current is driven by the *difference in the phases* of the correspondent wave functions. If the current provided by the generator is modified one has an *instantaneous voltage* that re-adjusts the phase to a new value of the current: when the equilibrium is attained again one has $V = 0$.

If a constant voltage V_0 is applied at the junction, Eqs. (18.29) and (18.30) show that the current oscillates at the frequency

$$\nu_0 = \frac{2e}{h}V_0 = 483.6\,\text{MHz}/\mu\text{V}, \tag{18.31}$$

usually undetected by a meter (see the sketch above). The oscillations of the current have been experimentally recorded with proper apparatus (see the book by *Buckel*).

18.6 The SQUID Device

The *SQUID* (from *S*uperconducting *QU*antum *I*nterference *D*evice) is based on the interference of Josephson currents running in junctions set in parallel, according to the schematic representation for a Nb ring given below:

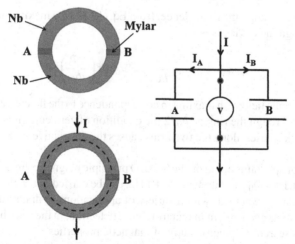

In a way resembling the interference phenomenon in optics, when the two currents have different phases they can interfere to zero. On the other hand, the phases are strongly sensible to the flux though the ring associated to an external magnetic field and the device can measure very fable fields.

A current generator provides the currents flowing in parallel along the two arms in the sketch above, while a voltmeter measures the voltage V. For identical junctions and zero magnetic field one has $V = 0$ and the two currents I_A and I_B, with phases ϕ_A and ϕ_B, sum up so that

$$I_T = I_A + I_B = I_0[sin\phi_A + sin\phi_B] = 2I_0 sin\left(\frac{\phi_A + \phi_B}{2}\right) cos\left(\frac{\phi_A - \phi_B}{2}\right).$$
(18.32)

Now it is shown that when the ring collects a flux equal to $\Phi_0/2$ (see Eq. (18.25)) the superconducting current goes to zero and a voltage appears at the voltmeter.

In the presence of a magnetic field, by moving along an internal line of the left side of the ring, following Eq. (18.24) the phase changes by

$$\delta_A = \phi_A - \frac{2e}{\hbar c}\int_A \mathbf{A} \cdot d\mathbf{l}.$$
(18.33)

Along the right side

$$\delta_B = \phi_B - \frac{2e}{\hbar c}\int_B \mathbf{A} \cdot d\mathbf{l}.$$
(18.34)

If one imagines to move all along the ring (for the side B in the opposite sense of side A) the whole phase difference must be zero (here we set $n = 0$ in Eq. (18.25)) and then

$$\phi_A - \phi_B = \frac{2e}{\hbar c}\int_{circ} \mathbf{A} \cdot d\mathbf{l} = \frac{2e}{\hbar c}\Phi,$$
(18.35)

Φ being the flux through the ring. Hence, from Eq. (18.32) the maximum current that can flow through the device is

$$I_{max} = 2I_0 cos\left|\frac{e\Phi}{\hbar c}\right| = 2I_0 cos\left(\frac{\pi\Phi}{\Phi_0}\right),$$
(18.36)

i.e. a function of the field with maxima in correspondence to the flux values $\Phi = n\Phi_0$. When the flux through the ring is $\Phi_0/2$ the condition of zero current is attained: the correction to the phases along the two arms causes the destructive interference of the two currents.

By means of apparatus based on the SQUID principle magnetic flux measurements of the order of $10^{-5}\Phi_0$ in a bandwidth of 1 Hz, can be carried out (fields of the order of 10^{-11} Oe can be detected) with a variety of applications in diagnostics (through the so called *biomagnetism*), in magneto-telluric detection, in the search of oil basins and in basic research for precise study of magnetic properties.

The SQUID has been used to prove the occurrence of the *Bohm-Aharonov* effect, i.e. the physical reality of the potential vector $\mathbf{A}$ in the regions where the magnetic field is absent.

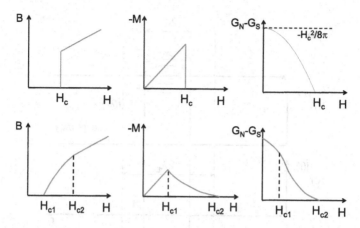

Fig. 18.6 Magnetization curves and *Gibbs free energies* (difference between normal N and superconducting S state) in type I (*top*) and type II (*bottom*) superconductors. For details see Sects. 18.9 and 18.10. H_c is the *thermodynamical critical field* (Sect. 18.1)

18.7 Type II Superconductors

Up to the years 1960 the superconducting materials at the highest T_c where the alloys, as Nb_3Sn with $T_c = 18.4$ K and critical field at 4.2 K around 20 T. The superconducting properties of the alloys are due their type II character, a behaviour predicted by *Abrikosov* and *Shubnikov* before the experimental discovery of that type of superconductors. The type II superconductors strongly differ from the ones implicitly described until now in regards of their response to magnetic field. In fact, in type II superconductors once that a certain value H_{c1} is attained the magnetic field penetrates in the material in form of *flux lines*, each carrying a quantum of flux, until a second critical field H_{c2} is reached. Above H_{c2} one has the transition to the ordinary state (see Fig. 18.6). The properties of these superconductors will be described at Sect. 18.9 in the framework of the *Ginzburg-Landau* theory. Here we only mention that the type II superconductors are the most suitable for the applications.[15]

In Fig. 18.7 a schematic one-dimensional representation of the properties of a flux line is given.

---

[15]The magnets with conventional (BCS) type II superconductors, as NbTi or Nb_3Sn, can reach up to 10 T and are the ones generally used in *NMR imaging* (see Sect. 6.3). The *magnetic levitation force* can be of the order of 10^4 N/m^2. The quality factor of superconducting *microwave cavities* can be increased up to 10^{10}, as in the ones currently used in particle accelerators. For extensive description of the applications, most for BCS superconductors, see the book by *Buckel*.

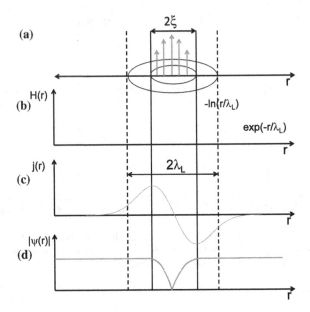

Fig. 18.7 Sketch of the structure of the vortex in terms of *London penetration length* λ_L and *coherence length* ξ (**a**); of the radial dependence of the magnetic field (**b**); of the density current (**c**) and of the order parameter (see Sect. 18.9.5) (**d**). It is noted that for $r \gg \lambda_L$ the material is completely superconducting. The condition $\xi \ll \lambda_L$ is the one for the realization of type II superconductors (see Sect. 18.10)

18.8 High-Temperature Superconductors

As described in the phenomenological overview (introductory section) since 1986 a new family of superconducting compounds has been discovered, now known with the acronym *HTcSC*. The core of these compounds is the CuO_2 plane: the electronic properties of the Cu^{2+} ions, with its localized magnetic moment, play the dominant role (see Appendix 17.1).[16]

At Sect. 13.3 it has been shown how the crystal field at octahedral coordination produced by the oxygen ions causes the structure of the e_g doublet (see Fig. 13.2). The elongation of the oxygen octahedron removes the doublet degeneracy and the generation of σ-bonds with the nearby oxygen in the plane is induced. Thus the planar atomic configuration can be sketched as in Fig. 18.8.

---

[16]This is true for the family of the cuprate superconductors. In MgB_2 the relatively high T_c (about 39 K) is still related to conventional BCS mechanism, in the presence of particular electronic band structure and phonon bath. In the iron-based superconductors discovered in 2008 (the *Fe-oxypnictides*) the magnetic ion is Fe and the pairing mechanism possibly involves magnetic excitations.

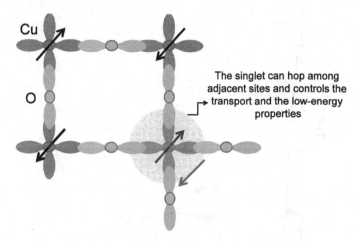

Fig. 18.8 Sketch of Copper 3d, Oxygen 2p orbitals and of the singlet induced by charge (hole) doping of the CuO_2 plane of high temperature superconductors

This is the planar CuO_2 structure occurring in the prototype La_2CuO_4, an insulating antiferromagnetic crystal that besides being the experimental realization of the 2DQHAF (see Appendix 17.1) can be considered the father of HTcSC. In fact, in this crystal the charge carriers (*holes*) are injected through Sr^{2+} for La^{3+} substitution, so that singlets at $S = 0$ can propagate onto the CuO_2 plane, as sketched in Fig. 18.8. As a result of this doping a complicate phase diagram (see Fig. 18.9) originates, which for the light charge doped region has been already reported at the Appendix 17.1 in another form.

The transport properties and the low-energy physics of the CuO_2 plane in the presence of the extra-holes due to the heterovalent substitutions can be discussed by referring to a model of charge hopping onto a square 2D lattice, a particular form of the Hubbard model (see Sect. 17.1). That simplified model takes into account the *electron-electron correlation* through a repulsive energy U between electrons on the same site and the *delocalization energy* t, namely the hopping matrix element between sites ith and $(i + 1)$th (see Sect. 12.7.3).

As already mentioned at Sect. 17.1, although apparently rather simple (and already being an oversimplification of the real physics in cuprates superconductors) the Hubbard Hamiltonian can be solved for arbitrary ratios of U/t only for the one-dimensional geometry. At half filling, namely for one electron per site, in the planar square lattice it is known that a metal-to-insulator transition occurs when the ratio U/t is increased above a certain value.

The parent compound, La_2CuO_4, is considered a *Mott insulator*,[17] the localization of the carriers being due to the *correlation energy*. The hopping that in principle one

---

[17]With reference to Chap. 12, the *Mott insulator* is somewhat opposite to the band insulator, its existence being due to the strong repulsive correlation, an effect that can hardly be described in the band-type theories.

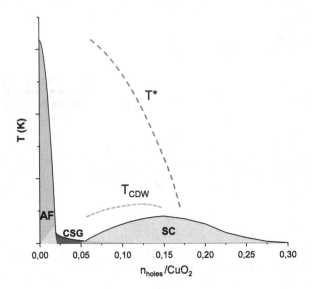

Fig. 18.9 Phase diagram of cuprates as a function of the hole doping. T^* schematically indicates the temperature below which anomalies in some quantities occur (effect commonly reported as *pseudo-gap opening*). T_{CDW} is the temperature below which a charge density wave phase is detected. On the left side the decrease of the Néel temperature T_N in the antiferromagnetic (AF) phase as a function of hole doping is reported (for details return to Appendix 17.1 and Fig. 17.9). CSG indicates a cluster spin-glass phase which develops in between the AF and SC phases and may also coexist with the latter one

has to consider in order to reduce the energy through the term t would require double occupation on a given site and then it would cost energy U larger than t. The array of electrons localized at the planar square lattice takes the *antiferromagnetic* (AF) *order*, with AF *exchange integral* J around 1500 K, through a super-exchange mechanism typical of oxides (see Sect. 17.2).

The AF spin alignment is the one that permits virtual hopping with a gain of energy given by $J = 4t^2/U$ (see Appendix 17.1) while for parallel spin (ferromagnetic alignment) no hopping at all is possible because of the Pauli principle.

At the opposite limit $t \gg U$ the hopping drives the system to a metallic-like state, with fermionic particles at pseudo-momentum $\hbar \mathbf{k}$, with the Fermi surface and Fermi-Dirac statistical behaviour.

When the electron vacancies are produced into the CuO_2 plane of the parent AF (*hole doping*) the AF order is rapidly destroyed by a few percent of holes (return to Appendix 17.1).

On increasing the doping extent x the superconducting phase is obtained, with the optimally doped condition for $x = 0.15$, as shown in Fig. 18.9.

In the framework of the scenario of electronic bands structure, the situation in the cuprates can be thought as generated from the copper d and oxygen p levels hybridisation, with the resulting *half-filled anti-bonding band*. Then the Coulomb correlation splits that band, causing the lower Hubbard band (*LHB*) and the upper Hubbard band (*UHB*), with energy gap U.

The undoped La_2CuO_4 corresponds to the filling of the LHB, with AF correlation. The charge doping generates electronic levels in the gap between LHB and UHB, allowing the itinerary of the carriers, as schematically indicated below:

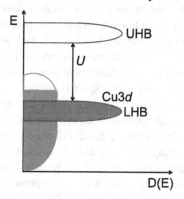

These qualitative considerations about the complex electronic structure in doped cuprates can be summarized by stating that superconductivity arises when vacant states are induced in 2D Heisenberg AF, the charge transport involving holes that jump in an antiferromagnetic background, with spin fluctuations.

Rather than trying to recall the many theories that have been attempted in order to describe HTcSC (the pairing mechanisms being nowadays still under debate), in the following the experimental and theoretical aspects that can be retained with a reasonable deal of confidence are summarized. In the next Sect. 18.9 the *Ginzburg-Landau theory*, that works very well (with some refinements and/or extensions) even for HTcSC, will be developed.

HTcSC's are *strongly type II*, with a relevant role of the granularity and of the anisotropy, so that the *effective mass*, the *London penetration length* (λ_L typically varying between 1500 and 5000 Å) and the *coherence length* ($\xi \sim 2 - 40$ Å) *have tensorial character*.[18]

The critical fields have been well established: for temperature close to zero H_{c1} is *of the order of a few hundreds of Oe* while H_{c2} is very high (possibly *around* $10^6 Oe$). The general properties regarding the behaviour under RF or microwave irradiation have been studied and are close to the ones for any good superconductor.

The *penetration of the magnetic field* in the intermediate range $H_{c1} < H < H_{c2}$ has been clearly evidenced and studied; the dynamics of the vortices (see Sect. 18.10) has been clarified to a large extent. The vortices have been found to move very easily for temperatures not so far below T_c and they control the transport properties.

---

[18]In anisotropic superconductors, in the reference frame aligned with the principal axes, there are three coherence lengths $\xi^{(i)}$ and three penetration lengths $\lambda_L^{(i)}$. These six lengths, not independent from each other, are commonly introduced in order to extend the isotropic Ginzburg-Landau theory by referring to $\lambda_L = [\lambda_L^{(1)} \lambda_L^{(2)} \lambda_L^{(3)}]^{1/3}$ and $\xi = [\xi^{(1)} \xi^{(2)} \xi^{(3)}]^{1/3}$ (see Sect. 18.9).

Only below a so-called *irreversibility temperature*[19] $T_{irr} < T_c$ the vortices can be considered fixed (in a way depending from impurities and/or defects) and thus no dissipation occurs.

In short, one could say that in spite of the uncertainties on the pairing mechanism and of the anomalous (and not yet understood) properties of the *underdoped compounds* (see Fig. 18.9) the Ginzburg-Landau approach works rather well and the main aspects of this novel family of superconductors can be satisfactorily described.

As regards the more controversial theoretical framework one can summarize the following. The carriers are *pairs in a singlet state*, the superconducting gap is of *d-character* (in the **k**-space there are nodes where the gap is zero). The pairing mechanisms should be considered strong, $\Delta(0)$ being about $7k_BT_c$. The *BCS theory* can be applied *only partially* and is not sufficient in order to describe the ordinary state as well as the superconducting state. The high T_c is due both to a large value of the term θ_D (see Eq. (18.8)) and to a large effective λ, possibly enhanced by the proximity to a metal-insulator crossover, as it occurs in the superconducting fullerides.

Likely the cuprates are on the verge of a so-called *quantum critical point*, namely in the proximity of crossovers driven by terms of the interactions rather than the usual transition driven by temperature. Therefore they can be expected to be much sensible to a certain number of parameters, including impurities and defects. Further elements about the properties of HTcSC are given in the subsequent section, in the form of the results obtained along the Ginzburg-Landau description.

18.9 Ginzburg-Landau (GL) Theory

18.9.1 The GL Functional

The GL theory for the superconductive state and the related superconducting transition is basically an extension of the Landau theory for the phase transition discussed at Sect. 15.2. The main success of the GL approach has been the one obtained by *Abrikosov*: the existence of the type II SC's was predicted before their experimental discovery. Furthermore the GL theory is the only practical approach to deal with the cases where spatial inhomogeneities play a crucial role (presence of magnetic field, thin films, boundary and proximity effects and to a large extent the fluctuations).

---

[19]Below T_{irr} one observes a difference between the diamagnetic susceptibility χ_{dia} (*Meissner effect*, see Sect. 18.3) measured after cooling in the presence of the magnetic field (*FC condition*) and the one measured at the same temperature after cooling in zero field (*ZFC condition*). In a certain temperature range below T_{irr} the ZFC diamagnetism is *time-dependent*, since the vortices are forced to penetrate inside the bulk, driving the relaxation process leading from the ZFC to the FC condition.

In the GL theory one has to reformulate the treatment given at Sect. 15.2 by applying it to a *complex order parameter* (due to gauge invariance[20]) of the form

$$\psi(\mathbf{r}) = |\psi(\mathbf{r})|e^{i\theta(\mathbf{r})}, \tag{18.37}$$

$\psi(\mathbf{r})^*\psi(\mathbf{r})$ being the density of pairs at point $\mathbf{r}$. The phase $\theta(\mathbf{r})$ is related to the superconducting current, as shown at Sect. 18.2 (Eq. (18.15) in particular).

We shall first assume $\psi(\mathbf{r}) = 0$ for $T > T_c$ and $\psi(\mathbf{r}) \neq 0$ for $T < T_c$, by neglecting for the moment the role of the superconducting fluctuations, that actually are significant in HTcSC (see Sect. 18.11). Then according to an expansion similar to the one given at Sect. 15.2, a *Gibbs free energy functional* (see note 8 at Sect. 15.3) is introduced in the form

$$G_S[\psi] = G_0 + \frac{1}{V}\int\left[a|\psi(\mathbf{r})|^2 + \frac{b}{2}|\psi(\mathbf{r})|^4 + \right.$$

$$\left. + \frac{1}{2m^*}\left|\left(-i\hbar\nabla + \frac{e^*\mathbf{A}}{c}\right)\psi\right|^2 + \frac{H^2(\mathbf{r})}{8\pi} - \mathbf{H}(\mathbf{r})\cdot\mathbf{M}(\mathbf{r})\right]d\mathbf{r}, \tag{18.38}$$

where the usual expression for the e.m. moment has been used in order to account for the presence of magnetic field into the gradient term, while the magnetic energy density is given by the last term (see Eqs. (15.7)–(15.10)).[21] Furthermore in Eq. (18.38) we have e^* and m^* to mean the application to the pairs ($e^* = +2e$ or $-2e$ according to the sign of the carriers, electrons or holes, and $m^* = 2m_{eff}$). Finally it is reminded that, according to Eq. (15.14), in Eq. (18.38) one has to assume $b(T) = b_0 > 0$ and

$$a(T) = a_0\left[\frac{T}{T_c} - 1\right] \equiv a_0\varepsilon. \tag{18.39}$$

For $T > T_c$ (normal state) the field is the external one and $M \simeq 0$. Thus in the thermodynamical potential the only magnetic contribution is the magnetic energy per unit volume for the "empty coil" creating the field.

Well below T_c, in the region of the sample where full superconductivity occurs, then $\mathbf{B} = 0$ while $\mathbf{M} = -(1/4\pi)\mathbf{H}_{ext}$ and the magnetic contribution

$$-\int \mathbf{H}(\mathbf{r})\cdot\mathbf{M}(\mathbf{r})d\mathbf{r}$$

---

[20]The phase and the magnetic vector potential $\mathbf{A}$ depend on the choice of the gauge but all the physical variables, including the magnetic field $\mathbf{B}$, are gauge-invariant (see Problem 18.14).

[21]Equation (18.38) should be considered a masterpiece of physical intuition. It has a possible justification in the light of the analogy with the Schrodinger equation by assigning to ψ the character of wavefunction. The BCS theory has clarified many aspects related to that equation. *Gor'kov* has shown that the GL theory can be derived from the BCS theory when the latter is generalized to include spatially varying situations, near the transition. ψ corresponds to the wavefunction of the centre of mass of the Cooper pair (see Sect. 18.2.3).

in Eq. (18.38) becomes $H_{ext}^2/4\pi$. When the magnetic terms are negligible $G[\psi] = F[\psi]$, the *Helmholtz free energy*. In Eq. (18.38) $\psi(\mathbf{r})$ and $\mathbf{A}(\mathbf{r})$ are both the unknown functions. The occurrence of superconductivity re-arranges the currents so that $G[\psi]$ is minimized.

Finally it is noted that if one uses the definition of velocity $v_{SC} = (1/m^*)(\hbar\nabla + 2e\mathbf{A}/c)$ according to Eq. (18.15), then the kinetic term involving gradient and field in Eq. (18.38) can be rewritten

$$\frac{1}{2m^*}\left[\hbar^2(\nabla|\psi|)^2 + \left(\hbar\nabla\theta + \frac{e^*\mathbf{A}}{c}\right)^2|\psi|^2\right] = \frac{1}{2}m^*v_{SC}^2n_c + \frac{\hbar^2}{2m^*}(\nabla|\psi|)^2, \quad (18.40)$$

namely the kinetic energy of the superconducting pairs and the energy term related to the gradient of the order parameter. In a superconducting block at the equilibrium, with constant order parameter, the London's scenario (Sect. 18.3) is recovered.

18.9.2 The GL Equations

In order to derive the equations controlling $\psi(\mathbf{r})$ and $\mathbf{A}(\mathbf{r})$ one has to minimize the GL functional (Eq. (18.38)). By deriving with respect to ψ^* while keeping ψ constant[22] one obtains the *first GL equation*:

$$a\psi + b|\psi|^2\psi + \frac{1}{2m^*}\left[-i\hbar\nabla + \frac{e^*\mathbf{A}}{c}\right]^2\psi = 0 =$$

$$= \frac{1}{2m^*}\left[\hbar^2\nabla^2\psi + \frac{2i\hbar e^*}{c}\mathbf{A}\cdot\nabla\psi - \frac{e^{*2}A^2}{c^2}\psi\right] - a\psi - b|\psi^2|\psi, \quad (18.41)$$

(being $div\mathbf{A} = 0$). By deriving with respect to the vector potential $\mathbf{A}$ one has

$$\frac{i\hbar e^*}{2m^*}\left[\psi^*\nabla\psi - \psi\nabla\psi^*\right] + \frac{e^{*2}}{m^*c}\mathbf{A}|\psi|^2 + \nabla\times(\nabla\times\mathbf{A})\frac{c}{4\pi} = 0, \quad (18.42)$$

and from $\mathbf{B} = \nabla\times\mathbf{A}$ and $\nabla\times\mathbf{B} = (4\pi/c)\mathbf{j}$, the current density turns out

$$\mathbf{j} = -\frac{i\hbar e^*}{2m^*}\left[\psi^*\nabla\psi - \psi\nabla\psi^*\right] - \frac{e^{*2}}{m^*c}\mathbf{A}|\psi|^2 = \left[\frac{e^*\hbar}{m^*}\nabla\theta - \frac{e^{*2}}{m^*c}\mathbf{A}\right]|\psi|^2 = e^*|\psi|^2v_{SC}, \quad (18.43)$$

---

[22]$\partial F/\partial\psi(\mathbf{r})^* = 0$ involves the *functional derivative*, the free energy being a function of infinitely many variables. By deriving with respect to $\psi(\mathbf{r})$ the complex conjugate of Eq. (18.41) would be obtained.

corresponding to $\mathbf{j} = -\partial F/\partial \mathbf{A(r)}$. Equation (18.43) is the *second GL equation*. By taking the curl of both members, since $rot\mathbf{V} = 0$ the London equation (Eq. (18.18)) is found.

Equations (18.41) and (18.43), when simultaneously solved yield the properties of the superconducting state. It is noted that $\mathbf{A(r)}$ is the microscopic vector potential due to $\mathbf{H}_{ext}$ and to the superconducting currents.

In the next Subsections we are going to apply the GL equations to particular situations, on one part deriving some of the properties already discussed in different context while novel aspects will emerge, particularly for type-II superconductors.

18.9.3 Uniform and Homogeneous SC and No Magnetic Field

In this case $|\psi|$ is site-independent, $\mathbf{V}|\psi| = 0$ and the density of the Cooper pairs n_c inside the SC is constant.[23] The second GL equation goes to zero while from Eq. (18.41), the phase being arbitrary and therefore set to zero, one has

$$a\psi + b\psi|\psi|^2 = 0, \qquad (18.44)$$

corresponding to the minimization of $F_{SC} = [a|\psi|^2 + (b/2)|\psi|^4]V$. Then $|\psi|^2 = -a/b$ and $\Delta F = -a^2/2b$ (see sketch below):

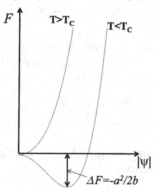

On the other hand the condensation energy per unit volume must correspond to $H_c^2/8\pi$, with H_c *thermodynamical critical field* (see Sect. 18.1 and Fig. 18.6). Thus, from $a|\psi|^2 + (b/2)|\psi|^4 = H_c^2/8\pi$, by using $|\psi|^2 = -a/b$, one derives

$$a = -\frac{1}{4\pi}\frac{H_c^2}{n_c} \quad \text{and} \quad b = \frac{1}{4\pi}\frac{H_c^2}{n_c^2}$$

---

[23]These conditions are also rather well verified in *nanoparticles*, namely particles of size much smaller than the coherence length (see Sect. 18.12), this case being often called *zero-dimensional condition*.

with n_c equilibrium density of the Cooper pairs (see Eq. (18.14)). Furthermore, near T_c, by returning to Eq. (18.39) one sees that[24]

$$n_c(T) = \frac{-a}{b} = \frac{a_0}{b}|\varepsilon|, \qquad (18.45)$$

and

$$H_c(T) \propto |\varepsilon|. \qquad (18.46)$$

The condensation energy is a measure of the gain in the free energy per unit volume when the transition from the normal to the superconducting state occurs. $VH_c^2/8\pi$ is usually very small, of the order of a few μeV per atom (see Problem 18.8).

18.9.4 Surface Effects (in Bulk SC and in the Absence of Field)

From the second GL equation for $\mathbf{A} = 0$ and $\mathbf{j} = 0$, one has $\psi^* \nabla \psi = \psi \nabla \psi^*$ and therefore the phase θ is site independent and it can be assumed zero. From the first GL equation, for $\mathbf{A} = 0$ one has

$$\frac{-\hbar^2}{2m^*} \nabla^2 \psi = -a\psi - b|\psi|^2 \psi. \qquad (18.47)$$

In order to discuss the situation occurring at the boundary between a bulk SC and the vacuum (or an ordinary metal) we shall refer to the 1D condition:

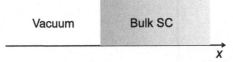

By change of variable, with $\psi = [a/b]^{1/2} p$ (p being a kind of normalized order parameter), Eq. (18.47) is rewritten

$$\frac{-\hbar^2}{2m^*|a|} \frac{d^2 p}{dx^2} + (1 - p^2)p = 0, \qquad (18.48)$$

resembling the one-dimensional non-linear Schrodinger equation. Then a dimensionless variable $\eta = (x/\xi)$, with

$$\xi^2(T) = \frac{\hbar^2}{2m^*|a|} \propto |\varepsilon|^{-1}, \qquad (18.49)$$

---

[24]Over the whole temperature range below T_c more appropriate temperature dependencies are $H_c(T) \propto [1 - (T/T_c)^2]$ and $|\psi|^2 = n_c \propto [1 - (T/T_c)^4]$.

is introduced (see Problem 18.10), so that Eq. (18.48) takes the form

$$\frac{d^2p}{d\eta^2} + (1 - p^2)p = 0,$$

of solution $p = tanh(\eta/\sqrt{2})$. Going back to the order parameter one writes

$$\psi(\eta\xi) = \left(\frac{|a|}{b}\right)^{1/2} tanh\left(\frac{\eta}{\sqrt{2}}\right) = \psi(\infty)tanh\left(\frac{\eta}{\sqrt{2}}\right)$$

with the behaviour illustrated below.

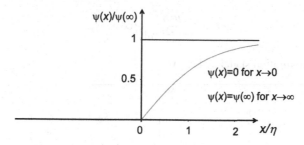

One observes that

$$\xi(T) = \frac{\hbar}{\sqrt{2m^*|a|}} \equiv \xi(0)|\varepsilon|^{-1/2} \qquad (18.50)$$

is a characteristic length measuring the variation of the order parameter on entering the bulk superconductor, thus it is the *coherence length* already defined at Sect. 18.2.2 and at Sect. 15.2.[25] Another definition of the coherence length shall be given in dealing with the fluctuations and the correlation function, at Sect. 18.11.

The critical exponent defining the divergence of $\xi(T)$ for $T \to T_c^-$ is therefore $1/2$, as it will turn out at Sect. 18.11, in the framework of the *Gaussian approximation*. An approximate expression for the correlation length can also be obtained as shown at Problem 18.10.

Finally we remark that by resorting to the expression of ξ the first GL equation can also be written in the form

$$-\left[\nabla - \frac{ie^*\mathbf{A}}{\hbar c}\right]^2 \psi + \frac{\psi}{\xi^2} + b\frac{2m^*}{\hbar^2}|\psi|^2\psi = 0, \qquad (18.51)$$

of frequent use, particularly in its linearized form.

---

[25]Return to the note n.10 of the present Chapter.

18.9.5 The London Penetration Length

From the second GL equation, by assuming for simplicity the homogeneity condition so that the variation in the order parameter can be neglected and $|\psi|^2 = |\psi_0|^2$, from

$$\mathbf{j} = \frac{-e^{*2}}{m^*c}\mathbf{A}|\psi_0|^2 \qquad (18.52)$$

and $\nabla^2\mathbf{A} = -(4\pi/c)\mathbf{j}$, one derives the equation for the penetration length already obtained from the London theory (Eq. (18.21)), leading to

$$\lambda_L^2 = \frac{m^*c^2}{4\pi e^{*2}|\psi_0|^2} \propto |\varepsilon|^{-1}, \qquad (18.53)$$

in the 1D configuration sketched below

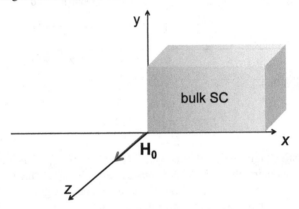

with

$$H_z(x) = H_0 exp[-x/\lambda_L]$$

$$A_y(x) = A_0 exp[-x/\lambda_L],$$

implying the boundary conditions at the surface (see Problem 18.4).

    As shown in Problem 18.7, from the second GL equation by circulating the current density along a close ring and imposing the periodicity condition to the phase because of the unicity of the order parameter, the flux quantization, already derived at Sect. 18.4, can be obtained.

    Since both λ_L and ξ diverge at T_c with $|a|^{-1/2}$, in the framework of the GL theory their ratio

$$\kappa = \frac{\lambda_L}{\xi} \qquad (18.54)$$

is *temperature independent.*[26] Furthermore, taking into account Eqs. (18.49) and (18.53), it is noted that

$$H_c(T)\lambda_L(T)\xi(T) = \frac{\Phi_0}{2\pi\sqrt{2}} = \text{constant} \tag{18.55}$$

Finally it should be remarked that the constants a and b, that in the framework of the GL theory control all the properties of a superconductor, according to the equations derived in this section can be obtained from the experiments.

18.10 The Parameter κ and the Vortex

The ratio $\kappa = \lambda_L(T)/\xi(T) = \sqrt{8}\pi\lambda_L^2(T)H_c(T)/\Phi_0$, with $\lambda_L^2 = m^*c^2/4\pi e^{*2}|\psi|^2$ and $\Phi_0 = hc/2e$, differentiates type I from type II superconductors:

$$\kappa < \frac{1}{\sqrt{2}} \quad \text{type I,} \quad \kappa > \frac{1}{\sqrt{2}} \quad \text{type II,}$$

with the magnetic phase diagrams sketched hereafter.

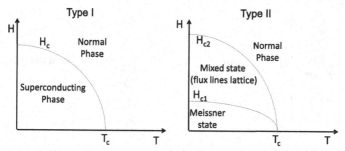

The relevant role of κ can be proved as follows. For a second-order phase transition (see Sect. 15.2) with small order parameter, from the first GL equation, by neglecting the contribution from currents, one writes

$$a\psi = \frac{1}{2m^*}\left[-i\hbar\nabla + \frac{2e\mathbf{A}}{c}\right]^2\psi, \tag{18.56}$$

namely the Schrodinger equation for a particle in magnetic field. The lowest-energy eigenvalue (see Eq. (A.13.1.4)) is $E_0 = \hbar\omega_c/2$, with $\omega_c = eH/mc$ the *cyclotron frequency.*

---

[26] κ is the only parameter that really plays the crucial role in the GL theory.

Non-zero solution and then superconductivity, are present for $H < H_{c2}$ with

$$\frac{e\hbar H_{c2}}{mc} = -a$$

and being $\xi^2 = (\hbar^2/2m^*|a(T)|)$,

$$\text{for } H_{c2}(T) = \frac{2m^*c}{e^*\hbar}|a(T)| = \frac{\Phi_0}{2\pi\xi^2(T)} \tag{18.57}$$

$$\text{or } H_{c2}(T) = \kappa\sqrt{2}H_c, \tag{18.58}$$

H_c being the *thermodynamic field* (see Sect. 18.1). Thus $\kappa = 1/\sqrt{2}$ represents the *threshold value* for having $H_{c2} > H_c$.

It should be remarked that the transition occurring at H_{c2} by varying the field or by varying the temperature, is a *second order one* since for any T below T_c the order parameter grows with continuity. At variance, for type I superconductors when on cooling one crosses a finite value of $H_c(T)$ the order parameter jumps from zero to a finite value and then the transition is *first order* (see Sect. 15.1).

According to the sketch in the previous page, H_{c1} can be defined as the field required to have the entrance in the material of just one vortex. For a vortex, schematized as below

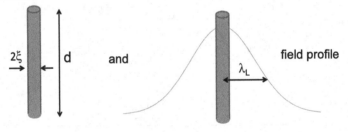

the energy cost is

$$(f_{SC} - f_0)V \simeq (H_{c2}^2/8\pi)\xi^2 d. \tag{18.59}$$

On the other hand the field penetrates only for a length of the order of λ_L and the energy gain therefore is of the order of $(H^2/8\pi)d\lambda_L^2$. Therefore the positive balance to have a vortex requires $\lambda_L \gg \xi$. Approximate expression for H_{c1} and H_{c2} are obtained at Problem 18.9[27]:

$$H_{c1} \simeq \frac{H_c}{\sqrt{2}\kappa}\ln\kappa \simeq \frac{\Phi_0}{4\pi\lambda_L^2} \tag{18.60}$$

---

[27]For details and for considerations on the effects of the magnetic field in other experimental conditions, see the book by *Poole, Farach* and *Creswick*. One could remark that an entire volume would not be sufficient in order to illustrate all the applications of the Ginzburg-Landau theory in a variety of circumstances, as remarked by the authors.

and

$$H_{c1}H_{c2} \simeq H_c^2 ln\kappa, \quad \text{or} \quad H_{c2} \simeq H_c\sqrt{2}\kappa. \tag{18.61}$$

Finally it is reminded that for superconductors arising from ordinary metals $\lambda_L \simeq 500 - 1000$ Å while $\xi \simeq 3000 - 5000$ Å. At variance, in HTcSC $\xi \simeq 2 - 40$ Å and $\lambda_L \sim 1000$ Å, both anisotropic. The alloying process in ordinary metals induces the so-called "*dirty*" *regime* by reducing the electronic mean free path, thus causing a much smaller effective coherence length. Therefore $\kappa \gg 1$ and II type superconductivity arises, with the increase of critical fields and critical currents.

In the field range $H_{c1} \ll H \ll H_{c2}$ the vortices density n (number per square cm) is approximately given by

$$n = \frac{< B_{int} >}{\Phi_0}, \tag{18.62}$$

where $< B_{int} >$ is an average internal field, being to a good approximation proportional to the applied field so that $n \simeq H_{ext}/\Phi_0$.

By referring to a single isolated vortex, for radial distance r much larger than the vortex core, according to Sect. 18.9.4, from the second GL equation and by taking the curl of both members of Eq. (18.52) and using the definition of the penetration length as in Eq. (18.53), one writes

$$\nabla \times \nabla \times \mathbf{B}(\mathbf{r}) = -\frac{\mathbf{B}(\mathbf{r})}{\lambda_L^2}, \tag{18.63}$$

as in the London scenario.

Now one has to consider that within the vortex, approximately a cylinder of radius ξ much smaller than λ_L, the external field penetrates and the flux is just a fluxon Φ_0. Thus in order to derive the field outside the vortex core we complete the Eq. (18.63) by writing

$$(\nabla \times \nabla \times \mathbf{B}(\mathbf{r}))\lambda_L^2 + \mathbf{B}(\mathbf{r}) = \mathbf{z}\Phi_0\delta(\mathbf{r}) \tag{18.64}$$

$\hat{z}$ being the direction of the external field, where the two-dimensional delta function reflects the singularity for $r = 0$ in the plane perpendicular to the field, meantime imposing the condition of the fluxon inside the core. By means of a lengthy mathematical procedure in cylindrical coordinates it can be shown that the solutions of this equation are modified *Bessel functions*. The field in the material, outside the vortex, is approximately[28]

$$B(r) \simeq \frac{\Phi_0}{2\pi\lambda_L^2}K_0\left(\frac{\sqrt{r^2 + 2\xi^2}}{\lambda_L}\right), \tag{18.65}$$

---

[28]It is reminded that for material strongly anisotropic in their superconducting properties, as the HTcSC, significant modifications to the expressions derived above have to be taken into account. For further details see the book by Poole, *Farach* and *Creswick*.

with $r = (x^2 + y^2)^{1/2}$ and where K_0 has the limits $-ln(x)$ for $x \ll 1$ and $\sqrt{\pi/2x}exp(-x)$ for $x \gg 1$. Then for $r \gg \lambda_L$ one has $B(r) \propto exp(-r/\lambda_L)$. On the other hand the superconducting current $|\mathbf{j}| \propto |\nabla \times \mathbf{B}| = (dB/dr)$ is flowing around the vortex in a sheet of the order of λ_L. Thus the main features involved in the Fig. 18.7 are now justified. The GL theory, particularly in the ingenious and extensive application carried out by *Abrikosov*, can be used to described other important aspects involving *proximity effects, interfaces* and *boundary conditions, flux flow and dissipation* (when the vortices are not sufficiently pinned by impurities or defects[29]). Furthermore the GL scenario is suited for the description of the superconducting fluctuations, as we shall see in the subsequent sections.

18.11 Effects of Superconducting Fluctuations

18.11.1 Introductory Remarks

As it has been pointed out at Sect. 18.2.2, in conventional superconductors (often reported as BCS superconductors) in the coherence volume, of the order of ξ^3, there is a large number of pairs. Thus it is not a surprise that the superconducting fluctuations, of nature similar to the ones described in general terms at Chap. 15, can play a significant role only in a very narrow temperature range around T_c, of difficult experimental detection in BCS superconductors. Correspondingly, one could say that mean field approaches to the superconducting transition yield reliable descriptions.

At variance, in HTcSC the coherence length is very small, the carrier density is reduced and the transition temperature is increased. Thus strong enhancement of the fluctuations occurs and several effects are detectable on approaching the transition. As a consequence of the superconducting fluctuations (SF) the temperature behaviour of the order parameter can be depicted as in Fig. 18.10.

A comprehensive description of the superconducting fluctuations and of the related effects (particularly the excess conductivity and the fluctuating diamagnetism above T_c) can be given by applying to the superconducting order parameter the basic recipe of the *Ginzburg-Landau theory*, through expansion of the free energy similar to the one used at Sect. 15.3 (see also Eq. (18.38)). Therefore the free energy density is written in the form[30]

---

[29] *Pinning* has technological importance in order to lock the vortices and avoid dissipation. For a current $\mathbf{j}$ flowing perpendicular to the field $\mathbf{H}$ the *Lorentz force* pushes the vortices along the $\mathbf{j} \times \mathbf{H}$ direction. The flux variation implies electric field parallel to $\mathbf{j}$ and then an effective electrical resistance. Also the *Magnus* sideway acting force can be involved in the vortex motions.

[30] Here we write α and β instead of a and b to emphasize that in this section we deal with the fluctuations around the equilibrium values.

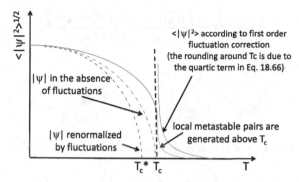

Fig. 18.10 The average value of the order parameter modulus (*solid line*) is different from zero even above T_c and smearing of the transition temperature also occurs. Above T_c^* one can say that local pairs, without long-range spatial coherence, are generated and they decay slower and slower on approaching the transition (see Sect. 15.2 for equivalence to other systems)

$$f = f_0 + \alpha(T)|\psi|^2 + \frac{\beta}{2}|\psi|^4 + \frac{1}{2m^*}|(-i\hbar\nabla + (e^*/c)\mathbf{A})\psi|^2, \qquad (18.66)$$

which includes spatial variations and the magnetic field through the potential vector
A. According to Eq. (18.39) $\alpha(T) = \alpha_0(T - T_c)/T_c \equiv \varepsilon\alpha_0$.

It is remarked that when the term in $|\psi|^4$ in Eq. (18.66) is neglected one operates in the so-called *Gaussian approximation* or *first-order fluctuation correction*. However, for $T \simeq T_c$ (see subsequently a more quantitative definition) where the fluctuations are strongly enhanced, that term is no longer negligible and it plays a crucial role, as it will be shown in describing the effects of fluctuations in *nanoparticles* (Sect. 18.12).

First in Eq. (18.66) we shall set $\mathbf{A} = 0$ and derive the collective amplitude and the decay time of the fluctuations in a way resembling what has been done at Sect. 15.3, for temperature above T_c.

The order parameter is expanded in the free-particle eigenfunctions:

$$\psi(\mathbf{r}) = \sum_{\mathbf{k}} \psi_{\mathbf{k}} e^{i\mathbf{k}\cdot\mathbf{r}}, \qquad (18.67)$$

equivalently to the Fourier expansion at Sect. 15.3. From Eq. (18.66) with $\mathbf{A} = 0$, by omitting the term in $|\psi|^4$, the free energy density becomes

$$f = f_0 + \sum_{\mathbf{k}}\left[\alpha + \frac{\hbar^2}{2m^*}k^2\right]|\psi_{\mathbf{k}}|^2, \qquad (18.68)$$

Now a thermodynamical average over the possible values of the order parameter has to be performed (see Eq. (15.19)):

$$< |\psi_{\mathbf{k}}|^2 > = \frac{\int |\psi_{\mathbf{k}}|^2 exp(-f/k_B T) d\psi_{\mathbf{k}} d\psi_{\mathbf{k}}^*}{\int exp(-f/k_B T) d\psi_{\mathbf{k}} d\psi_{\mathbf{k}}^*}. \tag{18.69}$$

Similarly to Eq. (15.20) one obtains

$$< |\psi_{\mathbf{k}}|^2 > = \frac{k_B T}{\alpha(1 + k^2 \xi^2)} \equiv \frac{< |\psi_{\mathbf{k}=0}|^2 >}{1 + k^2 \xi^2}, \tag{18.70}$$

where $\xi(T)$ is a *coherence length* analogous to the one in Eq. (18.49):

$$\xi(T) = \frac{\hbar}{\sqrt{2m^* \alpha}} = \xi_0 \left(\frac{T_c}{T - T_c} \right)^{1/2} = \xi_0 \varepsilon^{-1/2}, \tag{18.71}$$

with $\xi_0 = \xi(T = 0)$. $\xi(T)$ enters in the *correlation function for the fluctuations*

$$g(\mathbf{r} - \mathbf{r}') \equiv < \psi(\mathbf{r}) \psi^*(\mathbf{r}') > = \sum_{\mathbf{k}} < |\psi_{\mathbf{k}}|^2 > e^{i\mathbf{k}\cdot(\mathbf{r}-\mathbf{r}')}.$$

In fact, from Eq. (18.70) and by integration in the reciprocal space,[31] for $\mathbf{R} = \mathbf{r} - \mathbf{r}'$ one derives

$$g(\mathbf{R}) = \frac{m^* k_B T}{2\pi \hbar^2} \frac{e^{-R/\xi}}{R}, \tag{18.72}$$

as it could be expected in the light of Eq. (15.23). In the present context $\xi(T)$ is a measure of the distance over which the *fluctuations are coherent*. In a pictorial view $\xi(T)$ is the "size" of the metastable superconducting "droplets" formed above the bulk transition temperature.

As regards the time dependence of these fluctuating "droplets" one can start from the general equation for the deviation from the equilibrium of an order parameter (see Eq. (15.25)), by writing

$$\frac{\partial \psi}{\partial t} = \frac{-1}{\gamma} \frac{\partial f}{\partial \psi^*}.$$

From Eq. (18.66) under the condition $\mathbf{A} = 0$

$$-\gamma \frac{\partial \psi}{\partial t} = \alpha \psi - \frac{\hbar^2}{2m^*} \nabla^2 \psi + \beta |\psi|^2 \psi. \tag{18.73}$$

---

[31] The sum over $\mathbf{k}$ should be limited to a cut-off value of the order of the inverse coherence length ξ_0^{-1}, somewhat in analogy with the cut-off at the boundary of the Brillouin zone or at the Debye wave vector (see also footnote 33).

When the non-linear terms are neglected (as it is suited when the order parameter is small) the relaxation time of the uniform ($\mathbf{k} = 0$) mode, for which $\nabla^2\psi = 0$, turns out[32]

$$\tau_{GL} = \frac{\gamma}{\alpha} \propto \frac{T_c}{T - T_c}. \tag{18.74}$$

Thus, according to Eqs. (18.66) and (18.70), the so called *linearized time-dependent GL equation* can be written in the form

$$- \tau_{GL}\frac{\partial\psi}{\partial t} = (1 - \xi^2\nabla^2)\psi , \quad \text{for } T > T_c. \tag{18.75}$$

When ψ is expanded as in Eq. (18.67) one finds that the correlation function for each mode of the collective quantity decays exponentially in time:

$$< \psi_\mathbf{k}^*(0)\psi_\mathbf{k}(t) > = < |\psi_\mathbf{k}|^2 > e^{-t/\tau_k},$$

where

$$\tau_k = \frac{\tau_{GL}}{1 + \xi^2(T)k^2}, \tag{18.76}$$

the equivalent of the slowing down of the fluctuations described at Sect. 15.3.

A proper average of $\tau_\mathbf{k}$, with boundary of the order of ξ_0^{-1}, can be considered a sort of *lifetime of the Cooper pair* τ_{CP}. Furthermore, one can extend the use of this lifetime, controlled by τ_{GL}, also in the condensed state below T_c (notice that the extension of $\tau_\mathbf{k}$ in the form as in Eq. (18.76) to k-values far from $k = 0$ is somewhat arbitrary). An heuristic estimate of τ_{CP} can be obtained by resorting to Heisenberg uncertainty principle. In fact, by observing that the lifetime of a pair is equivalent to the time required to destroy a pair, implying a transition from the boundaries of the energy gap 2Δ, from $\Delta t \simeq \hbar/\Delta E$ and considering that $\Delta E \simeq 2\Delta$ one can write $\tau_{CP} \simeq \hbar/2\Delta$. Since $2\Delta \sim k_B T_c$, for T_c around $10\,\mathrm{K}$ τ_{CP} would turn out of the order of picoseconds.

18.11.2 Paraconductivity and Fluctuating Diamagnetism

As a consequence of the SF's one can remark that channels with no electrical resistance are being opened above T_c. Thus the conductivity for $T \to T_c^+$ is expected to

[32]The *Ginzburg-Landau (GL) correlation time* is often written $\tau_{GL} = (\gamma\hbar/\alpha)$ by adding $\hbar$ to the coefficient in Eq. (18.74) in order to get correspondence with the expression $\tau_{GL} = \hbar/8k_B(T - T_c)$ derived in the microscopic theories. It can be observe that in this way, without the non-linear term, Eq. (18.73) takes a form consistent with the time-dependent Schrodinger equation.

exhibit the raise of an extra-contribution, usually called paraconductivity. A sketch of a typical experimental evidence[33] is reported below:

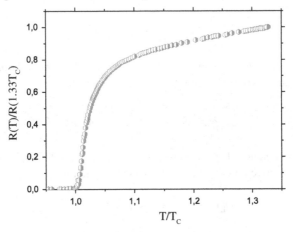

In the light of Eq. (18.43) the current density to be related to SF's is written (for $\mathbf{A} = 0$)

$$\mathbf{j} = \frac{e^*\hbar}{m^*}|\psi|^2\nabla\theta, \tag{18.77}$$

where $\nabla\theta$ has to be obtained from the linearized Eq. (18.73) for $\psi = |\psi|exp(i\theta)$. A term $e^*V\psi$ is added in order to account for an electric field $\mathcal{E}$. The imaginary part of that equation yields

$$\gamma\left[\frac{d\theta}{dt} + \frac{e^*V}{\hbar}\right] = \frac{\hbar^2}{2m^*}\nabla^2\theta, \tag{18.78}$$

implying the phase evolution $\theta(t) = \theta_0 - (e*/\hbar)\int_0^t Vdt$, as already discussed (see Eq. (18.30)).

For each mode $\mathbf{k}$ contributing to the extra current with a density of carriers $< |\psi_\mathbf{k}|^2 >$, with an acceleration given by $(dv_{SC}/d\tau_\mathbf{k}) = (\hbar/m^*)\nabla[\partial\theta/\partial\tau_\mathbf{k}] = e^*\mathcal{E}/m^*$, the extra-conductivity (responsible for the rounding in the resistivity above T_c) is written

$$\sigma_{SF} = \frac{\mathbf{j}}{\mathcal{E}} = \frac{e^{*2}}{m^*}\sum_\mathbf{k} < |\psi_\mathbf{k}|^2 > \tau_\mathbf{k}. \tag{18.79}$$

This form for σ_{SF} could be written on the basis of the analogy with the *Drude conductivity* (see Sect. 13.4)

$$\sigma_D = \frac{e^2 n_e \tau}{m}.$$

[33]A variety of data and extensive theory can be found in the book by *Larkin* and *Varlamov*. A comprehensive description of fluctuation effects, at a level comparable to the present text, is given at Chap. 8 of the book by *Tinkham*.

Equation (18.79) points out that paraconductivity involves an effective collision time and, instead of the electron density n_e, a kind of effective density of Cooper pairs averaged in the **k** space.

The fluctuating pairs associated to SF's also cause an extra-contribution to the magnetic susceptibility through a *negative* magnetization. This can somewhat be considered a precursor effect with respect to the *Meissner diamagnetism* described at Sect. 18.1.

For *evanescent magnetic field* (i.e. $H \rightarrow 0$), disregarding for the moment any field-related quenching of the pairs (see Sect. 18.12), from Eq. (18.66) in the Gaussian approximation, with $\mathbf{A} = \mathbf{H} \times \mathbf{r}/2$, the diamagnetic magnetization M_{dia} can be obtained by expanding the free energy in terms of the eigenfunctions for *Landau levels* E_{n,k_z} (see Appendix 13.1).

From the usual relation for the statistical average $< f > = -k_B T \ln Z$ in terms of the partition function

$$Z = \prod \frac{\pi k_B T}{E_{n,k_z}},$$

$< f >$ is evaluated *to the second order in H*, so that $M_{dia} = -d < f > /dH$ turns out *linear in* **H**. Finally the susceptibility could be found in the form

$$\chi_{dia} = -\frac{\pi k_B T}{6\Phi_0^2} \xi(T). \tag{18.80}$$

Instead of the lengthy procedure outlined above one can express an approximate expression of χ_{dia} equivalent to Eq. (18.80) by attributing to each Cooper pair a diamagnetic susceptibility as for atoms (see Sect. 4.5) with $< r^2 >$ substituted by $\xi^2(T)$, with the number of pairs per unit volume from the **k**-integration of $< |\psi_\mathbf{k}|^2 >$ (see Problem 18.11).

According to Eq. (18.80) the expected behaviour of the susceptibility would be as sketched below

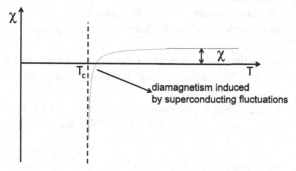

In reality the behaviour of the diamagnetic susceptibility for $T \rightarrow T_c^+$ is rather different from the one described above. First, for finite value of H one has the breakdown of the evanescent field approximation. For relatively large magnetic field

M_{dia} *is no longer linear in the field* and T_c *is decreased.* In other words, the field affects the fluctuating pairs and decreases the diamagnetic contribution to the magnetization.

Furthermore one often observes an *upturn* in the field dependence and $|M_{dia}|$ *decreases on increasing H.* The upturn in the magnetic field dependence of M_{dia} corresponds to the breakdown for finite field of the GL theory in the original formulation. The original theory has been integrated by *Gor'kov* to include short wave length (large k) fluctuations and non-locality effects and then the magnetization curves associated with the SF above T_c are fully accounted for (see the book by *Larkin* and *Varlamov*).

In Problem 18.12 it is shown how in the assumption of fluctuating superconducting droplets of size of the order or smaller than the coherence length (so that site-independent order parameter can be assumed, corresponding to *zero-dimensional condition*) then in a simple way the inversion in the field dependence of M_{dia} at the upturn field can be understood.

18.12 Superconducting Nanoparticles and the Zero-Dimensional Condition

A nice system that mimics the model of zero-dimensional condition, namely site-independent order parameter, is represented by a sample of independent nanoparticles each having diameter d smaller than the zero temperature coherence length $\xi(0)$. Then in Eq. (18.66) the condition of ψ site-independent is strictly valid and the gradient term can be set to zero. Meantime, by assuming spherical shape the potential vector is $A = H^2 d^2/10$ (see Problem 18.12), then from the complete GL functional including magnetic field and the term in $|\psi|^4$, the exact partition function and the field dependence of M_{dia} can be obtained also in the so-called *critical region*. In simple terms one can define as critical region the temperature range where the first-order perturbation theory breaks down and the term in $|\psi|^4$ in the GL functional cannot be neglected, the fluctuations having grown to high extent.

Typical temperature dependence of the diamagnetic susceptibility in nanoparticles is reported in Fig. 18.11.

It is noted that the first order correction theory (solid lines in the figures and corresponding to Eq. (18.80)) has to be abandoned close to the transition temperature and that χ_{dia} takes a finite value at T_c.

Going back to Eq. (18.66), for nanoparticles one can write the normalized wave function $\psi(\mathbf{r}) \equiv \psi_0 = \psi(V_{part})^{1/2}$, the GL functional being reduced to

$$f[\psi(\mathbf{r})] = f_0 + \alpha_0[T - T_c(0)]|\psi_0|^2 + \frac{\beta}{2}|\psi_0|^4 + \frac{e^{*2}A^2}{2m^*c^2}|\psi_0|^2, \qquad (18.81)$$

where $\alpha_0 = (\hbar^2/\xi_0^2 2m^*)$ while $A^2 = (H^2 d^2/10)$, as it is obtained in analogy to the evaluation of the moment of inertia of a sphere.

Fig. 18.11 Illustration of the temperature dependence of the diamagnetic susceptibility in the neighbourhood of the transition temperature in samples of independent Pb nanoparticles of diameter 750 Å (**a**) and 160 Å (**b**). The *solid lines* track the behaviour expected in the framework of the first-order fluctuation correction (Eq. (18.80)) [the experimental data have been taken at very small applied field, so that $T_c(H) \simeq T_c(0)$ (see Bernardi et al., Phys. Rev. B 74, 134509 (2006))

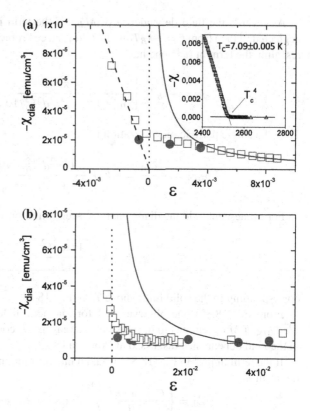

Thus one notices that the effect of the magnetic field is simply to renormalize the term in $|\psi_0|^2$. By remembering Sect. 18.9.3 one observess that the transition occurs at the temperature $T_c(H)$ where

$$\alpha_0\varepsilon + \frac{\pi^2\hbar^2H^2d^2}{10m^*\Phi_0^2} = 0, \qquad (18.82)$$

namely at

$$T_c(H) = T_c(0)\left[1 - \frac{4\pi^2H^2d^2\xi_0^2}{10\Phi_0^2}\right]. \qquad (18.83)$$

As regards the field dependence of M_{dia}, if the term in $|\psi_0|^4$ in Eq. (18.81) is neglected,[34] then from $f = -k_B T ln(\pi k_B T/\alpha_0 \varepsilon)$ (see Problem 18.12) again by simply rescaling the term in $|\psi|^2$ one has

$$f_0 = -k_B T ln \left[\frac{\pi k_B T}{\alpha_0 [\varepsilon + (4\pi^2 H^2 d^2 \xi_0^2/10\Phi_0^2)]} \right]. \tag{18.84}$$

Then the single particle magnetization is

$$M_{dia} = -\frac{\partial f_0}{\partial H} = -k_B T \frac{(4\pi^2 d^2 \xi_0^2/5\Phi_0^2)H}{\varepsilon + (2\pi^2 H^2 d^2 \xi_0^2/5\Phi_0^2)}. \tag{18.85}$$

For evanescent field the diamagnetic susceptibility turns out

$$\chi_{dia} = -k_B T \frac{4\pi^2 d^2 \xi_0^2}{5\Phi_0^2} \varepsilon^{-1}, \tag{18.86}$$

corresponding to the solid lines above T_c in Fig. 18.11.

From Eq. (18.85) one observes that for magnetic field of the order of $H_{up} \simeq \sqrt{\varepsilon} \Phi_0/\xi_0 d$ M_{dia} initiates to decrease on increasing H, corresponding to the upturn in the field dependence discussed at Problem 18.12.

If the term in $|\psi|^4$ in Eq. (18.81) is taken into account the partition function takes the form

$$Z_0 = \left(\frac{\pi^3 V_{part} k_B T}{2\beta} \right)^{1/2} |exp(x^2)(1 - erf(x))|,$$

with $x = \alpha(H)[V_{part}/2\beta k_B T]^{1/2}$ and the diamagnetic magnetization turns out

$$M_{dia} = \left[2\alpha/(\beta V_{part})^{1/2} \right] \left[\frac{k_B^{3/2} T_c T^{1/2} H\pi^2 \xi_0^2 d^2}{2.5\Phi_0^2} \right] \left[x - \frac{exp(x^2)}{\pi^{1/2}(1 - erf(x))} \right]$$

This expression accounts for susceptibility almost temperature independent within the critical region (see Fig. 18.11).

---

[34] To give a quantitative estimate of the temperature range where this assumption is appropriate, it is mentioned that the *Ginzburg-Levanyuk parameter* above which the first-order fluctuation correction is substantially valid, in zero-dimension is given by

$$G_i(0) \simeq 13.3[T_c(0)/T_F][\xi_0/d]^{3/2}$$

with T_F Fermi temperature. (For details see Chap. 2 of the book by *Larkin* and *Varlamov*.)

Problems

Problem 18.1 Give a qualitative description of the lack of electric resistance by considering the scattering event of a Cooper pair.

Solution: For an electric field not so strong to destroy the pairing each pair is accelerated in a way analogous to what illustrated at Sect. 13.4 for the electrons. When scattering occurs one electron changes its individual **k** and the second electron readjusts its one **k** in order to regain the wave function in the form (18.6), after the scattering event. Thus no change of the momentum of the pair occurs and the scattering is ineffective. The currents flows through the action of the field on the pairs as a whole and no change of state due to impurities or lattice vibrations is induced. The Cooper pair becomes the charge carrier introduced by Londons in their phenomenological equations.

Problem 18.2 From Eq. (18.5), derive Eq. (18.7).

Solution: The wave function can be written in the k-space in the form

$$\psi(\mathbf{r}) = \sum_{\mathbf{k}} g_{\mathbf{k}} e^{i\mathbf{k}\cdot\mathbf{r}}.$$

In the absence of interaction the electrons would set in a given **k** eigenstate and therefore only one coefficient $g_{\mathbf{k}}$ would be present in the summation. The effect of the interaction is to scatter the pair in other states and more than one coefficient will be different from zero. The states below E_F being occupied $g_{\mathbf{k}} = 0$ for $k < k_F$.

When $\psi(\mathbf{r})$ is substituted in Eq. (18.5) and both members are multiplied by $exp(-i\mathbf{k}' \cdot \mathbf{r})$ and then integrate over the volume V_C, one gets

$$\frac{\hbar^2 k^2}{2m} g_{\mathbf{k}} + \sum_{\mathbf{k}'} g_{\mathbf{k}'} V_{\mathbf{k},\mathbf{k}'} = (\Delta E + 2E_F) g_{\mathbf{k}},$$

where $V_{\mathbf{k},\mathbf{k}'} = (1/V_C) \int V(\mathbf{r}) exp[i(\mathbf{k} - \mathbf{k}') \cdot \mathbf{r}] d\mathbf{r}$ (note that $(1/V_C) \int exp[i(\mathbf{k} - \mathbf{k}') \cdot \mathbf{r}] d\mathbf{r} = \delta_{\mathbf{k},\mathbf{k}'}$). The scattering matrix element $V_{\mathbf{k},\mathbf{k}'}$ can be different from zero only when the two states are separated by an energy of the order of $\hbar\omega_D$ around the Fermi surface. Thus one can assume $V_{\mathbf{k},\mathbf{k}'} = -V$ (attractive potential) for $\hbar^2 k^2/2m \simeq \hbar^2 k'^2/2m < E_F + \hbar\omega_D$ and zero otherwise. Then

$$\frac{\hbar^2 k^2}{2m} g_{\mathbf{k}} + \sum_{\mathbf{k}'} g_{\mathbf{k}'} V_{\mathbf{k},\mathbf{k}'} = (\Delta E + 2E_F) g_{\mathbf{k}} \simeq \frac{\hbar^2 k^2}{2m} g_{\mathbf{k}} - V \sum_{\mathbf{k}'} g_{\mathbf{k}'} = (\Delta E + 2E_F) g_{\mathbf{k}},$$

with the sum limited to the energy shell specified above. Therefore

$$g_{\mathbf{k}} = \frac{A}{\Delta E + 2E_F - (\hbar^2 k^2/m)}$$

with $A = -V \sum_{\mathbf{k}'} g_{\mathbf{k}'}$ $\mathbf{k}$ independent, yielding

$$A = -V \sum_{\mathbf{k}} A \left[\Delta E + 2E_F - (\hbar^2 k^2/m) \right],$$

that can be rewritten

$$1 = V \sum_{\mathbf{k}} \frac{1}{2\zeta - \Delta E},$$

being ζ the departure of the energy from the Fermi energy, $\zeta = (\hbar^2 k^2/2m) - E_F$. By transforming the sum over an integral over the energy range E_F and $E_F + \hbar \omega_D$ (see Sect. 12.5 and Eq. (12.20)) and assuming that the density of states $D(E)$ in terms of ζ can be written $D(\zeta) \simeq D(0)$, one has

$$1 = \frac{1}{2} V D(0) ln \left(\frac{\Delta E - 2\hbar \omega_D}{\Delta E} \right)$$

giving

$$\Delta E = \frac{2\hbar \omega_D}{1 - exp(2/D(0)V)}$$

For small V, so that $exp(2/D(0)V) \gg 1$, Eq. (18.7) is found.

This sketch of treatment, with several approximations, can only be considered the starting point for the full BCS theory. Besides the books by *Grosso* and *Pastori Parravicini* and by *Annett* mentioned at Sect. 18.2, for presentations of the BCS picture of the superconducting state the books by *Goodstein* or by *Ibach* and *Luth* are also suggested.

Problem 18.3 Give an approximate estimate of the size of the Cooper pair by using the uncertainty principle.

Solution: An approximate estimate of the range ξ or the "size" of the pair, can be given as follows. Since the energies involved are of the order of Δ and therefore $\delta E \simeq \delta(p^2/2m) \simeq (\hbar k_F/m)\delta p \simeq v_F \delta p$, from

$$\xi \simeq \frac{\hbar}{\delta p} \simeq \frac{E_F}{k_F \Delta}$$

one obtains $\xi \sim 10^3$–10^4 Å, for the typical values of $\Delta = E_F(10^{-3}$–$10^{-4})$.

Since about 10^{19}–10^{20} electrons are involved in the pairing, in the "volume" of a pair, namely $\xi^3 \simeq 10^{-12} - 10^{-15}$ cm^3, one finds the center of mass of about 10^5–10^8 other Cooper pairs.

Problem 18.4 From the London equations evaluate the magnetic field and current density in a bulk superconductor occupying the half-space above the xy plane and inside an infinite superconductor slab according to the geometry sketched below.

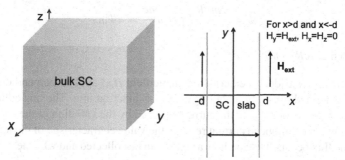

For x>d and x<-d
$H_y=H_{ext}$, $H_x=H_z=0$

Solution: For symmetry reasons $\mathbf{B}$ and $\mathbf{j}$ depend on z but not on x and y. Furthermore $j_z = 0$. Only components in the xy plane are expected for $rot\mathbf{j}$ and $\mathbf{B}$ (See Eq. (18.18)). Since one can set $B_x(z) = B_x(0)f(z)$, from $\partial^2 f/\partial z^2 = (1/\lambda_L)f$ one derives $f(z) = f(0)exp(-z/\lambda_L)$. The same z-dependence holds for $\mathbf{j}$ and $\mathbf{B}$. It is noted, however, that $\mathbf{j}$ and $\mathbf{B}$ are perpendicular, within the xy plane.

For the slab, field and current are functions of x only. Since the electric field is zero $\mathbf{j}$ is constant in time. From Eq. (18.20) $d^2 B_y(x)/dx^2 = B_y(x)/\lambda_L^2$, with the boundary condition $B_y(\pm d) = H_{ext}$, one has

$$B_y(x) = H_{ext} \frac{cosh(x/\lambda_L)}{cosh(d/\lambda_L)}.$$

For the current density, from

$$(4\pi/c)\mathbf{j} = rot\mathbf{B} = -\mathbf{k}\partial B/\partial x = -\mathbf{k}(1/\lambda_L)H_{ext}sinh(x/\lambda_L)/cosh(d/\lambda_L)$$

and from $\mathbf{B} = \mathbf{H}_{ext} + 4\pi\mathbf{M}$,

$$rot\mathbf{M} = \frac{\mathbf{j}}{c}.$$

Problem 18.5 In the assumption that the supercurrent along a ring can be considered equivalent to two electrons running friction-free along one of the Bohr orbit in the Hydrogen atom, prove that the flux quantization can be obtained from the Bohr-Sommerfeld condition (see Problem 1.4.4) for the angular momentum.

Solution: One can imagine that the current has been launched by increasing an external field perpendicular to the ring. The e.m. force $\mathcal{F} = (-1/c)d\Phi/dt$ corresponds to an electric field $\mathcal{E} = \mathcal{F}/2\pi R = -(1/2\pi Rc)d\Phi/dt$. The acceleration of the charge $2e$ yields $2e\mathcal{E} = 2mdv/dt$ and then

$$d\Phi = -(2\pi c/e)d(mvR).$$

From the Bohr condition $\int d(mvR) = n\hbar$ one has $\Phi = -n(hc/2e)$.

Analogous result is obtained by writing $\Phi = Li$, with L the inductance of the ring, the energy being $(1/2)Li^2 = (1/2)(2mv^2)$ and the current $i = 2ev/2\pi R$. Thus

$$\Phi = \frac{2mvR}{2e}2\pi = n\frac{hc}{2e} = n\Phi_0,$$

having used $2mvR = n\hbar$.

Problem 18.6 Consider an external magnetic field H_{ext} applied perpendicularly to the surface defined by a ring including an S-N-S junction in it. The flux collected is due to H_{ext} plus the one due to the current induced by the variation of the field itself. Show that when the induced current reaches the value characteristic of the junction a jump in the flux occurs. Discuss how a signal can be collected and why the sensitivity is so high.

Solution: The flux, including the term due to the induced current, is $\Phi = \Phi_{ext} + Li/c$. When the superconducting current reaches I_c:

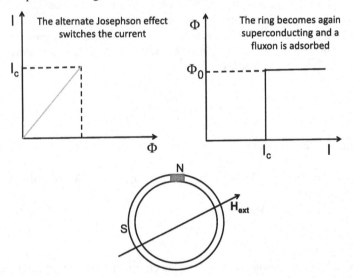

The change in the flux is very fast (it occurs in a time of the order of picosecond), then $d\Phi/dt$ is very large and another coil nearby the ring can collect the signal.

Problem 18.7 By resorting to the second GL equation and extracting the density current and circulating it along a close ring, prove that flux quantization can be derived.

Solution: From Eq. (18.43), since

$$\nabla\psi = i\psi\nabla\theta + e^{i\theta}\nabla|\psi(\mathbf{r})|$$

the current density turns out $\mathbf{j} = (\hbar e^*/m^*)|\psi|^2 - (e^{*2}/m^* c)|\psi|^2\mathbf{A}$. From the circulation along a ring, by taking into account that $\int_{circ}\nabla\theta\cdot d\mathbf{l} = n2\pi$ and that $\int_{circ}\mathbf{A}\cdot d\mathbf{l} = \Phi$, once again $\Phi = n\Phi_0$.

Problem 18.8 By estimating the condensation energy for each Nb atom show that the gain in free energy density in the superconducting phase is very small compared to the thermal and Fermi energies.

Solution: For Nb, with $T_c = 9.1\,\text{K}$, the critical field at zero temperature is around 2000 Oe. The condensation energy is about $H_c^2/8\pi \simeq 1.6 \times 10^5\,\text{erg/cm}^3$.

In Nb metal (bcc structure and lattice constant 3.3 Å) the volume of the cell is about $3.6 \times 10^{-23}\,\text{cm}^3$. The condensation energy per atom turns out $1.8\,\mu\text{eV}$. According to the BCS theory the condensation energy is of the order of $(k_B T_c)^2 D(E_F) \simeq k_B T_c(T_c/T_F)$.

Problem 18.9 Derive approximate expressions of the critical fields H_{c1} and H_{c2} from the estimate of the flux in correspondence to the first vortex and when the vortices are in contact.

Solution: When the first vortex is entering the sample H_{ext} is about H_{c1} and then $H_{c1}\pi\lambda_L^2 \simeq \Phi_0$ yielding $H_{c1} \simeq \Phi_0/\pi\lambda_L^2$. When H_{ext} is close to H_{c2} the effective area including a fluxon is now reduced to $\pi\xi^2$. Thus $H_{c2} \simeq (\Phi_0/\pi\xi^2) \simeq \sqrt{2}\kappa H_c$, $H_{c1} \simeq H_c/\sqrt{2}\kappa ln\kappa$ and $H_{c1}H_{c2} \simeq H_c^2 ln\kappa$ (Eq. (18.58)). It is reminded that, for isotropic superconductors, more correct expressions for the critical fields are $H_{c1} \simeq \Phi_0/4\pi\lambda_L^2$ and $H_{c2} \simeq \Phi_0/2\pi\xi^2$.

Problem 18.10 From Eq. (18.48) directly obtain an approximate estimate of the coherence length (for the deduction based on the fluctuation effects see Sect. 18.11.1).

Solution: From

$$\xi^2\frac{d^2f}{dx^2} + f - f^3 = 0,$$

by introducing $(g - 1) = f$ (with $f \ll 1$), to the first order $f - f^3 \simeq (1 + g) - (1 + 3g + \cdots) \simeq -2g$ and $d^2g/dx^2 \simeq 2g/\xi^2$ and $g(x) \simeq exp(\pm\sqrt{2}x/\xi(T))$ with $\xi^2(T) = \hbar^2/2m^*|a(T)|$.

Problem 18.11 By assuming for each pair due to superconducting fluctuations above T_c the diamagnetic susceptibility as for a system of two electrons at a distance given by the coherence length, express the diamagnetic susceptibility (for unit volume) to be attributed to the fluctuating diamagnetism.

Solution: From $\chi_{at} = -e^2 < r^2 > /mc^2$ with $< r^2 > \simeq \xi^2(T) = [\hbar^2/2m^*\alpha(T)]$, by evaluating the number density of Cooper pairs from

$$< |\psi|^2 > = \sum_{\mathbf{k}} < |\psi_{\mathbf{k}}|^2 > = \frac{1}{8\pi^3} \int_0^{1/\xi_0} 4\pi k^2 < |\psi_{\mathbf{k}}|^2 > dk$$

then Eq. (18.80) follows (it is noted that it is an approximate expression).

Problem 18.12 By referring to a model situation of fluctuating superconducting droplets of spherical shape and size d smaller than the coherence length (so that the order parameter can be considered site-independent) from Eq. (18.66), by considering that the potential vector is $A = H^2 d^2/10$ (analogously to the moment of inertia of a sphere), derive the effect of the magnetic field in correcting the transition temperature $T_c(0)$ and explain why the magnetization curves have an upturn in the field dependence of M_{dia} versus H.

Solution: The partition function for zero dimension is $Z_0 = \int d^2\psi_0 exp(-f[\psi_0]/k_B T)^{35}$ with

$f[\psi_0] = \alpha(T)|\psi_0|^2 + (e^{*2}/2m^*c^2)A^2|\psi_0|^2$. For $A = 0$ one would have $F_0 = -k_B T ln Z_0 = -k_B T ln[\pi k_B T/\alpha(T)]$. With $A \neq 0$ one can simple correct this result by writing

$$F_0 = -k_B T ln\left(\frac{\pi k_B T}{\alpha_0[\varepsilon + (4\pi\xi^2 < A^2 > /\Phi_0^2)]}\right),$$

with $\alpha_0 = \hbar^2/4m\xi_0^2$ (according to the GL coherence length, see Eq. (18.71)) while $\Phi_0 = hc/2e$. Then

$$T_c(H) = T_c(0)\left[1 - \frac{4\pi\xi^2 H^2 d^2}{10\Phi_0^2}\right]$$

and

$$M_{dia}(\varepsilon, T) = -\frac{\partial F_0}{\partial H} = -k_B TH \frac{(4\pi^2\xi_0^2 d^2/5\Phi_0^2)}{\varepsilon + (2\pi^2\xi_0^2 H^2 d^2/5\Phi_0^2)},$$

with an upturn field at $H_{up} \simeq \varepsilon\Phi_0/4\xi_0^2$ (for $d \simeq \xi$).

Problem 18.13 In the assumption that the first GL equation holds at all temperatures below $T_c(0)$, evaluate the temperature dependence of the upper critical field $H_{c2}(T)$.

Solution: Let us refer to the transition from the normal to the superconducting state occurring in the presence of an external field at $T_c(H) < T_c(0)$. For type II superconductors the transition is *second order* (see Sect. 18.10). Thus at the onset of the superconductivity the concentration of Cooper pairs $n_c = \psi^*\psi$ is small and the term $b|\psi|^2\psi$ in Eqs. (18.41) or (18.51), can be neglected. For the magnetic field inside

---

[35]This is the expression coming as functional integral $\int d^2\psi(\mathbf{r})exp(-f[\psi(\mathbf{r})]/k_B T)$ over the "field" $\psi(\mathbf{r})$. Thus $Z_0 = \int_0^\infty d|\psi_0|^2 exp(-\alpha|\psi_0|^2/k_B T) = \pi k_B T/\alpha(T)$.

the material one can assume $B \simeq H_{ext}$. By referring to the second Landau gauge for which $\mathbf{A} = H_{ext}(0, x, 0)$, Eq. (18.51) is written

$$\left(\frac{\partial^2}{\partial x^2} + \frac{\partial^2}{\partial z^2}\right)\psi + \left(\frac{\partial}{\partial y} - i\frac{2eH_{ext}x}{\hbar c}\right)\psi = -\xi^{-2}\psi$$

as already used for the Landau diamagnetism (Eq. (A.13.1.1)) and reminiscent of the Schrodinger equation for free particles in magnetic field. As for the derivation of the *Landau levels* (Appendix 13.1) the eigenfunctions are of the form

$$\psi(x) = \phi(x)e^{i(k_y y + k_z z)}$$

and by redefining $x' = x - (\hbar c/2eH_{ext})k_y$ one has

$$\left(-\frac{\hbar^2}{2}\frac{d^2}{dx'^2} + \frac{2e^2 H_{ext}^2}{c^2}x'^2\right)\phi(x') = \left(\frac{\hbar^2}{2\xi^2} - \frac{\hbar^2 k_z^2}{2}\right)\phi(x')$$

(see Eq. (A.13.1.2)). This equation is formally the one for a displaced harmonic oscillator (of unit mass) for which the eigenvalues are $E_n = (n + 1/2)\hbar\omega_c$ with $\omega_c = 2eH_{ext}/c$.

Thus one writes

$$\frac{\hbar^2}{2\xi^2} - \frac{\hbar^2 k_z^2}{2} = \left(n + \frac{1}{2}\right)\frac{2\hbar e H_{ext}}{c}.$$

The maximum allowed value of H_{ext} with non-zero solution is $H_{ext}^{max} = H_{c2}(T)$ corresponding to $n = 0$ an $k_z = 0$. Therefore

$$H_{c2}(T) \equiv H_{ext}^{max} = \frac{\hbar c}{2e}\frac{1}{\xi^2(T)} = \frac{\hbar c}{2e\xi^2(0)}\left(1 - \frac{T}{T_c}\right).$$

Problem 18.14 By considering a gauge transformation that adds $\nabla f(\mathbf{r})$ to the vector potential and that changes the phase of the order parameter, prove that the Ginzburg-Landau theory satisfies the gauge invariance.

Solution: From $\mathbf{A}(\mathbf{r}) = \mathbf{A}(\mathbf{r}) + \nabla f(\mathbf{r})$ by considering the momentum operator $\mathbf{p} = -i\hbar\nabla + (e^*\mathbf{A}/c)$ (see Eq. (18.38)), by changing the order parameter from $\psi(\mathbf{r})$ to $\psi(\mathbf{r})exp(i\theta(\mathbf{r}))$, one sees that

$$e^{i\theta(\mathbf{r})}\left(-i\hbar\nabla + \frac{e^*\mathbf{A}}{c}\right)\psi(\mathbf{r}) + \psi(\mathbf{r})e^{i\theta(\mathbf{r})}\hbar\nabla\theta(\mathbf{r}) =$$

$$= e^{i\theta(\mathbf{r})}\left[-i\hbar\nabla + e^*\left(\mathbf{A} + \frac{\hbar}{e^*}\nabla\theta(\mathbf{r})\right)\right]\psi(\mathbf{r}).$$

Thus, when the phase is changed, transforming $\psi(\mathbf{r})$ to $\psi(\mathbf{r})exp(i\theta(\mathbf{r}))$, and $\mathbf{A}$ transformed into $\mathbf{A} + (\hbar/e^*)\nabla\theta(\mathbf{r})$ the free energy is unchanged, as well as all the physical variables.

Specific References and Further Reading

1. G. Grosso and G. Pastori Parravicini, *Solid State Physics*, 2nd Edition, (Academic Press, 2013).
2. J.F. Annett, *Superconductivity, Superfluids and Condensates*, (Oxford University Press, Oxford, 2004).
3. W. Buckel, *Superconductivity- Fundamental and Applications*, (VCH Weinheim, 1991).
4. D.L. Goodstein, *States of Matter*, (Dover Publications Inc., 1985).
5. H. Ibach and H. Lüth, *Solid State Physics: an Introduction to Theory and Experiments*, (Springer Verlag, 1990).
6. C.P. Poole, H.A. Farach, R.J. Creswick, *Superconductivity*, (Academic Press, San Diego, 1995).
7. R.P. Feynman, R.B. Leighton and M. Sands, *The Feynman Lectures on Physics Vol. III*, (Addison Wesley, Palo Alto, 1965).
8. M. Tinkham, *Introduction to Superconductivity*, (Dover Publications Inc., New York, 1996).
9. A. Larkin and A.A. Varlamov, *Theory of Fluctuations in Superconductors*, (Oxford Science Publications, Clarendon Press, Oxford, 2005).
10. N.W. Ashcroft and N.D. Mermin, *Solid State Physics*, (Holt, Rinehart and Winston, 1976).
11. A. Barone and G. Paternó, *Physics and Applications of the Josephson Effect*, (John Wiley, New York, 1982).
12. J.G. Bednorz and K.A. Müller, *Early and Recent Aspects of Superconductivity*, (Springer-Verlag, Berlin, 1990).
13. G. Burns, *High Temperature Superconductivity - An Introduction*, (Academic Press Inc., 1992).
14. M.Cyrot and D.Pavuna, *Introduction to Superconductivity and High-T_c Materials*, (World Scientific, Singapore, 1992).
15. P.G. de Gennes, *Superconductivity of Metals and Alloys*, (Addison-Wesley, 1989).

Index

A

ABMR, 202
Abrikosov A.A., 540, 559, 564, 574
Absorption (and emission)
    coefficient, 44
Adiabatic approximation, 224, 354
Adiabatic demagnetization, 197, 210
Ag atom, 115, *see also* silver
Ag crystal, 393, *see also* silver
AgBr crystal, 346
AgMg crystal, 346
Al or aluminum crystal, 346, 382
Al_2O_3, 285
Alkali
    atoms, 1, 3, 61, 67, 131, 136
    crystals, 1, 393, 394
    hyperfine field, 161
    see also Li, Na, K, Rb, Cs and Fr
Alkali halides, 304
AlNi crystal, 346
Amit, D.J., 197
Ammonia maser, 281, 284
Ammonia molecule, *see* NH_3
    deuterated, 282
    in electric field, 281
Anharmonic potential, 469
Anharmonicity, 304–306, 308
    electrical, 305
    mechanical, 312
Annett, J.F., 545
Annihilation process, 24
Antibonding orbital, 245
Antiferroelectric, 446
    phase, 489
Antiferromagnetic state, 140
Antiferromagnets, 449, **513–517**
    one-dimensional (1D), 537

Antiproton (gyromagnetic ratio), 181
Antisymmetrical wavefunctions, 73, 74
Antisymmetry, 5, 6, 69, 74
As atom, 99
Aschcroft, N.W., 338
Aschcroft, N.W. and Mermin, N.D., 520
AsH_3 molecule, 282
Asymmetry parameter
    (of the electric field gradient), 167
Atomic diamagnetism, 140
Atomic orbitals, 11
Atomic polarizability, 123, 124
    see Stark effect
    of Hydrogen (ground state), 124
Atomic units, 145
Au crystal, 346
Auger effect, 70
Auto-correlation function, 151
Auto-ionizing states, 70

B

B^{3+} atom, 80
B_2 molecule, 252
Ba atom, 184
Balmer spectroscopic series, 15, 17, 174
Balzarotti, A., 86
Band of levels
    degenerate, 379
    in crystals, 354, 355, 364
    overlap, 380, 393
Band spectra (in molecule), 299
Bardeen, Cooper and Schriffer (BCS), 540
    theory, 540, 545, 547, 549, 564, 565, 587
Barn, 168
$BaTiO_3$, 277, 346
Be^{2+} atom, 73, 80

© Springer International Publishing Switzerland 2015
A. Rigamonti and P. Carretta, *Structure of Matter*,
UNITEXT for Physics, DOI 10.1007/978-3-319-17897-4

Be crystal, 393
BeCu crystal, 346
Bednorz, J.G. and Müller, K.A., 527, 541
Benzene molecule, 278, 280, 298, 355, 375,
     see also C_6H_6
Berezinskii-Kosterlitz-Thouless, 453
Bessel functions, 573
Bi atom, 181
     crystal, 393
Binary alloys, 446, 449
Black-body radiation, 44, 46, 52
Blakemore J.S., 368
Bloch equations, 204
     orbital, 354, 356, 375, 392
     oscillations, 366, 372, 408
     $T^{3/2}$ law, 518
     wave packet, 406, 418
Blundell, S.J., 145
Blundell, K.M., 470
Bohm-Aharonov effect, 558
Bohr atom, 7
     magneton, 26, 144
     model, 15, 26, 30, 173
     radius, 7, 145
     radius in positronium, 163, 414
Bohr-Sommerfeld condition, 15, 153
Bohr-van Leeuwen theorem, 383
Boltzmann statistics, 47, see also Maxwell-
     Boltzmann
Bonding (and anti-bonding) orbitals, 242,
     245
Born-Mayer repulsion, 263, 394, 400
Born-Oppenheimer separation, 224, 281,
     289, 353, 417
Born-Von Karmann, 48, 359
Born-Von Karmann boundary conditions, 48
Bose-Einstein condensate, 446
Bose-Einstein statistical distribution func-
     tion, 49
Bosons and bosonic particles, 327
Br_2, 252
Brackett series, 17
Bragg law, 345
     reflection, 374, 438
Branches, acoustic and optical, 417, 419,
     427, 439
Bravais lattice, 338, 346
Breit-Rabi diagram, 180
Brillouin cell and zone, 343–345, 350
     expansion, 514
     function, 138
     reduction to, 361, 423, 425
Broken symmetry, 448

Buckel, W., 557, 559
Budker, D., 40, 268
Bulk modulus, 399, 435
Burns, G., 338

C
C_2H_2 molecule, 328
C_2H_4 molecule, 275, 278
C_3H_3 molecule, 286
C_6H_6, 278, see also benzene
     C_6H_6 see benzene
     susceptibility, 281
C atom, 263, 274, 278, 286, see also carbon
     atom
Ca atom, 147
     crystal, 346
Calcium atom, 75
Canonical moment, 153
Carbon atom, 272, 274, 275, 277, 278, 286
Cellular method, 357
Central field approximation, 1–3
Central peak, of diffusive type, 461
Centrifugal distortion, 312
Centrifugal term, 7
Cesium maser, 285
CH, 263
CH_4, 274, 277
Charge transfer, 256
Chirality, 521
Chu, P.C.W., 541
Cini, M., 86
Cl, 263, 265
Cl^-, 263, 266
Clapeyron equation, 448, 449
Clausius-Mossotti relation, 307
Clebsch-Gordan coefficients, 92, 114
Cleeton, 282
Clementi-Raimondi rules, 112
Closed shells, 91
Cluster spin glass, 562
CO atom, 98
CO molecule, 254
     vibrational constant, 304
CO_2 molecule
     modes infrared active, 320
     normal modes, 318
     rotational levels, 332
Coarse-grain average, 449
Cochran, W., 428
Coexistence line, 446
Coherence length, 540, 549, 550, 563
Cohesive energy, 394, 401

for inert atoms crystals, 398
in KBr, 400
in molecular crystals, 398
Cole-Cole relationships, 499
Collapse transitions (in gels), 446
Compound doublets, 67
Compressibility, 399, *see also* bulk modulus
Conductivity, 405, *see also* electric transport
Configuration interaction CI, 259
Contact term, 160, 161, 163, 188, *see also*
  Fermi contact interaction
Cooper and Cooper pairs, 410, 457, 540, 545,
  554, 555, 568, 579, 584
  lifetime, 577
Copper, 213
  crystal, 348, 434
Copper oxides (cuprates), 511, 541
  Hg-based, 541
Correlation   diagram   (separated-united
  atoms), 231
Correlation effects, 113, 411
  electronic, 505–509
  energy, 561
Correlation length, 453
  in 2DQHAF, 524–530
Correlation time, 153
  distribution and effects on dielectric con-
    stants, 498–502
Correspondence principle, 23, 154, 292
Cosmological principle, 50
Coulomb gauge, 550
Coulomb integral, 71, 74, 76, 111
Covalent crystals, 391, 394
Cr^{3+} ion, 146
$CrBr_3$, 450
Creswick, R.J., 572
Critical
  behaviour, 445, 497
  current, 540
  divergences, 446
  exponent, 453, 489, 499, 525, 529, 531
  exponents, dynamical, 463
  Ising-like pseudospin, 500
  isochore, 448
  point, 362, 422, 445, 446, 454
  quantum, 525, 564
  region, 458, 514, 580
  temperature, 446
  variable, local, 466, 486, 488
  wave-vector, 459, 461
Critical fields
  H_{c1} and H_{c2}, 559, 563
  crystal field, 401

thermodynamic, 559, 572
Crystal field, 401
Crystallography (elementary), 338
Crystal momentum, 410
Crystal structures, 338
Cs atom, 162, *see also* Alkali atoms
  spin-orbit doublet, 67
CsCl crystal, 345
Cu crystal, *see* copper
Cu^{2+} ions, 560
$Cu_{1-x}Mn_x$, 511
$CuGeO_3$, 537
CuO_6 octahedron, 277
Curie law, 138
  susceptibility, 142, 488, 502
Curie-Weiss temperature, 514, 516
Current density, 405, 408
Cyclotron frequency, 153
Cyclotron motion and frequency, 360
Cylindrical coordinates, 228–231

D
d^2sp^3 hybridization, 277, 278, 393
1D Ising model, 508
D_2 molecule dissociation energy vs H_2
  and Raman spectra, 328
  and zero point energy, 302, 306
DBr molecule, 294
De Broglie, 345
De Broglie wavelength, 367, 372, 374, 385
De Haas-Van Alphen oscillations, 410
De Mille, D.P., 40, 268
Debye, 212
  correlation time, 482
  frequency and wave-vector, 406
Debye frequency, 576
  model for lattice vibration, 417, 419, 425,
    426, 428, 430, 466, 483, 541
  radius, 427
  temperature, 431, 432, 434, 439, 442
  temperature for elements, 431, 433, 434,
    441, 547, 548
Debye-type relaxor, 464
  relaxation time, 464
Debye-Waller factor, 433, 438
Degeneracy from dynamical equivalence, 6
Degeneracy, accidental and necessary, 11,
  126
Degree of ionicity, 262
Delocalization, 272
Density (of k-modes or of k-states), 48, 363
Density (of modes or of energy states), 364,
  547

Density functional theory, 113, 271
Density matrix, 142, 150, 205, 451, 492, 494
Depolarization, factor and field, 480
Determinantal eigenfunctions, 61
Deuterium, 22, 23
    quadrupole moment, 175
Diamagnetic levitation, 544
Diamagnetic susceptibility, 122, 141, 142,
        281, 288, 544, **564**, 587
    for inert gas atoms, 141, 142
    in superconducting nanoparticles, **580**
Diamagnetic suspension, 544
Diamagnetism (atomic), 141
    Landau, 409
    perfect, 543
    superdiamagnets, 544
Diamond, 347, 379, 391, 393, 394, 427, *see
        also* carbon
Diatomic
    crystal, 346, 423, 439, 442
    one-dimensional crystal, 423, 425
Dielectric measurements, 465
Dielectric response, ionic, **481, 482**
Dielectrics, 446
Diffuse (series lines), 67
Digonal hybridization, 277
Dilution model, **527–529**
Dipolar alphabet, 182
Dipolar field, 182
Dipole magnetic moment, 157
    field induced, 322
Dipole-dipole interaction, 164, 181
Dirac, 26, 28, 58, 90, 162, 174
    δ function, 42, 375
Disordered crystals, 446
Dispersion relation, 419
    vibrational, 419, 427
Dispersion relations, 363, 423, 424, 426, 517
Displacive ferroelectrics, 485, 492
Dissociation energy, 252
Distribution (of the Maxwellian velocities),
        53, 56
Dolinsek, J., 530
Doppler modulation, 435
    broadening, 53, 56
    first and second-order, 444
    second-order shift, 444
Double exchange (mixed valency), 516
Double excited states, 70
Doublet (spin-orbit), 61
    for alkali atoms, 90
Double-well, 469
2DQHAF, **524–530**

spin dilution and spin doping, 525
Drude model and conductivity, 405
Dy^{3+}, 116
Dynamical equivalence, 5, 6, 248
Dynamical matrix, 419, 421
Dynamical structure factor, 460–462, 465,
        472, 475
Dzyaloshinsky-Moriya, anisotropic ex-
        change, 513

E
Edward-Anderson order parameter, 517
Effective electron mass, 365, 385
Effective hyperfine field, 161
Effective nuclear charge, 5, 72, 78, 83, 142
Effective potential, 3, 5
e_g levels, 13, 404
Ehrenfest criterion, 452
Einstein
    model of crystal, 417, 425, 426, 431, 432,
        436, 441
    relations, 41, 44, 45
    relativity theory, 443
    temperature, 431
Electric and magnetic field effects in atoms,
        121
Electric dipole approximation, 41
    induced, 124, 319
    mechanism of transition, 41, *see also* se-
        lection rules
    of molecules, 230
    oscillating (in crystals), 405, 424
    quantum, associated to a pair of states,
        43
Electric field gradient, 169, *see also* quadru-
        pole interaction
Electric polarizability, 125
    for quantum oscillator, 128
    rotational, 283
Electric quadrupole (mechanism of transi-
        tion), 43, 118
Electric quadrupole moment, 166, *see also*
        quadrupole interaction
    of deuteron, 175
Electric quadrupole, selection rules, 114
Electric transport, 405
Electrical harmonicity, 320, 417
Electrical permeability, 143
Electro-optical properties, 346
Electromagnetic ranges, 36, 409
Electromagnetic symmetry, 144
Electromagnetic units, 145

Electron affinity, 80, 258
Electron states in crystals, 353, 379
Electron-electron repulsion, 74
   for atoms in the ground state, 141
   repulsive interaction, 2
Electronic charge transfer, 261
Electronic configuration, 33, 98
   in atoms, 33, 91
   in molecules, 228
Electronic spectra, 324
Eliashberg, 549
Ellipsoidal coordinates, 230
Empty lattice model, 363, 369
Energy functional, 78
Entropy
   from quadrupole levels, 221
   from rotational motion, 295, 300
   in spin systems, 210
   magnetic and lattice, 212
   of radiation, 50
   vibrational, 421
Entropy and specific heat, 177, 221, *see also*
   Specific heat
EPR, 178, 181, 204
Equipartition principle, 47
Eu^{2+} atom, 104
Eu^{3+} atom, 99
Evanescent field condition, 144, 218
Ewald procedure, 395
Exchange and correlation, 506, 507
Exchange degeneracy, 1, 92
Exchange frequency
   Hamiltonian, 75
Exchange integral, 74, 90, 98, 140, 200
   extended (in molecules), 255
   reduced, 255
Exchange interaction, 61, 510
Exchange symmetry, 6, 73, 74
Experimental probes, for critical dynamics,
   465, 486, 493
   2DQHAF, 525, 527, 561
Exponents, dynamical, 461

F
^{19}F nucleus, 117
F quantum number, 161
F_2^+, 166
F-center, 127
Fanfoni, M., 86
Farach, H.A., 572
Fe atom, 81, *see also* Iron
^{57}Fe nucleus, 117, 443

Fe^{2+} atom, 149, 411
Fe^{3+} atom, 142, 512
Fe-based superconductors and oxypnictides,
   541, 550
Fermi, 114
   contact interaction, 85
   energy, 441
   gas, 409, 411
   surface, 363, 410
   temperature, 434
   wavevector, 413
Fermi wave vector and energy or level, 411
Fermi-Dirac statistic or distribution, 368,
   386, 562
Fermions, 327
Ferrimagnetism, 513
Ferroelectric catastrophe, 481, 501
Ferroelectric state, 346
Ferroelectric transition, 318, 346
   in the mean field scenario, **485–492**
Ferroelectrics, 446
Ferromagnetic and antiferromagnetic, 446
Ferromagnetic metals, 409, 412
   ordered states, 508
Ferromagnetic or antiferromagnetic metals,
   409
Feynman, R.P., 39, 555
Fibonacci crystals, 338
FID, 208, 209
Fine structure, 28, 32, 33
Fine structure constant, 43, 57, 145, 174
Finite nuclear mass, 21
Finite size of the nucleus, 19
Finite width of the lines, 44
Fisher criterion, 452
Fluctuation-dissipation theorem, 150, 197,
   217, 460, 461, 470
Fluctuations, 445
   enhancement, 445, 446, 451
   quantum, 509, 517
   slowing-down, 445, 446, 460, 461, 463,
   486
   time-dependence, 460
Fluctuations of the e.m. field, 174
   of the magnetization, 452
Flux expulsion, 551
Flux flow, of vortices, 541
Flux quantization, 410
   in superconducting rings, **553–554**
Fluxon (superconductivity) and flux lines,
   410, 554, 573
Fock, 5, 112, 114
Fourier components, 373, 379

expansion, 343, 372
    transform, 419, 420
Fr atom, 61
Franck-Condon
    principle, 324, 326
Franck-Condon factor, 325
    principle, 324, 326
Free electron model, 365
Free energy, 178
Free energy density
    expansion, 455
Frenkel, 27
Friedman, R.S., 113
Frölich, 540
Frustration, **509–513**
Fullerene, 347
Functional, 458
Fundamental constants, 36

G

GaAs, 347, 393
Gallium specific heat, 387
    crystal, 388
    quadrupole moment, 388
γ emission (from ^{57}Fe), 117
Gamma-ray, 435–437
Gap (energy gap in crystals), 355, 372, 374,
    385, 392
GaSb, 393
Gauge invariance, 565
Gaussian approximation, 458, 569, 575
Gauss system, 143–145
Gaussian distribution (around the mean
    value), 197
Gd^{3+}, 146
Generalized moment, 25
Generalized susceptibility, 498
Gerlach, 115
Germanium atom, 108
    crystal, 347, 379, 391
Gibbs free energy, 451
    functional, 565
Ginzburg, 540
Ginzburg-Landau, 549
    correlation time, **577**
    equations, 566
    theory, 549, 550, 553, **564–571**, 571
    time-dependent, 577
Ginzburg-Levanyuk parameter, 582
Giulotto, L., 174
Glauber model, 492
Goldstein, H., 25

Goodstein, D.L., 584
Gor'kov, L., 565, 580
Graphite, 347
Grosso, G., 411, 545
Grotrian diagram, 62
Ground states of various atom, 98, 104
Group theory, 340
Group velocity, 359, 360, 365
Gunn diodes, 292
Gyromagnetic ratio, 31, 89, 93, 103, 144,
    158, 203

H

H atom, *see* Hydrogen
H^- atom, 258
H_2 molecule, 306
    comparison MO and, 257
    in the VB approach, 256
    mechanical anharmonicity, 305, 306
    Raman spectra, 328
    rotational constant, 294
    rotations and vibrations, 306
    specific heat, 330
    VB scenarios, 257
    vibration constant, 306
H_2O, 263, 273, 274
H_2^+, 223, *see also* Hydrogen molecule ion,
H_α line, 23, 174
$H_{\beta,\gamma,\delta}$ lines, 23, 174
Hahn, 208
Hahn echo, 208
Haken, H., 210, 329
Hall effect (fractional), 338
Hansch, 174
Harmonic approximation, 244, 302, 305,
    417, 419, 429
Harmonic potential, quasi-, 468, 487, 493
Hartree, 3, 5, 89, 112–114
Hartree-Fock theory, 354
HB, 263
HBr, 263
    rotational constant, 299
HCl, 261, 263
    and specific heat, 333
    rotational constant, 294
    rotational states, 333
    rotovibrational spectrum and deuterated
        molecule, 312
    Stark effect on rotational states, 301
    vibrational constant, temperature and
        specific heat, 314
He^+, 17, 86, 232

He$_2$, 249
He$_2^+$, 249
Heisenberg, 148, 235, 453, 487, 493, 508,
    519, 525, 537
    exchange frequency, 152
    Hamiltonian, 75, 148
    principle, 44, 232
    systems, 453
Heitler, 329
Helium atom, 2, 6, 69, 70, 74, 76, 80, 81, 90,
    110, 142, 254, 255, 328
Hellmann-Feynman theorem, 245
Helmholtz free energy, 210, 221, *see also*
    Free energy
Hermite polynomials, 129, 302, 303
Hertz, 145
Herzberg, G., 329
Heteronuclear molecules, 260, 305
HF, 263, 294, 309
Hg atom, 76
High-temperature (HT$_c$SC), **560–564**
Hole doping, 562
Holes, 385, 405, 408
Homogeneous functions, 531
Homonuclear molecules
    MO scenario, 248
Hopping integral, 506, 512
Hubbar bands (LHB, UHB), 562
Hubbard, 411
Hubbard Hamiltonian, 561. *see also* Stoner-
    Hubbard
Hückel, 279
Hückel criterium, 279
Hund rules, 97, 100, 104, 251
Hybrid band, 393
Hybrid orbitals, 83, 273, 346, 347
Hybridization, 391, *see also* Hybrid orbitals,
    d$^2 sp^3$
Hydrogen atom, 7
    fine and hyperfine structure, 31
    intergalactic, 56
    life-time of the $2p$ states, 51
    under irradiation, 56
Hydrogen bond, 396
Hydrogen molecule, 223, *see also* H$_2$ mole-
    cule
Hydrogen molecule ion, 223
Hydrogenic atoms, 1
    Darwin term in, 57
    expectation values, 204
    in weak and strong magnetic field, 166
    polarizability, 124
    quadrupole coupling constant, 167

Hyperfine quadrupole hamiltonian, 188, *see*
    also electric quadrupole and quadru-
    pole interaction,
Hyperfine structure, 31, 157
    for Na doublet, 176
    in Hydrogen, 172
    in Hydrogen molecule ion, 256

I

Incommensurate phases, 446, 498
Independent electron approximation, 71
Inert gas atoms, cohesive energy, 394
    diamagnetic susceptibility, 141, 142
Infrared radiation, 304
InSb, 347
Instability, 446
Intergalactic Hydrogen, 56
International system of units, 143
Interval rule, 94, 105, 162
Inversion doublet, 273, 282, 283
Inversion symmetry, 323
Inverted multiplet, 98
Ionic crystals, 391–394, 398
    cohesive energy, 394, 398
Iron crystal (bcc), 348, 411
Irreversibility temperature, 564
    FC and ZFC conditions, 553, 564
Ising model, 200, 508
Ising systems, 453
Ising-like pseudospin, 501
Isomer shift, 191, 438
Isotope effect, 540
Isotopic shift, 22, 68
Itinerant electrons, 413
    magnetic properties, 409, 411

J

J_1-J_2 model, 523
Jahn-Teller effect, 228
Jj scheme, 91–108
Johnston, D.C., 525
Josephson, 540
    effects, **554–557**
    junction, **554–557**

K

K atom, *see* Alkali atoms
K$^+$, 266
K$^+$ ion, 266
K$_\alpha$ line, 23, *see also* X-ray lines

K, momentum of the electron, role and properties, 358, 359
Kagomé lattice, **522**
Kamerlingh Onnes H., 539
κ parameter, **571–574**
KBr, 400
KBr crystal, 346
KCl, 263, 399, 400
KDP, 392
KDP-type ferroelectrics, 490, 497
KF, 263
Kimball, D.F., 40, 268
Kittel, C., 338
Klystron, 292
Kohn anomaly, 534
Koopmans theorem, 113
Kramers-Kronig relations, 461, 471, 483
Kronig and Penney model, 375

L
La atom, 104
La_2CuO_4, 507, **508, 511, 524, 527, 529, 561,** 563
Lamb, 28, 162, 174
Λ-doubling, 229
λ transition, 449
Landau, 551
    criterion, 452
    diamagnetism, 409
    gauge, 409
    levels, 409, 410
    statistical theory, 455, 460
Landau-Khalatnikov, 460
Lande' g factor, 32, 131
Lande' g nuclear, 158
Langevin function, 138
    susceptibility, 216, *see also* magnetic susceptibility
Laplace equation, 403
Larkin, A.I., 578, 580, 582
Larmor frequency, 95, 103, 122, 128, 409
    precession, 132, 202
Laser, 23, 185
Latent heats, 447, 449, 452, 541
Lattice vibrations, 417, 545
Laue X-ray diffraction, 337
LCAO molecular orbitals, 242
Lead crystal, 346, 391, *see also* Pb
Legendre polynomials, 71, 76, 111
Legendre transformation, 454
Lennard-Jones potential, 391, 396, 398
Li atom, 1, *see also* Alkali atoms

Grotrian diagram, 62
Li crystal, 392
Li^+ atom, 61
Li^{2+} ion, 15, 17
LiF crystal, 392
Life-time, 23, 435, 436, 438, 443
    of ^{57}Fe nucleus, 422
Ligand-field theory, 401
Light scattering, 465
LiH molecule, 263
LiH crystal, 346
Lindemann criterium, 433
Line at 21 cm, 163
Linear electric approximation, 305
Linear response theory, 140
Lithium
    bulk modulus, 389
    crystal, 348, 388
Lo Surdo, 124
Local density approximation, 113, 271
Local moments and delocalized electrons representation, 505
London
    equations, 544
    theory, **551–553**
London gauge, 551
London (Heinz and Fritz), 544
London interaction, 257, 396, 398
Long-wave length approximation, 41
Lorentz
    field, 480
    force, 121, 143, 144, 359, 574
    gauge, 25
    oscillator, 130
LS scheme, 91–92, 108, 109, 111, 114, 115, 130, 150
    in molecules, 130
LST relation, 484, 493
Lyman series, 16

M
Madelung constant, 395, 400
Magnetic anisotropy, 453
Magnetic dipole (mechanism of transitions), 43, 149, 169, 171, 174, 178, 194, 201
    selection rules, 114
Magnetic domains and domain walls, 519, 537
Magnetic field expulsion, 553
Magnetic field Hamiltonian, 129
Magnetic frustration, 509, 517
Magnetic moments, 25, 90, 95

effective, 95, 98, 136, 137
in field, 137
of nuclei, 25
Magnetic permeability, 143
Magnetic resonance, 201–209
Magnetic splitting, 26, 29
Magnetic susceptibility, 144
Magnetic temperature, 210
Magnetite, 512
Magnetization
curves, 146
field induced, 196
saturation, 138
spontaneous, 140, 519, 532
Magnetron, 292
Magnons, **509**, 517
Magnus force, 574
Manganites, 512
Martensitic materials, 446
Maser, 44, 283
Matrix Hamiltonian, 39
Maxwell-Boltzmann statistics, 47
Mc Millan, 549
Mean field exponents, 456
Mean field interaction, 121, 136
Mean field interaction or approximation
(MFA), 486, 487, 494, 500, 505, 519
bare susceptibility, 506
Mean free path, 405, 413
Mean free path of electrons, 413, 540, 549,
573
Mean square vibrational amplitude, 432
Mechanism of transition, *see* selection rules
Meissner effect, 540, 543, 549, 551, 564
Melting, 447
Mercury atom, 76, *see also* Hg
Mermin, N.D., 338, 519
Mermin-Wagner-Berezinsky theorem, 453,
519
Metals and metallic crystals, 391, 394
Mg crystal, 347
MgB_2, 541, 560
Miller indexes, 340, 345, 348
MKS system of units, 143
Mn, 346
Mn crystal, 516
MnF_2, 513
MnO, 510
MO-LCAO, 238–262
MO-LCAO-SCF, 238
Mobility, 405
Modes (of the radiation), 46
of vibration, *see* Vibrational motions

Molecular crystals, 391–392, 394, 398
Molecular orbitals (MO), 242, *see also* MO-
LCAO
Molecular velocities, 55
Moments, angular, magnetic and quadrupo-
lar, of the nuclei, 158
Monoatomic one-dimensional
crystal, lattice vibrations, 420
Monte Carlo simulations, 413, 462
Mori continuous fraction, 462
Morse potential, 306, 308–309
Moseley law, 53
Mössbauer effect, 53, 417, 435, 438, 443
Mössbauer spectroscopy, 438, 466
Motional broadening, 53
Mott
insulator, 527, 561
transition, 507
Müller, K.A., 527, 541
Multi-electron atoms, 1
Multiplets (quantum theory), 110
Muon, 22
life-time, 22
molecule, 243
Muonic atom, 1, 22, 23, 190, 445, 505, 539

N
N_2 molecule, 250
rotational constant, 294
rotovibrational structure and Raman
spectra, 328, 329
vibrational constant, 304
N_2^-, N_2^+, 252
Néel, temperature, 513, 514, 516, 527
antiferromagnetic order, 523
Na atom, 25, *see also* Sodium atom and alkali atoms
doublet, 63
doublet hyperfine, 176
quantum defect, 67
Stark effect, 127
Na crystal, 357
Na_2 molecule, vibrational constant, 304
NaCl crystal, 346, 392, 395, 399, 400, 493
NaCl molecule, 263
rotational constant, 292, 294
vibrational constant, 303, 313
$NaNbO_3$, 346
$NaNO_2$, 486
Nanoparticles (superconducting), 567, 575
Natural broadening, 44, 51
^{20}Ne nucleus, 117

Ne_2^+, 253
Nearly free electron model, 372, 374
Negative temperature, 46
Nesting and nesting wave-vector, 535
Neutron
    diffraction, 345, 349, 438
    spectroscopy, 427
Neutron diffraction, 329, 438
    inelastic scattering, 322
    spectroscopy, 427
Ni crystal, 346
Ni metal, 413
NMR, **193–204**
    imaging, 208
NO, 263
NO molecule, 268
Non-crossing rule (or non-intersecting rule),
    233
Non-linear σ model, 527, 529
Normal coordinates, 317
Normal modes
    in polyatomic molecules, 289, 319
    infrared active, 333
    spectroscopically independent, 319
Nuclear g-factor, 194
Nuclear magneton, 158
Nuclear moments, 157, 213, *see also* Nuclei
Nuclear motions
    in diatomic molecules (separation of ro-
        tational and vibrational motions), 289–
        294
    in molecules, 289, 333
Nuclear spin statistics (in homonuclear di-
    atomic molecules), 327, 330
Nuclear-size effects, 24
Nuclei, properties of, 158

O

O_2 molecule, 319
    Raman spectra and rotational lines, 328
    rotational constant, 294
    vibrational constant, 304
O_3, 251
Oblate rotator, 298
OCS molecule, rotational states in electric
    field, 332
Octahedral coordination
    crystal field, 404, 414
    of oxygen atoms, 404
Ohm law, 368, 405
Onsager
    field and reaction, 480–481

model, 508
Optical electron, 63, 70
Optical pumping, 133, 268
Order parameter, 445
    for superconductive transition, 458
Order-disorder ferroelectrics, 485, 490
Orientation electric polarizability, 297
Orientational electric polarizability, 296–
    297, **480–481**
Ornstein-Zernike expansion, 462
Ortho molecules and rotational states, 328
Ortho-Hydrogen, 330
Orthohelium, 70, 85
Overlap (band overlap), 393
Overlap integral, 239
Oxygen atom, 119
Oxygen octahedra, 346

P

P and R branches, 310
P crystal, 345
Palladium crystal, 348
Para molecules and rotational states, 327
Para-Helium, 70, 85
Para-Hydrogen, 327
Paraconductivity, 577–580
Paramagnetic susceptibility, 122, 138, 139,
    217, *see also* Magnetic susceptibility
    for Fermi gas, 409
Paramagnetism, 122, 141
    Van-Vleck, 141
Parity, 12
Partition function, 430, 433, 443
Paschen series, 16
Paschen-Back
    effect, 130
    on the Na doublet, 135, 136, 163, 176
    regime, 96, 132, 133
Pastori Parravicini, G., 411, 545
Pauli principle, 6, 74, 80, 89, 92, 119, 371,
    405
    susceptibility and, 389, 411, 412
    paramagnetism, 368, 383
Pauling entropy, 522
Pb crystal, 346, 379, *see also* Lead
Pd metal, 413
Penetration length, 552, 560, 563, **570–571**
Percolation threshold, 526
Periodic boundary conditions, 359, *see also*
    Born-Von Karmann
Periodical conditions, 48
Perturbation effects (in two levels system),
    37

Pfund series, 17
Phase diagrams, **446–454**
Phase transitions, 139, 140, 318
    paramagnetic-ferromagnetic, 448
Phonons, 430
Phosphorous atom, 102
Photon echoes, 208
Photon moment, 424
Photons (as bosonic particles), 47
Photons momentum, 436
Pinning, 574
Planck distribution function, 47, 429
Point groups, 340
Polarizability
    anisotropic, 323
    dipolar, ionic and electronic, 296
    electric, 296
    electronic, 479
    in HCl, 307
    in hydrogen, 397
    in molecule, 316
    of the harmonic oscillator, 315
    orientational, 296–297
    pseudo-orientational, 127
Polarization of the radiation and transitions,
    43, 132
Polarization, spontaneous, 346, 485, 488
Polyatomic molecules, 271
    normal modes, 316
Polymorphs and polymorphism, 446, 447
Poole, C.P., 572
Population inversion, 46
Positron, 22
Positronium, 22, 24
    hyperfine splitting, 163
    Zeeman effect, 130, 176
Potassium crystal, 385
Potential energy, 2, 5, 19, 25, 123, 224, 227,
    281, 289, 302, 308, 318, 354, 358,
    375, 397
Pound, 443, 444
Pre-dissociation, 228
Primitive
    cell and vectors, 339
    lattice, 341, 349–350
Principal (series lines), 67
Prolate rotator, 298
Proton magnetic moment, 194
Protonium, 24
Pseudo-gap, 562
Pseudo-potential, 373
Pseudo-spin

dynamics, in order-disorder ferro-
    electrics, **494–498**
formalism, theory, **490–492**
interaction, 75
Pt metal, 413
Purcell, E.M., 145

Q
Q-branch, 299, 311, 329
Quadrupole electric lens, 284
Quadrupole interaction and quadrupole cou-
    pling constant, 167
Quadrupole moment, 158
    of deuterium, 175
    of Gallium, 388
Quality factor, in microwave cavities, 559
Quantum critical (QC) regime, 526
Quantum defect, 62, 65, 68
Quantum disordered (QD) regime, 527
Quantum electrodynamics, 26
Quantum number F, 157
Quantum pressure (from electron gas), 382
Quantum rotator, 330
Quantum tunneling (in ferroelectrics), 490
Quasi-harmonic approximation, 493
Quasi-second order transitions, 489
Quenching of orbital momenta, 146, 415

R
Rabi, 180, 201, 268, 284
Rabi equation, 40
Radial equation and radial functions, 7
Radial probability density, 10, 65
Radiation damping, 46
Radiofrequency spectroscopy, 201
Radius of the first orbit (in Bohr atom), 7
Raman spectroscopy, 330
Random phase approximation, 140
Rare earth atom (electronic configuration
    and magnetic moments), 98
Rayleigh diffusion, 320, 328
Rb atom, 24, *see also* Alkali atoms
RbH molecule, vibrational frequency and
    dissociation energy, 309
Rebka, 443, 444
Reciprocal lattice, 337, 340, 343–345
    vectors (fundamental), 343
Recoil energy, 53, 436
Recoilless fraction, 438, 444
Reduced mass, 21, 22, 190
Relativistic
    effects, 438

mass, 173
shift, 186
terms, 28
transformation, 27
Relaxation mechanisms, 40, *see also* Resonance technique
Relaxational
behaviour, 469
critical, 486
modes, 486
Renormalization group, theory, 531
Renormalized classical (RC) regime, 526
Residual charge, 6
Residual first-order Doppler broadening, 186
Resistivity, 228, 337, 354, 358, 413, 539, 541, 543, *see also* Conductivity
Resonance absorption, 53, *see also* Mössbauer effect
Resonance integral, 239, 241, 354
Resonance technique, pulsed, 204
Response functions, **446–454**
Riemann zeta function, 49
RKKY interaction, **511–523**
Roothaan, 271
Rotational
constant, 292
frequency and motions, 291
spectroscopy (principles), 292
temperature, 294
Roto-vibrational
eigenvalues, 312
levels, 314
Roto-Vibrational eigenvalues and levels, 310
Rubidium atom, 68, 393
hyperfine field, 161
Rydberg atoms, 1, 21, 23, 25, 154, 445, 505, 539
Rydberg constant, 7, 69
Rydberg defect, 62

S
Sb crystal, 393
Scalar potential, 25, 41
Scaling concepts, 462
dynamical, 463, **525**
hypothesis, 532
universality, 508
Scattering
of electrons, 407
of photons, 321
Schottky anomaly, 178, 388

Screening cloud, 3
Second order transition, 531, 571
Selection rules, 41, 114, 126, 131, 132
electric dipole, 292, 323
for quantum magnetic number, 132, 176
Self-consistent field, 5
Semiconductors, 369
Semimetals, 393
Separated atoms scheme, 260
Shannon-Von Neumann entropy, 200
Sharp (series lines), 67
Shift (relativistic), 29, 31, 186
Shubnikov, L., 559
Si atom, 108
SI system of units, 143
Silicon crystal, 347, 379, 380, *see also* Tin
Silver, 213, 413, *see also* Ag crystal
Slater
determinant, 6, 61, 81
radial wavefunctions, 110
theory for multiplets, 7, 110
Slater, J.C., 5, 81, 110, 357
Slichter, C. P., 205
Sm atom, 98
Sn atom, 108
$SnCl_2 \cdot 2H_2O$, 500
Sodium atom, 53, 130, *see also* Na
hyperfine field, 176, 177, 186
Paschen Back effect, 135
Zeeman effect, 131
Soft modes, 484, 486, 492
optical modes, 502
Solid state lasers, 338
Sommerfeld
quantization, 15, 173
Sound velocity, 426, 439
sp^2 hybridization, 347, 391
sp^3 hybridization, 393
Space groups, 340
Spatial quantization, 15
Specific electronic charge (e/m), 147
Specific heat, 142, 200, 212, 295
Spectroscopic notations, 33
Spherical harmonics, 4, 7, 20, 43
addition theorem, 76, 403
Spherical model, 508
Spin, 26
echoes, 193, 208
eigenfunctions, 26, 84
Spin-density functional, 507
Spin-density waves, 512
Spin dynamics, critical, 525
Spin glass, 511, 516, 517, 521

freezing temperature, 516
Spin ice and ice, 522, 523
Spin-orbit interaction, 1, 23, 25, 26, 28, 30, 32, 52, 61, 67, 74, 76, 129, 130, 132, 149, 445, 505, 539
Spin-orbital, 26, 28
Spin-Peierls transition, 537
Spin-spin transitions, 196
Spin statistics, 193
Spin stiffness, 529
    doping dependence, 529
Spin temperature, 195, 196, 198, 212, 215
    negative, 213
Spin thermodynamics, 193, 197
Spin waves, 526, see also Magnons
Spin-exchange collisions, 268
Spin-orbit interaction, 90, 100, 102, 105, 160, 174
Spin-spin interaction, 92–97, 102
Spontaneous emission, 23, 44, 46
Squaring rule, 93
SQUID, 338, **557–559**
Sr crystal, 346
Stanley, H., 463, 473, 533
Stark effect, 301
    linear, 126
    on the Na doublet, 127
    quadratic, 124
Stationary states, 15
Statistical
    populations, 44, 193, 194
    temperature, 45
    weights, 45
Stefan-Boltzmann law, 49
Stern-Gerlach experiment, 115
Stimulated emission, 44, 46
Stirling approximation, 198
Stokes and anti-Stokes lines, 321
Stoner, 411
    criterium, 412
Stoner-Hubbard model, 464, 506
Sun, 52
    energy flow, 52
Superconducting fluctuations, **574–577**
    correlation function, 576, 577
    first order correction, 535
Superconducting gap, temperature dependence, 548
Superconducting transition, order, 540, 564
Superconductors
    coupling factor, 548
    density of states, 548
    high-temperature, 277, 404

strong, 564
surface effects, **568–569**
underdoped and overdoped, 527
zero-dimensional condition, 580
Superexchange, mechanism, **510**
Superfluidity, 446
Superselection rule, 80, 81
Susceptibility, 409, see also Landau diamagnetism
    magnetic, 368, 383
    negative, 409
    of Fermi gas, 409, see also Landau diamagnetism
    see Pauli susceptibility
Susceptibility, generalized, 497
    magnetic, 368, see also Magnetic susceptibility
    of a Fermi gas, see Pauli susceptibility
Svanberg, S., 186, 304

T
$T^{3/2}$ law, 518
T_1, 136, 137, 196, 205, 213, 219
T_2, 196, 208, 209, 217
T_1 and spin temperature, 219, 220
Tetrahedral or tetragonal hybridization, 274, 393
t_{2g} levels, 13
Thermal
    broadening, 53, 57, 436
    effects in crystals, 430, 438, 440
    energy in Debye crystal, 440
    properties (related to lattice vibrations), **430–433**
Thomas, 27, 114
Thomas and Frenkel semiclassical moment, 27, 30, 57
Thomas-Fermi method, 114
Thomson model, 18, 128
Tightly bound electron model, **375–378**
Time-dependent perturbation, **38–41**
Tin atom, **108, 109**
Tin crystal, 391, 393
Tinkham, M., 578
$t - J$ Hamiltonian, 508
TlBr crystal, 346
TlI crystal, 346
Townes, 283
Transition metal oxides, 510
Transition metals ions, in crystal field, **401–404**
Transition probabilities, **41–43**

for quadrupole interaction, 43
  magnetic, 43, 146, 165, 169
Translational
  invariance or translational symmetry,
    338–342, 339, 354, 428, 448
  operations, 340, 358
Trigonal hybridization, 275
Trouton rule, 447, 471
Tungsten
  crystal, 348
Tunnelling integral, 555
Two-level system, **37–40**, 178, 210, 248

U
^{92}U atom, 119
Ultrasound propagation, 427
Under-cooled and superheated liquid, 447
Unitary cell, **339–340**, 342, 343, 346, 380
United atoms scheme, 224, **231–233**, 263
Universality, 453, 508, 530
Universe (expansion), 47, 50, 51
Unsold theorem, 20, 90

V
V^{3+} atom, 104
Vacuum permeability, 145
Valence band, 385, 392, 393, 408
Valence bond, 237, 255, 256, **253–256**, 257–
    260, 391, see also VB
Van der Waals, 249, 392, 396, 446, 455, 530
Van Hove singularities, see Critical points
Van Vleck paramagnetism, 141
Vaporization, 447
Variable frequency laser, 162
Variational principle, procedure, 5, 78, 457
Varlamov, A.A., 578, 582
VB approach, 253, 255, 256, 260, 391
Vector potential, 25, 123, 159, 186
Vectorial model, 28, 81, 90, 111, 159
Venus, 285
Verbin, Y., 197
Vibrational
  model for lattice vibration, 426
  models of lattice vibrations, 419, 427
Vibrational frequency in H_2^+, 244
  in polyatomic molecules, 271
  lattice vibrations, 419
  motions in molecules, 289
  spectroscopy (principle), 292
  temperature, 295
Vibrational motions
  in crystals, 419

  in polyatomic molecules, 291
Vibrational spectra, 424
Vibrational temperature, 309, 315
Vibronic transitions, 324
Vilfan, M., 530
Virial theorem, 16, 165, 331
Volume shift, in Hydrogen and in muonic
    atoms, 190
Von Neumann-Wigner rule, 232
Vortex, 560, **571–574**, see also Fluxon

W
W crystal, see Tungsten
Wannier and Bloch functions, 506
Weak magnetic field (condition or regime),
    95, 121, 178
Weiss, 446
  mean field theory, 455
White, M., 534
Wiedemann-Franz law, 368
Wien law, 47, 49
Wigner crystallization, 413
Wigner-Eckart theorem, 96, 110, 130, 132,
    160, 189
Wigner-Seitz cell, 338, 339, 344
Williams, 283
Wilson, 531
Wolf, H.C., 210, 329

X
X-ray lines, 54–55
  diffraction, 392, 438, see also Bragg law
X-rays, 53
XY systems and model, 453, 508

Y
Yb^{3+} atom, 104
$YBa_2Cu_3O_{7-x}$ (YBCO), 541
Yellow doublet (for Na atom), 25, see also
    Na atom

Z
Zeeman effect, 130
  anomalous, 130
  in Hg, 149
  in positronium, 134
  levels, 194, 208
  normal, 130
  regime on hyperfine states, 163
  weak field, 130
Zeeman regime

on hyperfine states, 177
   weak field, 130
Zero-point energy, 243
Zero-temperature rotations, 327

Zn crystal, 347
Zone representation (reduced, extended, re-
   peated), 362, 369
Zumer, S., 530

Printed in the United States
By Bookmasters